Doing Calculus

All Night Long

(A Giant Step Backwards in Publishing)

Bruce E. Shapiro
California State University, Northridge
2014

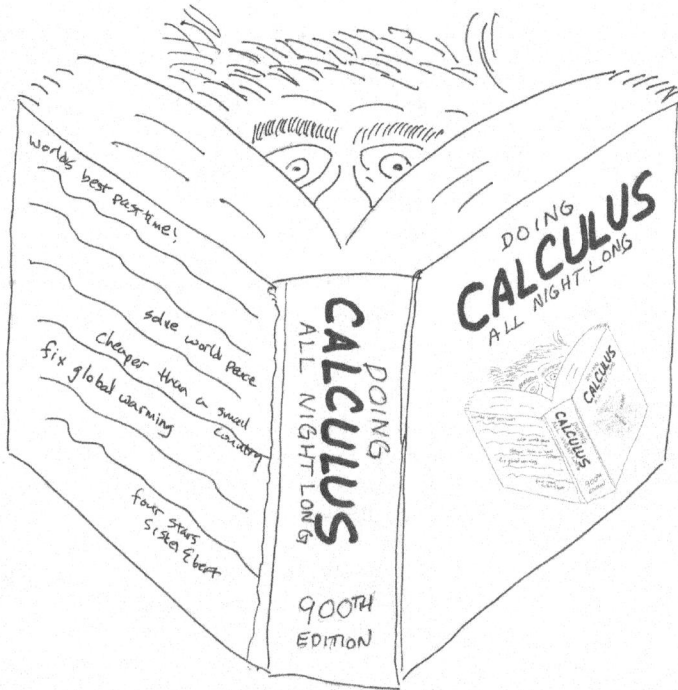

Sherwood
Forest

http://CalculusCastle.com

Doing Calculus All Night Long (A giant step backwards in publishing)
ISBN-13: 978-0-692-21536-4 (Sherwood Forest Books)
ISBN-10: 0692215360
Version 1.0, 11/8/14
Bruce E. Shapiro, Ph.D.
California State University, Northridge

This text was partially supported by the California State University, Northridge eText initiative.

Picture credits are given on page 471.

This document is provided in the hope that it will be useful but without any warranty, without even the implied warranty of merchantability or fitness for a particular purpose. The document is provided on an "as is" basis and the author has no obligations to provide corrections or modifications. The author makes no claims as to the accuracy of this document. In no event shall the author be liable to any party for direct, indirect, special, incidental, or consequential damages, including lost profits, unsatisfactory class performance, poor grades, confusion, misunderstanding, emotional disturbance or other general malaise arising out of the use of this document or any software described herein, even if the author has been advised of the possibility of such damage. It may contain typographical errors. While no factual errors are intended there is no surety of their absence.

This is not an official document. Any opinions expressed herein are totally arbitrary, are only presented to expose the student to diverse perspectives, and do not necessarily reflect the position of any specific individual, the California State University, Northridge, or any other organization.

Please report any errors, omissions, or suggestions for improvements to bruce.e.shapiro@csun.edu.

ISBN 978-0-692-21536-4

9 780692 215364

The Sherwood Forest imprint evokes the image of Robin Hood, who, in some legends, hid in Sherwood Forest while fighting to help the poor, as they suffered under the oppressive regime of the medieval English aristocracy.

A statue of Robin Hood stands in front of Nottingham Castle, his legendary home. The castle dates to the 17th century; the statue was built in 1952. (Photo by the author.)

Now as we enter the information age, modern university students suffer under the oppression of the expensive traditional publication model. Sherwood Forest Books aims to print low cost, affordable books in hard-copy and DRM-free electronic formats. In the logo, the roots of the tree sink down into the Earth, from which we all arise. The filament on the light bulb is a double helix, representing the DNA that binds all life on Earth and through which we grow, learn, interpret, and communicate our understanding of the world around us. This light of knowledge is spread through the easy and inexpensive dissemination of the printed word, like leaves on tree as they blow in the wind.

Contents

Calculus Made Simple

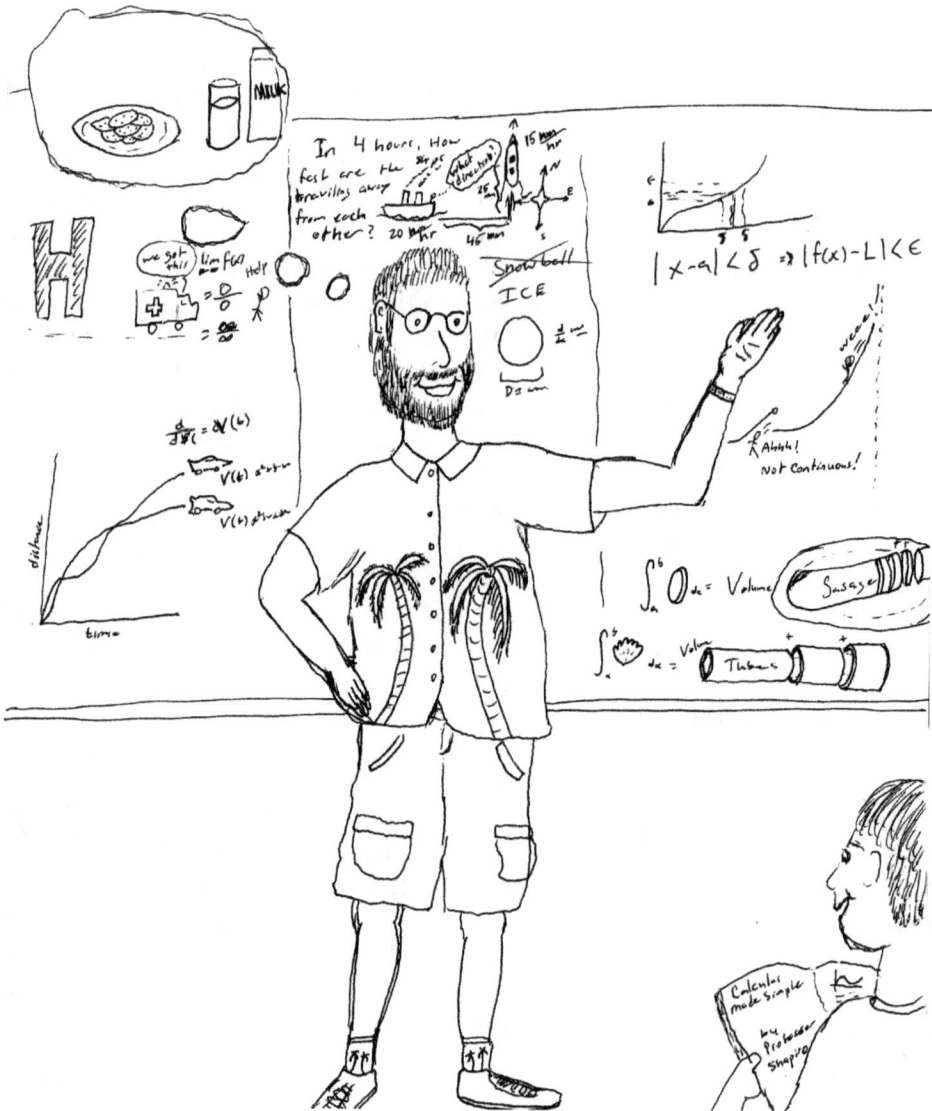

The Joy of Calculus

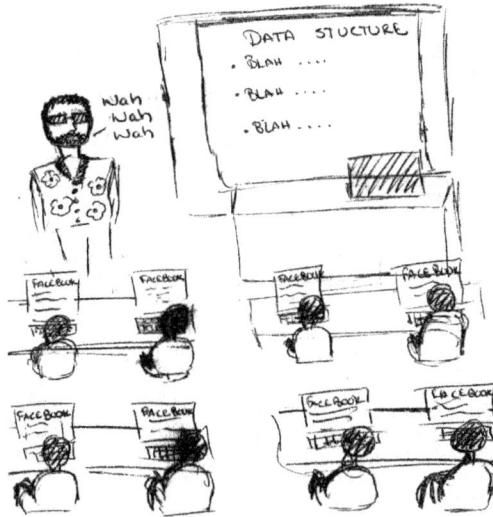

With thanks to the many hundreds of students who have sat patiently, rapt with attention, totally spellbound and engrossed, absorbing my many words of wisdom, through innumerable hours of totally captivating math lectures in boundless joy.[1]

A Preface, of Sorts

Calculus books are easy to find. There are so many that I wonder why we aren't drowning under the crush. A recent[2] search of the British Library using the keyword *calculus*, for example, returned 22,102 results[3]; a similar search of the US Library of Congress returned 5,698 listings[4]. Bookseller Biblio[5] had 64,702 textbooks listed under *calculus*, while Amazon[6] had 41,393 listings for *calculus* and 6,493 for *calculus textbook*. By comparison (at Amazon) there were only 31,808 listings for *Reader's Digest*, 5,289 for *pornography*, 24,064 for *soccer*, and 194,571 for *sex*. So calculus is more popular than Reader's Digest, porn, and soccer, but not sex.[7] However, there were 618,603 books on *mathematics*, so overall, math sells about 3.2 times better than sex – car dealers take note.[8]

---

[1] The better cartoons in the text were contributed by these students. The less clever ones I drew myself.

[2] This and the following searches were performed on 28 June 2014.

[3] http://explore.bl.uk

[4] http://www.loc.gov

[5] http://www.biblio.com, keyword *calculus* under category *textbook*. Note that many of the results for both Biblio and Amazon are copies of the same text from different sellers, and do not reflect different titles.

[6] http://www.amazon.com

[7] Some of the calculus references could be about dental tartar, kidney stones, the spider Oonopidae *Calclulus bicolor Purcell*, the Madeiran land snail Hygromiidae *Caseolus calculus*, or to *battlefield calculus*.

[8] Attentive readers should observe that this is a blatant misrepresentation of the statistics: all the data really tells us is that there are 3.2 times more book titles on the market; it says absolutely nothing about how many copies of each title gets sold. In fact, the way things are listed in these databases, 618,603 different people could be selling copies of the same math book. But who cares: it sounds a lot better the other way!

Math books sell better than sex. Who'd have thunk it? Year after year my students keep buying them. Precisely why, I don't know, since quite clearly they never read them. Sure, they are listed as "required." So are vitamins, and who takes those? These books are heavy and they are expensive. The only possible answer is that money is involved. Lots of it. Given that nearly a million students in US colleges and universities are enrolled in some form of calculus class at any given time, and that the typical college calculus book costs around \$200[9], that's around \$200 million in the book sales that *nobody ever reads*.

Table 1. Comparison of Calculus Books.

	Them	*Doing Calculus*
Pages	1000 - 1500	476
Cost	USD \$150-\$300	\$17.96
PDF	DRM Protected	No DRM
Solution Manual	Illegal Download	Write Your Own
Inexpensive Paperback	Only in Asia	Everywhere
Content	Detailed	Concise
Rigor	Precise	Descriptive
Different Versions	Many	All the same
Offers LMS*	Usually	No
Website	Often Useless	Totally Useless**
Test Bank Available	Usually	No
Snarky Comments	Rarely	Frequently
Silly Pictures	No	A Bunch
Helps Pay for my Retirement	No	Yes
Great Holiday Gift	No	Yes
Alien Conspiracy Theories	No	Yes
Can be used as a doorstop?	Easily	With Difficulty

*LMS=Learning Management System, some at extra cost.
**http://CalculusCastle.com

This raises two really interesting questions. First, where are all these books going? All these calculus books have gotta be affecting the Earth's gravitational potential. It's probably a big government secret, but I suspect that we are spiralling into the sun because we squandered our resources printing too many crappy calculus books. But enough about that. That's a problem for the physicists to deal with, since physicists like saving the universe and all that stuff. The second, and more important question is this: how can I get my hands on some of that cash?

Some mysterious and invisible force is motivating students to repeatedly throw (literally) tons of money at book stores and walk off with copies of these books, *sometimes paying hundreds of dollars merely to rent them for a few months*. It must be an alien conspiracy, because no self-respecting red-blooded human would sink so low. Pay \$800 every year for a smart telephone with technology that's out of date before you leave the store, well sure, that makes perfect sense, because the government and organized crime both need a new place to test the latest spyware. But buy a \$200 calculus book that never goes out of date, when its all posted on *Wikipedia* and *YouTube* anyway?

[9] Based on an average of the top ten best selling calculus textbooks on Amazon (7 Sept. 2014).

Me like dees book. You buy dees book. You reed dees book. Den mean croculus monster be nice and no eat you. Less you zeeba.

The aliens must run a cabal of crafty publishing houses that are secretly enticing these susceptible young minds to send all of their cash off-planet before one of them manages to prove $P = NP$ and makes everything we know about technology meaninglessly trivial.

In an effort to bring the Earth's orbit back into synch and save us from the terrors of global warming and so forth (as we crash into the sun) I offered to dig a big pit in my front yard and let students throw their money in there instead.

But the city government objected – something about underground high pressure gas lines and such. And the university administration insisted that student money is better spent (at least three times daily) on triple shot non fat soy no foam vanilla lattes with caramel drizzle at the campus coffee shops. Obviously the power of the cabal has spread beyond control.

In any case, I was told by the grand universal muckety-mucks that the only way I would ever get my hands on any of that $200 million would be to either win the lottery or write my own book.

Thus was born *Doing Calculus*.

I've also managed to collect lots of numbered bookmarks from the local *Seven-Eleven*.

The croculus monster in the figure explains quite clearly why you should buy *Doing Calculus* and not one of the other books. If you want something more specific, look at table 1.

But Why Should I Buy This Book?

If you've read this far, you probably aren't asking "why another damn book on calculus?" any more. You're probably asking "why isn't this guy locked up somewhere?"[10] Nevertheless, I'll answer the first question just before I hop back into my Tardis.[11] Read table 1. Listen to the croculus monster. *Doing Calculus* is a Giant Step Backwards in Publishing: there are no bells to lull you sleep, and no whistles to wake you up. Its just a book, which is all you should need. So buy it. If you are an instructor, tell your students to buy the book. Tell your friends to buy the book. Tell your mother to buy the book. Help me retire in luxury on the Nevada Riviera.[12]

How to Use This Book (For Instructors)

I'm not going to tell you how or what to teach.[13] Some instructors who are dead-set on using one of the door-stops like those listed in the references have found this material useful in supplementary instruction classes or discussion sections.

How to Use This Book (For Students)

Read it. Do the problems. Duh.

---

[10] As to question 2, you're the one who has read this far. I could say the same about you (unless you buy the book, of course, in which case you are clearly very intelligent).

[11] Photo by Zir of The Mark 2 fibreglass (Tom Yardley-Jones) Tardis as used in the 1980s for *Doctor Who*, on display at the BBC television center. Reproduced under the terms of the GNU Free documentation license (CC-By-2.5).

[12] When California sinks into the Pacific after **The Big One**.

[13] Nudge Nudge Wink Wink: this book originated as my lecture notes. That might tell you something.

Chapter 1

Slope and Rate of Change

Chapter Summary and Goal

This chapter will start with a discussion of slopes and the tangent line. This will rapidly lead to heuristic developments of limits and the derivative.

Student Learning Objectives

The student will:

1. Be able to distinguish between the slopes of secant lines and tangent lines.
2. Understand the concept of a limit.
3. Understand the relationship between slopes of a function and the derivative.
4. Learn the definition of a derivative.
5. Learn the relationship between tangent lines and the velocity problem.

The Mathematics of Change

Calculus is about change. More specifically, if gives us ways to explain, using mathematics, how a variable y might change when a variable x changes. Why do we care about this? Because change is a fact of life. We can use calculus to figure out how fast our local reservoir will empty during the next drought; to plan a flight path between Atlanta and London that uses the least fuel; to figure out how long it will take to pay off a mortgage; to figure out the dimensions of the largest rectangular tree-house you can build with a fixed amount of plywood; or to predict how fast a penny dropped off the top of the CN Tower in Toronto would be falling when it hit the ground. All of these problems involve figuring out how one variable changes in comparison to, or as mathematicians often like to say, **with respect to**, another variables.

You've probably already come across this concept in an algebra or pre-calculus class through the use of the **slope** of a line. Suppose we draw the plot of a non-vertical line

$$y = mx + b \tag{1.1}$$

where m is the slope and b is the y intercept (figure 1.1). If we pick *any* two points (x_1, y_1) and (x_2, y_2) on that line, then the slope is defined as

$$\text{slope} = \frac{\text{change in } y}{\text{change in } x} = \frac{\text{rise}}{\text{run}} = \frac{y_2 - y_1}{x_2 - x_1} \tag{1.2}$$

Calculus

Calculus is the **quantitative description** of how things change in comparison to one another. It has two branches:

a) **Differential Calculus**: calculation and application of the derivative to find rates of change.

b) **Integral Calculus**: calculate of areas under curves and inversion of the process of differentiation.

We have introduced the terms **rise** and **run** in equation 1.2 to refer to the **change in** x and **change in** y, respectively. These terms are commonly used throughout algebra and analytic geometry. Sometimes we also use the expressions Δy and Δx to refer to the rise and the run,

Figure 1.1: Slope of a line.

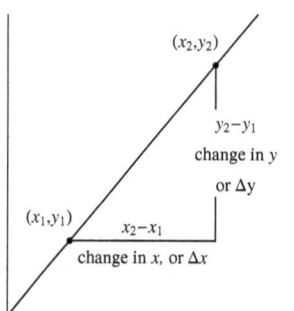

$$\Delta y = y_2 - y_1 \qquad \text{read this as "Delta y"} \qquad (1.3)$$
$$\Delta x = x_2 - x_1 \qquad \text{read this as "Delta x"} \qquad (1.4)$$

Here the Greek letter Δ is not a variable that is multiplied by the x or the y but part of the variable name, so that Δx is single complete variable name, and Δy is a single complete variable name.[1]

Tangent Lines

We want to extend the concept of the slope of a line to curves. Since curves can bend in any which-way they might choose, the idea of a single slope that applies to the entire curve doesn't make much sense. Instead, given any particular curve, and a particular point P, on that curve, we observe that if we look at it through a powerful enough magnifying glass, it looks more and more like a straight line as we get closer and closer to that point (see figure 1.2). Imagine, then, that we can figure out the slope m of this almost-nearly-straight line, and construct the line through P. This new line, which we have just drawn, has the *same slope as the curve at* P. We call this line the **tangent line to the curve at** P. The word tangent is derived from the latin verb *tangere*, to touch, as in, to touch, but not to cross or intersect. A tangent line (as we have described it above) just touches the curve at a single point, but does not cross it

---

[1] Technically, Δ is an operator that means "change in" so Δx means "change in x". It comes from the Liebniz notation that we will study in chapter 8.

Figure 1.2: Illustration of a point on a smooth curve, and two successive blow-ups under the magnifying glass, as you move from the left to the root, showing how the points in the neighborhood of the smooth curve look very much like a straight line if you look at the point under a magnifying glass and ignore the rest of the figure.

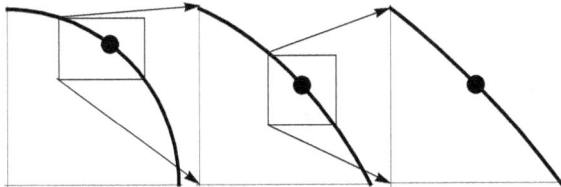

or intersect it (figure 1.3). The *slope of a curve at the point P* is defined to be the slope of the tangent line at P.

Definition 1.1. Slope of a Curve

The slope of a curve at a point P is the slope m of a line that is tangent to the curve at P.

Figure 1.3: Examples of tangent lines. The tangent line is just touching, but not intersecting, at the point of tangency.

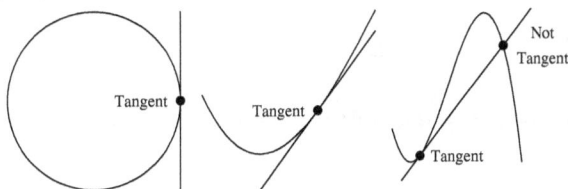

Let $f(x)$ be a smooth function. We want to develop an easy, methodical process for calculating the slope of the tangent line through a point P on the plot of $f(x)$. Let the coordinates of P be (x_1, y_1). Let $Q = (x_2, y_2)$ be another point on the curve, and construct the secant line through P and Q (see the curve in the left frame of figure 1.4.) Next, we imagine marching point Q

Figure 1.4: Using a sequence of secant lines to define the tangent line. The image on the left shows a single secant line between the points Q and P. Then, on the right, as the point Q approaches P, the secant line approaches the tangent line at P.

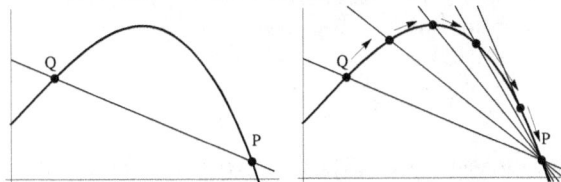

towards P, as indicated by the arrows in the right frame of figure 1.4. As Q gets **closer and closer** to P, the secant line gets closer and closer to the tangent line through P. We call this process of one point getting closer and closer to another point **taking a limit** or a **limiting**

process. We denote the limiting process by $Q \to P$, which we read as "Q goes to P," or "the limit as Q goes to P."

As Q is marching closer and closer to P, Q is trying to get as close as possible. Q may actually get all the way to P; but there are some cases where it may not actually get there. For example, we may have chosen P to be a hole in the function (a point $x = c$ where $f(x)$ is not defined, surrounded by an open neighborhood in which $f(x)$ is defined). It is often still possible to define the slope at a hole, even though the function is not defined there.

In order to do calculations, it is more convenient to talk about the coordinates (x, y) instead of P and Q. Since both P and Q lie on the function $f(x)$, we have

$$P = (x_1, y_1) = (x_1, f(x_1)) \tag{1.5}$$

$$Q = (x_2, y_2) = (x_2, f(x_2)) \tag{1.6}$$

Since x is the **independent variable** and y is the **dependent variable**, instead of saying Q approaches P, we say that x_2 approaches x_1, which we write as $x_2 \to x_1$. Instead of saying "the limit as Q goes to P" we say "the limit as x_2 goes to x_1." At the same time, we observe that $y_2 \to y_1$ because $Q \to P$.

Since the slope of the secant is

$$m_{\text{secant}} = \frac{f(x_2) - f(x_1)}{x_2 - x_1} \tag{1.7}$$

we can then define the **slope of the tangent line at** x_1 as the limit of m_{secant} as $x_2 \to x_1$. We will call this slope $f'(x_1)$.

$$f'(x_1) = \lim_{x_2 \to x_1} \frac{f(x_2) - f(x_1)}{x_2 - x_1} \tag{1.8}$$

If we define the fixed point $x_1 = a$ and let $x = x_2$ then (1.8) becomes

$$f'(a) = \lim_{x \to a} \frac{f(x) - f(a)}{x - a} \tag{1.9}$$

Equation 1.9 gives the slope of the tangent line at the point $x = a$. It is a number, and is called the derivative of $f(x)$ at the point $x = a$.

If we instead define $h = x_2 - x_1$ in (1.8) then $x_2 = x_1 + h$. So when $x_2 \to x_1$, h must go to zero; it is the horizontal distance between the coordinates of P and Q. Substituting,

$$f'(x_1) = \lim_{x_1 + h \to x_1} \frac{f(x_1 + h) - f(x_1)}{h} \tag{1.10}$$

Since there is only one x coordinate (that of P) the index is no longer needed, and the equation for the **slope of the tangent line** at x becomes

$$f'(x) = \lim_{h \to 0} \frac{f(x + h) - f(x)}{h} \tag{1.11}$$

We call the slope of the tangent line at x the **derivative** of $f(x)$ and denote it by $f'(x)$. When the limit (1.11) exists we say that $f(x)$ is **differentiable at** x. We emphasize here that while the slope of a line is a number, and the slope of a curve at a particular point, the derivative of a function is another function. ***The derivative $f'(x)$ gives the slope of $f(x)$ as a continually changing function of x.***

Definition 1.2. The Derivative

The **derivative of a function** $f(x)$ is the **function** $f'(x)$ defined by

$$f'(x) = \lim_{h \to 0} \frac{f(x+h) - f(x)}{h} \qquad (1.12)$$

The **derivative at the point** $x = a$ is a **number** $f'(a)$ that may be calculated either by by setting $x = a$ in the formula for $f'(x)$ or by calculating a limit such as

$$f'(a) = \lim_{h \to 0} \frac{f(a+h) - f(a)}{h} \qquad (1.13)$$

or

$$f'(a) = \lim_{x \to a} \frac{f(x) - f(a)}{x - a} \qquad (1.14)$$

Example 1.1. Find the slope of the tangent line to $y = x^2$ at the point $(1,1)$ by simulating the process of $Q \to P$ empirically, and use this slope to calculate the equation of the tangent line at $(1,1)$.

Solution. By an *empirical* calculation we mean we want to experimentally calculate the values of the slop as Q marches toward P. Let $Q = (x, y) = (x, x^2)$ be any other point on the curve of $y = x^2$. Then the slope is

$$m = \lim_{x_2 \to x_1} \frac{y_2 - y_1}{x_2 - x_1} = \lim_{x \to 1} \frac{x^2 - 1}{x - 1} = \lim_{x \to 1} \frac{(x-1)(x+1)}{x-1} = x + 1 \qquad (1.15a)$$

We will arbitrarily pick a sequence of values of x that approach 1 from the left, and see what the slope appears to be approaching:

x	$m = 1 + x$
0.9	1.9
0.99	1.99
0.999	1.999
0.9999	1.999
0.9999999999	1.9999999999

as $x \to 1$ from the left, it would appear that $m \to 2$. What about if $x \to 1$ from the right? We can repeat the empircal calculation:

x	$m = 1 + x$
1.1	2.1
1.01	2.01
1.001	2.001
1.0001	2.0001
1.0000000001	2.0000000001

It would appear that $m \to 2$ as $x \to 1$ from the right as well. Thus the tangent line through $(1,1)$ is (using the point-slope form of the equation of a line),

$$y - 1 = 2(x - 1) = 2x - 2 \qquad (1.15b)$$

Bringing the 1 to the right hand side gives the equation in the more standard slope-intercept form of $y = 2x - 1$. $\qquad \square$

As it turns it, we could have simply plugged $x = 1$ into the last step of equation (1.15a)

$$m = \lim_{x \to 1} \frac{x^2 - 1}{x - 1} = \lim_{x \to 1}(x + 1) = 2 \qquad (1.16)$$

We will see why we can do this when we discuss the limit laws (such as theorem 3.9) in chapter 2. In this case, since $f(x) = x + 1$ is a polynomial, theorem 3.9 allows us to simply substitute the value of x into the formula for $f(x)$. We *could not* have plugged $x = 1$ into $g(x) = \dfrac{x^2 - 1}{x - 1}$, even though $g(x) = f(x)$ for all $x \neq 1$, to calculate the limit because that would have led to the rather perplexing conundrum of fraction equal to $0/0$. We will also discus $0/0$ limits in chapter 2.

Average and Instantaneous Rate of Change

Suppose that the amount or quantity of something is a function of time: position, altitude of an airplane, the odometer on your car, amount of flour in a cannister on your kitchen counter, amount of gasoline in your car. We want to be able to describe that quantity $y = f(t)$ changes with time. Suppose there is an amount $y_1 = f(t_1)$ at time t_1, and an amount $y_2 = f(t_2)$ at time t_2.

Let $\Delta t = t_2 - t_1$ be the change in t, and $\Delta y = y_2 - y_1$ be the corresponding change in y over the timespan Δt starting at t_1.

Then the **average rate of change** of y is the total change in y divided by Δt, and we will denote this by v_{average}:

$$v_{\text{average}} = \frac{\Delta y}{\Delta t} = \frac{f(t_2) - f(t_1)}{t_2 - t_1} \tag{1.17}$$

If y is the *position*, for example, then v is the *average velocity* or *speed*, and is measured in km/sec. If y is the amount of gasoline in your car, it may be measured in gallons/day. If y is the number of gallons in a tank of water, then dy/dt is rate at which you use the water.

We observe that **the average rate of change is the slope of the secant line** to $f(t)$ from time $t = t_1$ to time $t = t_2$.

We use the term **instantaneous rate of change** to refer to the limit (see figure 1.5)

$$v_{\text{instantenous}}(t_1) = \lim_{t_2 \to t_1} \frac{f(t_2) - f(t_1)}{t_2 - t_1} = \lim_{\Delta t \to 0} \frac{\Delta y}{\Delta t} \tag{1.18}$$

which is precisely the derivative at t_1. Hence

$$v_{\text{instantaneous}}(t) = f'(t) = \lim_{a \to t} \frac{f(a) - f(t)}{a - t} = \lim_{h \to 0} \frac{f(t + h) - f(t)}{h} \tag{1.19}$$

The instantaneous rate of change at any time $t = a$ is the slope of the tangent line at $t = a$, and is equal to the value of the derivative $f'(a)$.

Example 1.2. In the absence of air resistance, the distance an object falls when it is dropped is related to the time since it has fallen by the equation

$$y = \frac{1}{2}gt^2 \tag{1.20}$$

where $g = 9.80665$ m/sec^2 is the Earth standard acceleration due to gravity. Estimate the velocity that a US penny has when it hits the ground if it is dropped off the top of the tallest building in Los Angeles (the US Bank tower, 310 Meters).

Solution. We first solve for the time it takes to hit the ground; solving (1.20) for t,

$$t = \sqrt{\frac{2y}{g}} = \sqrt{\frac{2 \times 310}{9.80665}} = 7.95125 \text{ sec} \tag{1.21a}$$

Figure 1.5: Average and instantaneous rate of change. The function $f(t)$ is plotted on the dependent axis and time on the independent axis. The instantaneous rate of change at any particular time, such as at the time points A and C (illustrated) is the slope of the tangent line at that point. The average rate of change is measured over and interval between two points, and is the slope of the secant line connecting the values of the function at the two endpoints of the interval $[A, B]$.

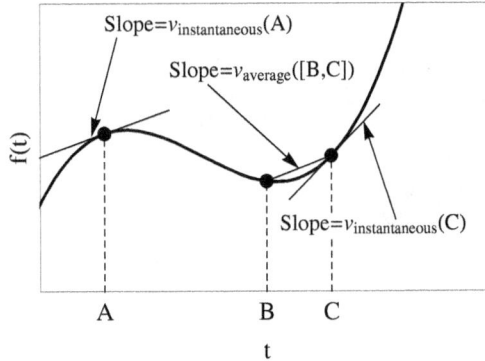

Next, we need to figure out the instantaneous velocity at $t = 7.95125$. The average velocity between t_1 and t_2 is

$$v_{\text{average}} = \frac{gt_2^2/2 - gt_1^2/2}{t_2 - t_1} = \frac{g}{2} \cdot \frac{(t_2 - t_1)(t_2 + t_1)}{t_2 - t_1} = \frac{g(t_2 + t_1)}{2} \tag{1.21b}$$

The instantaneous velocity is

$$v(t_1) = \lim_{t_2 \to t_1} \frac{g(t_2 + t_1)}{2} \tag{1.21c}$$

The instantaneous velocity at $t = 7.95125$ is then

$$v(7.95125) = \lim_{t \to 7.95125} \frac{g(t + 7.95125)}{2} \tag{1.21d}$$

Calculating this limit empirically from the left we obtain the following values:

t	$g(t + 7.95125)/2$
7.9	77.7238
7.95	77.9690
7.951	77.9739
7.9512	77.9749
7.95125	77.9751

Here is a table as we approach from the right

t	$g(t + 7.95125)/2$
8	78.2142
7.96	78.018
7.952	77.9788
7.9513	77.9754
7.95126	77.9752

Doing Calculus

Truncating to three decimals we estimate that the instantaneous velocity is 77.975 meters/second.

In fact, due to air resistance, the speed would probably only be about half that fast (because the equation we started with is actually incorrect). $\square$

Exercises

Empirically estimate the slope of each of the following functions at the specified point.

1. $y = x^3 - 4$ at $x = 2$ ans: 12

2. $y = \sin x$ at $x = \dfrac{\pi}{3}$ ans: 0.5

3. $y = x(x - 1)$ at $x = 0.5$ ans: 0

4. $y = x^4 - x^2$ at $x = 1$ ans: 2

5. $y = \sqrt{x}$ at $x = 2$ ans: 0.353553

6. $y = \dfrac{1}{\sqrt{x}}$ at $x = 4$ ans: -0.0625

Use an empirical estimate of the slope to find the equation of the tangent line to each of the following functions at the specified point.

7. $y = x^4 - x^2$ at $x = 1$ ans: $y = 2x - 2$

8. $y = x - x^3$ at $x = 2$ ans: $y = 16 - 11x$

9. $y = \tan x$ at $x = \dfrac{\pi}{4}$ ans: $y = 2x - 0.571$

10. $y = x \sin\left(\dfrac{1}{x}\right)$ at $x = \dfrac{1}{x}$
    ans: $y = \pi x - 1$

11. Suppose that a ball is thrown straight up into the air with an initial velocity of 50 ft/sec so that its height in feet after t seconds is given by $y = 50t - 16t^2$.

    (a) Find the average velocity over the period $1 \le t \le 1.5$

    (b) Find the average velocity over the period $1 \le t \le 1.1$

    (c) Find the average velocity over the period $1 \le t \le 1.01$

    (d) Estimate the instantaneous velocity at $t = 1$

12. The position of a particle moving in a straight line is given by $t^2 - 4t + 8$ meters, where t is measured in seconds.

    (a) Estimate the average velocity over the interval $[3, 4]$

    (b) Estimate the average velocity over the interval $[3.5, 4]$

    (c) Estimate the average velocity over the interval $[4, 5]$

    (d) Estimate the average velocity over the interval $[4, 4.5]$

    (e) Estimate the instantaneous velocity at $t = 4$.

Chapter 2

Limits

Chapter Summary and Goal

Limits in math are about getting closer and closer – as close as you possibly can – without necessarily getting all the way there. The classic example is Zeno's paradox. In order to run a mile, you have to first run a half the mile. Then you have run have the rest, which is a quarter mile, to get to the 3/4 mile mark. Then you have to run half the rest, which is 1/8 mile, to get to the 7/8 mile mark. Then you have run half the rest – well you get it. When you describe it this way, it seems like you never get there. How may miles do you actually run?

$$S = \frac{1}{2} + \frac{1}{4} + \frac{1}{8} + \frac{1}{16} + \frac{1}{32} + \frac{1}{64} + \cdots \tag{2.1}$$

This is a sum of an infinite number of terms and you have to run all of those distances to get where you are going, and yet you know the total is only a mile. How is this possible?

The technical answer is that if we define the function

$$S(n) = \frac{1}{2} + \frac{1}{4} + \frac{1}{8} + \frac{1}{16} + \cdots + \frac{1}{2^n} = 1 - \frac{1}{2^n} \tag{2.2}$$

we can make $S(n)$ as close to 1 as we like by letting n be as big as we like.

This is an example of a type of *limit* – when we can make the value of the function as close to some number L as we like (in this case L) – by choosing the value of the argument in a suitable range (in this case n sufficiently large), and we would write $\lim_{n \to \infty} S(n) = 1$.

We are going to start our investigation of limits (in this chapter) by looking at functions of a continuous variable x, rather than functions of an integer. We will return to limits of functions of integers like $S(n)$ when we study sequences and series.

Student Learning Objectives

The student will:

1. Understand the concept of a limit.
2. Understand the distinction between one-sided limits and two-sided limits.
3. Be able to locate vertical asymptotes in a function.
4. Be able to evaluate simple limits by examination or calculation.

The Concept of a limit

We can ask the following question: What happens to $f(x)$ as x gets very close to – but is not quite equal to – a? Think of pulling up to a stop sign: you are allowed to get as close as you like to the line drawn across the street but you are not allowed to touch or cross it.

When we do this we say that we are *approaching a. We are never, ever, allowed to actually touch the value of a when we approach a.* We can get as close we like. But we can't touch it. Like the line on the street. If we touch it, we have failed. A buzzer should go off in our heads and give us an imaginary traffic ticket.

We represent the process of approaching a symbolically by $x \to a$.

We can now restate our original question as this:

What happens to $f(x)$ as $x \to a$?

If $f(x)$ keeps getting closer to some number L in this process ($f(x)$ is allowed to actually reach L, but it is not required to) then we say that $f(x)$ *approaches* L as $x \to a$. We also write this as $f(x) \to L$, and express the whole process symbolically as

$$f(x) \to L \text{ as } x \to a \tag{2.3}$$

Another way to write this statement is as

$$\lim_{x \to a} f(x) = L \tag{2.4}$$

which is read as *the limit of $f(x)$ as x approaches a is L.*

When the function is **nice and smooth** our natural intuition will usually lead us to the correct answer to the following questions:

1. What happens to $f(x)$ as $x \to a$?
2. Is there a unique number L such that $f(x) \to L$ as $x \to a$?

However, sometimes our intuition will lead us very far astray, and for this reason we will need to be more precise in our calculation as time proceeds forward. For now we will remain somewhat ad-hoc and use definition 2.1 to explain what we mean by nice and smooth. This will allow us to begin to fine tune our intuitive fibers. In subsequent discussion more precise definitions of these terms will be presented.

Definition 2.1. Nice and Smooth Function

We say a function is nice and smooth if we can a plot of it without lifting the pencil from the paper.

Example 2.1. Consider the nice and smooth function $f(x) = x^2 - x + 2$. What happens to $f(x)$ as $x \to 2$?

Solution. We will make an educated guess at the answer, by attempting to approximate the limit experimentally. We will list a table of numbers that get closer and closer to 2, like 1.9, 1.99, 1.999, 1.9999, etc., and see what happens to $f(x)$ at each value of x. We also need to approach from above, and we do this by making a second table of numbers, like 2.1, 2.01, 2.001, 2.0001, etc.

For our educated guess to be considered a limit, the sequence of values of $f(x)$ that we calculate in each table must appear to be approaching the same number. This is what our tables look like (you should fill in the blanks yourself):

x (on left)	$f(x)$ (fill in value)	x (on right)	$f(x)$ (fill in value)
1.9		2.1	
1.99		2.01	
1.999		2.001	
$\vdots$			
$2 - h$		$2 + h$	

If your calculations were correct the numbers in each table should have seemed to have been approaching a value of 4. The drawing on the left side of figure 2.1 illustrates what happens: we are approaching the position $x = 2$ on the x-axis from both directions. In each case as we move along the curve of $f(x)$, we *approach* the dot, which corresponds to a *limit* of 4. In fact, the actual value of $f(x) = 4$ does not really matter; we are only

Figure 2.1: Two functions whose limit as $x \to 2$ is 4. The value of $f(4)$ is irrelevant to the value of the limit.

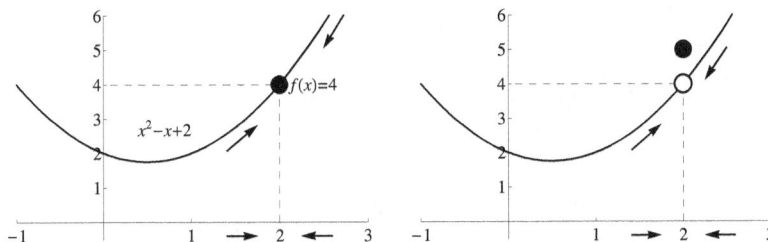

asking what happens as we get *very close* to $x = 4$. But in the table of numbers we **never actually include the value** $x = 2$, the numbers only keep getting closer and closer to 2. We would have obtained the same result if the picture looked like the one on the right hand side of figure 2.1. In the drawing on the right

$$f(x) = \begin{cases} x^2 - x + 2, & x \neq 2 \\ 5, & x = 2 \end{cases} \qquad (2.5a)$$

whereas in the drawing on the left

$$f(x) = x^2 - x + 2 \text{ for all } x \qquad (2.5b)$$

In both case, $\lim_{x \to 2} f(x) = 2$. $\square$

Definition 2.2. Limit.

If we can make the value of $f(x)$ as close to L as we like by choosing x sufficiently close to a (but not precisely equal to a), as we approach a from both sides (the left and the right), then we say that *the limit of $f(x)$ as x approaches a equals L*, which we may write (equivalently) as either

$$\lim_{x \to a} f(x) = L \qquad (2.6)$$

or

$$f(x) \to L \text{ as } x \to a \qquad (2.7)$$

Example 2.2. Demonstrate numerically that $\lim_{x \to 1} \dfrac{x-1}{x^2-1} = 0.5$

Solution. Doing a numerical computation as $x \to 1$ from both sides of $x = 1$ we produce the following table.

x	$f(x)$		x	$f(x)$
0.9	0.526316		1.1	0.47619
0.95	0.512821		1.05	0.487805
0.99	0.502513		1.01	0.497512
0.995	0.501253		1.005	0.498753
0.999	0.50025		1.001	0.49975
0.9995	0.500125		1.0005	0.499875
0.9999	0.500025		1.0001	0.499975

The first column and the second column **appear** to be approaching the same number, which is around 0.5. Thus we **estimate** that $f(x) \to 5$. ☐

Example 2.3. Demonstrate numerically that $\lim_{x \to 1} f(x)$ does not exist for

$$f(x) = \begin{cases} \dfrac{x+1}{x^2-1} & \text{if } x \neq 1 \\ 1.5 & \text{if } x = 1 \end{cases} \tag{2.9}$$

Solution. Doing a numerical computation as $x \to 1$ from both sides of $x = 1$ we produce the following table.

x	$f(x)$		x	$f(x)$
0.9	-10.		1.1	10.
0.95	-20.		1.05	20.
0.99	-100.		1.01	100.
0.995	-200.		1.005	200.
0.999	-1000.		1.001	1000.
0.9995	-2000.		1.0005	2000.
0.9999	-10000.		1.0001	10000.

The values approaching 1 from the left are getting large and negative; the values approaching 1 from the right are getting large and positive. Neither column appears to be converging. Thuse we conlude that the limit does not exist. ☐

Example 2.4. Use the calculator to guess $\lim_{x \to 0} \dfrac{\sin x}{x}$

Solution. The table of values we calculate is shown here.

x	$f(x)$
1	0.8414709848078965
0.1	0.9983341664682815
0.01	0.9999833334166665
0.001	0.9999998333333416
0.0001	0.9999999983333334
$\vdots$	
$x \to 0$	1

Thus we guess that $\lim_{x \to 0} \dfrac{\sin x}{x} = 1$ ☐

Unfortunately, we can't always use the guessing method, even with a computer, because of numerical errors. This is demonstrated in the following example.

Example 2.5. Estimate $\lim\limits_{x \to 0} \dfrac{\sqrt{x^2 + 4} - 2}{x^2}$

Solution.   As before we calculate a the values of the function at a sequence of numbers on both sides of $x = 0$. The results are illustrated in the following table.

x	$f(x)$	x	$f(x)$
1.	0.236068	-1.	0.236068
0.5	0.246211	-0.5	0.246211
0.1	0.249844	-0.1	0.249844
0.05	0.249961	-0.05	0.249961
0.01	0.249998	-0.01	0.249998
0.001	0.25	-0.001	0.25
0.0001	0.25	-0.0001	0.25
0.00001	0.25	-0.00001	0.25
10^{-6}	0.250022	-10^{-6}	0.250022
10^{-7}	0.222045	-10^{-7}	0.222045
10^{-8}	0.	-10^{-8}	0.
10^{-9}	0.	-10^{-9}	0.
10^{-10}	0.	-10^{-10}	0.

Starting at about 10^{-6} and smaller a numerical error arises, and beyond 10^{-8} the calculator actually gives an answer of 0. The correct answer is actually 0.25. $\square$

Another problem with estimating the limit numerically is that the answer can be misleading. For example, if a function is highly oscillatory, the numbers we choose might actually happen to hit right on a sequence that converges.

Example 2.6. Estimate $\lim\limits_{x \to 0} \sin \dfrac{\pi}{x}$

Solution. Our first sequence of guesses pretty clearly suggests that

$$\lim_{x \to 0} \sin \frac{\pi}{x} = 0 \qquad (2.13a)$$

as we can see here,

x	$f(x)$	x	$f(x)$
1	0	-1	0
0.5	0	-0.5	0
0.1	0	-0.1	0
0.05	0	-0.05	0
0.01	0	-0.01	0
0.005	0	-0.005	0
0.001	0	-0.001	0
0.005	0	-0.005	0
0.0001	0	-0.0001	0
0.00005	0	-0.00005	0
0.00001	0	-0.00001	0

Now here's another sequence of numbers approaching zero from both directions:

x	$f(x)$		x	$f(x)$
0.3	-0.86603		-0.3	0.86603
0.09	-0.34202		-0.09	0.34202
0.07	0.78183		-0.07	-0.78183
0.03	-0.86603		-0.03	0.86603
0.009	-0.34202		-0.009	0.34202
0.007	0.43388		-0.007	-0.43388
0.003	-0.86603		-0.003	0.86603
0.0009	-0.34202		-0.0009	0.34202
0.0007	0.97493		-0.0007	-0.97493

Now the numbers are all over the place. What is going on here? The answer can

Figure 2.2: Plot of $y = \sin \frac{\pi}{x}$ illustrating how the numbers were chosen in example 2.6. The numbers approaching from the left (marked with a symbol $\otimes$) were used in the first calculation, which approached zero; the numbers approaching from the right (marked with symbol $\odot$) were used in the second calculation. The arrows indicate the x values, and the symbols the $f(x)$ values.

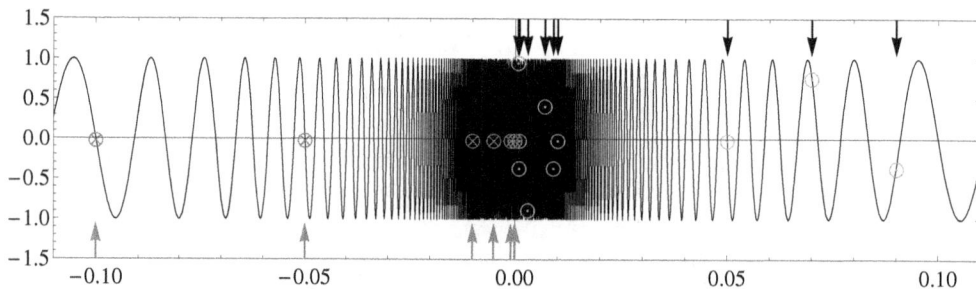

be found in figure 2.2. The function is highly oscillatory, and getting infinitely more oscillatory as $x \to 0$. There is no limit - we just happened to hit on a bunch of zeros with a lucky guess the first time through.   □

Left and Right Limits

The rules for calculating limits get more complicated when a function is not nice and smooth. A typical example is the **unit step function** which has a unit jump from $y = 0$ to $y = 0$ as x increases across the origin from negative values to positive values.

Definition 2.3. Unit Step Function

The unit step function (figure 2.3) at the origin is defined by

$$U(x) = \begin{cases} 0 & \text{if } x < 0 \\ 1 & \text{if } x \geq 0 \end{cases} \tag{2.14}$$

You need to lift up your pencil to draw the curve past the origin, so the function is not nice and smooth, at least not at the point $x = 0$. The difficulty is illustrated in example 2.7.

Example 2.7. Estimate $\lim_{x \to 0} U(x)$ numerically.

Figure 2.3: A unit step function at the origin.

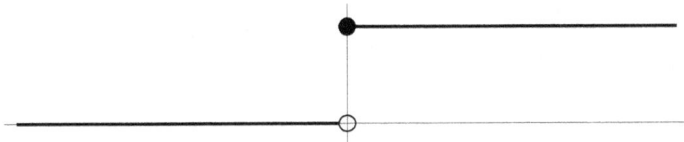

Solution. Following our usual numerical process, we'll fill in a table. Here's what happens.

x	$U(x)$		x	$U(x)$
-1.0	0		1.0	1
-0.1	0		0.1	1
-0.01	0		0.01	1
-0.001	0		0.001	1
-0.0001	0		0.0001	1
-0.00001	0		0.00001	1

As $x \to 0$ from the left, it appears that $U(x) \to 0$.

As $x \to 0$ from the right, it appears that $U(x) \to 1$.

Since these numbers are different, then we conclude that $\lim\limits_{x \to 0} U(x)$ does not exist (DNE is a common shorthand for this).   □

This example suggests that it may be useful to define a more general concept of a limit: a limit from just the left, or a limit from just the right. For the regular limit to exist, both of these one-sided limits must exist, and must be equal to the same number.

Definition 2.4. Left Hand Limit

If we can make the value of $f(x)$ as close to L as we like by choosing x as close to a as we like *on the left* (i.e., $x < a$) (but never equal to a), then we write

$$\lim_{x \to a^-} f(x) = L \tag{2.16}$$

or alternatively,

$$f(x) \to L \text{ as } x \to a^- \tag{2.17}$$

We say this: *"The limit as x approaches a from the left of $f(x)$ is L."*

Definition 2.5. Right Hand Limit

If we can make the value of $f(x)$ as close to L as we like by choosing x as close to a as we like *on the right* (i.e., $x > a$) (but never equal to a), then we write

$$\lim_{x \to a^+} f(x) = L \tag{2.18}$$

or alternatively,

$$f(x) \to L \text{ as } x \to a^+ \tag{2.19}$$

We say this: *"The limit as x approaches a from the right of $f(x)$ is L."*

Doing Calculus

Thus for the limit $\lim_{x \to a} f(x)$ to exist, and have some value L, all of the following must be true:

1. $\lim_{x \to a^-} f(x)$ must exist; and

2. $\lim_{x \to a^+} f(x)$ must exist; and

3. For some number L_1, $\lim_{x \to a^-} f(x) = L_1$; and

4. For some number L_2, $\lim_{x \to a^+} f(x) = L_2$; and

5. $L_1 = L_2 = L$

This is stated more concisely in theorem 2.1.

Theorem 2.1. One Sided Limit Theorem

The limit $\lim_{x \to a} f(x) = L$ if and only if $\lim_{x \to a-} f(x) = L$ and $\lim_{x \to a^+} f(x) = L$

Remark

The value of $f(a)$ is irrelevant in computing any of the limits as $x \to a$, $x \to a^-$, or $x \to a^+$. It does not even have to be defined.

If and Only If Statements. The "if and only if" statement in theorem 2.1 is a type of statement that we will use frequently as we advance further in mathematics. It comes from the study of mathematical logic. It is commonly represented by the symbol " $\iff$ ", or written in shorthand as the word iff (an if with the letter f repeated). So if P and Q are statements about something, like "John likes the apples" (P) and "The apples in the store are fresh" (Q), then $P \iff Q$ is equivalent to "John likes the applies if and only if the apples in the store are fresh."

What does this " $\iff$ " really mean, though? The double arrow gives us a hint. It comes from putting together two single arrows, each of which means "implies." The statement $Q \implies P$ is read as "Q implies P." Another way of reading $Q \implies P$ is "If Q, then P".

In the previous paragraph, $Q \implies P$ means that "If the apples in the store are fresh then John likes the apples."

The statement $P \iff Q$ means that both of the following are true:

1. $P \implies Q$, i.e., if P is true, then Q is true.

2. $Q \implies P$, i.e., if Q is true, then P is true.

Thus $P \iff Q$ means that both of the following are true:

1. If John likes then apples, then the applies in the store are fresh; and

2. If the apples in the store are fresh, then John likes the apples.

Taking this back to theorem 2.1 we had a statement

$$\lim_{x \to a} f(x) = L \iff \lim_{x \to a^-} f(x) = L \text{ and } \lim_{x \to a^+} f(x) = L \tag{2.20}$$

This means that both of the following statements are true.

1. If $\lim\limits_{x \to a} f(x) = L$ then $\lim\limits_{x \to a^-} f(x) = L$ and $\lim\limits_{x \to a^+} f(x) = L$.

2. If $\lim\limits_{x \to a^-} f(x) = L$ and $\lim\limits_{x \to a^+} f(x) = L$ are both true, then $\lim\limits_{x \to a} f(x) = L$

Example 2.8. Evaluate each of the following limits if figure 2.4.

(a) $\lim\limits_{x \to 0^-} f(x)$

(b) $\lim\limits_{x \to 0^+} f(x)$

(c) $\lim\limits_{x \to 0} f(x)$

(d) $\lim\limits_{x \to 1^-} f(x)$

(e) $\lim\limits_{x \to 1^+} f(x)$

(f) $\lim\limits_{x \to 1} f(x)$

(g) $\lim\limits_{x \to 2^-} f(x)$

(h) $\lim\limits_{x \to 2^+} f(x)$

(i) $\lim\limits_{x \to 2} f(x)$

(j) $\lim\limits_{x \to -1^-} f(x)$

(k) $\lim\limits_{x \to -1^+} f(x)$

(l) $\lim\limits_{x \to -1} f(x)$

Figure 2.4: Function to be evaluated for example 2.8.

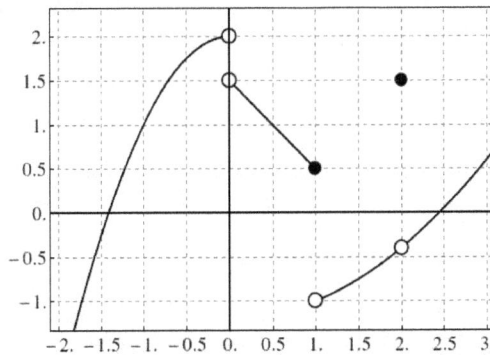

Solution.

(a) $\lim\limits_{x \to 0^-} f(x) = 2$; the fact that $f(x)$ is undefined at $x = 0$ does not matter, only the value that f is approaching, from the left, as $x \to 0$ from the left.

(b) $\lim\limits_{x \to 0^+} f(x) = 1.5$; the argument is the same as in part (a). Only the value of f as it is approaching from the right matters, not the fact that $f(0)$ is undefined.

(c) $\lim\limits_{x \to 0} f(x)$ DNE (does not exist) because the limit from the left (part (a)) and the limit from the right (part(b)) are different.

(d) $\lim\limits_{x \to 1^-} f(x) = 0.5$. Same argument as in part (a). Even though $f(1) = .5$ this does not matter.

(e) $\lim\limits_{x \to 1^+} f(x) = -1$. The value of $f(1)$ does not matter. The function is approaching -1, even though the value of $f(1) = .5 \neq -1$, the limit from the right is still -1.

(f) $\lim\limits_{x \to 1} f(x)$ DNE (does not exist) because the limit from the left is .5 and the limit from the right is -1, and these are not equal.

(g) $\lim\limits_{x \to 2^-} f(x) = -.5$

(h) $\lim\limits_{x \to 2^+} f(x) = -.5$

(i) $\lim\limits_{x \to 2} f(x) = -.5$ because the limits from the right and the left are equal at this point.

(j) $\lim\limits_{x \to -1^-} f(x) = 1$; the function is nice and smooth here, so there is no confusion in any of the three limits.

(k) $\lim\limits_{x \to -1^+} f(x) = 1$

(l) $\lim\limits_{x \to -1} f(x) = 1$

$\square$

Infinite Limits

Sometimes a function $f(x)$ will take on larger and larger values as $x \to a$. In some cases these values become unbounded, and we the limit has a value of ∞ or $-\infty$.

Definition 2.6. Infinite Limits

Suppose that $f(x)$ is defined for all x in some interval that includes a (except possibly at a itself). If the values of $f(x)$ can me made as large as we like by taking x sufficiently close to a (but not equal to a) then

$$\lim_{x \to a} f(x) = \infty \tag{2.22}$$

or equivalently

$$f(x) \to \infty \text{ as } x \to a \tag{2.23}$$

These are called infinite limits, and occur when functions have denominators that approach zero. For example, consider the function $y = \dfrac{x}{(x-5)^2}$, which is illustrated on the left in figure 2.5. If we set the denominator equal to zero in the function, we found points in the domain of the function where it is undefined because we are attempting to divide by zero.

$$(x-5)^2 = 0 \implies (x-5) = 0 \implies x = 5 \tag{2.24}$$

We say in this case that $\lim\limits_{x \to 5} \dfrac{x}{(x-5)^2} = \infty$

Figure 2.5: Both of the functions $y = x/(x-5)^2$ (shown on the left) and $y = x/(x-5)$ (shown on the right) have vertical asymptotes at $x = 5$.

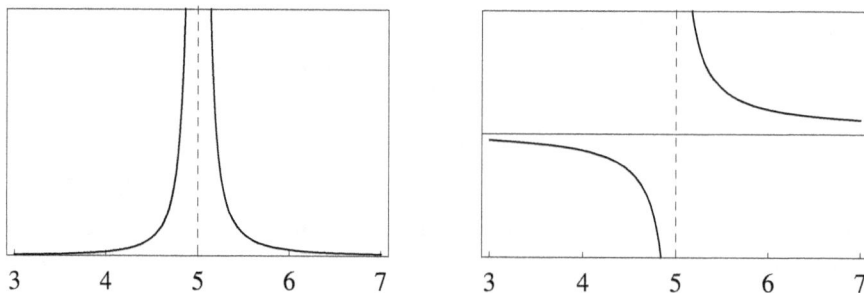

We can make a similar definition for limits that have values of $-\infty$.

Definition 2.7. Infinite Limits $(-\infty)$

Suppose that $f(x)$ is defined for all x in some interval that includes a (except possibly at a itself). If the values of $f(x)$ can me made as large (negative) as we like by taking x sufficiently close to a (but not equal to a) then

$$\lim_{x \to a} f(x) = -\infty \qquad (2.25)$$

or equivalently

$$f(x) \to -\infty \text{ as } x \to a \qquad (2.26)$$

We also allow limits from the left and limits from the right to be infinite. If any of the limits are infinite (either to ∞ or $-\infty$) – from the left, as $x \to a^-$; the right, as $x \to a^+$; or as $x \to a$ – then we say the function has a vertical asymptote at $x = a$.

Definition 2.8. Vertical Asymptote

The line $x = a$ is called a ***vertical asymptote*** of $f(x)$ if any of the following are true:

$$\begin{array}{lll}
\lim_{x \to a} f(x) = \infty & \lim_{x \to a^-} f(x) = \infty & \lim_{x \to a^+} f(x) = \infty \\
\lim_{x \to a} f(x) = -\infty & \lim_{x \to a^-} f(x) = -\infty & \lim_{x \to a^+} f(x) = -\infty
\end{array} \qquad (2.27)$$

For example the function

$$f(x) = \frac{x}{x - 5} \qquad (2.28)$$

(see figure 2.5) has a limit from the left of $-\infty$ and a limit from the right of ∞:

$$\lim_{x \to 5^+} \frac{x}{x - 5} = \infty \qquad (2.29)$$

$$\lim_{x \to 5^-} \frac{x}{x - 5} = -\infty \qquad (2.30)$$

Exercises

1. The function $f(x)$ is plotted in the following figure.

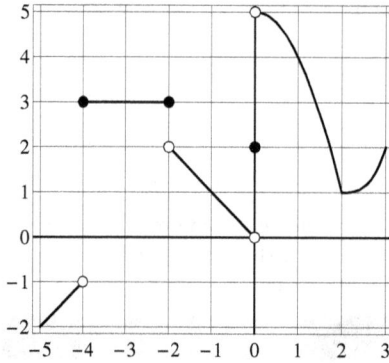

Based on this figure, find each of the following limits.

(a) $\lim\limits_{x \to -4^-} f(x)$ (g) $\lim\limits_{x \to 0^-} f(x)$

(b) $\lim\limits_{x \to -4^+} f(x)$ (h) $\lim\limits_{x \to 0^+} f(x)$

(c) $\lim\limits_{x \to -4} f(x)$ (i) $\lim\limits_{x \to 0} f(x)$

(d) $\lim\limits_{x \to -2^-} f(x)$ (j) $\lim\limits_{x \to 2^-} f(x)$

(e) $\lim\limits_{x \to -2^+} f(x)$ (k) $\lim\limits_{x \to 2^+} f(x)$

(f) $\lim\limits_{x \to -2} f(x)$ (l) $\lim\limits_{x \to 2} f(x)$

2. The function $f(x)$ is plotted in the following figure.

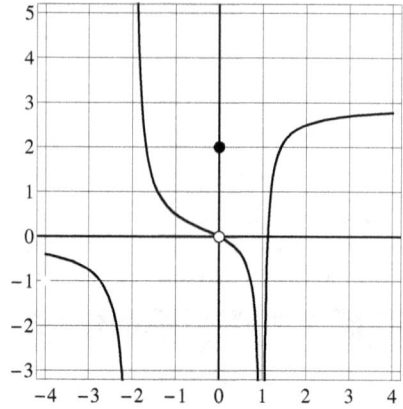

Based on this figure, find each of the following limits.

(a) $\lim\limits_{x \to -2^-} f(x)$ (f) $\lim\limits_{x \to 0} f(x)$

(b) $\lim\limits_{x \to -2^+} f(x)$ (g) $\lim\limits_{x \to 1^-} f(x)$

(c) $\lim\limits_{x \to -2} f(x)$

(d) $\lim\limits_{x \to 0^-} f(x)$ (h) $\lim\limits_{x \to 1^+} f(x)$

(e) $\lim\limits_{x \to 0^+} f(x)$ (i) $\lim\limits_{x \to 1} f(x)$

Estimate each of the following limits empirically.

3. $\lim\limits_{x \to 5} (x^2 + 1)$ Ans: 26

4. $\lim\limits_{x \to 5} \dfrac{x^2 - 5}{x + 5}$ Ans: 2

5. $\lim\limits_{x \to 5} \dfrac{x^2 - 25}{x - 5}$ Ans: 10

6. $\lim\limits_{x \to -5} \dfrac{x^2 - 25}{x - 5}$ Ans: 0

7. $\lim\limits_{h \to 0} \dfrac{\dfrac{1}{3 + h} - \dfrac{1}{3}}{h}$ Ans: -1/9

8. $\lim\limits_{h \to 0} \dfrac{\sqrt{4 + h} - 2}{h}$ Ans: 1/4

Chapter 3

Limit Laws

Chapter Summary and Goal

The calculation of limits would be brutally tedious if we had to get out our calculators every time we needed to know a limit. In fact, as we saw in example 2.6, we can't always trust our calculators.

Fortunately there are better ways to calculate limits. We will study these *limit laws* in this chapter.

Student Learning Objectives

The student will:

1. Be able to understand, justify, and explain how to use the limit laws.
2. Be able to calculate limits using the limit laws.
3. Be able to understand and apply the squeeze theorem.

Rules for Calculating Limits

Theorem 3.1. The Limit of a Constant is a Constant

If c is any constant,
$$\lim_{x \to a} c = c \qquad (3.1)$$

For example,
$$\lim_{x \to 12} 17 = 17 \qquad (3.2)$$

Theorem 3.2. Limit of a Linear Function

If a is any number, then
$$\lim_{x \to a} x = a \qquad (3.3)$$

For example,
$$\lim_{x \to 12} x = 12 \tag{3.4}$$

Theorem 3.3. Sum andDifference Rule

If the limits $\lim_{x \to a} f(x)$ and $\lim_{x \to a} g(x)$ both exist then

$$\lim_{x \to a}[f(x) + g(x)] = \lim_{x \to a} f(x) + \lim_{x \to a} g(x) \tag{3.5}$$
$$\lim_{x \to a}[f(x) - g(x)] = \lim_{x \to a} f(x) - \lim_{x \to a} g(x) \tag{3.6}$$

For example,
$$\lim_{x \to 3}(4x - 5) = \lim_{x \to 3}(4x) - \lim_{x \to 3} 5 \tag{3.7}$$

Theorem 3.4. Constant Multiple Rule

If c is any constant and the limit $\lim_{x \to a} f(x)$ exists then

$$\lim_{x \to a}[cf(x)] = c \lim_{x \to a} f(x) \tag{3.8}$$

Example 3.1. Find $\lim_{x \to 3}(4x - 5)$

Solution.

$$\lim_{x \to 3}(4x - 5) = \lim_{x \to 3}(4x) - \lim_{x \to 3} 5 \qquad \text{(theorems 3.3 and 3.1)} \tag{3.9a}$$
$$= 4 \lim_{x \to 3}(x) - 5 \qquad \text{(theorem 3.4)} \tag{3.9b}$$
$$= 4(3) - 5 = 7 \qquad \text{(theorem 3.2)} \tag{3.9c}$$

$\square$

Theorem 3.5. Limit of a Product

The limit of a product is the product of the limits: if both of the limits $\lim_{x \to a} f(x)$ and $\lim_{x \to a} g(x)$ exist, then

$$\lim_{x \to a}[f(x)g(x)] = \left(\lim_{x \to a} f(x)\right)\left(\lim_{x \to a} g(x)\right) \tag{3.10}$$

Example 3.2. Find $\lim_{x \to 5}(x + 7)(x + 3)$

Solution.

$$\lim_{x \to 5}(x + 7)(x + 3)$$
$$= \left(\lim_{x \to 5}(x + 7)\right)\left(\lim_{x \to 5}(x + 3)\right) \qquad \text{(theorem 3.5)} \tag{3.11a}$$
$$= (5 + 7)(5 + 3) \qquad \text{(theorems 3.2, 3.1)} \tag{3.11b}$$
$$= 90 \tag{3.11c}$$

□

> ## Theorem 3.6. Limit of a Quotient
>
> If the limits $\lim\limits_{x \to a} f(x)$ and $\lim\limits_{x \to a} g(x)$ both exist, and the limit $\lim\limits_{x \to a} g(x) \neq 0$, then
>
> $$\lim_{x \to a} \frac{f(x)}{g(x)} = \frac{\lim\limits_{x \to a} f(x)}{\lim\limits_{x \to a} g(x)} \tag{3.12}$$

Example 3.3. Find $\lim\limits_{x \to 5} \dfrac{x+7}{x+3}$

Solution.

$$\lim_{x \to 5} \frac{x+7}{x+3} = \frac{\lim_{x \to 5}(x+7)}{\lim_{x \to 5}(x+3)} = \frac{5+7}{5+3} = \frac{12}{8} = \frac{3}{2} \tag{3.13a}$$

□

> ## Theorem 3.7. Limit of a Power
>
> The limit of a power is the power of the limit: if the limit $\lim\limits_{x \to a} f(x)$ exists, then
>
> $$\lim_{x \to a} (f(x))^n = \left(\lim_{x \to a} f(x) \right)^n \tag{3.14}$$
>
> for any positive integer n.

Example 3.4. Find $\lim\limits_{x \to 5} (x+12)^3$

Solution.

$$\lim_{x \to 5} (x+12)^3 = \left(\lim_{x \to 5} (x+12) \right)^3 = (5+12)^3 = 4913 \tag{3.15a}$$

□

> ## Theorem 3.8. Limit of a Root
>
> The limit of a power is the power of the limit: if the limit $\lim\limits_{x \to a} f(x)$ exists, then
>
> $$\lim_{x \to a} \sqrt[n]{f(x)} = \sqrt[n]{\lim_{x \to a} f(x)} \tag{3.16}$$
>
> for any positive integer n.

Example 3.5. Find $\lim\limits_{x \to 5} \sqrt{x+11}$

Solution.

$$\lim_{x \to 5} \sqrt{x+11} = \sqrt{\lim_{x \to 5}(x+11)} = \sqrt{5+11} = \sqrt{16} = 4 \tag{3.17a}$$

□

The limit laws allow us to considerably simplify the calculation of limits. For example, when $f(x)$ is either a polynomial or a quotient of polynomials, we can substitute in directly. The see this, we observe that a polynomial is a sum of terms of the form $c_j x^j$, where j is a non-negative integer, as in

$$f(x) = c_0 + c_1 x + c_2 x^2 + \cdots + c_n x^n \tag{3.18}$$

We have seen (by combing theorems 3.3, 3.4, and 3.7) that we can write

Table of Limit Laws

$\lim_{x \to a} C = C$	(3.19)		$\lim_{x \to a} x = a$	(3.20)	
$\lim_{x \to a} [f(x) \pm g(x)] = \lim_{x \to a} f(x) \pm \lim_{x \to a} g(x)$	(3.21)		$\lim_{x \to a} [Cf(x)] = C \lim_{x \to a} f(x)$	(3.22)	
$\lim_{x \to a} [f(x)g(x)] = \left(\lim_{x \to a} f(x) \right) \left(\lim_{x \to a} g(x) \right)$	(3.23)		$\lim_{x \to a} \dfrac{f(x)}{g(x)} = \dfrac{\lim_{x \to a} f(x)}{\lim_{x \to a} g(x)}$	(3.24)	
$\lim_{x \to a} (f(x))^n = \left(\lim_{x \to a} f(x) \right)^n$	(3.25)		$\lim_{x \to a} \sqrt[n]{f(x)} = \sqrt[n]{\lim_{x \to a} f(x)}$	(3.26)	

$$\lim_{x \to a} f(x) = \lim_{x \to a} \left(c_0 + c_1 x + c_2 x^2 + c_3 x^3 + \cdots + c_n x^n \right) \tag{3.27}$$

$$= \lim_{x \to a} c_0 + \lim_{x \to a} c_1 x + \lim_{x \to a} c_2 x^2 + \cdots + \lim_{x \to a} c_n x^n \tag{3.28}$$

$$= c_0 + c_1 \lim_{x \to a} x + c_2 \lim_{x \to a} x^2 + \cdots + c_n \lim_{x \to a} x^n \tag{3.29}$$

$$= c_0 + c_1 a + c_2 a^2 + \cdots + c_n a^n \tag{3.30}$$

$$= f(a) \tag{3.31}$$

Similarly, a rational function $r(x) = f(x)/g(x)$ is the quotient of two polynomials, so by the theorem 3.6 and equation 3.31, so long as $g(a) \neq 0$,

$$\lim_{x \to a} r(x) = \lim_{x \to a} \frac{f(x)}{g(x)} \tag{3.32}$$

$$= \frac{\lim_{x \to a} f(x)}{\lim_{x \to a} g(x)} \tag{3.33}$$

$$= \frac{f(a)}{g(a)} \tag{3.34}$$

$$= r(a) \tag{3.35}$$

Thus we have the following direct substitution rule.

Theorem 3.9. Direct Substitution.

If $f(x)$ is a rational function or polynomial and x is in the domain of $f(x)$ then $\lim_{x \to a} f(x) = f(a)$.

Example 3.6. Find $\lim_{x \to 2} (-3x^2 + 6x - 7)$

Solution. Since $f(x)$ is a polynomial we can use direct substitution;

$$\lim_{x \to 2} (-3x^2 + 6x - 7) = -3(2^2) + 6(2) - 7 = -7 \tag{3.36a}$$

$\square$

Example 3.7. Find $\lim_{x \to 2} \dfrac{x^3 + 2x^2 + 1}{5 - 3x}$

Solution. This is a rational function so we will try direct substitution. The only thing we have to check for is to make sure the denominator is not zero.

$$\lim_{x \to 2} \frac{x^3 + 2x^2 + 1}{5 - 3x} = \frac{2^3 + 2(2^2) + 1}{5 - 3(2)} = \frac{8 + 8 + 1}{5 - 6} = -17 \tag{3.37a}$$

$\square$

When the denominator is zero in a rational function, we cannot use direct substitution. For example, the following calculation using direct substitution is somewhat perplexing.

$$\lim_{x \to -1} \frac{x^2 + x}{x^2 - 3x - 4} \to \text{ with direct substitution} \tag{3.38a}$$

$$\to \frac{(-1)^2 + (-1)}{(-1)^2 - 3(-1) - 4} \tag{3.38b}$$

$$\to \frac{0}{0} \tag{3.38c}$$

The question is, what does a limit of $0/0$ mean? We cannot immediately say that the limit is undefined! In fact, sometimes these limits are quite well defined. Theorem 3.10 tells us what to do: we replace $f(x)$ with another function $g(x)$ that is equal to $f(x)$ everywhere except at $g(x)$, but which does not have a denominator that is equal to zero.

Theorem 3.10. Limit of Equivalent Function

If $f(x) = g(x)$ for all $x \neq a$ then

$$\lim_{x \to a} f(x) = \lim_{x \to a} g(x) \tag{3.39}$$

There are a lot of tricks to using theorem 3.10, but the most common one is factoring.

Example 3.8. Find $\displaystyle\lim_{x \to -1} \frac{x^2 + x}{x^2 - 3x - 4}$

Solution. As we saw in equations 3.38a, simple substitution gives $0/0$. So instead we try to simplify the function by factoring it.

$$\lim_{x \to -1} \frac{x^2 + x}{x^2 - 3x - 4} = \lim_{x \to -1} \frac{x(x + 1)}{(x - 4)(x + 1)} \tag{3.40a}$$

$$= \lim_{x \to -1} \frac{x}{x - 4} \tag{3.40b}$$

$$= \frac{1}{5} \tag{3.40c}$$

$\square$

The following example demonstrates how we can get a finite answer by subtracting $\infty - \infty$.

Example 3.9. Find $\displaystyle\lim_{t \to 0} \left(\frac{1}{t} - \frac{1}{t^2 + t} \right)$

Solution. This is the difference of two limits, each of which appears to be infinite, so it looks like we are subtracting ∞ from ∞, which doesn't seem to make any more sense than dividing zero by zero. However, we are not allowed to write

$$\lim_{t \to 0} \left(\frac{1}{t} - \frac{1}{t^2 + t} \right) = \lim_{t \to 0} \frac{1}{t} - \lim_{t \to 0} \frac{1}{t^2 + t} \tag{3.41a}$$

because, in fact, neither of these limits exists (each term has a different left and right limit, one going to ∞ and the other to $-\infty$). Theorem 3.3 only applies when both limits exist. Instead, we use a different method.

Here the trick is to put things over a common denominator, and then factor, and cancel common factors.

$$\lim_{t \to 0} \left(\frac{1}{t} - \frac{1}{t^2 + t} \right) = \lim_{t \to 0} \left(\frac{t^2 + t - t}{t(t^2 + t)} \right) \tag{3.41b}$$

$$= \lim_{t \to 0} \frac{t^2}{t(t)(t+1)} \tag{3.41c}$$

$$= \lim_{t \to 0} \frac{1}{t+1} = 1 \tag{3.41d}$$

$\square$

Example 3.10. Find $\lim\limits_{x \to 2} \dfrac{\sqrt{5x + 6} - 4}{x - 2}$

Solution. This is a 0/0 form, but there is nothing to factor and nothing to put over a common denominator, so we need a new trick. We rationalize the numerator by multiplying the entire fraction by 1.

$$\lim_{x \to 2} \frac{\sqrt{5x + 6} - 4}{x - 2} = \lim_{x \to 2} \frac{\sqrt{5x + 6} - 4}{x - 2} \times \left(\frac{\sqrt{5x + 6} + 4}{\sqrt{5x + 6} + 4} \right) \tag{3.42a}$$

$$= \lim_{x \to 2} \frac{(5x + 6) - 16}{(x - 2)(\sqrt{5x + 6} + 4)} \tag{3.42b}$$

$$= \lim_{x \to 2} \frac{(5x - 10)}{(x - 2)(\sqrt{5x + 6} + 4)} \tag{3.42c}$$

$$= \lim_{x \to 2} \frac{5(x - 2)}{(x - 2)(\sqrt{5x + 6} + 4)} \tag{3.42d}$$

$$= \lim_{x \to 2} \frac{5}{\sqrt{5x + 6} + 4} \tag{3.42e}$$

$$= \frac{5}{\sqrt{16} + 4} = \frac{5}{8} \tag{3.42f}$$

$\square$

Squeeze Theorem

Our final limit law encapsulates the following intuitive result: if a function is trapped or squeezed between two other functions as $x \to a$ then it can only go to one place. And that is the the same place that the functions doing the squeezing do. It's like you are on a luge or a water slide. You can only go where the slide takes you.

Theorem 3.11. Squeeze Theorem

If $f(x) \leq g(x) \leq h(x)$ when x is near a and

$$\lim_{x \to a} f(x) = L = \lim_{x \to a} h(x) \tag{3.43}$$

then

$$\lim_{x \to a} g(x) = L \tag{3.44}$$

Example 3.11. Show that

$$\lim_{x \to 0} \left(x^2 \sin \frac{1}{x} \right) = 0 \tag{3.45}$$

by squeezing it between $f(x) = -x^2$ and $h(x) = x^2$

Solution. Note that we **cannot say** (you should explain this as an exercise)

$$\lim_{x \to 0} \left(x^2 \sin \frac{1}{x} \right) = \left(\lim_{x \to 0} x^2 \right) \left(\lim_{x \to 0} \sin \frac{1}{x} \right) \tag{3.46a}$$

However, we do observe that that since

Figure 3.1: Plot of $f(x) = x^2 \sin(1/x)$ demonstrating that it is bounded about by x^2 and below by $-x^2$.

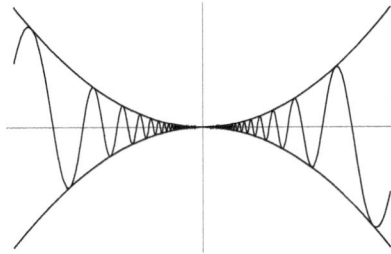

$$-1 \leq \sin \frac{1}{x} < 1, \tag{3.46b}$$

then we also know that (multiply through by x^2),

$$-x^2 \leq x^2 \sin \frac{1}{x} < x^2 \tag{3.46c}$$

Since the limits of the functions the left and right are both zero,

$$\lim_{x \to 0} (-x^2) = 0 \tag{3.46d}$$

$$\lim_{x \to 0} x^2 = 0 \tag{3.46e}$$

The function in the middle gets squeezed between the two zeros, and hence we must have $x^2 \sin \frac{1}{x} \to 0$ as $x \to 0$. $\qquad\square$

---

[1]AKA the *Two Policemen Theorem*: If you are walking between two policemen who are going to the same place as each other, you will go wherever they are going. Sometimes called the *Two Policemen and a Drunk Theorem*: no matter how much the drunk stumbles, the policemen will eventually get to the station. In some countries, it is called the *Three Policemen Theorem*, because the drunk is also a policeman. Other names are the *pinching theorem* and the *sandwich theorem*.

Exercises

Use the limit laws to find each of the following limits.

1. $\lim\limits_{x \to 3} 7x$ Ans: 21

2. $\lim\limits_{x \to 5} (12x + 19)$ Ans: 79

3. $\lim\limits_{x \to 4} (x^2 - 14x + 15)$ Ans: -25

4. $\lim\limits_{x \to 2} \dfrac{x - 3}{x + 3}$ Ans: -1/5

5. $\lim\limits_{x \to 3} \dfrac{x^2 - 9}{x^2 + 9}$ Ans: 0

6. $\lim\limits_{x \to 3} \dfrac{x^2 - 9}{x - 3}$ Ans: 6

7. $\lim\limits_{x \to 4} \sqrt{16 - x^2}$ Ans: 0

8. $\lim\limits_{x \to 3} \sqrt[4]{x^5 + 3x + 4}$ Ans: 4

9. $\lim\limits_{x \to 4} \dfrac{\sqrt{x^2 - 16}}{\sqrt{x - 4}}$ Ans: $2\sqrt{2}$

10. $\lim\limits_{x \to 4} \dfrac{x - 4}{x^2 - 2x - 8}$ Ans: 1/6

11. $\lim\limits_{x \to 4} \dfrac{x - 2}{x^2 + 3x - 10}$ Ans: 1/9

12. $\lim\limits_{x \to 3} \dfrac{5x^2 - 45}{x^3 + 2x^2 - 9x - 18}$ Ans: 1

13. $\lim\limits_{x \to 2} \dfrac{12 - 6x}{x^3 - 8}$ Ans: $-\dfrac{1}{2}$

14. $\lim\limits_{x \to 1} \dfrac{\sqrt{x + 3} - 2}{x - 1}$ Ans: 1/4

15. $\lim\limits_{x \to 4} \dfrac{\sqrt{36(x + 5)} - 18}{x - 4}$ Ans: 1

16. $\lim\limits_{h \to 0} \dfrac{(7 + h)^2 - 49}{h}$ Ans: 14

17. $\lim\limits_{h \to 0} \dfrac{(2 + h)^3 - 8}{h}$ Ans: 12

18. $\lim\limits_{h \to 0} \dfrac{\dfrac{3}{4 + h} - \dfrac{3}{4}}{h}$ Ans: -3/16

Royal Solver

Chapter 4

Limits of Trigonometric Functions

Chapter Summary and Goal

In this chapter we will apply the limit laws to various combinations of trigonometric and algebraic functions. The most important of these limits are

$$\lim_{\theta \to 0} \frac{\sin \theta}{\theta} = 1, \qquad \lim_{\theta \to 0} \frac{1 - \cos \theta}{\theta} = 0 \qquad (4.1)$$

Student Learning Objectives

The student will:

1. Learn how to apply the limit laws to trigonometric functions.
2. Learn how to use basic geometry to derive limit laws.
3. Learn how to apply limit rules for trigonometric functions.

Two Fundamental Limits

Our first goal is to calculate $\lim_{\theta \to 0} \dfrac{\sin \theta}{\theta}$. We will do this using a geometric argument that allows us to bound the ratio $\dfrac{\sin \theta}{\theta}$ both above and below, i.e., we will find functions $m(\theta)$ and $M(\theta)$, using only geometry, such that

$$
\begin{array}{ccccc}
m(\theta) & \leq & \dfrac{\sin \theta}{\theta} & \leq & M(\theta) \\
\downarrow & & \downarrow & & \downarrow \\
\ell & \leq & L & \leq & \ell
\end{array}
\qquad (4.2)
$$

for θ near zero. Will then take the limit as $\theta \to 0$ and apply the squeeze theorem. This will work so long as $\lim_{\theta \to 0} m(\theta) = \lim_{\theta \to 0} M(\theta) = \ell$ for some number ℓ.

We start by considering an arc $\overset{\frown}{AOB}$ of a unit circle, centered at O, and passing through the points A and and B, as illustrated in figure 4.1. Define θ as the central angle,

$$\theta = \angle BOC \qquad (4.3)$$

Figure 4.1: Geometry used in derivation of the $\lim\limits_{\theta\to 0}\dfrac{\sin\theta}{\theta}$.

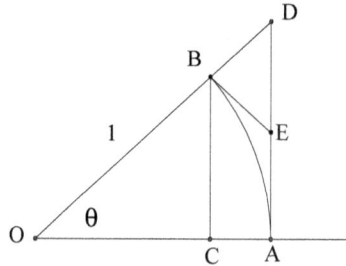

Define point C as the foot of the perpendicular line dropped from B to $\overline{OA}$. Then $\triangle OBC$ is a right triangle with altitude

$$|\overline{BC}| = \sin\angle BOC = \sin\theta \tag{4.4}$$

The length of the arc of the circle from A to B is

$$|\widehat{AB}| = \overbrace{(\text{radius})}^{\text{this is } |\overline{OB}|=1} \times \underbrace{(\text{angle in radians})}_{\text{this is }\theta} = \theta \tag{4.5}$$

Examining the figure, we observe that

$$|\overline{BC}| < |\widehat{AB}| \tag{4.6}$$

hence

$$\sin\theta < \theta \tag{4.7}$$

because

$$\sin\theta = |\overline{BC}| < |\widehat{AB}| = \theta \tag{4.8}$$

Dividing both sides of this by θ,

$$\frac{\sin\theta}{\theta} < 1 \tag{4.9}$$

Next, we construct a vertical line that is tangent to the circle at A and define D as the point where that line intersects $\overrightarrow{OB}$, as illustrated. We then construct a tangent line to the circle through B and define its intersection with $\overline{AD}$ as the point E.

The the arc length of the circle is shorter than the length of the path from A to B through E (see the figure):

$$\theta = |\widehat{AB}| < |\overline{BE}| + |\overline{AE}| \tag{4.10}$$

Since $\overline{BE}$ is a tangent line, then $\overline{BD}$ is perpendicular to $\overline{BE}$. Hence $\triangle DEB$ is a right triangle with hypotenuse $\overline{DE}$. Since the hypotenuse is the longest side of a right triangle,

$$|\overline{BE}| < |\overline{DE}| \tag{4.11}$$

and combining the last two results gives

$$\theta < |\overline{AE}| + |\overline{DE}| = |\overline{AD}| = \overbrace{|\overline{OA}|}^{=1}\tan\theta = \tan\theta = \frac{\sin\theta}{\cos\theta} \tag{4.12}$$

where we have used the definition of the tangent in the last step. Therefore

$$\theta < \frac{\sin\theta}{\cos\theta} \tag{4.13}$$

If we cross multiply,

$$\cos\theta < \frac{\sin\theta}{\theta} \tag{4.14}$$

Combining this with inequality (4.9) gives

$$\cos\theta < \frac{\sin\theta}{\theta} < 1 \tag{4.15}$$

Taking the limit as $\theta \to 0$,

$$\lim_{\theta\to 0}\cos\theta \quad < \quad \lim_{\theta\to 0}\frac{\sin\theta}{\theta} \quad < \quad \lim_{\theta\to 0}1 \tag{4.16}$$
$$\downarrow \qquad\qquad\qquad\qquad\qquad\qquad \downarrow$$
$$1 \qquad\qquad\qquad\qquad\qquad\qquad 1$$

Therefore by the squeeze theorem we have the following result.

Theorem 4.1. First Fundamental Trigonometric Limit

$$\lim_{\theta\to 0}\frac{\sin\theta}{\theta} = 1 \tag{4.17}$$

Example 4.1. Find $\lim\limits_{x\to 0}\dfrac{\sin 4x}{x}$

Solution. To apply result 4.17 the argument of the sin should be the same as the quantity in the denominator. We can always make this happen by multiplying by one.

$$\lim_{x\to 0}\frac{\sin 4x}{x} = \lim_{x\to 0}\frac{\sin 4x}{x}\times\frac{4}{4} = 4\lim_{x\to 0}\frac{\sin 4x}{4x} = 4\times 1 = 4 \tag{4.18a}$$

The reason why this next-to-last step works is because $\lim\limits_{x\to 0}f(x) = \lim\limits_{4x\to 0}f(x)$ for any function $f(x)$. This follows from the observation that $4x \to 0$ as $x \to 0$. □

Example 4.2. Find $\lim\limits_{x\to 0}\dfrac{\tan 4x}{\sin 3x}$

Solution. Now we have to extend the trick used in the previous example to the argument of each sin function.

$$\lim_{x\to 0}\frac{\tan 4x}{\sin 3x} = \lim_{x\to 0}\overbrace{\frac{\sin 4x}{\cos 4x}}^{\text{def. of tan}}\times\frac{1}{\sin 3x} \tag{4.19a}$$

$$= \lim_{x\to 0}\frac{\sin 4x}{\cos 4x}\times\overbrace{\frac{4x}{4x}}^{\text{mult. by 1}}\times\frac{1}{\sin 3x}\times\overbrace{\frac{3x}{3x}}^{\text{mult. by 1}} \tag{4.19b}$$

$$= \lim_{x\to 0}\underbrace{\frac{\sin 4x}{4x}}_{\to 1}\times\underbrace{\frac{1}{\cos 4x}}_{\to 1}\times\underbrace{\frac{3x}{\sin 3x}}_{\to 1}=\frac{4}{3}\times\underbrace{\frac{4x}{3x}}_{=4/3} \tag{4.19c}$$

In the last step we have made use of the fact that

$$\lim_{x\to 0}\frac{x}{\sin x} = \frac{1}{\left(\lim\limits_{x\to 0}\dfrac{x}{\sin x}\right)} = 1 \tag{4.19d}$$

Equation 4.19d follows because $\lim(1/f) = \lim 1\lim f = 1/\lim f$ whenever $\lim f$ exists. □

The second important limit we want to find provides an example of how we can use this formula.

Example 4.3. Find $\displaystyle\lim_{\theta \to 0} \frac{\cos\theta - 1}{\theta}$.

Solution. The trick is to multiply the equation by one, and use a trigonometric identity to change resulting form of the numerator.

$$\lim_{\theta \to 0} \frac{\cos\theta - 1}{\theta} =$$

$$\lim_{\theta \to 0} \frac{\cos\theta - 1}{\theta} \times \frac{\cos\theta + 1}{\cos\theta + 1} \qquad\qquad \text{multiply by 1} \qquad (4.20a)$$

$$= \lim_{\theta \to 0} \frac{\cos^2\theta - 1}{\theta(\cos\theta + 1)} \qquad\qquad \text{diff. of squares} \qquad (4.20b)$$

$$= \lim_{\theta \to 0} \frac{-\sin^2\theta}{\theta(\cos\theta + 1)} \qquad\qquad \text{trig identity} \qquad (4.20c)$$

$$= -\lim_{\theta \to 0} \left(\frac{\sin\theta}{\theta} \times \frac{\sin\theta}{\cos\theta + 1} \right) \qquad\qquad \text{rearrange} \qquad (4.20d)$$

$$= -\left(\lim_{\theta \to 0} \frac{\sin\theta}{\theta} \right) \left(\lim_{\theta \to 0} \frac{\sin\theta}{\cos\theta + 1} \right) \qquad\qquad \text{lim. of product} \qquad (4.20e)$$

$$= -(1) \times \frac{0}{1 + 1} = 0 \qquad\qquad (4.20f)$$

□ This limit arises so frequently we also summarize it as theorem 4.21.

Theorem 4.2. Second Fundamental Trigonometric Limit

$$\lim_{\theta \to 0} \frac{\cos\theta - 1}{\theta} = 0 \qquad\qquad (4.21)$$

Exercises

Calculate each of the following limits

1. $\displaystyle\lim_{\theta \to 0} \frac{\sin 3\theta}{\sin 4\theta}$ (ans: 3/4)

2. $\displaystyle\lim_{t \to 0} \frac{\sin 5t}{\tan \pi t}$ (ans: 5/π)

3. $\displaystyle\lim_{x \to 0} \frac{x}{\sin 4x}$ (ans: 1/4)

4. $\displaystyle\lim_{x \to 0} \frac{x + 3\tan x}{2x}$ (ans: 2)

5. $\displaystyle\lim_{x \to 0} \frac{x^2 \tan x}{x + 7\sin^2 x}$ (ans: 1/8)

6. $\displaystyle\lim_{x \to 0} x^2 \cot(5x^2)$ (ans: 1/5)

7. $\displaystyle\lim_{x \to 0} x \cot(5x^2)$ (ans: ∞)

8. $\displaystyle\lim_{x \to 0} \frac{2\tan x}{\sin x + x\cos x}$ (ans: 1)

9. $\displaystyle\lim_{x \to 0} \frac{1 - x^2}{\cos x}$ (ans: 1)

10. $\displaystyle\lim_{x \to 0} \frac{x^2}{\sin(\pi x)\sin(2\pi x)}$ (ans: $1/2\pi^2$)

11. $\displaystyle\lim_{x \to 1} \frac{(x + 3)(1 - \cos(1 - x))}{x^2 - 2x + 1}$ (ans: 2)

Doing Calculus

Chapter 5

Asymptotes

Chapter Summary and Goal

An **asymptote** occurs when the plot of curve becomes very close to a second curve in the infinite reaches of the coordinate plane. Most commonly the term asymptote is used when the second curve is a line, as in when the curve becomes extremely close to a line for points extremely far away from the origin.

Asymptotes fall into three classes, and they are closely connected to limits:

1. **Vertical Asymtotes**, when the plot of $f(x)$ becomes infinitesimally close to a vertical line as $x \to \pm\infty$;
2. **Horizontal Asymptotes**, when the plot of $f(x)$ becomes infinitesimally close to a vertical line, as $y \to \pm\infty$; and
3. **Oblique Asymptotes**, when the plot of $f(x)$ either becomes infinitesimally close to another line that is neither horizontal nor vertical, or becomes infinitesimally close to another curve, as $x \to \infty$. Oblique asymptotes are sometimes called **slant asymptotes**.

Each of these classes of asymptotes is defined by a particular type of limit, and we will learn how to identify these asymptotes in this chapter.

Student Learning Objectives

The student will:

1. Understand the distinction between the different types of asymptotes.
2. Be able to determine the locations of all horizontal and vertical asymptotes.
3. Understand the concept of limits at infinity, and be able to calculate limits at infinity using the rules of limits.

Vertical Asymptotes

We have already seen vertical asymptotes in definition 2.8, which we repeat here as definition 2.8. Typically they occur in functions with denominators at points where the denominator is equal to zero.

Definition 5.1. Vertical Asymptote

The line $x = a$ is called a ***vertical asymptote*** of $f(x)$ if any of the following are true:

$$\lim_{x \to a} f(x) = \infty \qquad \lim_{x \to a^-} f(x) = \infty \qquad \lim_{x \to a^+} f(x) = \infty$$
$$\lim_{x \to a} f(x) = -\infty \qquad \lim_{x \to a^-} f(x) = -\infty \qquad \lim_{x \to a^+} f(x) = -\infty \tag{5.1}$$

Example 5.1. Find the vertical asymptotes of $f(x) = \dfrac{x + 5}{(x - 3)(x - 7)}$

Solution. We observe that the domain of $f(x)$ is all real numbers except for the points where the denominator is zero, namely, $x = 3$ and $x = 7$. At these points, $f(x)$ is undefined. Each of these lines is a vertical asymptotes. For example, to see what happens as $x \to 3^-$, we let $x = 3 - \epsilon$ for some small number ϵ, and then consider the limit as $\epsilon \to 0^+$.

$$f(3 - \epsilon) = \frac{(3 - \epsilon) + 5}{((3 - \epsilon) - 3)((3 - \epsilon) - 7)} \tag{5.2a}$$

$$= \frac{8 - \epsilon}{-\epsilon(-\epsilon - 4)} = \frac{8 - \epsilon}{\epsilon(\epsilon + 4)} \to \infty \tag{5.2b}$$

as $\epsilon \to 0^+$. Thus

$$\lim_{x \to 3^-} f(x) = \infty \tag{5.2c}$$

This automatically tells us that $x = 3$ is a vertical asymptote, and that $f(x)$ become infinitely large as it approaches from the left (see figure 5.1). To see what happens as we approach the same asymptote from the right, we let $x = 3 + \epsilon$ and then also consider what happens as as $\epsilon \to 0^+$:

$$f(3 + \epsilon) = \frac{(3 + \epsilon) + 5}{((3 + \epsilon) - 3)((3 + \epsilon) - 7)} \tag{5.2d}$$

$$= \frac{8 + \epsilon}{\epsilon(\epsilon - 4)} = \to -\infty \tag{5.2e}$$

so $f(x)$ becomes infinitely large and negative as we approach from the right, i.e.,

$$\lim_{x \to 3^+} f(x) = -\infty \tag{5.2f}$$

A similar analysis can be performed at $x = 5$, to show that $x = 5$ is an asymptote. Although both of the following are true,

$$\lim_{x \to 5^-} f(x) = -\infty \text{ and } \lim_{x \to 5^+} f(x) = \infty \tag{5.2g}$$

either equation by itself is sufficient to demonstrate that $x = 5$ is also an asymptote. $\square$

Horizontal Asymptotes

The definition of a limit at infinity is based on our intuition: we say that

$$\lim_{x \to \infty} f(x) = L \tag{5.3}$$

if it is possible to make $f(x)$ arbitrarily close to L by choosing x sufficiently large (figure 6.3). We may write $f \to L$ as $x \to \infty$, and say "f goes to L as x goes to infinity."

Figure 5.1: The function in example 5.1 has two vertical asymptotes, illustrated by the dashed lines here.

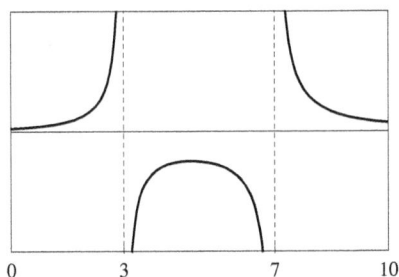

$$0 \qquad\quad 3 \qquad\qquad\quad 7 \qquad\quad 10$$

We can similarly define

$$\lim_{x \to -\infty} f(x) = L \tag{5.4}$$

if it is possible to make $f(x)$ arbitrarily close to L by choosing x sufficiently large, but in the negative x direction. We may write $f \to L$ as $x \to -\infty$, and say "f goes to L as x goes to minus infinity." We will formalize these definitions in chapter 6.

Definition 5.2. Horizontal Asymptote

The line $y = L$ is called a **horizontal asymptote** of $f(x)$ if $f(x) \to L$ as either $x \to \infty$ or $x \to -\infty$.

There is no restriction in the definitions of either the limit at infinity or the horizontal asymptote that the function never actually cross the asymptote prior to reaching the limit. An example of a function that crosses the asymptote infinitely many number of times is illustrated in figure 5.2.

Figure 5.2: Example of a horizontal asymptote at $y = L$; in this case the function $f(x) = 1 - \sin(x)/x$ crosses the asympote infinitely many times.

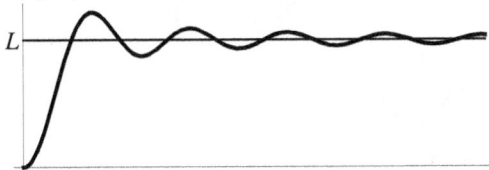

Example 5.2. Find $\displaystyle\lim_{x \to \infty} \frac{7x^2 - 3x + 8}{9x^2 + 4x}$.

Solution. We will divide by the lowest power of the denominator, and simplify. Since both the numerator and denominator are quadratic, we divide each by x^2. This is equivalent to multiplying by one, so it does not change the result of the limit.

$$\lim_{x \to \infty} \frac{7x^2 - 3x + 8}{9x^2 + 4x} = \lim_{x \to \infty} \frac{7x^2 - 3x + 8}{9x^2 + 4x} \times \frac{1/x^2}{1/x^2} \tag{5.5a}$$

$$= \lim_{x \to \infty} \frac{\dfrac{7x^2 - 3x + 8}{x^2}}{\dfrac{9x^2 + 4x}{x^2}} \tag{5.5b}$$

$$= \lim_{x \to \infty} \frac{7 - \dfrac{3}{x} + \dfrac{8}{x^2}}{9 + \dfrac{4}{x}} \tag{5.5c}$$

$$= \frac{\lim_{x \to \infty} 7 - \lim_{x \to \infty} \dfrac{3}{x} + \lim_{x \to \infty} \dfrac{8}{x^2}}{\lim_{x \to \infty} 9 + \lim_{x \to \infty} \dfrac{4}{x}} \tag{5.5d}$$

$$= \frac{7 - 0 + 0}{9 + 0} = \frac{7}{9} \tag{5.5e}$$

$\square$

> ### Heuristic 5.1. Rule for Finding Limits at Infinity of Fractions
>
> Divide by the highest power in the denominator.

Example 5.3. Find $\lim_{x \to \infty} \dfrac{3x^3 + 2x}{x + 7}$.

Solution. Applying the rule of dividing by the highest power of the denominator, we divide the numerator and denominator by x^3.

$$\lim_{x \to \infty} \frac{3x^3 + 2x}{x + 7} = \frac{3x^3 + 2x}{x + 7} \times \frac{(1/x^3)}{(1/x^3)} \tag{5.6a}$$

$$= \lim_{x \to \infty} \frac{3 + \dfrac{2}{x^2}}{\dfrac{1}{x^2} + \dfrac{7}{x^3}} \tag{5.6b}$$

$$= \frac{\lim_{x \to \infty} 3 + \lim_{x \to \infty} \dfrac{2}{x^2}}{\lim_{x \to \infty} \dfrac{1}{x^2} + \lim_{x \to \infty} \dfrac{7}{x^3}} \tag{5.6c}$$

$$= \frac{3 + 0}{0 + 0} = \infty \tag{5.6d}$$

$\square$

Example 5.4. Find $\lim_{x \to \infty} \dfrac{8x + 14}{x^2 + 6}$.

Solution. Following the same procedure, we divide numerator and denominator by x^2.

$$\lim_{x \to \infty} \frac{8x + 14}{x^2 + 6} = \lim_{x \to \infty} \frac{8x + 14}{x^2 + 6} \times \frac{(1/x^2)}{(1/x^2)} \tag{5.7a}$$

$$= \lim_{x \to \infty} \frac{\dfrac{8}{x} + \dfrac{14}{x^2}}{1 + \dfrac{6}{x^2}} \tag{5.7b}$$

$$= \frac{\lim_{x \to \infty} \dfrac{8}{x} + \lim_{x \to \infty} \dfrac{14}{x^2}}{\lim_{x \to \infty} 1 + \lim_{x \to \infty} \dfrac{6}{x^2}} \tag{5.7c}$$

$$= \frac{0+0}{1+0} = 0 \tag{5.7d}$$

$\square$

The last three examples illustrate the following useful rule for rational functions.

Theorem 5.1. Limit of a Rational Function at Infinity

If $a_n, b_m \neq 0$, where n and m are the largest exponents of two polynomials $a(x)$ and $b(x)$, then

$$\lim_{x \to \infty} \frac{a(x)}{b(x)} = \lim_{x \to \infty} \frac{a_n x^n + a_{n-1} x^{n-1} + \cdots + a_0}{b_m x^m + b_{m-1} x^{m-1} + \cdots + b_0} \tag{5.8a}$$

$$= \begin{cases} 0, & \text{if } n < m \\ a_n/b_n, & \text{if } n = m \\ \infty, & \text{if } n > m \end{cases} \tag{5.8b}$$

Example 5.5. Find $\displaystyle\lim_{x \to \infty} \frac{36x^7 - 12x^3 + 8}{6x^7 + x^5}$

Solution. Since both the numerator and denominator are polynomials of degree 7, we read off the coefficients. The limit is $36/6 = 6$. $\square$

Figure 5.3: A function with both horizontal and vertical asymptotes, as discussed in example 5.6.

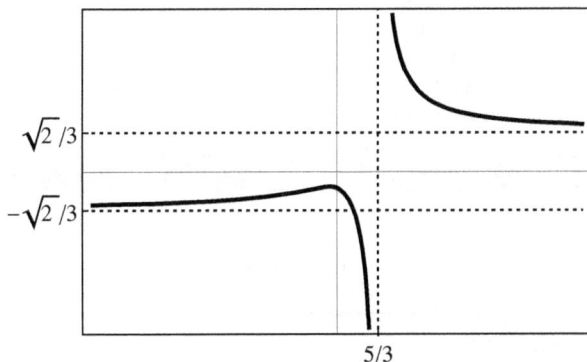

Example 5.6. Find $\displaystyle\lim_{x \to \infty} \frac{\sqrt{2x^2 + 1}}{3x - 5}$

Solution. This is not a quotient of polynomials so we resort to the earlier method of dividing by x (x raised to the highest power in the denominator).

$$\lim_{x \to \infty} \frac{\sqrt{2x^2 + 1}}{3x - 5} = \lim_{x \to \infty} \frac{\sqrt{2x^2 + 1}}{3x - 5} \times \frac{(1/x)}{(1/x)} \tag{5.10a}$$

$$= \lim_{x \to \infty} \frac{\dfrac{1}{x}\sqrt{x^2 \left(2 + \dfrac{1}{x^2}\right)}}{3 - \dfrac{5}{x}} \tag{5.10b}$$

$$= \lim_{x \to \infty} \frac{\frac{\sqrt{x^2}}{x}\sqrt{2 + \frac{1}{x^2}}}{3 - \frac{5}{x}} = \lim_{x \to \infty} \frac{\sqrt{2 + \frac{1}{x^2}}}{3 - \frac{5}{x}} \tag{5.10c}$$

$$= \frac{\lim_{x \to \infty} \sqrt{2 + \frac{1}{x^2}}}{\lim_{x \to \infty} 3 - \lim_{x \to \infty} \frac{5}{x}} = \frac{\sqrt{2 + 0}}{3 + 0} = \frac{\sqrt{2}}{3} \tag{5.10d}$$

Thus this function has a horizontal asymptote a $y = \frac{\sqrt{2}}{3}$.   □

Example 5.7. Repeat the previous example, but as $x \to -\infty$.

Solution. The solution is identical **except** for the implicit assumption that we made in (5.10c), that

$$\lim_{x \to \infty} \frac{\sqrt{x^2}}{x} = \lim_{x \to \infty} \frac{|x|}{|x|} = \lim_{x \to \infty} 1 = 1 \tag{5.11a}$$

because now we have

$$\lim_{x \to -\infty} \frac{\sqrt{x^2}}{x} = \lim_{x \to -\infty} \frac{|x|}{-|x|} = \lim_{x \to \infty} -1 = -1 \tag{5.11b}$$

The rest of the limit proceeds as before, but with a minus sign, so that

$$\lim_{x \to -\infty} \frac{\sqrt{2x^2 + 1}}{3x - 5} = -\frac{\sqrt{2}}{3} \tag{5.11c}$$

□

An important observation that can be made from the previous example is summariz,ed in the following box. When we "pull" an x into a square root as $\sqrt{x^2}$, when the square root represents a single positive square root (as in a function definition) then we must take into account the sign of x. For limits as $x \to -\infty$, the sign is negative, and this explains a change in signs whose origin might otherwise be somewhat perplexing.

Theorem 5.2. Incorporating x under the square root.

$$x = \begin{cases} \sqrt{x^2} & \text{if } x \geq 0, \text{ e.g., as } x \to \infty \\ -\sqrt{x^2} & \text{if } x < 0, \text{ e.g., as } x \to -\infty \end{cases} \tag{5.12}$$

Example 5.8. Find $\lim_{x \to \infty} (\sqrt{x^2 + 1} - x)$

Solution. This has the form of $\infty - \infty$, so the answer is not immediately obvious. The trick here is to multiply by 1 in a form to create a difference of squares.

$$\lim_{x \to \infty} (\sqrt{x^2 + 1} - x) = \lim_{x \to \infty} (\sqrt{x^2 + 1} - x) \times \left(\frac{\sqrt{x^2 + 1} + x}{\sqrt{x^2 + 1} + x} \right) \tag{5.13a}$$

$$= \lim_{x \to \infty} \frac{x^2 + 1 - x^2}{\sqrt{x^2 + 1} + x} = \lim_{x \to \infty} \frac{1}{\sqrt{x^2 + 1} + x} = 0 \tag{5.13b}$$

□

Don't depend on your intuition too much for these limits; the last example may seem to make sense because it sort of should be something like $\sqrt{\infty^2} - \infty \to 0$ But if we use the same logic on the following example to conclude the result sort of should be $-\infty + \sqrt{\infty^2} \to 0$ we would be wrong!

Example 5.9. Find $\lim_{x \to -\infty} \left(x + \sqrt{x^2 + x}\right)$

Solution. Using the "multiply by one" trick

$$\lim_{x \to -\infty} \left(x + \sqrt{x^2 + x}\right)$$

$$= \lim_{x \to -\infty} \left(x + \sqrt{x^2 + x}\right) \times \frac{x - \sqrt{x^2 + x}}{x - \sqrt{x^2 + x}} \tag{5.14a}$$

$$= \lim_{x \to -\infty} \frac{x^2 - (x^2 + x)}{x - \sqrt{x^2 + x}} \tag{5.14b}$$

$$= \lim_{x \to -\infty} \frac{-x}{x - \sqrt{x^2 + x}} \times \frac{(1/x)}{(1/x)} \tag{5.14c}$$

$$= \lim_{x \to -\infty} \frac{-1}{1 - \dfrac{\sqrt{x^2}}{x} \sqrt{1 + \dfrac{1}{x}}} \tag{5.14d}$$

$$= \lim_{x \to -\infty} \frac{-1}{1 + \dfrac{\sqrt{x^2}}{|x|} \sqrt{1 + \dfrac{1}{x}}} \tag{5.14e}$$

$$= \lim_{x \to -\infty} \frac{-1}{1 + \dfrac{\sqrt{x^2}}{\sqrt{x^2}} \sqrt{1 + \dfrac{1}{x}}} \tag{5.14f}$$

$$= \lim_{x \to -\infty} \frac{-1}{1 + \sqrt{1 + \dfrac{1}{x}}} = -\frac{1}{2} \tag{5.14g}$$

Thus the function $f(x) = x + \sqrt{x^2 + x}$ has a horizontal asymptote at $x = -1/2$. □

Example 5.10. Find $\lim_{x \to \infty} \left(\sqrt{x^2 + ax} - \sqrt{x^2 + bx}\right)$ where a and b are positive real constants.

Solution. We use the same "trick" as in the previous example.

$$\lim_{x \to \infty} \left(\sqrt{x^2 + ax} - \sqrt{x^2 + bx}\right) =$$

$$= \lim_{x \to \infty} \left(\sqrt{x^2 + ax} - \sqrt{x^2 + bx}\right) \times \frac{\sqrt{x^2 + ax} + \sqrt{x^2 + bx}}{\sqrt{x^2 + ax} + \sqrt{x^2 + bx}} \tag{5.15a}$$

$$= \lim_{x \to \infty} \frac{x^2 + ax - x^2 - bx}{\sqrt{x^2 + ax} + \sqrt{x^2 + bx}} \tag{5.15b}$$

$$= \lim_{x \to \infty} \frac{(a - b)x}{\sqrt{x^2 + ax} + \sqrt{x^2 + bx}} \times \frac{(1/x)}{(1/x)} \tag{5.15c}$$

$$= \lim_{x \to \infty} \frac{(a - b)}{\sqrt{1 + (a/x)} + \sqrt{1 + (b/x)}} \tag{5.15d}$$

$$= \frac{\lim_{x \to \infty} (a - b)}{\lim_{x \to \infty} \sqrt{1 + (a/x)} + \lim_{x \to \infty} \sqrt{1 + (b/x)}} = \frac{a - b}{2} \tag{5.15e}$$

This function has a horizontal asymptote at $y = (a - b)/2$, as illustrated in figure 5.4. □

Figure 5.4: The horizontal asymptote to the function calculated in example 5.10 is indicated by the dashed line.

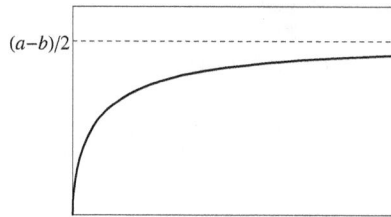

Oblique Asymptotes

An **Oblique asymptote** (or **slant** asymptote) occurs when one function becomes arbitrarily close to another as $x \to \pm\infty$.[1] Typically this means a rational function approaches $\pm\infty$ because the power of the numerator is larger than the power of the denominator. The **degree of the slant** is difference of the powers. If the numerator is one degree higher than the denominator, then the oblique asymptote is linear. To find the equation of the oblique asymptote, we do long division.

Figure 5.5: Oblique asymptote calculated in example 5.11.

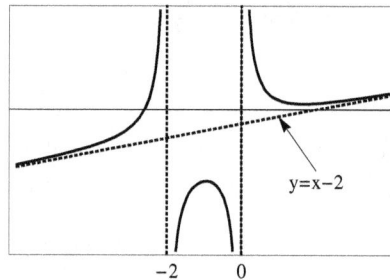

Example 5.11. Find the oblique asymptotes of $y = \dfrac{x^3 - 4x + 7}{x^2 + 2x}$

Solution.   We observe that the power of the numerator is one higher than the power of the denominator so the function has an oblique asymptote. By long division,

$$
\begin{array}{r}
x - 2 \\
x^2 + 2x \overline{\smash{\big)}\ x^3 \quad\quad - 4x + 7} \\
\underline{-x^3 - 2x^2} \\
-2x^2 - 4x \\
\underline{2x^2 + 4x}
\end{array}
\tag{5.16a}
$$

Therefore an oblique asymptote occurs at $y = x - 2$. We also observe that $f(x)$ has vertical asymptotes at $x = -2$ and $x = 0$. Except at $x = 0$ and $x = -2$, $f(x)$ is continuous.

---

[1] Some texts restrict the concept of oblique asymptotes to straight lines; here we are including more general functions, and for oblique asymptotes, we do not require that the asymptote itself be a line.

To find out the limits from the left and right at the asymptotes we consider what happens as x approaches each asymptote very close. To see what happens as $x \to 0^+$ we let $x \approx \epsilon$, where $\epsilon > 0$ is a very small number. Then

$$\lim_{x \to 0^+} \frac{x^3 - 4x + 7}{x(x+2)} \approx \frac{\epsilon^3 - 4\epsilon + 7}{\epsilon(\epsilon+2)} \approx \frac{7}{2\epsilon} > 0 \to \infty \tag{5.16b}$$

Similarly, to examine what happens when $x \to 0^-$, we let $x \approx -\epsilon$ where $\epsilon > 0$ and small:

$$\lim_{x \to 0^-} \frac{x^3 - 4x + 7}{x(x+2)} \approx \frac{(-\epsilon)^3 - 4(-\epsilon) + 7}{(-\epsilon)((-\epsilon)+2)} \approx -\frac{7}{2\epsilon} < 0 \to -\infty \tag{5.16c}$$

As $x \to -2^+$ we let $x \approx -2 + \epsilon$, where $\epsilon > 0$,

$$\lim_{x \to -2^+} \frac{x^3 - 4x + 7}{x(x+2)} \approx \frac{(-2+\epsilon)^3 - 4(-2+\epsilon) + 7}{(-2+\epsilon)((-2+\epsilon)+2)} \tag{5.16d}$$

$$\approx \frac{(-8 + 12\epsilon - 6\epsilon^2 + \epsilon^3) + 8 - 4\epsilon + 7}{(-2+\epsilon)(\epsilon)} \tag{5.16e}$$

$$\approx -\frac{7}{2\epsilon} < 0 \to -\infty \tag{5.16f}$$

As $x \to -2^-$ we let $x \approx -2 - \epsilon$, where $\epsilon > 0$,

$$\lim_{x \to -2^-} \frac{x^3 - 4x + 7}{x(x+2)} \approx \frac{(-2-\epsilon)^3 - 4(-2-\epsilon) + 7}{(-2-\epsilon)((-2-\epsilon)+2)} \tag{5.16g}$$

$$\approx \frac{(-8 - 12\epsilon - 6\epsilon^2 - \epsilon^3) + 8 + 4\epsilon + 7}{(-2-\epsilon)(-\epsilon)} \tag{5.16h}$$

$$\approx \frac{7}{2\epsilon} > 0 \to \infty \tag{5.16i}$$

The function is plotted in figure 5.5.   □

Example 5.12. Find all the asymptotes of $y = \dfrac{x^5 - 2x^3}{x^2 - 4}$

Solution. The denominator is a difference of squares and can be factored as

$$x^2 - 4 = (x-2)(x+2) \tag{5.17a}$$

and we see that there are vertical asymptotes at $x = \pm 2$. Since the numerator has a highest power of 5 and the denominator has a highest power of 2, there is an oblique asymptote of degree 5-2 = 3. To find the equation of this cubic asymptote we use long division:

$$
\begin{array}{r}
x^3 + 2x \\
x^2 - 4 \overline{) x^5 - 2x^3} \\
-x^5 + 4x^3 \\
\hline
2x^3 \\
-2x^3 + 8x \\
\hline
8x
\end{array}
\tag{5.17b}
$$

The equation of the oblique asymptote is then

$$y = x^3 + 2x \tag{5.17c}$$

The asymptotes are plotted as the dotted lines in figure 5.6.

Figure 5.6: The function $y = (x^5 - 2x^3)/(x^2 - 4)$ has an oblique cubic asymptote of $x^3 + 2x$ (see example 5.12), indicated by the dotted curve in the figure. This function also has two vertical asymptotes at $x = \pm 2$.

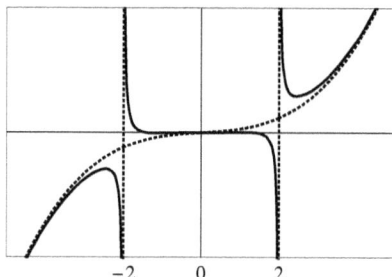

We conclude by noting that the function $g(x)$ is an oblique asymptote if $f(x)$ can be made arbitrarily close to $g(x)$ by choosing x sufficiently large.

Definition 5.3. Oblique Asymptotes

A function $g(x)$ is an **Oblique Asymptote** of a function $f(x)$ if

$$\lim_{x \to \infty} |f(x) - g(x)| = 0 \qquad (5.18)$$

An interesting consequence of this is that if g is a oblique asymptote of f then f is an oblique asymptote of g.

Exercises

Use the limit laws to find each of the following limits at infinity.

1. $\lim\limits_{x \to \infty} \dfrac{2x + 7}{3x + 8}$

2. $\lim\limits_{x \to \infty} \dfrac{8x - 17}{16x}$

3. $\lim\limits_{x \to \infty} \dfrac{8x}{16x - 14}$

4. $\lim\limits_{x \to \infty} \dfrac{x^3}{2x^3 + x^2}$

5. $\lim\limits_{x \to \infty} \dfrac{x^2}{4x^3 + x^2 + x}$

6. $\lim\limits_{x \to \infty} \dfrac{\sqrt{x^2 - 8}}{x - 8}$

7. $\lim\limits_{x \to \infty} \dfrac{4x^2 - 5x}{6x + 12x^2 + 7}$

8. $\lim\limits_{x \to \infty} \dfrac{3x^3 + 2x^2 + x}{4x^4 + 3x^3 + 2x^2 + x}$

9. $\lim\limits_{x \to \infty} \dfrac{\sqrt{x + 4} - \sqrt{x}}{\sqrt{x}}$

10. $\lim\limits_{x \to \infty} \left(\dfrac{x + 1}{\sqrt{x}} - \dfrac{x}{\sqrt{x + 1}} \right)$

11. $\lim\limits_{x \to \infty} x \left(\left(3 + \dfrac{1}{x} \right)^2 - 9 \right)$

Find the oblique asymptotes of each of the following functions.

12. $\dfrac{4x^2 - 5x}{6x + 7}$
ans: $y = -\dfrac{53}{25} + \dfrac{4x}{5}$

13. $\dfrac{4x^2 - 36x + 6}{4x}$
ans: $y = x - 9$

14. $\dfrac{x^3 + 3x^2 - 4x - 12}{x^2 - x - 2}$
ans: $y - 4 + x$

15. $\dfrac{x^4 - 16}{x - 9}$
ans: $y = x^3 + 9x^2 + 82x + 738$

Chapter 6

Formal Theory of Limits

Chapter Summary and Goal

In this chapter we will approach the concept of a limit more formally. We will dissect the definition using mathematical logic and explain what each term means. We will then demonstrate how to prove whether or not a limit exists, and how to find out the value of that limit, using this formal process.[1]

Student Learning Objectives

The student will:

1. Learn the formal definition of a limit.
2. Be able to prove limits using the formal definition.

A Precise Mathematical Approach to Limits

In definition 2.2 we said the following.

> ### Definition 6.1. Limit.
>
> If we can make the value of $f(x)$ as close to L as we like by choosing x sufficiently close to a (but not precisely equal to a), then we say that *the limit of $f(x)$ as x approaches a equals L*, written as
>
> $$\lim_{x \to a} f(x) = L \qquad (6.1)$$

This is what a mathematician means when he or she makes a statement like definition 6.1 (see figure 6.1):

If whenever you pick out any small number (call it ϵ)
  Constructive Explanation: Project horizontal line segments at $y = L \pm \epsilon$ from the y axis to the curve of $y = f(x)$. These will define the top and bottom of a bounding box about the line $y = L$. See figure 6.1.A.

Then I can find another small number (call it δ) that depends on ϵ
  Such That

---

[1]The material in this chapter is not used in the remainder of the book and may be skipped without any loss in flow.

Whenever $|x - a| < \delta$

> *Note:* this is how mathematicians say x is sufficiently close to a.
>
> *Constructive Explanation:*
>
> 1. Project the arrows drawn in figure 6.1.A down to the x axis, as shown in figure 6.1.B.
> 2. Measure the distances m and M from a where they intersect the axis.
> 3. One of these distances will be smaller than the other.
> 4. Define $\delta = min(m, M)$ (the smaller of the two distances in step figure 6.1.B).
> 5. Project arrows back up from $a \pm \delta$ to $f(x)$ and then back to the y axis, as shown in figure 6.1.C.

Then $|f(x) - L| < \epsilon$

> *Note:* This is how mathematicians say $f(x)$ *is as close to L as we like*)
>
> *Constructive Explanation:* Observe that the arrows you projected all the way back to the y axis formed a *smaller* box than the original box. Thus you have found an interval about a, say $(a - \delta, a + \delta)$, such that whenever x is in that interval, then $f(x)$ is in the original interval $(L - \epsilon, L + \epsilon)$, except possibly at the point $x = a$.

Figure 6.1: Demonstration of $\lim_{x \to a} f(x) = L$. In frame (A), you give me ϵ. In frame (B), I find the values of x that pull f out of the bounding box. In frame (C) I have chosen δ as the maximum of m, M.

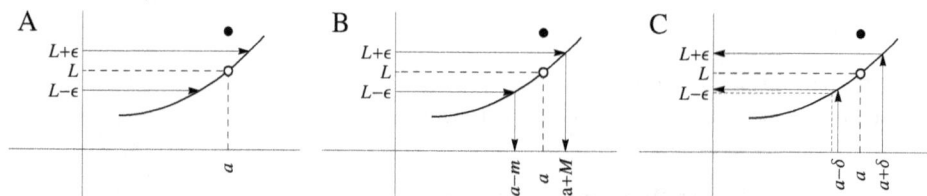

Definition 6.2. Limit, Precise Definition.

Let $f(x)$ be a function that is defined on some open interval containing a, except possibly at the number a itself (this means that f does not have to be defined at a, although its allowed to be defined there).

Then we say *the limit of $f(x)$ as x approaches a is L* if: given any $\epsilon > 0$, there exists a $\delta > 0$, such that:

$$|x - a| < \delta \implies |f(x) - L| < \epsilon \tag{6.2}$$

and we write

$$\lim_{x \to a} f(x) = L \tag{6.3}$$

Application of the precise definition of a limit is a two step process: analysis and proof.

1. ***Analysis.*** During the analysis you translate the equation from limit form (an equation that looks something like 6.3) into its epsilon and delta form (something that looks like equation 6.2). Then you figure out either a value of δ that will ensure that (6.2) is true. Almost always δ will depend on ϵ.

Figure 6.2: Structure of a typical limit proof

Let $\epsilon > 0$ be given.
Let $\delta =$ (insert formula for something that may
depend on ϵ here)
Then

$$|x - a| < \delta \implies \cdots$$
$$\implies \cdots$$
$$\implies |f(x) - L| < \epsilon$$

and therefore we may conclude that
$$\lim_{x \to a} f(x) = L \quad \square$$

2. **Proof**. In the proof you actually demonstrate that (6.2) is true. The first line of the proof step is invariably something like *"let $\epsilon > 0$ be given"* and the second line is usually something like *"define $\delta = \cdots$"* followed by some formula that depends on ϵ. The remaining steps must be logically justified, one after the other, and form a progression demonstrating that if you now assume that $|x - a| < \delta$ then $|f(x) - L| < \epsilon$. This is often an orderly (an annotated) reversal of the steps in the analysis (figure 6.2).

Example 6.1. Show that $\lim_{x \to 3}(4x - 5) = 7$

Solution. <u>Analysis Step</u>. By syntactically comparing the expressions $\lim_{x \to 3}(4x - 5) = 7$ and $\lim_{x \to a}(f(x)) = L$ we are able to make the following associations

$$f(x) = 4x - 5 \tag{6.4a}$$
$$a = 3 \tag{6.4b}$$
$$L = 7 \tag{6.4c}$$

We need to demonstrate that (6.2) is true for this example. For reference, we copy equation 6.2 here:
$$|x - a| < \delta \implies |f(x) - L| < \epsilon \tag{6.4d}$$

Substituting equations 6.4a to 6.4c into equation 6.4d tells us that what we realy need to demonstrate is this:

$$|x - 3| < \delta \implies |(4x - 5) - 7| < \epsilon \tag{6.4e}$$

There was no right way to approach this equation. Ultimately, we want a formula for $\delta(\epsilon)$ (δ as a function of ϵ) such that (6.4e) works, i.e., whenever

$$|x - 3| < \delta \tag{6.4f}$$

then (simplifying the right hand side of equation 6.4e),

$$|4x - 12| < \epsilon \tag{6.4g}$$

which is equivalent to

$$4|x - 3| < \epsilon \tag{6.4h}$$

which in turn is equivalent to

$$|x - 3| < \frac{\epsilon}{4} \tag{6.4i}$$

Now we need to make an observation. We want (6.4f) to imply (6.4g). But (6.4g) is entirely equivalent to (6.4i), and (6.4i) looks a whole lot like (6.4f). In fact, if we make the totally arbitrary definition $\delta = \epsilon/4$ then they will be identical.

This gives us an idea for our formula for δ:

$$\delta = \frac{\epsilon}{4} \tag{6.4j}$$

In fact, we could use any δ that is smaller than this; the $\epsilon/4$ is the limiting case.

<u>Proof Step.</u> Let $\epsilon > 0$ be given, and define $\delta = \dfrac{\epsilon}{4}$. Then if

$$|x - 3| < \delta \tag{6.4k}$$

by substituting the value of $\delta = \epsilon/4$,

$$|x - 3| < \frac{\epsilon}{4} \tag{6.4l}$$

Multiply by 4

$$|4x - 12| < \epsilon \tag{6.4m}$$

Since $12 = 5+7$, a little rearrangement gives us

$$|(4x - 5) - 7| < \epsilon \tag{6.4n}$$

The sequence of steps from equations 6.4k through 6.4n tells us that whenever $|x-3| < \delta$ then $(4x - 5) - 7| < \epsilon$. Since we didn't put any restrictions whatsoever on the value of ϵ (except to say that it was positive), this holds for any $\epsilon > 0$.

This is exactly the definition of the limit $\lim\limits_{x \to 3}(4x - 5) = 7$ $\square$

Nonlinear functions are somewhat more difficult to work with, as we illustrate in the following example.

Example 6.2. Show that $\lim\limits_{x \to 4} \sqrt{x} = 2$

Solution. <u>Analysis Step.</u> Comparing to (6.2) we have the following associations:

$$f(x) = \sqrt{x}, \quad L = 2, \quad a = 4 \tag{6.5a}$$

and we want to show that for any $\epsilon > 0$, we can find a $\delta > 0$, such that

$$|x - 4| < \delta \implies |\sqrt{x} - 2| < \epsilon \tag{6.5b}$$

This time, instead of taking apart the right hand side of the implication, we will work with the left.This means that starting with $|x - 4| < \delta$ we have a *goal* of demonstrating that $|\sqrt{x} - 2| < \epsilon$. But by the definition of absolute value, $|x - 4| < \delta$ is equivalent to

$$-\delta < x - 4 < \delta \tag{6.5c}$$

We an factor the center part of this inequality into a difference of squares:

$$-\delta < \left(\sqrt{x} - 2\right)\left(\sqrt{x} + 2\right) < \delta \tag{6.5d}$$

For values of x near 4 we observe that

$$\sqrt{x} + 2 \approx \sqrt{4} + 2 = 2 \tag{6.5e}$$

$$\sqrt{x} - 2 \approx \sqrt{4} - 2 = 0 \tag{6.5f}$$

So it is safe to divide through the $\sqrt{x} + 2$, because it is nowhere near zero.

$$\frac{-\delta}{\sqrt{x}+2} < \sqrt{x} - 2 < \frac{\delta}{\sqrt{x}+2} \tag{6.5g}$$

which is equivalent to

$$\left|\sqrt{x} - 2\right| < \frac{\delta}{\sqrt{x}+2} \tag{6.5h}$$

The right hand side looks a lot like our goal, which is $\left|\sqrt{x} - 2\right| < \epsilon$:

$$\left|\sqrt{x} - 2\right| < \frac{\delta}{\sqrt{x}+2} < \cdots \text{we want} \cdots < \epsilon \tag{6.5i}$$

Proof Step. Let $\epsilon > 0$ be given and define $\delta = \epsilon$.

Then suppose that

$$|x - 4| < \delta = \epsilon \tag{6.5j}$$

Since $x > 0$ near $x = 4$,

$$2 + \sqrt{x} > 2 > 1 \tag{6.5k}$$

and therefore

$$|x - 4| < \delta = \epsilon < \epsilon(2 + \sqrt{x}) \tag{6.5l}$$

Dividing by $2 + \sqrt{x}$,

$$\frac{|x - 4|}{2 + \sqrt{x}} < \epsilon \tag{6.5m}$$

Using the definition of absolute value and then factoring,

$$-\epsilon < \frac{\left(\sqrt{x} - 2\right)\left(\sqrt{x} + 2\right)}{2 + \sqrt{x}} < \epsilon \tag{6.5n}$$

Since the $2 + \sqrt{x}$ is positive it is never zero and can be safely cancelled, leaving us with

$$-\epsilon < \sqrt{x} - 2 < \epsilon \tag{6.5o}$$

or

$$\left|\sqrt{x} - 2\right| < \epsilon \tag{6.5p}$$

as required.

$\square$

Example 6.3. Prove that $\lim_{x \to 5} x^3 = 125$

Solution. Analysis. We want to find δ such that

$$|x - 5| < \delta \implies |x^3 - 125| < \epsilon \tag{6.6a}$$

Working backwards from the ϵ inequality, we have

$$|x^3 - 125| < \epsilon \iff -\epsilon < x^3 - 125 < \epsilon \tag{6.6b}$$

$$\iff -\epsilon < (x - 5)(x^2 + 5x + 25) < \epsilon \tag{6.6c}$$

$$\iff -\frac{\epsilon}{x^2 + 5x + 25} < x - 5 < \frac{\epsilon}{x^2 + 5x + 25} \tag{6.6d}$$

We want to somehow make the last line look like $|x - 5| < \delta$. Since

$$\frac{1}{x^2 + 5x + 25} \approx \frac{1}{75} \text{ for } x \text{ near } 5 \tag{6.6e}$$

this suggests that some $\delta < \epsilon/75$ should work, say $\delta = \epsilon/100$.

Proof. Let $\epsilon > 0$ be given. Choose $\delta = \epsilon/100$.

Then if

$$|x - 5| < \delta = \frac{\epsilon}{100} \tag{6.6f}$$

then

$$-\frac{\epsilon}{100} < x - 5 < \frac{\epsilon}{100} \tag{6.6g}$$

Focusing on the right-half inequality,

$$x < 5 + \frac{\epsilon}{100} \tag{6.6h}$$

and therefore

$$x^2 + 5x + 25 < \left(5 + \frac{\epsilon}{100}\right)^2 + 5\left(5 + \frac{\epsilon}{100}\right) + 25 \tag{6.6i}$$

$$= 25 + \frac{\epsilon}{10} + \frac{\epsilon^2}{10,000} + 25 + \frac{\epsilon}{100} + 25 \tag{6.6j}$$

$$= 75 + \epsilon(.11 + .0001\epsilon) \tag{6.6k}$$

Since ϵ is a very small number, the quantity $\epsilon(.11 + .0001\epsilon)$ is also small. We are safe in saying that

$$\epsilon(.11 + .0001\epsilon) < 1 < 25 \tag{6.6l}$$

and therefore

$$x^2 + 5x + 25 < 75 + 25 = 100 \tag{6.6m}$$

Therefore near $x = 5$

$$\frac{1}{100} < \frac{1}{x^2 + 5x + 25} \tag{6.6n}$$

and therefore near $x = 5$ (from 6.6f)

$$|x - 5| < \delta = \frac{\epsilon}{100} < \frac{\epsilon}{x^2 + 5x + 25} \tag{6.6o}$$

Since we are near $x = 5$, we don't have to worry about the sign of x, only of $x - 5$. The last equation is equivalent to

$$-\frac{\epsilon}{x^2 + 5x + 25} < x - 5 < \frac{\epsilon}{x^2 + 5x + 25} \tag{6.6p}$$

The denominator is positive near $x = 5$ so we can multiply through by $x^2 + 5x + 25$, to give

$$-\epsilon < (x - 5)(x^2 + 5x + 25) < \epsilon \tag{6.6q}$$

$$\implies -\epsilon < x^3 - 125 < \epsilon \tag{6.6r}$$

$$\implies |x^3 - 125| < \epsilon \tag{6.6s}$$

Thus for any $\epsilon > 0$, whenever $|x - 5| < \delta$ then $|x^3 - 125| < \epsilon$. This is the strict precise of $\lim\limits_{x \to 5} x^3 = 125$. □

Example 6.4. Prove that $\lim\limits_{x \to 4}(x^2 - 8x) = -16$.

Solution. Dissecting the limit formula gives $L = -16$, $a = 4$ and $f(x) = x^2 - 8x$. Thus we need to show that for any $\epsilon > 0$, there exists a a $\delta > 0$ such that

$$|x - 4| < \delta \implies |x^2 - 8x - (-16)| < \epsilon \tag{6.7a}$$

The expression on the right hand side of (6.7a) is equivalent to

$$|(x - 4)^2| < \epsilon \tag{6.7b}$$

Taking the square root of both sides gives $|x - 4| < \epsilon$. This suggests that we try $\delta = \sqrt{\epsilon}$.

<u>Proof.</u> Let $\epsilon > 0$ be given. Then define $\delta = \sqrt{\epsilon}$.

$$|x - 4| < \delta = \sqrt{\epsilon} \implies |x - 4|^2 < \epsilon \tag{6.7c}$$
$$|x^2 - 8x + 16 < \epsilon \tag{6.7d}$$
$$|(x^2 - 8x) - (-16)| < \epsilon \tag{6.7e}$$

The last line gives the required inequality. $\square$

Precise Definitions of Left and Right Hand Limits

The left and right limits can also be precisely defined. Since we are only approaching from a single direction, there is no need for absolute values in the δ equation.

Definition 6.3. Left Hand Limit, Precise Definition.

If for every $\epsilon > 0$ there exists some $\delta > 0$ such that

$$a - \delta < x < a \implies |f(x) - L| < \epsilon \tag{6.8}$$

then $\lim\limits_{x \to a^-} f(x) = L$

Definition 6.4. Right Hand Limit, Precise Definition.

if for every $\epsilon > 0$ there exists some $\delta > 0$ such that

$$a < x < a + \delta \implies |f(x) - L| < \epsilon \tag{6.9}$$

then $\lim\limits_{x \to a^+} f(x) = L$

Formal proofs for left and right hand limits are similar to formal proofs for normal limits although the absolute value restriction is removed on the δ condition.

Example 6.5. Prove that $\lim\limits_{x \to 2^-} \sqrt{2 - x} = 0$.

Solution. This limit is easily verified using the limit laws, so we know already that we have the correct answer. The limit from the right does not exist because the function is undefined for $x > 2$, so the limit only makes sense in terms of an approach from the left.

Comparing variables in the equation with $\lim\limits_{x \to a^-} f(x) = L$ we have $a = 2$, $L = 0$ and

$f(x) = \sqrt{2 - x}$. Thus the statement we need to prove is this: For every $\epsilon > 0$, there exists some $\delta > 0$ such that (compare with eq. 6.8

$$2 - \delta < x < 2 \implies |\sqrt{2 - x} - 0| < \epsilon \tag{6.10a}$$

Analysis Step: Look at the right hand side of the $\implies$. We want

$$|\sqrt{2 - x}| < \epsilon, \text{ which is true when} \tag{6.10b}$$
$$\sqrt{2 - x} < \epsilon, \text{ which is true when} \tag{6.10c}$$
$$2 - x < \epsilon^2, \text{ which is true when} \tag{6.10d}$$
$$2 - \epsilon^2 < x, \text{ which looks similar to eq. } 6.10a \tag{6.10e}$$

So to make things work we an try to let $\delta = \epsilon^2$ (or smaller).

Proof Step: Let $\epsilon > 0$ be given. Then define $\delta = \epsilon^2$

Suppose that $2 - \delta < x < 2$ (this is the left hand side of eq. 6.10a so it is what we need to assume). Since $\delta = \epsilon^2$,

$$2 - \epsilon^2 < x < 2 \tag{6.10f}$$

The first part of this inequality gives us

$$2 - \epsilon^2 < x \implies 2 - \epsilon^2 - x < 0 \tag{6.10g}$$
$$\implies 2 - x < \epsilon^2 \tag{6.10h}$$
$$\implies \sqrt{2 - x} < \epsilon \tag{6.10i}$$

Since $\epsilon > 0$,

$$|\sqrt{2 - x}| < \epsilon| \tag{6.10j}$$

and thus

$$|\sqrt{2 - x} - 0| < \epsilon \tag{6.10k}$$

which is precisely the right hand side of equation 6.10a. Thus, given any arbitrary $\epsilon > 0$, we are able to find a δ (specifically, by setting $\delta = \epsilon^2$), such that the right hand side of equation 6.10a can be derived from the left hand side. This is the formal proof of the limit. $\square$

Precise Definition of Infinite Limits

A function has an infinite limit ($f \to \infty$) as $x \to a$ if we can make f arbitrarily large whenever x is sufficiently close to a. This means that if you pick some really big number (say M) then I can find some really little number (say ϵ) and draw a "circle" (of radius ϵ) about a, such that, whenever you come inside the circle, the value of f will be bigger than M.

Definition 6.5. Infinite Limit, Precise Definition.

Let f be defined on an open interval that contains a (except possibly at a). Then

$$\lim_{x \to a} f(x) = \infty \tag{6.11}$$

means that for every $M > 0$ (no matter how large) there is a positive number $\delta > 0$ (that is usually very small) such that

$$|x - a| < \delta \implies f(x) > M \tag{6.12}$$

and

$$\lim_{x \to a} f(x) = -\infty \tag{6.13}$$

means that for every $N < 0$ (no matter how large and negative) there is a positive number $\delta > 0$ (that is usually very small) such that

$$|x - a| < \delta \implies f(x) < N \tag{6.14}$$

Example 6.6. Prove that $\lim_{x \to 0} \dfrac{1}{x^2} = \infty$.

Solution. <u>Analysis.</u> Let $M > 0$ be given. We need to find $\delta > 0$ such that

$$|x - 0| < \delta \implies \frac{1}{x^2} > M \tag{6.15a}$$

We observe that

$$\frac{1}{x^2} > M \iff x^2 < \frac{1}{M} \iff |x| < \frac{1}{\sqrt{M}} \tag{6.15b}$$

<u>Proof.</u> Reverse the order of the steps. Let $M > 0$ be given.

Then choose $\delta = \dfrac{1}{\sqrt{M}}$. Then if

$$|x| < \delta = \frac{1}{\sqrt{M}} \tag{6.15c}$$

by rearrangement we have

$$\frac{1}{x^2} > M \tag{6.15d}$$

and therefore $f(x) > M$ as required. $\qquad \square$

Precise Definition of Asymptotic Limits

By an asymptotic limit here we are referring to the type of limit that leads to a horizontal asymptote. Most texts call this a "limit at infinity" which leads to some confusion with the "infinite limits" we just discussed. Infinite limits correspond to vertical asymptotes; asymptotic limits correspond to horizontal asymptotes.

Figure 6.3: Illustration of the concept of a limit as $x \to \infty$: $f(x)$ can be made arbitrarily close to L (within a distance ϵ) by choosing x to be sufficiently large (e.g., larger than some large N).

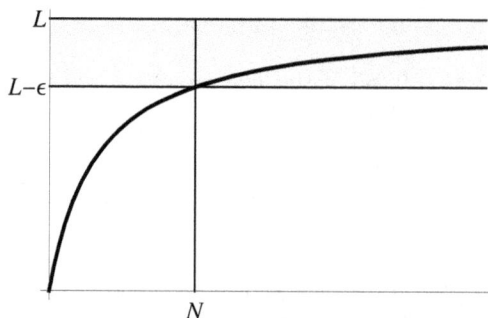

Definition 6.6. Asymptotic Limit (Limit at Infinity)

Let $f(x)$ be defined on an open interval (a, ∞). Then we write

$$\lim_{x \to \infty} f(x) = L \tag{6.16}$$

if, given any (very small number) $\epsilon > 0$, it is possible to find a (usually very large number) N such that for all $x > N$,

$$|f(x) - L| < \epsilon \tag{6.17}$$

and we say that $f(x)$ has a horizontal asymptote at $y = L$.

Definition 6.7. Asymptotic Limit (Limit at Negative Infinity)

Let $f(x)$ be defined on an open interval $(-\infty, b)$. Then we write

$$\lim_{x \to -\infty} f(x) = L \tag{6.18}$$

if, given any (very small number) $\epsilon > 0$, it is possible to find a (usually very large in magnitue and negative number) N such that for all $x < N$,

$$|f(x) - L| < \epsilon \tag{6.19}$$

and we say that $f(x)$ has a horizontal asymptote at $y = L$.

Example 6.7. Show that $\displaystyle\lim_{x \to \infty} \frac{1}{x} = 0$

Solution. We need to prove that $1/x \to 0$ as $x \to \infty$. To do this we must show that for any $\epsilon > 0$, there exists an N such that for all $x > N$

$$|1/x| = |1/x - 0| = |f(x) - L| < \epsilon \tag{6.20a}$$

This means that we need to pick x bigger than $1/\epsilon$. If we let $N = 2/\epsilon > 1/\epsilon$ then this will be guaranteed to occur. Here is how the proof goes:

Let $\epsilon > 0$ be given. Then define $N = 2/\epsilon$. For all $x > N$,

$$\frac{1}{x} < \frac{1}{N} = \frac{\epsilon}{2} < \epsilon \tag{6.20b}$$

Thus for any $\epsilon > 0$, there exists some $N > 0$ (the one we found was $N = 2/\epsilon$, but there are lots of others), such that for all $x > N$,

$$|f(x) - L| = \left|\frac{1}{x} - 0\right| < \epsilon \tag{6.20c}$$

Thus $\lim\limits_{x \to \infty} \dfrac{1}{x} = 0$. $\square$

In theorem 6.1 we generalize the last example to functions of the form $1/x^r$.

Theorem 6.1. Power Limits

Let r be any rational number. Then

$$\lim_{x \to \pm\infty} \frac{1}{x^r} = 0 \tag{6.21}$$

Proof. We proved the case $r = 1$ in example 6.7 (for $x \to \infty$). We skip the case of $x \to -\infty$. The general result follows from the application of limit laws. If r is a rational number, then there exist some non-zero integers such that $r = p/q$. So we may write

$$\lim_{x \to \pm\infty} \frac{1}{x^r} = \lim_{x \to \pm\infty} \frac{1}{x^{p/q}} = \lim_{x \to \pm\infty} \left(\frac{1}{x}\right)^{p/q} = \left(\lim_{x \to \pm\infty} \frac{1}{x}\right)^{p/q} \tag{6.22}$$

$$\tag{6.23}$$

$$= 0^{p/q} = \sqrt[q]{0^p} = \sqrt[q]{0} = 0 \tag{6.24}$$

$\square$

Exercises

1. Let $f(x) = 5x + 8$. Prove that $\lim_{x \to 1} = 13$ using the formal theory of limits. This means that *for any $\epsilon > 0$ you must find a $\delta > 0$ such that whenever $|x - 1| < \delta$ then $|5x + 8 - 13| < \epsilon$.*

   (a) Explain why the statement in italics is what you need to find.
   (b) Suppose somebody tells you what ϵ is. Find a δ (as a function of ϵ) that will work.

2. Let $f(x) = x^2 - 12x$.

   (a) Find $\lim_{x \to 6} f(x)$ using the limit laws from the previous chapters.
   (b) Formulate a statement in terms of ϵ and δ of what you need to prove to show this formally.
   (c) Suppose somebody tells you what ϵ is. Find a δ (as a function of ϵ) that will work.

Prove each of the following limits using the formal theory.

3. $\lim_{x \to 2} 3x = 6$

4. $\lim_{x \to 5} (4x - 7) = 13$

5. $\lim_{x \to 2} (x^2 - 4) = 0$

6. $\lim_{x \to 3} (x + 2x) = 6$

7. $\lim_{x \to 2} \dfrac{x^2 - 4}{x - 2} = 4$

8. $\lim_{x \to 2} \dfrac{x^2 + 4}{x - 2} = 4$

9. $\lim_{x \to 3} \sqrt{9 - 3x} = 0$

10. $\lim_{x \to 10^-} f(x) = 1$ for $f(x) = \begin{cases} 1, & \text{if } x \leq 10 \\ 2, & \text{if } x > 10 \end{cases}$

11. $\lim_{x \to 3+} f(x) = 9$ for $f(x) = \begin{cases} 2, & \text{if } x \leq 3 \\ x^2, & \text{if } x > 3 \end{cases}$

12. Prove the squeeze theorem (theorem 3.11) using the formal definition of a limit. Suppose that

    (a) $g(x) \leq f(x) \leq h(x)$
    (b) $h(x) \to L$ as $x \to a$
    (c) $h(x) \to L$ as $x \to a$

    Prove the theorem as follows:

    (a) Write equations in terms of ϵ and δ for each of 12b and 12c, assuming the have the same ϵ but different δ's.
    (b) Write a statement in terms of ϵ and δ for what you need to prove for f. Use the same ϵ as for g and h.
    (c) Chose δ in the f statement as the minimum of the δ's for g and h

Chapter 7

Continuity

Chapter Summary and Goal

In this chapter we will explore what we meant earlier in the text by our ad-hoc use of the term *nice and smooth function*. The mathematical term for a *nice and smooth function* is a *continuous function*. The concept of continuity is defined formally in this chapter in terms of limits.root

Student Learning Objectives

The student will:

1. Understand the concept of continuity.
2. Be able to identify the intervals of continuity of a function.
3. Be able to locate points of discontinuity on a function.
4. Understand and apply the intermediate value theorem.

Continuous Functions

We we say that a function is continuous, we a making a formal (precise) mathematical statement that means the function is, in a certain sense, nice and smooth (you may want to refer back the ad-hoc definition of "nice-and-smooth" we gave earlier in definition 2.1).[1] This formal concept of continuity is closely related to the concept of a limit. We will first examine the continuity of a function at a single point $x = a$, and then expand this definition to the continuity of the function over an entire interval. In simple terms, a function is continuous at $x = a$ if you can draw a plot of the function through the point $x = a$ without lifting your pencil up off the paper (figure 7.1). The consequences of this are the following:

1. $f(a)$ must be defined, i.e., $f(x)$ is defined at $x = a$. Otherwise, you would have to lift up your pencil to draw the hole at $f(x)$.

2. $\lim\limits_{x \to a} f(x)$ must exist. Specifically, the limit from the left and the limit from the right must be the same, because if there were a jump from the left to the right, you would have to lift up your pencil (see theorem 2.1).

---

[1] Traditionally the term "smooth" is reserved for differentiable functions, not continuous functions, so we are breaking from tradition a bit by using it here.

Figure 7.1: Examples of continuity and discontinuity. At $x = a$, $\lim\limits_{x \to a} f(x) \neq f(a)$ so f is not continuous. At $x = b$, $\lim\limits_{x \to b^-} f(x) = f(b)$ but $\lim\limits_{x \to b^+} \neq f(b)$, so f is also not continuous at b. At $x = c$, the limits exist from both directions and are equal to $f(c)$, so f is continuous at c.

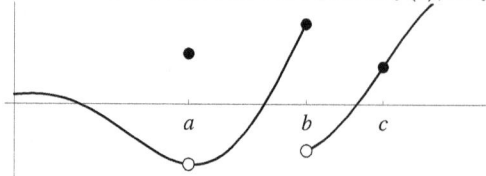

3. $\lim\limits_{x \to a} f(x) = f(a)$. Otherwise, if there were a jump, either from the limit from the left to $f(a)$, or from $f(a)$ to the limit from the right, then you would have to lift up your pencil.

In fact, item (3) in the above does not make any sense unless items (1) and (2) are already true, so , in a sense, stating the first two items is somewhat superfluous. We therefore take it as our definition of **continuity at a point** $x = a$.

Definition 7.1. Continuity at a Point

A function $f(x)$ is said to be **continuous at the point** $x = a$ if

$$\lim_{x \to a} f(x) = f(a) \tag{7.1}$$

If $f(x)$ is not continuous at a we say that $f(x)$ **is discontinuous at** a.

We also have a special name when a function is continuous at every point on an interval.

Definition 7.2. Continuity on an Interval

A function is **continuous on an interval** (a, b) if it is continuous at every point in the interval.

Here is the distinction between the two definitions. When $f(x)$ is continuous at a, you can draw $f(x)$ through the point at $(a, f(a))$ without lifting your pencil. When f is continuous on (a, b) you can draw $f(x)$ through the interval (a, b) without lifting your pencil.

Example 7.1. Where is $f(x) = \dfrac{x^2 - x - 2}{x - 2}$ continuous?

Solution. We observe that

$$\lim_{x \to 2^-} f(x) = \lim_{x \to 2^+} f(x) = 3 \tag{7.2a}$$

and therefore

$$\lim_{x \to 2} f(x) = 3 \tag{7.2b}$$

However, $f(x)$ is undefined at $x = 2$, because the denominator is equal to zero. We write

$$\lim_{x \to 2} f(x) \neq f(2) \tag{7.2c}$$

because $f(x)$ is undefined, and conclude that $f(x)$ is not continuous at $x = 2$.

For all other x, since $f(x)$ is a rational function with non-zero denominator, we can use direct substitution to find the limit (theorem 3.9). Thus at any point $a \neq 2$, $\lim\limits_{x \to a} f(x) = f(a)$ and therefore $f(x)$ is continuous at that a. $\square$

Figure 7.2: Removable singularity of the function analyzed in examples 7.1 and 7.2.

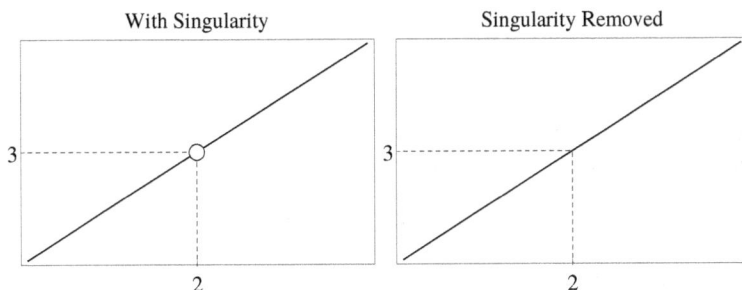

Sometimes we can remove a discontinuity at $x = a$, when it amounts to nothing more than a hole in the function at a. These are often caused by a special type of division by zero called a removable singularity.

Definition 7.3. Removable Singularity

We call the point $x = a$ a **removable singularity** of $f(x)$ if $f(x)$ is not defined at $x = a$ but $\lim\limits_{x \to a} f(x) = L$ for some finite number L. Then we can define a new function $F(x)$ with the singularity removed by

$$F(x) = \begin{cases} f(x), & x \neq a \\ L, & x = a \end{cases} \tag{7.3}$$

Example 7.2. Fix the removable singular in $y = \dfrac{x^2 - x - 2}{x - 2}$ we encountered in example 7.1.

Solution. We observed that

$$f(x) = \frac{x^2 - x - 2}{x - 2} = \frac{(x+1)(x-2)}{x-2} = x + 1 \tag{7.4a}$$

except at $x = 2$. As we found in the earlier example, $f(x) \to 3$ as $x \to 3$.

But we also saw that for all $x \neq 2$, the new function $g(x) = x + 1 = f(x)$. Since $g(x)$ is a line, it is continuous at all x. (This follows from theorem 3.9.) By definition 7.3, $g(x)$ removes the singularity in $f(x)$, as illustrated in See figure 7.2. $\square$

In example 7.1 we calculated the set of all points where $f(x)$ is continuous. This has a name: it is called the interval of continuity. The interval of continuity in example 7.1 is $(-\infty, 2) \cup (2, \infty)$[2] The interval of continuity for $g(x)$ in example 7.2 is $(-\infty, \infty)$.

---

[2] The symbol $\cup$ represents set union.

Definition 7.4. Interval of Continuity

The **interval of continuity** is the set of all points in the domain of a function $f(x)$ at which the function is continuous.

Example 7.3. Find the interval of continuity of the function

$$f(x) = \begin{cases} \dfrac{2}{x^2} & \text{if } x \neq 0 \\ 1 & \text{if } x = 0 \end{cases} \tag{7.5}$$

Solution. This is a rational function; the only point where there is any concern is where the denominator is zero. This occurs at $x = 0$. We have

$$\lim_{x \to 0} f(x) = \infty \neq 1 = f(0) \tag{7.6a}$$

which means that $f(x)$ is not continuous at $x = 0$, but it is continuous everywhere else. So the interval of continuity is $\{(-\infty, 0), (0, \infty)\}$.   □

We can also define a concept of directional continuity. This occurs when the function has a different limit from the right than from the left, and the function value at $x = a$ is equal to the limit from one of the two directions.

Definition 7.5. Directional Continuity.

We say that $f(x)$ is **continuous from the right at** a (or **right continuous at** a) if

$$\lim_{x \to a^+} f(x) = f(a) \tag{7.7}$$

and **continuous from the left at** a (or **left continuous at** a) if

$$\lim_{x \to a^-} f(x) = f(a) \tag{7.8}$$

Figure 7.3: The function (7.9) is right continuous at $x = 2$.

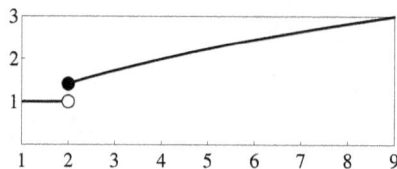

Example 7.4. The function

$$f(x) = \begin{cases} 1, & \text{if } x < 2 \\ \sqrt{x}, & \text{if } x \geq 2 \end{cases} \tag{7.9}$$

is right-continuous at $x = 2$.

Solution. This is illustrated in figure 7.3. To verify this algebraically we calculate

$$\lim_{x \to 2^+} f(x) = \lim_{x \to 2^+} \sqrt{x} = \sqrt{2} = f(2) \tag{7.10a}$$

Doing Calculus

It is not left-continuous, because

$$\lim_{x \to 2^-} f(x) = \lim_{x \to 2^-} 1 = 1 \neq \sqrt{2} = f(2) \tag{7.10b}$$

$\square$

The interval of continuity is always a subset of a function's of domain. The domain of $f(x) = \sqrt{x}$ is $[0, \infty)$. Since $f(x)$ is continuous everywhere in $(0, \infty)$, and is right-continuous at $x = 0$, the interval of continuity is $[0, \infty)$.

Remark 7.1. Notation for the interval of continuity:

Notation	Continuity at a	Continuity at b
(a, b)	None	None
$(a, b]$	None	From the left
$[a, b)$	From the right	None
$[a, b]$	From the right	From the left

Example 7.5. Find the interval of continuity for $f(x) = 1 - \sqrt{2 - x^2}$

Solution. The domain is limited by the square root, whose argument must be non-negative.

$$2 - x^2 \geq 0 \tag{7.11a}$$

$$2 \geq x^2 \tag{7.11b}$$

$$x^2 \leq 2 \tag{7.11c}$$

$$-\sqrt{2} \leq x \leq \sqrt{2} \tag{7.11d}$$

Therefore the domain of $f(x)$ is $[-\sqrt{2}, \sqrt{2}]$.

To see where the function is continuous, we need to consider (a) every point in the interior of the interval $(-\sqrt{2}, \sqrt{2})$; and (b) to see if $\lim_{x \to a} f(x) = f(a)$, where $a = \pm\sqrt{2}$.

In the interior of the domain, we can apply the limit laws:

$$\lim_{x \to a} f(x) = \lim_{x \to a} (1 - \sqrt{2 - x^2}) \tag{7.11e}$$

$$= \lim_{x \to a} 1 - \lim_{x \to a} \sqrt{2 - x^2} \qquad \text{(limit of a sum)} \tag{7.11f}$$

$$= 1 - \lim_{x \to a} \sqrt{2 - x^2} \qquad \text{(limit of a constant)} \tag{7.11g}$$

$$= 1 - \sqrt{\lim_{x \to a} (2 - x^2)} \qquad \text{(limit of a square root)} \tag{7.11h}$$

$$= 1 - \sqrt{2 - a^2} \qquad \text{(substitution law)} \tag{7.11i}$$

$$= f(a) \qquad \text{(definition of $f(x)$)} \tag{7.11j}$$

Thus $f(x)$ is continuous at every point in $(-\sqrt{2}, \sqrt{2})$.

We must examine each endpoint individually to see if they are included. At $x = -\sqrt{2}$, we must approach from the right.

$$\lim_{x \to -\sqrt{2}} f(x) = \lim_{x \to -\sqrt{2}} (1 - \sqrt{2 - x^2}) = (1 - \sqrt{2 - (-\sqrt{2})^2}) = 1 = f(-\sqrt{2}) \tag{7.11k}$$

Thus $f(x)$ is right continuous at the left endpoint. Similarly, at $x = \sqrt{2}$, we should approach from the left.

$$\lim_{x \to \sqrt{2}} f(x) = \lim_{x \to \sqrt{2}} (1 - \sqrt{2 - x^2}) = (1 - \sqrt{2 - (\sqrt{2})^2}) = 1 = f(\sqrt{2}) \qquad (7.11l)$$

Thus $f(x)$ is left continuous at the right endpoint. Therefore the interval of continuity is $[-\sqrt{2}, \sqrt{2}]$. $\qquad\qquad\qquad\qquad\qquad\qquad\qquad\qquad\qquad\qquad\qquad\qquad\qquad\qquad\qquad\qquad\square$

Continuity Rules

We will proceed to summarize a number of standard continuity rules. All of these rules are direct consequences of the limit laws discussed in chapter 3. Our first result, for example, follows from theorem 3.3. Suppose that $f(x)$ and $g(x)$ are continuous at a. Then $f(x) \to f(a)$ and $g(x) \to g(a)$, so that

$$\lim_{x \to a} (f(x) + g(x)) = \lim_{x \to a} f(x) + \lim_{x \to a} g(x) = f(a) + g(a) \qquad (7.12)$$

Thus if we define a new function $h(x) = f(x) + g(x)$,

$$\lim_{x \to a} h(x) = f(a) + g(a) = h(a) \qquad (7.13)$$

making h continuous at a. A similar argument can be used to prove that the difference $f(x) - g(x)$ is continuous.

Theorem 7.1. Continuity of Sum and Difference

If $f(x)$ and $g(x)$ are continuous at $x = a$ then their sum $f(x) + g(x)$ and their difference $f(x) - g(x)$ are continuous at a.

Example 7.6. Show that $y = x + \sqrt{x}$ is continuous at $x = 4$.

Solution. This is the sum of two functions $f(x) = x$ and $f(x) = \sqrt{x}$.

Since

$$\lim_{x \to 4} f(x) = \lim_{x \to 4} x = 4 = f(4) \qquad (7.14a)$$

we know that $f(x)$ is continuous at 4. Similarly, since

$$\lim_{x \to 4} g(x) = \lim_{x \to 4} \sqrt{x} = 2 = g(4) \qquad (7.14b)$$

we also know that g is continuous at 4. Hence $f(x) + g(x)$ is continuous at 4. $\qquad\square$

From the constant multiple rule (theorem 3.4), if $f(x) \to f(a)$ then

$$\lim_{x \to a} (cf(x)) = c \lim_{x \to a} = cf(a) \qquad (7.15)$$

As with the sum or product, if we define a new function $g(x) = cf(x)$, then

$$\lim_{x \to a} g(x) = cf(a) = g(a) \qquad (7.16)$$

Therefore $g(x) = cf(x)$ is continuous at a.

Theorem 7.2. Continuity of a Constant Multiple

If $f(x)$ is continuous at $x = a$ that $cf(x)$ is continuous at $x = a$.

Example 7.7. Show that $y = 10\sqrt{x}$ is continuous at $x = 9$

Solution. If we define $f(x) = \sqrt{x}$ then $y = 10f(x)$. Since

$$f(x) = \sqrt{x} \to \sqrt{9} = 3 = f(9), \tag{7.17a}$$

we know that $f(x)$ is continuous at $x = 9$.Therefore by the constant multiple rule $10\sqrt{x}$ is continuous at $x = 9$. □

Similarly we have product and quotient rules. Their proofs are similar and are left to the exercises.

Theorem 7.3. Continuity of a Product

If $f(x)$ and $g(x)$ are continuous at $x = a$ then their product $f(x)g(x)$ is continuous at $x = a$.

Theorem 7.4. Continuity of Roots

All roots are continuous on their domain. Let n be any positive integer. Then

$$f(x) = x^{1/n} = \sqrt[n]{x} \tag{7.18}$$

is continuous on $[0, \infty)$ if n is an even integer, and is continuous on $(-\infty, \infty)$ if n is an odd integer.

Example 7.8. The function $y = x\sqrt{x}$ is continuous at $x = 16$.

Solution. Let $f(x) = x$ and $g(x) = \sqrt{x}$. Then

$$f(x) = x \to 16 = f(16) \text{ as } x \to 16 \tag{7.19a}$$
$$g(x) = \sqrt{x} \to \sqrt{16} = 4 = g(16) \text{ as } x \to 16 \tag{7.19b}$$

Thus both f and g are continuous at $x = 16$. By the product rule, $f(x)g(x)$ is continuous at $x = 16$. □

Theorem 7.5. Continuity of a Quotient

If $f(x)$ and $g(x)$ are continuous at $x = a$ and $g(a) \neq 0$, then their quotient $\dfrac{f(x)}{g(x)}$ is continuous at $x = a$

Example 7.9. The function $y = \dfrac{x + \sqrt{x}}{x - \sqrt{x}}$ is continuous at $x = 25$.

Solution. Let $f(x) = x + \sqrt{x}$ and $g(x) = x - \sqrt{x}$. By the direct substitution rule for limits (theorem 3.9) both x and $\sqrt{x}$ are continuous at $x = 25$. Furthermore, $g(25) = 25 - \sqrt{25} = 20 \neq 0$, so theorem 7.5 applies. Hence $f(x)$ is continuous at $x = 25$. □

The next results all follow from the direct substitution rule for limits (theorem 3.9).

> ### Theorem 7.6. Continuity of Polynomials
>
> Let $p_n(x)$ be a polynomial, where n is any positive integer:
>
> $$p_n(x) = a_0 + a_1 x + a_2 x^2 + \cdots + a_n x^n \qquad (7.21)$$
>
> where $a_0, a_1, \ldots, a_n$ are any constants. Then the interval of continuity of $p_n(x)$ is $(-\infty, \infty)$.

Example 7.10. Find the interval of continuity of $y = x^{73} - 2x^{28} + 19$.

Solution. Since y is a polynomial, by theorem 7.6 the interval of continuity is $(-\infty, \infty)$. □

Since a rational function is a quotient of polynomials, by theorems 7.6 and 7.5, the interval of continuity of a rational function will be all real numbers except for the roots of the denominator. We state this explicitly as a separate continuity theorem.

> ### Theorem 7.7. Continuity of Rational Functions
>
> Let p_n and q_d be polynomials where n and d are any positive integers, and define
>
> $$r(x) = \frac{p_n}{q_d} = \frac{a_0 + a_1 x + a_2 x^2 + \cdots + a_n x^n}{b_0 + b_1 x + b_2 x^2 + \cdots + b_d x^d} \qquad (7.23)$$
>
> where $a_0, a_1, \ldots,$ and $b_0, b_1, dots$ are any constants. Then $r(x)$ is continuous at all x except at the roots of $q_d(x)$, i.e., the points where the denominator $q_d(x) = 0$.

Example 7.11. Find the interval of continuity of $y = \dfrac{7x^3 - 3x^2 + 4}{12x^2 + 6x - 3}$.

Solution. This is a rational function, so by theorem 7.7 it is continuous everywhere except where the denominator is zero. So from the quadratic equation, the possible points of discontinuity occur at

$$x = \frac{-6 \pm \sqrt{6^2 - 4 \cdot 12 \cdot -3}}{2 \cdot 12} = \frac{-6 \pm \sqrt{180}}{24} = -\frac{1}{4} \pm \frac{\sqrt{5}}{4} \qquad (7.24a)$$

The interval of continuity is thus

$$\left\{ \left(-\infty, -\tfrac{1}{4} - \tfrac{\sqrt{5}}{4}\right), \left(-\tfrac{1}{4} - \tfrac{\sqrt{5}}{4}, -\tfrac{1}{4} + \tfrac{\sqrt{5}}{4}\right), \left(-\tfrac{1}{4} + \tfrac{\sqrt{5}}{4}, \infty\right) \right\}$$

i.e., the entire real line with the two roots excluded. □

> ### Theorem 7.8. Continuity of Trigonometric Functions
>
> All trigonometric functions are continuous on their domain.

Example 7.12. Find the domain of continuity of $y = x^{1/3} + \sin x$

Solution. Let a be any real number. By theorem 7.4, $\sqrt[3]{x}$ is continuous at $x = a$. Since the domain of $\sin x$ is all real numbers, by theorem 7.8, it is also continuous at $x = a$. By theorem 7.1 their sum $x^{1/3} + \sin x$ is continuous at $x = a$ and therefore its interval of continuity is all real numbers $(-\infty, \infty)$. □

Example 7.13. Find the interval of continuity of $y = f(x) + g(x) - h(x)$, where f, g, and h are defined by the the under-brackets as follows:

$$y = \underbrace{\sqrt{x}}_{f(x)} + \underbrace{\frac{x+1}{x-1}}_{g(x)} - \underbrace{\frac{x+1}{x^2+1}}_{h(x)} \tag{7.26}$$

Solution. We consider the function as a sum of three terms $y = f(x) + g(x) + h(x)$. The first term $f(x) = \sqrt{x}$ is a root. By theorem 7.4, roots are continuous at every point in their domain. So the first term is continuous on $[0, \infty)$.

The second term $g(x) = \dfrac{x+1}{x-1}$ is a rational function. By theorem 7.7, a rational function is continuous at every point except where the denominator is zero. Thus the second term is continuous on $(-\infty, 1) \cup (1, \infty)$.

The third term $h(x) = \dfrac{x+1}{x^2+1}$ is also a rational function. In this case, the denominator can never be zero. Hence the third term is continuous on $(-\infty, \infty)$.

By theorem 7.1 the interval of continuity is the intersection of the intervals of continuity of the intervals for f, g and h, which is $[0, 1) \cup (1, \infty)$. These intervals are illustrated in figure 7.4. □

Figure 7.4: The interval of continuity of the function in example 7.12 is determined as the intersection of the intervals of continuity of the individual terms of the sum, as illustrated here. The intersection is given by the line on the top.

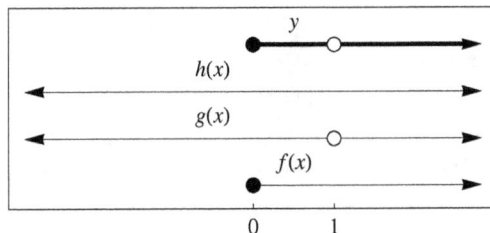

Limits of Continuous and Composite Functions

The concept of continuity allows us to add one new limit law to our toolbox: *The limit of a continuous function is the function of the limit*.

Theorem 7.9. Limit of Function is Function of Limit

If $f(x)$ is continuous at b and
$$\lim_{x \to a} g(x) = b$$
then
$$\lim_{x \to a} f(g(x)) = f(b)$$
i.e.,
$$\lim_{x \to a} f(g(x)) = f\left(\lim_{x \to a} g(x)\right)$$

Example 7.14. Find $\lim_{x \to \infty} \sin(1/x)$

Solution.   Since $\lim_{x \to \infty} (1/x) = 0$ and since $\sin x$ is continuous at $x = 0$, then by theorem 7.9
$$\lim_{x \to \infty} \sin(1/x) = \sin(0) = 0. \qquad \square$$

A consequence of this limit law is the following.

Theorem 7.10. Continuity of Composite Functions

If f is continuous at a and f is continuous at $g(a)$ then its composition
$$(f \circ g)(x) = f(g(x)) \tag{7.29}$$
is continuous at $x = a$, i.e.,
$$h(x) = f(g(x)) \tag{7.30}$$
is continuous at $x = a$.

Example 7.15. Show that $h(x) = \sin(x^2)$ is continuous at $x = \sqrt{\pi}$

Solution.   Let $f(x) = \sin(x)$ and $g(x) = x^2$.

Then $h(x) = \sin(x^2) = \sin(g(x)) = f(g(x))$.

Since $g(x)$ is a polynomial it is continuous at all x, and in particular it is continuous at $x = \sqrt{\pi}$.

Since $f(x)$ is a trigonometric function it is continuous at all points in its domain, and the domain of $\sin x$ is all real numbers. In particular, $\sin x$ is continuous at $g(\sqrt{\pi}) = (\sqrt{\pi})^2 = \pi$.

Therefore by theorem 7.10, $h(x) = f(g(x)) = \sin(x^2)$ is continuous at $x = \sqrt{\pi}$. $\square$

Intermediate Value Theorem

The intermediate value theorem (see figure 7.5) tells us that in travelling from $(x, y) = (a, p)$ to $(x, y) = (b, q)$, a *continuous* function must pass through every possible y value between p and q – there are no leaps, jumps, or skips. It is possible to pass through other values outside this range, but no value inside the range can be missed.

Figure 7.5: Two smooth (continuous) curves are shown that join the points (a, p) to (b, q). Each one of these curves passes through every y value in the range $q \leq y \leq p$. One of the curves passes through some y values more than once; this is not ruled out by the intermediate value theorem. The figure also illustrates how the function may take on values outside of the y box defined by $[p, q]$; this is also not ruled out.

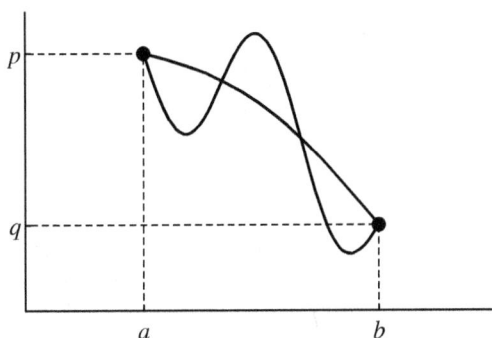

Theorem 7.11. Intermediate Value Theorem.

Suppose that $f(x)$ is continuous on the closed interval $[a, b]$ such that $f(a) \neq f(b)$. Let y be any number between $f(a)$ and $f(b)$. Then there exists some number $c \in (a, b)$ such that $f(c) = y$.

Example 7.16. Show that the function

$$f(x) = 4x^3 - 6x^2 + 3x - 2 \tag{7.32}$$

has a root in the interval $[1, 2]$.

Solution. We can demonstrate this by calculating the value of the function at each endpoint and showing that it has different signs. We calculate that

$$f(1) = 4 - 6 + 3 - 2 = -1 \tag{7.33a}$$
$$f(2) = 4(8) - 6(4) + 3(2) - 2 = 12 \tag{7.33b}$$

Since $f(1) = -1 > 0$ and $f(2) = 12 < 0$. Since $f(x)$ must must pass through every point between -1 and 12, it must also pass through 0. If there is some point where $f(x) = 0$ then there is some number x that corresponds to this y value. We call this x value c. This is the c value referred to in the intermediate value theorem. This is how we state the conclusion of the intermediate value theorem: *there must be some point c in $(1, 2)$ such that $f(c) = 0$.* This value of c corresponds to the root of $f(x)$, i.e., the root is at $x = c$. $\square$

Figure 7.6: Examples of functions that are continuous (left) and discontinuous (right) at $x = a$ in terms of the formal approach to continuity.

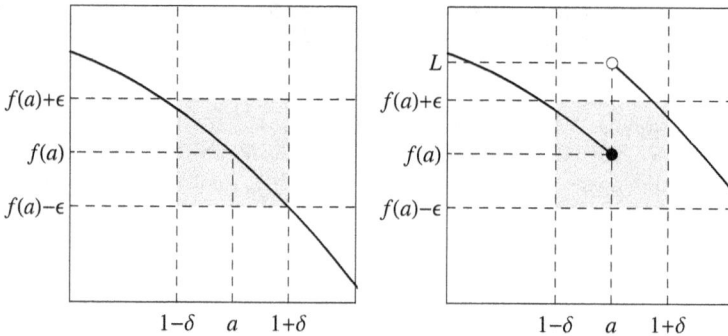

Formal Approach to Continuity

It is possible to look a continuity in the same formal way we look at limits. When we say that $f(x)$ is continuous at $x = a$ that means that (1) $f(x)$ is defined at $x = a$; and (2) $f(x) \rightarrow f(a)$ as $x \rightarrow a$. Thus saying the $f(x)$ is continuous at $x = a$ is equivalent to saying that

$$\lim_{x \to a} f(x) = f(a) \tag{7.34}$$

since the right-hand side does not even make sense unless $f(x)$ is defined at $x = a$.

In terms of the language of "getting really close to $x = a$", we say the $f(x)$ is continuous at $x = a$ if $f(x)$ can be as close as we like to $f(a)$ whenever x is sufficiently close to $x = a$.

Imagine drawing a box around the point $P = (a, f(a))$.

Let the height of the box be $\pm\epsilon$ and the width be $\pm\delta$ (see figure 7.6), centered around P. If the function is continuous, no matter what the height of the box (however tiny), it is possible to find some width (δ) so the the limit from both the left and right equal $f(a)$. If the function is not continuous (see the figure on the right), then for some very tiny ϵ, there is not possible to make the box wide enough so that the limit from the right and the left both equal $f(a)$.

Theorem 7.12. Continuity (Formal Definition)

We say that $f(x)$ is continuous at $x = a$ if (1) $f(x)$ is defined at $x = a$ and (2) given any $\epsilon > 0$ there exists a $\delta > 0$ (δ may depend on ϵ) such that

$$|x - a| < \delta \implies |f(x) - f(a)| < \epsilon \tag{7.35}$$

Proving that a function is continuous using the formal definition is identical to proving that the limit is $f(a)$ using the formal theory of limits.

Example 7.17. Show that $f(x) = x^3$ is continuous at $x = 3$ using the formal approach to continuity.

Solution. Reformulating in terms of limits, this is equivalent to showing that $\lim_{x \to 5} x^3 = 125$. This was done in example 6.3. □

Differentiability and Continuity

If $f(x)$ is differentiable at a point $x = a$ then the limit

$$f'(a) = \lim_{h \to 0} \frac{1}{h}(f(a+h) - f(a)) \tag{7.37}$$

exists. But

$$0 = f'(a) \cdot 0 = f'(a) \overbrace{\lim_{h \to 0} h}^{\text{this } = 0} \tag{7.38}$$

$$= \left(\lim_{h \to 0} \frac{1}{h}(f(a+h) - f(a)) \right) \left(\lim_{h \to 0} h \right) \tag{7.39}$$

$$= \lim_{h \to 0} \left[\left(\frac{1}{h}(f(a+h) - f(a)) \right) \cdot (h) \right] \tag{7.40}$$

$$= \lim_{h \to 0} (f(a+h) - f(a)) \tag{7.41}$$

$$= \lim_{h \to 0} f(a+h) - \lim_{h \to 0} f(a) \tag{7.42}$$

$$= \left(\lim_{h \to 0} f(a+h) \right) - f(a) \tag{7.43}$$

This tells as that $\lim_{h \to 0} f(a + h) = f(a)$. Therefore if we define $x = a + h$,

$$\lim_{x \to a} f(x) = \lim_{h \to 0} f(a+h) = f(a) \tag{7.44}$$

This is the definition of continuity.

> **Theorem 7.13. Differentiability Implies Continuity**
>
> If $f(x)$ is differentiable at $x = a$ then $f(x)$ is continuous at $x = a$.

The converse of Theorem 7.13 is not true. As an example, consider $y = |x - 5|$ at $x = 5$.

Exercises

1. Use the following figure for exercise 1.

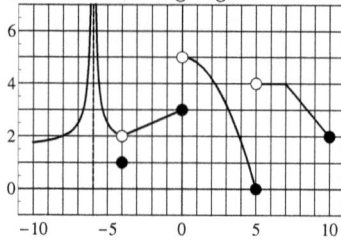

Identify whether $f(x)$ is (a) Left Continuous; (b) Right continuous and (c) Continuous at each of the following locations:

(a) $x = -6$ (e) $x = 5$
(b) $x = -4$ (f) $x = 7$
(c) $x = 0$ (g) $x = 10$
(d) $x = 3$

Find the interval of continuity of each of the following functions.

2. $f(x) = \dfrac{x^2 + 2x - 1}{5 - 3x}$

3. $g(x) = x^{100} - 2x^{37} + 75$

4. $h(x) = \sqrt{x} + \dfrac{x+1}{x-1} + \dfrac{x+1}{x^2+1}$

5. $p(x) = \dfrac{x^2 + 2x + 17}{x^2 - 1}$

6. $q(x) = \dfrac{\sin x}{2 + \cos x}$

7. $y = \cos \sqrt{x}$

8. $y = \sqrt{\sin x}$

9. $f(x) = \dfrac{\sin \sqrt{x}}{\sqrt{x}}$

10. $f(x) = \begin{cases} x, & \text{if } x \leq 1 \\ x^2, & \text{if } x > 1 \end{cases}$

11. $f(x) = \begin{cases} 1 - x, & \text{if } x \leq 0 \\ 1 - x^2, & \text{if } x \geq 0 \end{cases}$

12. $f(x) = \begin{cases} 1 - x^2, & \text{if } x < 0 \\ \sec x, & \text{if } 0 \leq x \leq \pi \\ \pi^2 - 1 - x^2, & \text{if } x > \pi \end{cases}$

13. Prove theorem 7.3. Show that the product of two functions is continuous. Hint: Assume that $f(x)$ and $g(x)$ are continuous at $x = a$ and apply theorem 3.5.

14. Prove theorem 7.5. Show that the quotient of two functions is continuous. Hint: Assume that $f(x)$ and $g(x)$ are continuous at $x = a$ and that $g(a) \neq 0$. Apply theorem 3.6.

15. Prove theorem 7.6. Show that polynomials are continuous everywhere.

Find each of the following functions, find the real constant a that will make $f(x)$ continuous.

16. $f(x) = \begin{cases} ax - 3, & x < 2 \\ ax^2, & x \geq 2 \end{cases}$

17. $f(x) = \begin{cases} ax - 1, & x < -8 \\ x^2 + 2x - 3, & x \geq -8 \end{cases}$

18. Show that $y = \dfrac{2x^2 + 2x - 24}{x - 3}$ has a removable discontinuity at $x = 3$ and find a value of $f(3)$ that removes the discontinuity.

Use the intermediate value theorem to prove that the following functions have roots in the indicated intervals.

19. $y = x - 3$, $[2,5]$

20. $y = x^2 - 2x - 8$, $[3,5]$

21. $y = x^2 - 3x - 9$, $[-3, 0]$

22. $y = x^3 \cos x + x$, $[\pi/4, \pi/2]$

Chapter 8

Calculus Without Limits

Chapter Summary and Goal

Here we will present the concept of infinitesimals. This approach was taken by Leibniz to develop the calculus without using limits.

Student Learning Objectives

The student will:

1. Understand the concept of infinitesimals.
2. Develop calculus from using hyperreals.
3. Calculate simple derivatives using the Leibniz Derivative.

Calculus Without Limits

The concept of a limit was formalized in the 19th century - more than 200 years after Newton and Leibniz simultaneously developed the calculus – principally by two famous mathematicians (Cauchy and Weierstrass) who were the giants of their day.

> **Concept of an Infinitesimal**
>
> An infinitesimal is so small that no matter how many you collect, you can never fill up a tea spoon.

Both Newton and Leibniz required that users think "outside the box" and hypothesized the existence of infinitesimal numbers, or numbers that are so small they could never amount to anything. The existence of such numbers violates a basic premise of math known as the Principle of Archimedes.

> **Principle of Archimedes**
>
> You can fill a bathtub with a teaspoon.

Newton sidestepped this conundrum by only considering only ratios of infinitesimals, and claiming that the ratio would remain finite. It is the ratio of these infinitesimals that became the derivative. It was the attempt to formally show that this ratio does, in fact, remain finite, that led to the development of the concept of a limit in the 19th centuries.

Leibniz, on the other hand, developed a whole new system of mathematics based on infinitesimals. His approach led to identical results but was largely ignored by the community until the mid 20th century, when the mathematics of hyperreal numbers was formally developed by Abraham Robinson and others.

More precisely, the **Principle of Archimedes** says that for any two nonzero numbers x and M, such that $x < M$, if you add x to itself enough times you will get a number larger than M, i.e., for some positive n,

$$\underbrace{x + x + \cdots + x}_{n \text{ times}} > M \tag{8.1}$$

All real numbers and all of traditional mathematics obeys the principle of Archimedes.

Axiom 8.1. Principle of Archimedes: Formal Mathematical Statement

For any real numbers x and M such that $x < M$, there exists some integer n such that

$$nx > M \tag{8.2}$$

This axiom is so ingrained into our mathematical subconscious that we rarely even bother to think about it; we tend to accept it as a physical truth or law, without question. When we turn to infinitesimals, however, we must abandon the principle of Archimedes. An infinitesimal is a new type of number than can be added to a real number but is not itself a real number.

Definition 8.1. Infinitesimal

An infinitesimal is a number h that is smaller than the smallest real number.

Given any real number M,
$$nh < M \tag{8.3}$$
for all integers n.

We will denote this using the symbol $\ll$. We will read

$$p \ll q \tag{8.4}$$

as p **is infinitely smaller than** q **but non-zero**, p **is infinitesimal with respect to** q, or p **is lower order than** q. In more colloquial terms, this means that p **is a whole lot smaller than** q, in the sense that the size of p can be ignored in comparison to the size of q, for all practical calculations.

Theorem 8.1. Infinitesimally Smaller

If h is any infinitesimal and x is any real number then

$$h \ll x \tag{8.5}$$

Among themselves, infinitesimals have the same properties as real numbers. You can add, subtract, multiply, and divide them to make new infinitesimals:

$$h + h = 2h \tag{8.6}$$

$$h + h + h = 3h \tag{8.7}$$

$$h \times h = (h)^2 \tag{8.8}$$

and so forth.

Numbers like h are known as **first order infinitesimals**. The basic property of first order infinitesimals is this: you can keep adding them together as much as you like, but their sum will never add up to even the smallest real number.

How Big are Infinitesimals?

No matter how many of them you add together, they will never add up to even the smallest real number.

We are allowed to combine real numbers together with infinitesimals. When we add an infinitesimal to a real number we get a a **hyperreal** number, such as

$$x = 17 + 3h \tag{8.9}$$

Along with very small numbers, we allow very big numbers, which we call infinite hyperreals. An **infinite hyperreal** is a number X is that is larger than any real number x.

Theorem 8.2. Properties of Infinitesimals

Let h be any infinitesimal, a be any real number. and X an infinite hyperreal. Then:

	h	a	X
$\square \cdot h$	infinitesimal	infinitesimal	indeterminate
$\square + h$	infinitesimal	hyperreal	infinite hyperreal
$h/\square$	indeterminate	infinitesimal	infinitesimal
$\square/h \ (h \neq 0)$	indeterminate	infinite	infinite hyperreal
$1/\square$	infinite	finite real	infinitesimal

When we multiply two first order infinitesimals together, as in (8.8), the result h^2, is called, a **second order infinitesimal**. Every second order infinitesimal is smaller than any first order infinitesimal.

$$h^2 \ll h \tag{8.10}$$

Furthermore, we can keep adding second order infinitesimals together as long as you like and you will never reach a first order infinitesimal. Thus

$$3h + h^2 > 3h \tag{8.11}$$

but there is no integer n such that

$$3h + nh^2 > 4h \tag{8.12}$$

We can continue to define 4th order, 5th order, etc., and all higher order infinitesimals to any order. We can imagine an infinite succession of all orders of ininitesimals, with

$$\cdots \ll h^n \ll h^{n-1} \ll \cdots h^3 \cdots \ll h^2 \ll h \ll x \tag{8.13}$$

for any real number x.

Definition 8.2. Hyperreal Numbers

A hyperreal number is the sum of a real number and an infinitesimal number.

Since the sum of any real number x and any infinitesimal h is a hyperreal that is very close to x we will define the **rounding operator** $R(x)$ to be the nearest real to x.

Definition 8.3. Standard Part or Rounding Operator

Let x be any finite hyperreal. Then the **Standard Part** of x is given by

$$R(x) = \text{nearest real number to } x \tag{8.14}$$

Theorem 8.3. Properties of the Rounding Operator

Let x and y be hyperreal. Then

$$R(x + y) = R(x) + R(y) \tag{8.15}$$
$$R(xy) = R(x)R(y) \tag{8.16}$$

Example 8.1. Find $R(h)$

Solution. Since $0 + h$ is a hyperreal,

$$R(h) = R(0 + h) \tag{8.17}$$
$$= R(0) \tag{8.18}$$
$$= 0 \tag{8.19}$$

$\square$

Example 8.2. Find $R(12 - 3h + 18h^2)$

Solution.

$$R(12 - 3h + 18h^2) = R(12) - R(3h) + 18(h2) \tag{8.20a}$$
$$= 12 - 3R(h) + 18R(h)R(h) \tag{8.20b}$$
$$= 12 - 3(0) + 18(0)(0) \tag{8.20c}$$
$$= 12 \tag{8.20d}$$

$\square$

Hyperreals and Slopes

Consider a the slope of a secant line to a function $f(x)$ that intersects f at two points that are infinitesimally far apart:

$$(a, f(a)) \text{ and } (a + h, f(a + h))$$

The slope is

$$m = \frac{y_2 - y_1}{x_2 - x_1} = \frac{f(a + h) - f(a)}{h} \tag{8.21}$$

We will define the derivative of $f(x)$ at $x = a$ as $R(m)$.

Definition 8.4. Leibniz Derivative

$$f'(x) = R\left(\frac{f(x + h) - f(x)}{h}\right) \tag{8.22}$$

Example 8.3. Calculate $f'(x)$ for $f(x) = 5x^2 + 3x + 15$.

Solution. We have

$$f(x + h) = 5(x + h)^2 + 3(x + h) + 15 \tag{8.23a}$$

$$= 5x^2 + 10xh + 5h^2 + 3x + 3h + 15 \tag{8.23b}$$

Hence

$$f(x + h) - f(x) = 5x^2 + 10xh + 5h^2 + 3x + 3h + 15$$
$$- 5x^2 - 3x - 15 \tag{8.23c}$$

$$= 10xh + 5h^2 + 3h \tag{8.23d}$$

$$= h(10x + 5h + 3) \tag{8.23e}$$

and consequently the derivative is

$$f'(x) = R\left(\frac{f(x + h) - f(x)}{h}\right) \tag{8.23f}$$

$$= R(10x + 5h + 3) = 10x + 3 \tag{8.23g}$$

$\square$

Thus we can derive a formula for the derivative without using limits at all. In fact, virtually every known fact about calculus can be derived using the Leibniz definition and without EVER calculating any limits.

Leibniz did not use the symbol h. Instead, he defined an infinitesimal in x using an operator called the delta-operator, written as Δx, read as "delta-x." This is somewhat confusing because it looks like two numbers Δ and x multiplied together but it is not. The number Δx is a single number, that is in fact an infinitesimal. This is the origin of the notation Δx for a very small number, used by most calculus books. Leibniz's heritage is that we use the notation dy/dx to represent a derivative, even though most mathematicians do not even understand where the origin of the notation, and just think of Δx as a "small" number, but not as an infinitesimal.

Example 8.4. If $f(x) = x^n$, find a formula for $f'(x)$.[1]

Solution. Using the binomial expansion (A.17) for $(x + h)^n$ for $n \geq 3$, and substituting into equation 8.22, gives,

$$f'(x) = R\left(\frac{f(x + h) - f(x)}{h}\right) = R\left(\frac{(x + h)^n - x^n}{h}\right) \tag{8.24a}$$

$$= R\left(\frac{x^n + nx^{n-1}h + \frac{n(n-1)}{2}x^{n-2}h^2 + \cdots + nxh^{n-1} + h^n - x^n}{h}\right) \tag{8.24b}$$

$$= R\left(\frac{nx^{n-1}h + \frac{n(n-1)}{2}x^{n-2}h^2 + \cdots + nxh^{n-1} + h^n}{h}\right) \tag{8.24c}$$

$$= R\left(nx^{n-1} + \frac{n(n-1)}{2}x^{n-2}h + \cdots + nxh^{n-2} + h^{n-1}\right) \tag{8.24d}$$

$$= nx^{n-1} \tag{8.24e}$$

$\square$

---

[1]In chapter 9 we will re-derive this formula using a factoring technique instead of the binomial theorem.

Example 8.5. Find the derivative of $y = \sqrt{x}$.

Solution. Substituting $f(x) = \sqrt{x}$ and $f(x+h) = \sqrt{x+h}$ into equation 8.22 gives

$$f'(x) = R\left(\frac{f(x+h) - f(x)}{h}\right) = R\left(\frac{\sqrt{x+h} - \sqrt{x}}{h}\right) \tag{8.25a}$$

$$= R\left(\frac{\sqrt{x+h} - \sqrt{x}}{h} \cdot \frac{\sqrt{x+h} + \sqrt{x}}{\sqrt{x+h} + \sqrt{x}}\right) \tag{8.25b}$$

$$= R\left(\frac{x + h - x}{h(\sqrt{x+h} + \sqrt{x})}\right) = R\left(\frac{h}{h(\sqrt{x+h} + \sqrt{x})}\right) \tag{8.25c}$$

$$= R\left(\frac{1}{\sqrt{x+h} + \sqrt{x}}\right) = \frac{1}{\sqrt{x} + \sqrt{x}} = \frac{1}{2\sqrt{x}} \tag{8.25d}$$

□

Example 8.6. Repeat example 8.5 by assuming that the result of example 8.4 holds for all real exponents.

Solution. In equation 8.24e we derived

$$(x^n)' = nx^{n-1} \tag{8.26a}$$

In example 8.5 we had $f(x) = \sqrt{x} = x^{1/2}$. If we assume that equation 8.24e holds for $n = 1/2$ then

$$f'(x) = \frac{1}{2}x^{1/2-1} = \frac{1}{2}x^{-1/2} = \frac{1}{2\sqrt{x}} \tag{8.26b}$$

as before.   □

Increments and Linearization

We have seen that the derivative is the standard part of the hyperreal quotient

$$\frac{f(x+h) - f(x)}{h} \tag{8.27}$$

where h is some infinitesimal. We will often write this fraction as $\Delta y / \Delta x$, i.e.,

$$\frac{\Delta y}{\Delta x} = \frac{f(x+h) - f(x)}{h} \tag{8.28}$$

where Δx represents an infinitesimal change in x (e.g., the "run") and Δy represents the corresponding change in y (e.g., the "rise"). We make the association

$$h = \Delta x \tag{8.29}$$

and

$$\Delta y = f(x+h) - f(x) \tag{8.30}$$

The question we ask here is the following: is Δy an infinitesimal? We will answer this question by considering two cases: either $h = 0$ or $h \neq 0$.

If $h = 0$, then $\Delta y = f(x+h) - f(x) = f(x) - f(x) = 0$. Thus $\Delta y = 0$, and therefore infinitesimal.

What happens when $h \neq 0$? Then since

$$\frac{\Delta y}{\Delta x} \approx R\left(\frac{\Delta y}{\Delta x}\right) = f'(x) \tag{8.31}$$

the left hand side and the right hand side only differ by some infinitesimal number, say ϵ. We can write this as

$$\frac{\Delta y}{\Delta x} = f'(x) + \epsilon \qquad (8.32)$$

Multiplying the equation through by Δx gives

$$\Delta y = f'(x)\Delta x + \epsilon \Delta x \qquad (8.33)$$

As we showed above, (8.33) still holds when $\Delta = 0$, by choosing $\epsilon = 0$. Thus we have proven the increment theorem.

Theorem 8.4. Increment Theorem

Let $y = f(x)$ is differentiable at x, and let h be infinitesimal. Then

$$\Delta y = f(x + h) - f(x) \qquad (8.34)$$

is also infinitesimal, and furthermore, there exists some infinitesimal ϵ such that

$$\Delta y = h f'(x) + \epsilon h \qquad (8.35)$$

In equation 8.35, the first term $h f'(x)$ is a first order infinitesimal, while the second term ϵh is a second order infinitesimal. In most practical calculations, such as the definition of the differential, it is ignored:

$$\epsilon h \ll h f'(x) \implies \Delta y \approx h f'(x) \qquad (8.36)$$

Setting $h = \Delta x$ gives us the definition of a differential.

Definition 8.5. Differential

Let $y = f(x)$ be a differentiable function and Δx an infinitesimal.
Then the **differential of y at x is**

$$dy = f'(x)\Delta x \qquad (8.37)$$

One common numerical application of differentials is to use them to rewrite (8.34) as

$$f(x + h) = f(x) + \Delta y \qquad (8.38)$$

If we approximate the infinitesimal Δy by differential $dy = f'(x)h$ (for some small increment h in x) then we have

$$f(x + h) \approx f(x) + f'(x)h \qquad (8.39)$$

The numerical application is as follows: if we know $f(a)$ then for a small perturbation in x away from a, say to $a + h$, we can approximate $f(a + h)$ by

$$f(a + h) \approx f(a) + f'(a)h \qquad (8.40)$$

where h is now a very small real number. If we define the point $x = a + h$, then $h = x - a$, so that

$$f(x) \approx f(a) + f'(a)(x - a) \qquad (8.41)$$

This is equivalent to approximating the curve by a straight line with slope $f'(a)$ through the point $(a, f(a))$. To see this, consider the point-slope equation of a line. The equation of a line through the point (x_1, y_1) with slope m is given by

$$y = y_1 + m(x - x_1) \qquad (8.42)$$

The tangent line at the point $(a, f(a))$ has slope $f'(a)$. If we substitute $m = f'(a)$ and $(x_1, y_1) = (a, f(a))$ into equation 8.42 we arrive precisely at equation 8.41. This tangent line is called the **linearization** of f at a.

Method 8.1. Linearization

If $f(x)$ is differentiable at $x = a$ then the **linearization** of f near a is

$$f(x) = f(a) + f'(a)(x - a) \qquad (8.43)$$

Example 8.7. Find the linearization of $y = x^{1/3}$ about $x = 8$.

Solution. The linearization is given by

$$y = f(a) + f'(a)(x - a) = f(8) + f'(8)(x - 8) \qquad (8.44a)$$

where $f(8) = 8^{1/3} = 2$ By the power law (Example 8.4),

$$f'(x) = \frac{1}{3}x^{-2/3} \qquad (8.44b)$$

and therefore

$$f'(8) = \frac{1}{3}8^{-2/3} = \frac{1}{12} \qquad (8.44c)$$

Thus the linearization is

$$y = f(8) + f'(8)(x - 8) = 2 + \frac{1}{12}(x - 8) = \frac{1}{12}x + \frac{4}{3} \qquad (8.44d)$$

$\square$

Example 8.8. Use the linearization method of equation 8.43 to approximate $\sqrt{4.2}$

Solution. Let $f(x) = \sqrt{x}$ and $a = 4$. Then $f(a) = \sqrt{4} = 2$ and

$$f'(x) = \frac{1}{2\sqrt{x}} \implies f'(a) = \frac{1}{2\sqrt{4}} = 0.25 \qquad (8.45a)$$

Thus

$$\sqrt{4.2} = f(4.2) \qquad (8.45b)$$
$$\approx f(a) + f'(a)(x - a) \qquad (8.45c)$$
$$= 2 + 0.25(4.2 - 4) = 4.05 \qquad (8.45d)$$

$\square$

Exercises

Use the Liebniz definition of the derivative (definition 8.4) to find general formulas for each of the following.

1. $y = x^2$
2. $y = x^3$
3. $y = Cx$, where C is a constant.
4. $y = ax + b$, where a and b are constants
5. $y = x^2 - 2x + 3$
6. $y = 1/\sqrt{x}$
7. $y = x^{1/3}$
8. $y = 1/x$

Use definition 8.4 to prove each of the following:

9. The constant multiple rule,

$$(Cf(x))' = Cf'(x),$$

for any constant C.

10. The sum rule,

$$(f(x) + g(x))' = f'(x) + g'(x).$$

In exercises 11 through 13, find the linearizations of the functions at points specified.

11. $y = x^2$ near $x = 5$
12. $y = \sqrt{x}$ near $x = 36$
13. $y = x^{1/4}$ near $x = 16$

14. Approximate $\sqrt{63.}$ using a linear approximation.

15. Approximate $\sqrt[3]{25.}$

16. Use differentials to estimate how many liters of paint are needed to cover the walls of a room that is 4 by 3 meters and 2.5 meters tall with a 1 mm thick layer of paint. You should include the ceiling by can ignore the area of doors and windows.

17. One side of a right triangle is 10 inches long and the opposite angle is 30 degrees, measured with an accuracy of two degrees. Use differentials to estimate the error in computing the length of the hypotenuse. (Note: you may look up the derivatives of trigonometric functions which we have not yet derived yet.)

Chapter 9

Basic Formulas for Derivatives

Chapter Summary and Goal

In this chapter we will use the basic definition of a derivative as given in definition 1.2 (or the equivalent definition 8.4) to develop a table of derivatives of some of the basic functions that we know from algebra. We repeat that earlier definition here for reference.

Definition 9.1. The Derivative

The derivative $f'(x)$ is defined by the following limit, if it exists:

$$f'(x) = \lim_{h \to 0} \frac{f(x+h) - f(x)}{h} \tag{9.1}$$

Student Learning Objectives

The student will:

1. Use the definition of the derivative to calculate new differentiation formulas.
2. Calculate derivatives of simple functions from algebra.
3. Learn to apply the basic rules of differentiation such as the constant multiple rule.
4. Understand linearity of the derivative and the sum/difference rule.
5. Understand the power rule.
6. Be able to apply the basic rules in combination with one another.

Calculation of the Derivative

We derived equation 9.1 by considering the limit of the slopes of secant lines as they approach a tangent (you may want to refer to back to figure 1.4 at this point). If $f'(x)$ is the derivative as a function of x, then $f'(c)$ is the derivative of f at the point $x = c$, and $f'(c)$ is **the slope of the tangent line to the function** $f(x)$ at $x = c$.

If the derivative exists at a point $x = c$, we say that $f(x)$ is **differentiable at** c. If the derivative exists on any interval (a, b) then we say $f(x)$ is **differentiable on** (a, b). If the limit does not exist at any point c, then we say **the derivative does not exist at** c, or that $f(x)$ is **not differentiable at** c. If the derivative does not exist anywhere on an interval (a, b) in the domain of f, then we say that $f(x)$ **is not differentiable on** (a, b).

Denoting the derivative by $f'(x)$ is known as the **Newton notation** for the derivative. Another commonly used notation for the derivative is the **Liebniz notation**. If $y = f(x)$ then we write

$$\frac{dy}{dx} = f'(x) \tag{9.2}$$

If we think of Δy and Δx as small increments in the y and x directions, representing the incremental rise and run that make up the slope, respectively (figure 9.1), then it is helpful to think of the Leibniz notation as representing the limit

$$\frac{dy}{dx} = \lim_{\Delta x \to 0} \frac{\Delta y}{\Delta x} \tag{9.3}$$

This is no different from definition 1.2 because

$$\Delta x = h \tag{9.4}$$
$$\Delta y = f(x + h) - f(x) \tag{9.5}$$

(You might want to compare this to how we defined increments and differentials in theorem 8.4 and definition 8.5.)   In the Liebniz notation we denote the derivative at the point $x = c$

Figure 9.1: As $Q \to P$, the ratio of the rise over the run (equation 9.3), $\Delta y/\Delta x \to dy/dx$.

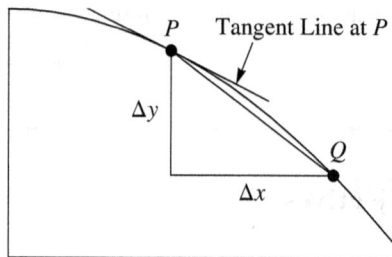

by the left hand side of the following equation:

$$\left. \frac{dy}{dx} \right|_{x=c} = y'(c) \tag{9.6}$$

For the most part we will use the notations y', f', and dy/dx interchangeably.[1]

The **second derivative**, denoted by $f''(x)$ is the derivative of $f'(x)$, if it exists. If $y = f(x)$ then we may also write

$$f''(x) = y'' = \frac{d^2 y}{dx^2} \tag{9.7}$$

Second derivatives commonly arise in physical problems as a measure of curvature and acceleration.

The **third derivative**, denoted by $f'''(x)$ is the derivative of $f''(x)$, if it exists. If $y = f(x)$ then we may also write

$$f'''(x) = y'' = \frac{d^3 y}{dx^3} \tag{9.8}$$

---

[1]Strictly speaking, we only use dy/dx when $y = f(x)$. If $y = f(t)$ then we use y', f' and dy/dt interchangeably, as the variable x would not exist in the equation, and so forth.

Beyond $n = 3$, the prime notation is rarely used for differentiation; instead, a small raised exponent surrounded by parenthesis is used. This is sometimes uses for y'' and y''' as well. Thus we define the notation

$$\frac{d^2y}{dx^2} = f^{(2)}(x) = y^{(2)} = y'' = \frac{d}{dx}\frac{dy}{dx} \tag{9.9}$$

$$\frac{d^3y}{dx^3} = f^{(3)}(x) = y^{(3)} = y''' = \frac{d}{dx}\frac{d^2y}{dx^2} \tag{9.10}$$

$$\vdots$$

$$\frac{d^ny}{dx^n} = f^{(n)}(x) = y^{(n)} = \left(y^{(n-1)}\right)' = \frac{d}{dx}\left(\frac{d^{n-1}y}{dx^{n-1}}\right) \tag{9.11}$$

Derivative of a Constant Function

To find the derivative y' of function $y = f(x) = C$ we substitute $f(x) = C$ into (9.1),

$$f'(x) = \lim_{h \to 0} \frac{f(x+h) - f(x)}{x} = \lim_{h \to 0} \frac{C - C}{h} = 0 \tag{9.12}$$

Theorem 9.1. Derivative of a Constant

The derivative of a constant is zero:

$$\frac{d}{dx}(C) = 0 \tag{9.13}$$

Derivative of $y = x$

To find the derivative y' of $y = x$ we substitute $f(x) = x$ into (9.1),

$$f'(x) = \lim_{h \to 0} \frac{f(x+h) - f(x)}{x} = \lim_{h \to 0} \frac{x + h - x}{h} = 1 \tag{9.14}$$

Theorem 9.2. The derivative of $y=x$ is 1:

$$\frac{d}{dx}(x) = 1 \tag{9.15}$$

The Constant Multiple Rule

To find the derivative of the function $g(x) = Cf(x)$, where C is a known constant and we know the derivative $f'(x)$ of $f(x)$, we substitute $g(x)$ into (9.1).

$$[Cf(x)]' = g'(x) = \lim_{h \to 0} \frac{g(x+h) - g(x)}{h} \tag{9.16a}$$

$$= \lim_{h \to 0} \frac{Cf(x+h) - Cf(x)}{h} \tag{9.16b}$$

$$= \lim_{h \to 0} \frac{C(f(x+h) - f(x))}{h} \tag{9.16c}$$

$$= C \underbrace{\lim_{h \to 0} \frac{f(x+h) - f(x)}{h}}_{\text{this is } f'(x)} = Cf'(x) \quad \square \tag{9.16d}$$

The last line follows from limit theorem 3.4, which allows us to pull a multiplicative constant out of the limit.

Theorem 9.3. The Constant Multiple Rule

The derivative of a constant times a function is a constant times the derivative, or more succinctly, we are allowed to "pull a multiplicative constant out of the derivative:"

$$\frac{d}{dx}(Cf(x)) = C\frac{df}{dx} \tag{9.17}$$

or

$$[Cf(x)]' = Cf'(x) \tag{9.18}$$

The general principle that we have used here – that the derivative of a function stays the same even if we change the name (e.g., from $Cf(x)$ to $g(x)$ and then back again to $Cf(x)$) – is going to be used so often that we will state it as a rule.

Theorem 9.4. Equality of Derivatives

If $f(x) = g(x)$ on an interval (a, b) where $f'(x)$ and $g'(x)$ are defined, then $f'(x) = g'(x)$ for all x in (a, b).

The Sum and Difference Rules

We are often faced in mathematics with functions that are the sum or difference of two simpler functions, such as $f(x) = x^2 + x$, or $g(x) = \cos x - \tan x$. Here we ask the question of how the derivatives of the sum (or difference) is related to the component function, e.g., if we know the derivative of $u(x) = \sqrt{x}$ and we know the derivative of $v(x) = x^3$, how can we make use of that information to determine the derivative of $u(x) + v(x) = \sqrt{x} + x^3$?

We approach the problem as before: suppose that $u(x)$ and $v(x)$ are two functions, and that we know how to compute (i.e., we know formulas for) $u'(x)$ and $v'(x)$. Then we define a new function $f(x) = u(x) + v(x)$, and begin by calculating $f'(x)$ using definition 1.2.

$$f'(x) = \lim_{h \to 0} \frac{f(x+h) - f(x)}{h} \tag{9.19a}$$

$$= \lim_{h \to 0} \frac{\overbrace{(u(x+h) + v(x+h))}^{f(x+h)} - \overbrace{(u(x) + v(x))}^{f(x)}}{h} \tag{9.19b}$$

$$= \lim_{h \to 0} \frac{u(x+h) \overbrace{-u(x) + v(x+h)}^{\text{switch order of these terms}} -v(x)}{h} \tag{9.19c}$$

$$= \lim_{h \to 0} \left(\frac{u(x+h) - u(x)}{h} + \frac{v(x+h) - v(x)}{h} \right) \tag{9.19d}$$

$$\overbrace{= \lim_{h \to 0} \underbrace{\frac{u(x+h) - u(x)}{h}}_{u'(x)} + \lim_{h \to 0} \underbrace{\frac{v(x+h) - v(x)}{h}}_{v'(x)}}^{\text{from theorem 3.3}} \tag{9.19e}$$

$$= u'(x) + v'(x) \tag{9.19f}$$

This result is called the **sum rule**; a similar result, the **difference rule**, is obtained by finding the derivative $u'(x) - v'(x)$.

Theorem 9.5. Sum Rule

The derivative of the sum is the sum of the derivatives

$$(u(x) + v(x))' = u'(x) + v'(x) \tag{9.20}$$

Theorem 9.6. Difference Rule

The derivative of the difference is the difference of the derivatives

$$(u(x) - v(x))' = u'(x) - v'(x) \tag{9.21}$$

Derivative of a Line

A line with slope m and y intercept b has an equation $y = mx + b$, where m and b is a constant. Thus we can use several of the last two results to calculate its derivative:

$$y' = \frac{d}{dx} (mx + b) \tag{9.22a}$$

$$= \frac{d}{dx} (mx) + \frac{d}{dx} (b) \qquad \text{by result 9.20} \tag{9.22b}$$

$$= m \frac{d}{dx} (x) \qquad \text{by results 9.17 and 9.13} \tag{9.22c}$$

$$= m \qquad \text{by result 9.15} \tag{9.22d}$$

This leads us to a very important observation: the **derivative of a line is its slope**.

Theorem 9.7. Derivative of a Line

The derivative of a line $y = mx + b$ is the slope m:

$$\frac{d}{dx} (mx + b) = m \tag{9.23}$$

Linearity

When taken together, theorems 9.3, 9.5 and 9.6 tell us that for any functions f and g and for any constants a and b (that may be either positive or negative),

$$(af(x) + bg(x))' = af'(x) + bg'(x) \tag{9.24}$$

It is sometimes convenient to think of $f'(x)$ (or $\dfrac{d}{dx} f(x)$) in terms of operators. An operator is a special kind of function that works on functions (figure 9.2). Linear operators are a special class of operators.

Figure 9.2: The derivative as an operator. D is a function that takes a function as input and returns a different function as output.

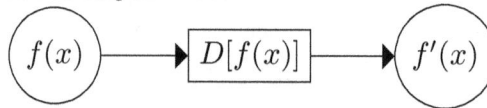

Definition 9.2. Linear Operators

A **linear operator** is any function $L[f]$ with input domain the space of all functions and whose output is another function such that for any two function $f(x)$ and $g(x)$, and for any two constants a and b,

$$L[af(x) + bg(x)] = aL[f(x)] + bL[g(x)] \tag{9.25}$$

We can define an operator

$$D[f(x)] = \frac{d}{dx} f(x) \tag{9.26}$$

whose input is the function $f(x)$ and whose output is the function $f'(x)$. From equation 9.24, if f and g are any two functions and a and b are any two numbers, then

$$D[af(x) + bg(x)] = \frac{d}{dx} (() \, af(x) + b(g(x))) \tag{9.27}$$

$$= a\frac{d}{dx} f(x) + b\frac{d}{dx} g(x) \tag{9.28}$$

$$= aD[f(x)] + bD[g(x)] \tag{9.29}$$

Thus by definition 9.2, D is linear.

Theorem 9.8. Linearity of the Derivative

The derivative is a linear operator.

Power Laws

To motivate the general rule for the derivative of x^n we will first explicitly calculate the derivatives of x^2 and x^3.

If we define $f(x) = x^2$ then

$$f(x + h) = (x + h)' = x' + 2xh + h' \tag{9.30}$$

Substituting (9.30) into result 9.1,

$$\frac{d}{dx}\left(x^2\right) = \lim_{h \to 0} \frac{f(x+h) - f(x)}{h} = \lim_{h \to 0} \frac{(x^2 + 2xh + h^2) - x^2}{h} \tag{9.31a}$$

$$= \lim_{h \to 0} \frac{2xh + h^2}{h} \tag{9.31b}$$

$$= \lim_{h \to 0} \frac{h(2x + h)}{h} \tag{9.31c}$$

$$= \lim_{h \to 0} (2x + h) \tag{9.31d}$$

$$= 2x \tag{9.31e}$$

Similarly, if we define $f(x) = x^3$ then

$$g(x+h) = (x+h)^3 = x^3 + 3x^2h + 3xh^2 + h^3 \tag{9.32}$$

Substituting (9.32) into result 9.1,

$$\frac{d}{dx}\left(x^3\right) = \lim_{h \to 0} \frac{f(x+h) - f(x)}{h} \tag{9.33a}$$

$$= \lim_{h \to 0} \frac{(x^3 + 3x^2h + 3xh^2 + h^3) - x^3}{h} \tag{9.33b}$$

$$= \lim_{h \to 0} \frac{3x^2h + 3xh^2 + h^3}{h} \tag{9.33c}$$

$$= \lim_{h \to 0} \frac{h(3x^2 + 3xh + h^2)}{h} \tag{9.33d}$$

$$= \lim_{h \to 0} (3x^2 + 3xh + h^2) \tag{9.33e}$$

$$= 3x^2 \tag{9.33f}$$

If we observe that $x^0 = 1$ and $x^1 = 1$, then we can make the following associations:

x^0	has derivative	$0 = 0 \times x^{0-1}$
x^1	has derivative	$1 = x^0 = 1 \times x^{1-1}$
x^2	has derivative	$2x = 2x^{2-1}$
x^3	has derivative	$3x^2 = 3x^{2-1}$

The generalization of this pattern is know as the power law.

Theorem 9.9. Power Law

Let n be any real number. Then

$$\frac{d}{dx}(x^n) = nx^{n-1} \tag{9.34}$$

Example 9.1. Find the derivative of $f(x) = x^7$.

Solution.

$$f'(x) = \frac{d}{dx}(x^7) = 7x^{7-1} = 7x^6 \tag{9.35a}$$

$\square$

Example 9.2. Find the derivative of $f(x) = 2300x^{1978}$.

Solution.

$$f'(x) = \frac{d}{dx}\left(2300x^{1978}\right) = 2300 \times 1978x^{1978-1} = 4,549,400x^{1977} \qquad (9.36a)$$

□

Proof. (Theorem 9.9 rule integer exponents.[2]) The general result follows from the following algebraic factorization rule (see (A.14)),

$$y^n - x^n = (y - x)(y^{n-1} + y^{n-2}x + \cdots + yx^{n-2} + x^{n-1}) \qquad (9.37a)$$

If we let $y = x + h$ in (9.37a) then

$$(x+h)^n - x^n = ((x+h) - x) \times$$
$$((x+h)^{n-1} + (x+h)^{n-2}x + \cdots + (x+h)x^{n-2} + x^{n-1}) \qquad (9.37b)$$
$$= h((x+h)^{n-1} + (x+h)^{n-2}x + \cdots + (x+h)x^{n-2} + x^{n-1}) \qquad (9.37c)$$
$$\frac{(x+h)^n - x^n}{h} = (x+h)^{n-1} + (x+h)^{n-2}x + \cdots + (x+h)x^{n-2} + x^{n-1} \qquad (9.37d)$$

Therefore the derivative becomes

$$\frac{d}{dx}(x^n) = \lim_{h \to 0} \frac{f(x+h) - f(x)}{h} \qquad (9.37ea)$$

$$= \lim_{h \to 0} \frac{(x+h)^n - x^n}{h} \qquad (9.37eb)$$

$$= \lim_{h \to 0} \left[(x+h)^{n-1} + (x+h)^{n-2}x + \cdots \right.$$
$$\left. + (x+h)x^{n-2} + x^{n-1}\right] \qquad (9.37ec)$$

$$= \underbrace{x^{n-1} + x^{n-1} + \cdots + x^{n-1}}_{n \text{terms}} \qquad (9.37ed)$$

$$= nx^{n-1} \qquad (9.37ee)$$

This verifies rule 9.34 for all positive integers.

The demonstration of why the the power law works with negative integer exponents requires the use of the quotient rule, which will be derived in chapter 10.

The demonstration of why the power law works for real numbers requires the use of exponential and logarithmic functions. The derivatives of these functions will be discussed in chapter 15

□

The derivative of a polynomial is found by applying the sum (or difference) rule, constant multiple rule, and power law successively.

Example 9.3. Find the derivative of $y = 3x^5 - 5x^2 + 7$

Solution.

$$y' = \left(3x^5 - 5x^2 + 7\right)' \qquad (9.6)$$

$$= \left(3x^5\right) + \left(-5x^2\right)' + (7)' \qquad \text{by result } 9.20 \qquad (9.7)$$

$$= 3\left(x^5\right) - 5\left(x^2\right)' + 0 \qquad \text{by results } 9.17 \text{ and } 9.13 \qquad (9.8)$$

---

[2]See also example 8.4 for a derivation using the Binomial theorem.

$$= 3 \times 5x^{5-1} - 5 \times 2x^{2-1} \qquad\qquad \text{by result } 9.34 \qquad (9.9)$$

$$= 15x^4 - 10x \qquad\qquad\qquad\qquad\qquad\qquad\qquad (9.10)$$

$\square$

Example 9.4. Find the points on the curve $y = 36x^2 - 20x^3 + 3x^4$ where the tangent line is horizontal.

Solution. When the tangent line is horizontal, its slope is zero, and hence its derivative, is zero, the derivative is zero. Hence the points where the tangent is horizontal satisfy

$$0 = f'(x) = \frac{d}{dx}\left(36x^2 - 20x^3 + 3x^4\right)$$
$$= 72x - 60x^2 + 12x^3$$
$$= 12x(6 - 5x + x^2)$$
$$= 12x(x - 3)(x - 2)$$

Thus the tangent line is horizontal at $x = 0$, $x = 2$, and $x = 3$. The corresponding *points* are $(0,0)$, $(2,32)$, and $(3,27)$, and are illustrated in figure 9.3. $\square$

Figure 9.3: $f(x)$ used in example 9.4, showing the points where the tangent line is horizontal.

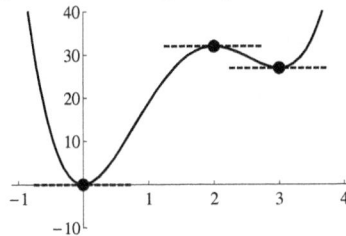

Example 9.5. Find the equations of all the tangent lines to the curve $y = x^3 - 4x$ which are parallel to the line $y = 8x + 1$

Solution. The line $y = 8x + 1$ has a constant slope of 8, so we need to find all the points on the curve of $y = x^3 - 4x$ where the tangent line has slope of 8. Then we need to find the equations of tangent lines at each of these points.

The x values of the points where the slope is 8 are the solutions of $y' = 8$,

Figure 9.4: $f(x)$ used in example 9.5, showing the points (short dashes) where the tangent line is parallel to the line $y = 8x + 1$ (longer dashes). The parallel tangent lines are shown as dashed lines.

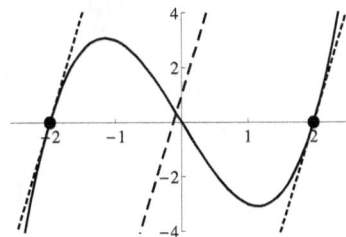

$$8 = (x^3 - 4x)' = 3x^2 - 4 \qquad\qquad\qquad (9.11a)$$

$$12 = 3x^2 \tag{9.11b}$$

$$4 = x^2 \tag{9.11c}$$

Therefore $x = \pm 2$; the corresponding points are (2,0) and (-2,0).

The equations of the lines can be found using the point-slope form, $y - y_1 = m(x - x_1)$. The lines are $y = 8x - 16$ and $y = 8x + 16$ as illustrated in figure 9.4.

$\square$

Derivatives of Sines and Cosines

The following trigonometric angle sum formulas will be useful for deriving these formulas.

$$\sin(A + B) = \sin A \cos B + \cos A \sin B \tag{9.12}$$

$$\cos(A + B) = \cos A \cos B - \sin A \sin B \tag{9.13}$$

Rule: Derivatives of $\sin x$ and $\cos x$

$$\frac{d}{dx} \sin x = \cos x \tag{9.14}$$

$$\frac{d}{dx} \cos x = -\sin x \qquad \text{(exercise)} \tag{9.15}$$

To get a formula for $\dfrac{d}{dx} \sin x$, we let $f(x) = \sin x$. Then

$$\frac{d}{dx} \sin x = f'(x) = \lim_{h \to 0} \frac{f(x + h) - f(x)}{h} \tag{9.16a}$$

$$= \lim_{h \to 0} \frac{\sin(x + h) - \sin x}{h} \tag{9.16b}$$

$$= \lim_{h \to 0} \frac{\sin x \cos h + \cos x \sin h - \sin x}{h} \tag{9.16c}$$

$$= \lim_{h \to 0} \frac{(\sin x \cos h - \sin x) + \cos x \sin h}{h} \tag{9.16d}$$

$$= \lim_{h \to 0} \frac{\sin x(\cos h - 1) + \cos x \sin h}{h} \tag{9.16e}$$

$$= \lim_{h \to 0} \left(\frac{\sin x(\cos h - 1)}{h} + \frac{\cos x \sin h}{h} \right) \tag{9.16f}$$

$$= \lim_{h \to 0} \frac{\sin x(\cos h - 1)}{h} + \lim_{h \to 0} \frac{\cos x \sin h}{h} \tag{9.16g}$$

$$= \underbrace{\left(\lim_{h \to 0} \sin x \right)}_{\to \sin x} \underbrace{\left(\lim_{h \to 0} \frac{(\cos h - 1)}{h} \right)}_{\to 0} + \underbrace{\left(\lim_{h \to 0} \cos x \right)}_{\to \cos x} \underbrace{\left(\lim_{h \to 0} \frac{\sin h}{h} \right)}_{\to 1} \tag{9.16h}$$

$$= \cos x \tag{9.16i}$$

Exercises

Use the definition of a derivative (9.1) to derive formulas for each of the expressions in exercises 1 through 4.

1. $f(x) = \sqrt{x}$ Ans: $\dfrac{1}{2\sqrt{x}}$

2. $f(x) = x^{-3/2}$ Ans: $-\dfrac{3}{2x^{5/2}}$

3. $f(x) = \dfrac{1}{\sqrt{x}}$ Ans: $-\dfrac{1}{2x^{3/2}}$

4. $f(x) = \dfrac{1}{x}$ Ans: $-\dfrac{1}{x^2}$

Use the definition of a derivative (9.1) and any trigonometric identities you know to find y' when y is given by the expressions in exercises 5 through 8.

5. $y = \cos x$ ans: $y' = -\sin x$

6. $y = \cos(2x)$ ans: $y' = -2\sin 2x$

7. $y = \sin^2(3x)$ ans: $y' = 3\sin(6x)$

8. $y = x\sin x$ ans: $y' = \sin x + x\cos x$

Use the rules of differentiation to find $f'(x)$ if y is given by the expressions in exercises 9 through 20.

9. $y = 3x - 14$ ans: 3

10. $y = 42x^5 - 6x^4 + 3x^2 + 7$
    ans:$210x^4 - 24x^3 + 6x$

11. $y = x^3 + x^2 + x + 14$
    ans: $3x^2 + 2x + 1$

12. $y = -x^5 + 12x^4 - 3x^2$
    ans: $-5x^4 + 48x^3 - 6x$

13. $y = x^{3/2}$ ans: $\dfrac{3x^2}{2}$

14. $y = 7x^{-5}$ ans: $-35x^{-6}$

15. $y = 8x^6 + 6x^{-8}$ ans: $48(x^5 - x^{-7})$

16. $y = \dfrac{2x^5 - 12}{4x^3}$ ans: $9x^{-4} + x$

17. $y = 5x^{1/2} - x^2$ ans: $\frac{5}{2\sqrt{x}} - 2x$

18. $y = \pi\sqrt{x} - \dfrac{3}{\sqrt{x}}$ ans: $\dfrac{\pi x + 3}{2x^{3/2}}$

19. $y = (x+2)(x-3)$ ans: $2x - 1$

20. $y = \sqrt[3]{x^{99}}$ ans: $33x^{32}$

Find the points on the curve given in exercises 21 through 24 where the tangent line is horizontal.

21. $y = 6x^2 - 3x$
    ans: $\left(\frac{1}{4}, -\frac{3}{8}\right)$

22. $y = -12x^2 - 4x^3 + 3x^4$
    ans: (-1,0), (0,0), (2,-32)

23. $y = x^4 - x^2$
    ans: $(0,0)$, $\left(-\frac{1}{\sqrt{2}}, -\frac{1}{4}\right)$, $\left(\frac{1}{\sqrt{2}}, -\frac{1}{4}\right)$

24. $y = x^4 - 4x^3 - 20x^2$
    ans: (-2, -32), (0,0), (5,-375)

Find the points on the following curves where the tangent line is parallel to the given line.

25. $y = x^2 + 3x + y$, $y = 9x - 2$
    Ans: $x = 3$

26. $y = x^3 + 3x^2 - 4x$, $4y = 7x + 4$
    Ans: $x = 2/3$, $x = -8/3$

Chapter 10

Product and Quotient Rules

Chapter Summary and Goal

In this chapter we will generalize the formulas for derivatives that we learned in the previous chapter so that we can find derivatives of products fg and quotients f/g in terms of the individual derivatives f' and g'.

Student Learning Objectives

The student will:

1. Know when and how to apply the product rule to calculate the derivative $(fg)'$ in terms of the individual derivatives f' and g'.
2. Know when and how to apply the quotient rule to calculate the derivative $(f/g)'$ in terms of the individual derivatives f' and g.

Product Rule

The product rule gives us a formula for the derivative of the product $f(x)g(x)$ when we already know how to calculate the individual derivatives $f'(x)$ and $g(x)$. Here we begin by deriving the product rule using the definition of the derivative, equation 9.1. Understanding this and the following derivations is important because students will frequently find that the methods used here can also be used to derive other formulas. Following this we will and present several examples.

Let's assume that we are given two functions $f(x)$ and $g(x)$ and that we know how to calculate $f'(x)$ and $g'(x)$. Our immediate goal is to find the derivative of the function

$$p(x) = f(x)g(x) \tag{10.1}$$

This "trick" of defining a new function in terms of the known functions allows us to apply the definition of a derivative to p immediately.

$$(f(x)g(x))' = p'(x) = \lim_{h \to 0} \frac{p(x+h) - p(x)}{h} = \lim_{h \to 0} \frac{f(x+h)g(x+h) - f(x)g(x)}{h} \tag{10.2a}$$

$$= \lim_{h \to 0} \frac{f(x+h)g(x+h) \overbrace{- f(x+h)g(x) + f(x+h)g(x)}^{\text{this is adding zero}} - f(x)g(x)}{h} \tag{10.2b}$$

$$= \lim_{h \to 0} \left(\frac{f(x+h)g(x+h) - f(x+h)g(x)}{h} \right.$$

$$\left. + \frac{f(x+h)g(x) - f(x)g(x)}{h} \right) \qquad (10.2c)$$

$$= \lim_{h \to 0} \frac{f(x+h)g(x+h) - f(x+h)g(x)}{h}$$

$$+ \lim_{h \to 0} \frac{f(x+h)g(x) - f(x)g(x)}{h} \qquad (10.2d)$$

$$= \lim_{h \to 0} \left(f(x+h) \frac{g(x+h) - g(x)}{h} \right) + \lim_{h \to 0} \left(g(x) \frac{f(x+h) - f(x)}{h} \right) \qquad (10.2e)$$

$$= \overbrace{\lim_{h \to 0} (f(x+h))}^{\to f(x)} \overbrace{\lim_{h \to 0} \left(\frac{g(x+h) - g(x)}{h} \right)}^{=g'(x)}$$

$$+ \overbrace{\lim_{h \to 0} (g(x))}^{=g(x)} \left(\overbrace{\lim_{h \to 0} \frac{f(x+h) - f(x)}{h}}^{=f'(x)} \right) \qquad (10.2f)$$

$$= f(x)g'(x) + g(x)f'(x) \quad \square \qquad (10.2g)$$

Theorem 10.1. Product Rule

If $f(x)$ and $g(x)$ are differentiable then

$$(f(x)g(x))' = f(x)g'(x) + g(x)f'(x) \qquad (10.3)$$

Example 10.1. Let $p(x) = (x^2 + 3x)(4x^{17} + 6x^{43})$. Find $p'(x)$.

Solution. If we write $p(x) = f(x)g(x)$, where

$$f(x) = x^2 + 3x \qquad (10.4a)$$
$$g(x) = 4x^{17} + 6x^{43} \qquad (10.4b)$$

By the rules of differentiation (power, sum, and constant multiple rule),

$$f'(x) = 2x + 3 \qquad (10.4c)$$
$$g'(x) = 4(17)x^{16} + 6(43)x^{42} = 68x^{16} + 258x^{42} \qquad (10.4d)$$

Therefore the product rule (Theorem 10.1) gives

$$p'(x) = f(x)g'(x) + g(x)f'(x) \qquad (10.4e)$$
$$= (x^2 + 3x)(68x^{16} + 258x^{42}) + (4x^{17} + 6x^{43})(2x + 3) \qquad (10.4f)$$

$$\square$$

The following example demonstrates that the product rule gives the same result as the power rule for derivatives of $x^{p+q} = (x^p)(x^q)$:

Example 10.2. Find the derivative of $y = (6x^3)(7x^4)$ first by using the product rule, and then by first multiplying out the product and using the power rule.

Solution. Using the product rule,

$$\frac{d}{dx} \left(6x^9 \right) \left(7x^4 \right) = 6x^9 \frac{d}{dx} 7x^4 + 7x^4 \frac{d}{dx} 6x^9 \qquad (10.5a)$$

$$= \left(6x^3\right)\left(7(4)x^3\right) + \left(7x^4\right)\left(6(3)x^2\right) \tag{10.5b}$$

$$= 168x^{3+3} + 126x^{4+2} = 294x^6 \tag{10.5c}$$

If we multiply the product of monomials out first, and then differentiate with the power rule,

$$\frac{d}{dx}\left(6x^3\right)\left(7x^4\right) = \frac{d}{dx}\left((6)(7)(x^3)(x^4)\right) = \frac{d}{dx}\left(42x^{3+4}\right) \tag{10.5d}$$

$$= 42\frac{d}{dx}x^7 = 42(7)x^6 = 294x^6 \tag{10.5e}$$

$\square$

Example 10.3. Find the derivative of $(3x+4)(8x^{16} - x^3)$.

Solution. Using the product rule,

$$\frac{d}{dx}\left((3x+4)(8x^{16} - x^3)\right)$$

$$= (3x+4)(8x^{16} - x^3)' + (3x+4)'(8x^{16} - x^3) \tag{10.6a}$$

$$= (3x+4)(128x^{15} - 3x^2) + 3(8x^{16} - x^3) \tag{10.6b}$$

$\square$

Example 10.4. Find the derivative of $y = x^2 \cos x$.

Solution. Using the product rule,

$$y' = (x^2 \cos x)' = x^2(\cos x)' + (x^2)' \cos x = -x^2 \sin x + 2x \cos x \tag{10.7a}$$

$\square$

Example 10.5. Find the equation of the line tangent to $y = x \sin x$ at $x = \dfrac{\pi}{2}$.

Solution. Using the point-slope formula for a line with slope m through a point $(\pi/2, f(\pi/2))$,

$$y = f(\pi/2) + m(x - \pi/2) \tag{10.8a}$$

At $x = \pi/2$ we have

$$f(\pi/2) = \frac{\pi}{2} \sin \frac{\pi}{2} = \frac{\pi}{2} \tag{10.8b}$$

and therefore, since the slope of the tangent line at $x = \pi/2$ is the derivative at $\pi/2$, we can use $m = f'(\pi/2)$, to give us

$$y = \frac{\pi}{2} + f'\left(\frac{\pi}{2}\right)\left(x - \frac{\pi}{2}\right) \tag{10.8c}$$

From the product rule,

$$f'(x) = \frac{d}{dx}(x \sin x) = x\frac{d}{dx}\sin x + (\sin x)\frac{d}{dx}x \tag{10.8d}$$

$$= x \cos x + \sin x \tag{10.8e}$$

$$f'\left(\frac{\pi}{2}\right) = \frac{\pi}{2} \cdot \cos \frac{\pi}{2} + \sin \frac{\pi}{2} = 1 \tag{10.8f}$$

Hence the equation of the tangent line is $y = x$.

$\square$

The Quotient Rule

The quotient rule gives us a formula for the derivative of the quotient $(f(x)/g(x))$ in terms of the individual derivatives $f'(x)$ and $g'(x)$ and the original functions $f(x)$ and $g(x)$. Our immediate goal is find the derivative of

$$p(x) = \frac{f(x)}{g(x)} \tag{10.9}$$

so we proceed by applying the definition of a derivative to p.

$$\frac{d}{dx}\frac{f(x)}{g(x)} = p'(x) = \lim_{h \to 0} \frac{p(x+h) - p(x)}{h} \tag{10.10a}$$

$$= \lim_{h \to 0} \frac{\dfrac{f(x+h)}{g(x+h)} - \dfrac{f(x)}{g(x)}}{h} \tag{10.10b}$$

$$= \lim_{h \to 0} \frac{\dfrac{g(x)f(x+h) - f(x)g(x+h)}{g(x)g(x+h)}}{\dfrac{h}{1}} \tag{10.10c}$$

$$= \lim_{h \to 0} \frac{g(x)f(x+h) - f(x)g(x+h)}{hg(x)g(x+h)} \tag{10.10d}$$

$$= \lim_{h \to 0} \frac{g(x)f(x+h) \overbrace{-f(x)g(x) + f(x)g(x)}^{\text{adding zero}} -f(x)g(x+h)}{hg(x)g(x+h)} \tag{10.10e}$$

$$= \lim_{h \to 0} \left(\frac{g(x)f(x+h) - f(x)g(x)}{hg(x)g(x+h)} + \frac{f(x)g(x) - f(x)g(x+h)}{hg(x)g(x+h)} \right) \tag{10.10f}$$

$$= \lim_{h \to 0} \frac{g(x)f(x+h) - f(x)g(x)}{hg(x)g(x+h)} + \lim_{h \to 0} \frac{f(x)g(x) - f(x)g(x+h)}{hg(x)g(x+h)} \tag{10.10g}$$

$$= \lim_{h \to 0} \frac{g(x)}{g(x)g(x+h)} \frac{(f(x+h) - f(x))}{h}$$

$$\qquad\qquad + \lim_{h \to 0} \frac{f(x)}{g(x)g(x+h)} \frac{g(x) - g(x+h)}{h} \tag{10.10h}$$

$$= \lim_{h \to 0} \overbrace{\frac{1}{g(x+h)}}^{=1/g(x)} \overbrace{\lim_{h \to 0} \frac{(f(x+h) - f(x))}{h}}^{=f'(x)}$$

$$\qquad\qquad + \underbrace{\lim_{h \to 0} \frac{f(x)}{g(x)g(x+h)}}_{=f(x)/g^2(x)} \underbrace{\lim_{h \to 0} \frac{g(x) - g(x+h)}{h}}_{=g'(x)} \tag{10.10i}$$

$$= \frac{1}{g(x)} \times f'(x) + \frac{f(x)}{(g(x))^2} \times \left(-g'(x)\right) = \frac{g(x)f'(x) - f(x)g'(x)}{(g(x))^2} \quad \square \tag{10.10j}$$

Theorem 10.2. Quotient Rule

If $f(x)$ and $g(x)$ are differentiable, then wherever $g(x) \neq 0$,

$$\left[\frac{f(x)}{g(x)}\right]' = \frac{g(x)f(x)' - f(x)g(x)'}{g(x)^2} \tag{10.11}$$

Example 10.6. Find $q'(x)$ for $q(x) = \dfrac{4x^3 - 3x}{2x + 17}$.

Solution. Define $f(x)$ and $g(x)$ by $q = f/g$ where

$$f(x) = 4x^3 - 3x \tag{10.12a}$$

$$g(x) = 2x + 17 \tag{10.12b}$$

Differentiating f and g,

$$f'(x) = 12x^2 - 3 \tag{10.12c}$$

$$g'(x) = 2 \tag{10.12d}$$

Thus by the quotient rule

$$q'(x) = \frac{gf' - fg'}{g^2} \tag{10.12e}$$

$$= \frac{(2x + 17)(12x^2 - 3) - (4x^3 - 3x)(2)}{(2x + 17)^2} \tag{10.12f}$$

$$= \frac{16x^3 + 204x^2 - 51}{(2x + 17)^2} \tag{10.12g}$$

□

Example 10.7. Find the derivative of $y = \dfrac{x^2 + x - 2}{x^3 + 6}$.

Solution. Using the quotient rule,

$$y' = \frac{d}{dx} \frac{x^2 + x - 2}{x^3 + 6} \tag{10.13a}$$

$$= \frac{(x^3 + 6)(x^2 + x - 2)' - (x^2 + x - 2)(x^3 + 6)'}{(x^3 + 6)^2} \tag{10.13b}$$

$$= \frac{(x^3 + 6)(2x + 1) - (x^2 + x - 2)(3x^2)}{(x^3 + 6)^2} \tag{10.13c}$$

$$= \frac{x^4 - 2x^3 + 6x^2 + 12x + 6}{(x^3 + 6)^2} \tag{10.13d}$$

□

Example 10.8. Find the equation of the line tangent to the curve $y = \dfrac{x}{1 + x^2}$ at $x = 1$.

Solution. The point-slope equation of any-line[1] is

$$y - y_1 = m(x - x_1) \tag{10.14a}$$

where m is the slope and (x_1, y_1) is a point on the line.

In this case, $m = f'(1)$ and the particular point on the curve we must use is $(1, f(1)) = (1, 1/2)$ (because $y = 1/2$ when $x = 1$ and we are asked to find the tangent at $x = 1$). Hence the equation of the tangent line is

$$y - \frac{1}{2} = m(x - 1) \tag{10.14b}$$

---

[1]Except for a vertical line, of course. For a vertical line, the slope is undefined, and the equation would be $x = x_1$.

To get the slope $y'(1)$ we first need to find $y'(x)$, and for that we use the quotient rule.

$$y' = \frac{(1+x^2)x' - x(1+x^2)'}{(1+x^2)^2} \tag{10.14c}$$

$$= \frac{1+x^2 - x(2x)}{(1+x^2)^2} \tag{10.14d}$$

$$= \frac{1-x^2}{(1+x^2)^2} \tag{10.14e}$$

Hence the slope is

$$m = y'(1) = 0 \tag{10.14f}$$

Therefore the tangent line is $y = 1/2$.   □

Example 10.9. At what points on the hyperbola $(x-1)y = 4x$ is the tangent line parallel to the line $4x + y = 0$?

Solution. Since the line $y = -4x$ has a slope of $m = -4$, we are asked to find the points on the curve of $(x-1)y = 4x$ that have slope of -4; equivalently, we are asked to find the points where $y'(x) = -4$. Solving for y and using the quotient rule,

$$y' = \frac{d}{dx}\frac{4x}{x-1} = \frac{(x-1)(4x)' - (4x)(x-1)'}{(x-1)^2} \tag{10.15a}$$

$$= \frac{4(x-1) - 4x}{(x-1)^2} = \frac{-4}{(x-1)^2} \tag{10.15b}$$

Thus the points where $y'(x) = -4$ satisfy

$$-4 = \frac{-4}{(x-1)^2} \tag{10.15c}$$

Solving for x,

$$(x-1)^2 = 1 \tag{10.15d}$$

$$x^2 - 2x + 1 = 1 \tag{10.15e}$$

$$x^2 - 2x = 0 \tag{10.15f}$$

$$x(x-2) = 0 \tag{10.15g}$$

$$x = 0, 2 \tag{10.15h}$$

Since $y = 4x/(x-1)$, at $x = 0$, $y = 0$, and at $x = 2$, $y = 8$.

So the requested points are (0,0) and (2,8).   □

Example 10.10. Find the derivative of $y = \tan x$.

Solution. From the definition of the tangent and the quotient rule, we obtain

$$(\tan x)' = \frac{d}{dx}\frac{\sin x}{\cos x} = \frac{(\cos x)(\sin x)' - (\sin x)(\cos x)'}{\cos^2 x} \tag{10.16a}$$

$$= \frac{\cos^x + \sin^2 x}{\cos^2 x} = \frac{1}{\cos^2 x} = \sec^2 x \tag{10.16b}$$

□

Example 10.11. Find an equation of the tangent line to $f(x) = \sin x \tan x$ at $x = \pi/3$.

Solution. Using the point-slope form for the equation of a line with slope m through the point (a, b), the equation is

$$y = b + m(x - a) \tag{10.17a}$$

We are interested in the tangent line at $x = \pi/3$; this means $a = \pi/3$, and

$$b = f(a) = \sin\frac{\pi}{3}\tan\frac{\pi}{3} = \frac{\sqrt{3}}{2} \cdot \sqrt{3} = \frac{3}{2} \tag{10.17b}$$

Thus the equation is $y = \frac{3}{2} + m\left(x - \frac{\pi}{3}\right)$. The slope is the derivative of $f'(a) = f'(\pi/3)$. Differentiating and using the product rule,

$$f'(x) = \frac{d}{dx}(\sin x \tan x) = \sin x \frac{d}{dx}\tan x + \tan x \frac{d}{dx}\sin x \tag{10.17c}$$

$$= \sin x \sec^2 x + \tan x \cos x = \frac{\sin x}{\cos^2 x} + \frac{\sin x \cos x}{\cos x} \tag{10.17d}$$

$$= \sec x \tan x + \sin x \tag{10.17e}$$

$$f'\left(\frac{\pi}{3}\right) = \sec\frac{\pi}{3}\tan\frac{\pi}{3} + \sin\frac{\pi}{3} = 2 \cdot \sqrt{3} + \frac{\sqrt{3}}{2} = \frac{5\sqrt{3}}{2} \tag{10.17f}$$

Thus the equation of the line is

$$y = \frac{3}{2} + \frac{5\sqrt{3}}{2}\left(x - \frac{\pi}{3}\right) = \frac{5\sqrt{3}}{2}x - \frac{5\sqrt{3}}{2} \cdot \frac{\pi}{3} + \frac{3}{2} = \frac{5\sqrt{3}}{2}x + \frac{3}{2} - \frac{5\pi}{2\sqrt{3}} \tag{10.17g}$$

$\square$

Example 10.12. Find the derivative of $y = \sec x$.

Solution. From the definition of the secant and the quotient rule,

$$(\sec x)' = \frac{d}{dx}\frac{1}{\cos x} = \frac{(\cos x)(1)' - (1)(\cos x)'}{\cos^2 x} \tag{10.18a}$$

$$= \frac{\sin x}{\cos^2 x} = \frac{1}{\cos x} \times \frac{\sin x}{\cos x} = \sec x \tan x \tag{10.18b}$$

$\square$

Table 10.1. Summary of Differentiation Formulas

$C' = 0$	(10.19)	$(Cf(x))' = Cf'(x)$	(10.20)
$x' = 1$	(10.21)	$(f(x) \pm g(x))' = f'(x) \pm g'(x)$	(10.22)
$(x^n)' = nx^{n-1}$	(10.23)	$(fg)' = fg' + f'g$	(10.24)
$(\sqrt{x})' = \dfrac{1}{2\sqrt{x}}$	(10.25)	$\left(\dfrac{f}{g}\right)' = \left(\dfrac{gf' - fg'}{g^2}\right)$	(10.26)
$(\sin x)' = \cos x$	(10.27)	$(\cos x)' = -\sin x$	(10.28)
$(\tan x)' = \sec^2 x$	(10.29)	$(\sec x)' = \sec x \tan x$	(10.30)
$(\cot x)' = -\csc^2 x$	(10.31)	$(\csc x)' = -\csc x \cot x$	(10.32)

Exercises

Use the rules in table 10.1 to find $f'(x)$ if y is given by the following equations.

1. $y = (x+2)(x+3)$ $\qquad$ Ans: $5+2x$
2. $y = (x+99)\sqrt{x}$ $\qquad$ Ans: $\frac{3(33+x)}{2\sqrt{x}}$
3. $y = (\tan x)\sqrt{x}$ $\qquad$ Ans: $\frac{2x\sec^2 x + \tan x}{2\sqrt{x}}$
4. $\dfrac{1+x}{1-x}$ $\qquad$ Ans: $\frac{2}{(x-1)^2}$
5. $\dfrac{1+\dfrac{x}{x+1}}{\dfrac{x+1}{x+2}+1}$ $\qquad$ Ans: $\frac{1}{(x+1)^2} - \frac{4}{(2x+3)^2}$
6. $\dfrac{1+\sqrt{x}}{1-\sqrt{x}}$ $\qquad$ Ans: $\frac{1}{(\sqrt{x}-1)^2\sqrt{x}}$
7. $\dfrac{x+\sqrt{x}}{x-\sqrt{x}}$ $\qquad$ Ans: $-\frac{1}{(\sqrt{x}-1)^2\sqrt{x}}$
8. $y = x\tan x + \sec x$
   Ans: $\tan x + (x+\sin x)\sec^2 x$
9. $y = \dfrac{1+\cos x}{1+\sin x}$ $\qquad$ Ans: $-\frac{\sin x + \cos x + 1}{(\sin x + 1)^2}$
10. $y = \sec x \tan x$
    Ans: $\sec x \left(\tan^2 x + \sec^2 x\right)$
11. $y = x\cos x \tan x$ $\qquad$ Ans: $x\cos x + \sin x$
12. $y = x\sin x$ $\qquad$ Ans: $y' = \sin x + x\cos x$

In exercises 13 and 14, verify the trigonometric derivatives using the quotient rule.

13. $(\cot x)' = -\csc^2 x$
14. $(\csc x)' = -\cot x \csc x$

In exercises 15 through 20, find the points on the following curves where the tangent line is horizontal.

15. $y = (x+1)(x+2)$ $\qquad$ Ans: $x = -3/2$
16. $y = x^2 + 2x - 3$ $\qquad$ Ans: $x = -1$
17. $y = x^3 - x$ $\qquad$ Ans: $x = \pm 1/\sqrt{3}$
18. $y = \dfrac{x^2}{1+x^2}$ $\qquad$ Ans: 0
19. $y = \dfrac{x^2}{x+1}$ $\qquad$ Ans: x=-2, 0
20. $y = \dfrac{(x-1)^2}{x+1}$ $\qquad$ Ans: $x = -3, 1$

In exercises 21 through 22, find the points on the given curves where where the tangent line is parallel to the specified line.

21. $y = x^2 + 2x - 4$, $y = x + 3$
    Ans:(-1/2, -19/4)
22. $y = 2x^3 - 3x^2$, $y = 36x - 7$
    Ans:(-2,28), (3, 27)

Prove the stated theorems in exercises 23 through 25.

23. Prove the generalized product rule: if f, g, and h are all differentiable, then $(fgh)' = f'gh + fg'h + fgh'$.
24. Prove that the product rule is consistent with the power rule: show that
$$(x^{p+q})' = (x^p)(x^q)' + (x^p)'(x^q)$$
25. Prove the power law for negative exponents using quotient law, i.e., assume that the power law is true for positive integer exponents. Then write x^{-n} as $1/x^n$ and use the quotient law to prove that power law for x^{-n}.

Chapter 11

The Chain Rule

Chapter Summary and Goal

The chain rule tells us how to calculate the derivative of a function of a function, which is more precisely called the composition of two functions. For example, in the previous chapters, we learned how to calculate the derivatives of

$$f(x) = \sqrt{x} \tag{11.1}$$

and

$$g(u) = \sin(u) \tag{11.2}$$

The chain rule will now give us a way to calculate the derivative of

$$h(x) = g(f(x)) = \sin(\sqrt{(x)}) \tag{11.3}$$

based on information about the derivative of $\sin x$ and $\sqrt{x}$. More generally we will learn how to calculate the derivative of $(f \circ g)(x) = f(g(x))$ in this chapter.

Student Learning Objectives

The student will:

1. Learn to recognize when the chain rule should be used.
2. Learn to use the chain rule to calculate the derivative of a function.
3. Be able to apply the chain rule using both the Leibnitz and Newton notations.
4. Understand the proof of the chain rule.

Unravelling the Chain

In this chapter we will explore why the different forms of the chain rule are really identical; where the chain rule comes from; and how to apply the chain rule to find derivatives of a function.

The chain rule gives us a generalized method for finding the derivative of a function of a function, such as $\sqrt{1 + x + x^2}$ or $\cos \sqrt{u}$. We know, for example that

$$\frac{d}{du}\sqrt{u} = \frac{1}{2\sqrt{u}}, \tag{11.4}$$

and that

$$\frac{d}{dx}(1 + x + x^2) = 1 + 2x, \tag{11.5}$$

but how can we combine this information together to find $\dfrac{d}{dx}\sqrt{1 + x + x^2}$? To state this particular example more succinctly, let us write $f(x) = \sqrt{x}$ and $g(x) = 1 + x + x^2$. Then since

$$\sqrt{1 + x + x^2} = f(g(x)) \tag{11.6}$$

what we are really asking for is a way to calculate

$$\frac{d}{dx}\sqrt{1 + x + x^2} = \frac{d}{dx}f(g(x)) \tag{11.7}$$

Rather than approaching these problems on a case-by-case basis, we will examine the general problem of finding a method to calculate $\dfrac{d}{dx}f(g(x))$. Then we will also be able to solve other problems besides $(d/dx)\sqrt{1 + x + x^2}$, like $(d/du)\cos\sqrt{u}$.

Theorem 11.1. Chain Rule (Newton Form)

$$(f(g(x)))' = f'(g(x))g'(x) \tag{11.8}$$

where

$$f'(g(x)) = f'(u)|_{u=g(x)} \tag{11.9}$$

In the Newton Form the formula for chain rule looks very different:

Theorem 11.2. Chain Rule (Leibniz Form)

$$\frac{dy}{dx} = \frac{dy}{dz}\frac{dz}{dx} \tag{11.10}$$

Proof. To see where the chain rule comes from it is easiest to appeal to the Leibniz Derivative (definition 8.4). We want to find the derivative $u'(x)$ of the function $u(x) = f(g(x))$ where the function $f(x)$ and $g(x)$ are both differentiable. Then from 8.4,

$$f'(x) = R\left(\frac{f(x + h) - f(x)}{h}\right) \tag{11.11a}$$

$$g'(x) = R\left(\frac{g(x + h) - g(x)}{h}\right) = R\left(\frac{\delta}{h}\right) \tag{11.11b}$$

where R is the Standard Part operator (definition 8.3), h is an infinitesimal change in x, and

$$\delta = g(x + h) - g(x) \tag{11.11c}$$

is the infinitesimal change in $g(x)$ corresponding to a change h in x (see the increment theorem, theorem 8.4). Since $u(x) = f(g(x))$ we have

$$u(x + h) = f(g(x + h)) \tag{11.11d}$$

and therefore

$$u'(x) = R\left(\frac{u(x + h) - u(x)}{h}\right) \tag{11.11e}$$

$$= R\left(\frac{f(g(x+h)) - f(g(x))}{h}\right) \tag{11.11f}$$

$$= R\left(\frac{f(g(x+h)) - f(g(x))}{h} \cdot \frac{g(x+h) - g(x)}{g(x+h) - g(x)}\right) \tag{11.11g}$$

$$= R\left(\frac{f(g(x+h)) - f(g(x))}{g(x+h) - g(x)} \cdot \frac{g(x+h) - g(x)}{h}\right) \tag{11.11h}$$

$$= R\left(\frac{f(g(x+h)) - f(g(x))}{g(x+h) - g(x)}\right) \cdot R\left(\frac{g(x+h) - g(x)}{h}\right) \tag{11.11i}$$

Using (11.11c) in the first term and definition 8.4 in the second term

$$u'(x) = R\left(\frac{f(g(x) + \delta) - f(g(x))}{\delta}\right) \cdot \left(\frac{dg}{dx}\right) \tag{11.11j}$$

$$= R\left(\frac{f(z + \delta) - f(z)}{\delta}\right) \cdot \left(\frac{dg}{dx}\right) \qquad \text{where } z = g(x) \tag{11.11k}$$

$$= \left(\frac{df}{dz}\right)\bigg|_{z=g(x)} \cdot \left(\frac{dg}{dx}\right) \tag{11.11l}$$

Because z and $g(x)$ mean the same thing in the last line, we sometimes write this as

$$\frac{df}{dx} = \frac{df}{dz} \cdot \frac{dz}{dx} \tag{11.11m}$$

Equation 11.11m is true for any intermediate variable z even if we are not originally considering f to be a composite function.[1] Furthermore, equation 11.11l can be written in the Newton form by observing that since $u(x) = f(g(x))$ then $u' = f'(g(x))$. Thus

$$f'(g(x)) = f'(z)|_{z=g(x)} \cdot g'(x) = f'(g(x))g'(x) \tag{11.11n}$$

Thus we have the two different forms for the chain rule, given by theorems 11.1 and 11.2 in (11.11m) and (11.11n).

□

One way to think of the chain rule is in terms of an **inside function** and an **outside function**. Consider the functions $\sqrt{\sin x}$ and $\sin\sqrt{x}$. In the following table we've describe them in terms of their inside functions and outside functions. The actual variable names have been suppressed to avoid getting confused. Read the small square □ as the word "something," as in "square root of something."

Example of Function	Outside Function	Inside Function
$y = \sqrt{\sin x}$	$\sqrt{\square}$	$\sin(\square)$
$y = \sin\sqrt{x}$	$\sin(\square)$	$\sqrt{\square}$

The function $f(g(x))$ in this notation has an outside function $f(\square)$ (f of something) and an inside function $g(\square)$ (g of something). The chain rule can be thought of in this way:

$$f'(g(x)) = (\text{Derivative of Outside Function Evaluated at Inside Function})$$
$$\times (\text{Derivative of Inside Functions}) \tag{11.12}$$

---

[1] The reason why this works is because we are forcing f to be a composition of functions by introducing an intermediate variable.

The derivative of $y = \sqrt{\sin x}$ is then

$$= \left(\text{Derivative of } \sqrt{\Box} \text{ at } \sin x\right) \times \left(\text{Derivative of } \sin x\right) \qquad (11.13)$$

$$= \left(\frac{1}{2\sqrt{\Box}} \text{ with } \Box = \sin x\right) \times \cos x = \frac{\cos x}{2\sqrt{\sin x}} \qquad (11.14)$$

Similarly, the derivative of $\sin \sqrt{x}$ is then

$$= \left(\text{Derivative of } \sin \Box \text{ at } \sqrt{x}\right) \times \left(\text{Derivative of} \sqrt{x}\right) \qquad (11.15)$$

$$= \left(\cos \Box \text{ with } \Box = \sqrt{x}\right) \times \frac{1}{2\sqrt{x}} = \frac{\cos \sqrt{x}}{2\sqrt{x}} \qquad (11.16)$$

Example 11.1. Find $f'(x)$ for $f(x) = \sqrt{x^2 + 1}$.

Solution. The outside function is $g(u) = \sqrt{u}$ and the inside function is $h(x) = x^2 + 1$. Then

$$g'(u) = \frac{1}{2\sqrt{u}} \qquad (11.17a)$$

$$g'(h(x)) = \frac{1}{2\sqrt{h(x)}} = \frac{1}{2\sqrt{x^2 + 1}} \qquad (11.17b)$$

$$h'(x) = 2x \qquad (11.17c)$$

$$f'(x) = g'(h(x))h'(x) = \frac{2x}{2\sqrt{x^2 + 1}} = \frac{x}{\sqrt{x^2 + 1}} \qquad (11.17d)$$

$\Box$

Example 11.2. Find $f'(x)$ for $f(x) = \sqrt{3x^2 + 2}$.

Solution. The inside function is

$$u(x) = 3x^2 + 2 \qquad (11.18a)$$

Then

$$f(x) = \sqrt{3x^2 + 2} = \sqrt{u} \qquad (11.18b)$$

Hence

$$f'(x) = \frac{df}{dx} = \frac{df}{du}\frac{du}{dx} \qquad (11.18c)$$

Since

$$\frac{df}{du} = \frac{d}{du}\sqrt{u} = \frac{d}{du}u^{1/2} = \frac{1}{2}u^{-1/2} = \frac{1}{2\sqrt{u}} \qquad (11.18d)$$

and

$$\frac{du}{dx} = \frac{d}{dx}\left(3x^2 + 2\right) = 6x \qquad (11.18e)$$

we end up with

$$f'(x) = \frac{1}{2\sqrt{u}} \times (6x) = \frac{3x}{\sqrt{3x^2 + 2}} \qquad (11.18f)$$

$\Box$

Example 11.3. Find dy/dx for $y = \cos(x^2)$.

Solution. Let $u = x^2$. Then $y = \cos u$ and

$$\frac{dy}{dx} = \frac{dy}{du}\frac{du}{dx} = \left(\frac{d}{du}\cos u\right) \times \left(\frac{d}{dx}x^2\right) \qquad (11.19a)$$

$$= (-\sin u) \times (2x) \qquad (11.19b)$$

$$= -2x\sin(x^2) \qquad (11.19c)$$

$\Box$

Example 11.4. Let $y = (\cos x)^2$. Find dy/dx.

Solution. Let $u = \cos x$. Then $y = u^2$, and

$$\frac{dy}{dx} = \frac{dy}{du}\frac{du}{dx} = \left(\frac{d}{du}u^2\right) \times \left(\frac{d}{dx}\cos x\right) = -2u\sin x = -2\cos x\sin x \qquad (11.20a)$$

$\square$

Example 11.5. Find $f'(x)$ if $f(x) = (x^3 - 1)^{100}$

Solution. Let $u = x^3 - 1$. Then $y = u^{100}$ and

$$y' = \frac{dy}{du}\frac{du}{dx} = 100u^{99}(3x^2) = 300x^2(x^3 - 1)^{99} \qquad (11.21a)$$

$\square$

It is useful to write a general formula for the derivative of

$$y = (g(x))^n \qquad (11.22)$$

To do this we can substitute $u = g(x)$. This gives

$$y = u^n \qquad (11.23)$$

Differentiating and applying the chain rule,

$$\frac{dy}{dx} = \frac{dy}{du}\frac{du}{dx} = \left(nu^{n-1}\right)g'(x) = n(g(x))^{n-1}g'(x) \qquad (11.24)$$

Theorem 11.3. Generalized Power Rule

If $u = g(x)$ is differentiable then

$$\frac{d}{dx}(u^n) = nu^{n-1}\frac{du}{dx} \qquad (11.25)$$

or equivalently

$$\frac{d}{dx}(g(x))^n = n(g(x))^{n-1}g'(x) \qquad (11.26)$$

Example 11.6. Differentiate $y = (4x^7 - 3x^6)^{427}$.

Solution. Applying theorem 11.3, we bring down the 427 to the front of the quantity in brackets, subtract one from the exponent, and then differentiate the quantity in brackets:

$$\frac{d}{dx}\left((4x^7 - 3x^6)^{427}\right) = 427(4x^7 - 3x^6)^{426}\frac{d}{dx}(4x^7 - 3x^6) \qquad (11.27a)$$

$$= 427(4x^7 - 3x^2)^{426}(28x^6 - 18x^5). \qquad (11.27b)$$

$\square$

Example 11.7. Differentiate $f(x) = \dfrac{1}{\sqrt{x^2 + 99}}$.

Solution. This is also a generalized power, so we can apply theorem 11.3 as before. We begin by converting the square root in the denominator to an exponent.

$$\frac{d}{dx}\frac{1}{\sqrt{x^2 + 99}} = \frac{d}{dx}(x^2 + 99)^{-1/2} = -\frac{1}{2}(x^2 + 99)^{-3/2}\frac{d}{dx}(x^2 + 99) \qquad (11.28a)$$

$$= -\frac{x}{(x^2 + 99)^{3/2}}. \qquad (11.28b)$$

$\square$

Example 11.8. Let $y = \sin(x \cos x)$. Find y'.

Solution. If we define $u = x \cos x$, then $y = \sin u$. Hence

$$y' = \frac{dy}{dx} = \frac{dy}{du}\frac{du}{dx} \tag{11.29a}$$

$$= \left(\frac{d}{du}\sin u\right) \times \left(\frac{d}{dx}[x \cos x]\right) \tag{11.29b}$$

By the product rule,

$$\frac{d}{dx}x \cos x = (x)(\cos x)' + (\cos x)(x)' = -x \sin x + \cos x \tag{11.29c}$$

and thus

$$y' = \cos u(-x \sin x + \cos x) = [\cos(x \cos x)](-x \sin x + \cos x). \tag{11.29d}$$

$\square$

Example 11.9. Find dy/dx for $y = \dfrac{x}{\sqrt{7 - 3x}}$.

Solution.   For this problem we have to first apply the chain rule, then apply the quotient rule. We have a quotient f/g where $f = x$ and $g = \sqrt{7 - 3x}$. Hence $f' = 1$ and by the chain rule:

$$\frac{d}{dx}\sqrt{7 - 3x} = \frac{1}{2}(7 - 3x)^{-3/2}\frac{d}{dx}(7 - 3x) = \frac{-3}{2(7 - 3x)^{3/2}} \tag{11.30a}$$

By the quotient rule,

$$\frac{dy}{dx} = \frac{d}{dx}\frac{f}{g} = \frac{gf' - fg'}{g^2} \tag{11.30b}$$

$$= \frac{(\sqrt{7 - 3x})(1) - (x)(-3/[2(7 - 3x)^{3/2}]}{7 - 3x} \tag{11.30c}$$

$$= \frac{(7 - 3x)^2 + 3x/2}{(7 - 3x)^{5/2}}. \tag{11.30d}$$

$\square$

Example 11.10. Find $f'(x)$ if $f(x) = \sin(\cos(\tan x))$.

Solution. In this example we apply the chain rule twice. The first time, the sine is the outside function and $\cos \tan x$ as the inside function:

$$f'(x) = \cos(\cos(\tan x))\frac{d}{dx}\cos(\tan x) \tag{11.31a}$$

Next, the cosine is the outside function and the $\tan x$ is the inside function:

$$f'(x) = \cos(\cos(\tan x))(-\sin(\tan x))\frac{d}{dx}\tan x \tag{11.31b}$$

$$= -\cos(\cos(\tan x))(\sin(\tan x))\sec^2 x. \tag{11.31c}$$

$\square$

Exercises

Use the chain rule to find the derivative of each of the functions in exercises 1 through 21.

1. $f(x) = (2x + 4)^{-2}$

2. $f(u) = \sin(u^4)$

3. $g(\theta) = \cos^5 \theta$

4. $f(x) = \sqrt{7x - 3}$

5. $f(x) = (2x^2 + 5)^4(12x^3 - 8x)^{10}$

6. $y = x \tan \sqrt{x}$

7. $f(t) = \dfrac{t^3}{(3t - 1)^2}$

8. $f(x) = \cos(2\pi x - 3\pi)$

9. $g(x) = \sqrt{4 - 2x^3}$

10. $f(x) = \sin(\cos x))$

11. $f(x) = \sin(\cos(\sin(x)))$

12. $f(x) = \sqrt{\tan(\sqrt{x})}$

13. $y = \sqrt{1 + \sqrt{1 + \sqrt{x}}}$

14. $y = \dfrac{\sqrt{2x + 1}}{3x + 7}$

15. $y = \dfrac{5x - 3}{\sqrt{7x - 4}}$

16. $y = \sqrt{\dfrac{2x + 3}{4x + 5}}$

17. $y = \dfrac{x^2}{(2x^3 + 7)^3}$

18. $y = \dfrac{x^2 + 4}{x^3 + x}$

19. $y = \dfrac{\sqrt{x}}{3 + x}$

20. $y = \sin(\sqrt{2x + 3})$

21. $y = \dfrac{(1 + x)^2}{(x + x^2)^3}$

Use the chain rule to find the second derivative of each of the functions in exercise 22 through 24.

22. $f(x) = \sqrt{1 + x^2}$

23. $f(x) = \dfrac{1 + x}{1 - x}$

24. $y = \sin^2 \pi t$

25. In this problem, you will prove the power rule for rational exponents, i.e., you will show that when $y = x^{p/q}$, for any integers p, q, then $y' = (p/q)x^{(p/q)-1}$.

     (a) Let $y = x^{p/q}$. Show that $y^q = x^p$.

     (b) Calculate $\dfrac{d}{dx}y^q$ using the chain rule.

     (c) Calculate $\dfrac{d}{dx}x^p$.

     (d) Set the last two derivatives equal, solve for y', and simplify.

26. Derive the quotient rule from the product rule as $(u \cdot (v)^{-1})'$ using the chain rule, where $(v)^{-1} = 1/v$.

(Answers on following page)

Answers to Selected Exercises

1. $-\dfrac{4}{(2x+4)^3}$

2. $4u^3 \cos\left(u^4\right)$

3. $-5\sin(\theta)\cos^4(\theta)$

4. $\dfrac{7}{2\sqrt{7x-3}}$

5. $16x(2x^2+5)^3(12x^3-8x)^{10}$ $+10(2x^2+5)^4(36x^2-8)(12x^3-8x)^9$

6. $\tan\sqrt{x} + \frac{1}{2}\sqrt{x}\sec^2\sqrt{x}$

7. $\dfrac{3(t^3-t^2)}{(3t-1)^3}$

8. $2\pi\sin(2\pi x)$

9. $-\dfrac{3x^2}{\sqrt{4-2x^3}}$

10. $-\sin(x)\cos(\cos(x))$

11. $-\cos(x)\sin(\sin(x))\times$ $\cos(\cos(\sin(x)))$

12. $\dfrac{\sec^2\sqrt{x}}{4\sqrt{x}\sqrt{\tan\sqrt{x}}}$

13. $\dfrac{1}{8\sqrt{\sqrt{\sqrt{x}+1}+1}}\times$ $\dfrac{1}{\sqrt{\sqrt{x}+1}\sqrt{x}}$

14. $\dfrac{4-3x}{\sqrt{2x+1}(3x+7)^2}$

15. $\dfrac{35x-19}{2(7x-4)^{3/2}}$

16. $-\dfrac{1}{(4x+5)^2}\sqrt{\dfrac{4x+5}{2x+3}}$

17. $-\dfrac{14(x^4-x)}{(2x^3+7)^4}$

18. $\dfrac{-x^4-11x^2-4}{x^2(x^2+1)^2}$

19. $\dfrac{3-x}{2\sqrt{x}(x+3)^2}$

20. $\dfrac{\cos(\sqrt{2x+3})}{\sqrt{2x+3}}$

21. $\dfrac{-4x-3}{x^4(x+1)^2}$

22. $\dfrac{1}{(x^2+1)^{3/2}}$

23. $-\dfrac{4}{(x-1)^3}$

24. $2\pi^2\cos(2\pi t)$

Chapter 12

Implicit Differentiation

Chapter Summary and Goal

Implicit differentiation utilizes the rules of differentiation, such as the chain rule, the product rule, and the quotient rule, to find y' when the function y is defined implicitly. This may happen, for example, when we cannot solve explicitly for y, as in the equation $y = x \sin y$.

Student Learning Objectives

The student will:

1. Learn the distinction between explicit and implicit function definitions.
2. Learn to differentiate an an equation implicitly.
3. Learn to solve for y' algebraically after differentiating both sides of the equation.
4. Learn how to find a derivative implicitly.

Implicit Functions

Functions may be defined in two ways: **implicitly** and **explicitly**. The normal way of defining functions, as in the equation

$$y = 3x^2 + \sin x \tag{12.1}$$

has the form $y = f(x)$. We can separate out the x and y dependence so that all of the x variables are on one side of the equation, and the y variable stands alone on the other side of the equation. Sometimes we come across a function definition that is not written in this format, such as

$$F(x, y) = y - x^3 + 3x^2 + 12 = 0 \tag{12.2}$$

or

$$G(x, y) = x^3 + y^3 - xy = 0 \tag{12.3}$$

Both (12.2) and (12.3) are examples of **Implicit Function Definitions**. Implicit functions are normally written in the form of some $F(x, y) = 0$, although we will not make much use of that fact in this section.

With equation 12.2 it is possible to solve for y, as in

$$y = x^3 - 3x^2 - 12 \tag{12.4}$$

111

When this occurs, we say that *the function $F(x, y) = 0$ can be explicitly solved for some function $y = f(x)$.*

However, there are many situations when when this is not possible; equation 12.3 is such an example. In these situations is not possible to solve the function $G(x, y) = 0$ explicitly for some function function $y = g(x)$.

This does not mean that such a function does not exist, however. It only means that we can't find a formula for it. In fact, due to a very important theorem from advanced calculus (theorem 12.1), such a function almost always exists in small pieces, except at points where the tangent line is vertical.

Theorem 12.1. Implicit Function Theorem

If $F(x, y) = 0$ is a smooth function of x and y and (a, b) is a point where the tangent line is not vertical, then $F(x, y)$ may be solved implicitly for for some function $y = f(x)$ *in the neighborhood* of (a, b).[1]

Our take-away message from the implicit function theorem is that many implicitly defined functions can be broken down into explicit functions and points where the tangent line is vertical. One simple example that you should already be familiar with is the unit circle

$$x^2 + y^2 = 1 \tag{12.5}$$

There are two points where the tangent lines are vertical: (1,0) and (-1, 0). Away from these points, there are functions that describe the points on the circle. Every point on the top half of the circle (not counting the end points) lies on the function

$$f(x) = \sqrt{1 - x^2} \tag{12.6}$$

while every point on the bottom semi-circle lies on the function

$$g(x) = -\sqrt{1 - x^2} \tag{12.7}$$

Thus the circle can be broken down into a combination of (a) functions and (b) points where the tangent line is vertical (see figure 12.1). Another example is given by the Folium of Descartes (see the right side of figure 12.1).

Differentiating Implicitly Defined Functions

The basic idea of **implicit differentiation** is this: differentiate both sides of the equation, set the derivatives equal, and solve for y'.

Once we apply the rules of differentiation to each side of the equation, the remainder of the process is nothing more than solving an algebraic equation, where the unknown is y'. We begin by illustrating this with an example.

Example 12.1. Find y' for the Folium of Descartes $x^3 + y^3 = 10xy$.

Solution. We differentiate both sides of the equation,

$$\frac{d}{dx}\left(x^3 + y^3\right) = \frac{d}{dx}\left(10xy\right) \tag{12.8a}$$

---

[1]This does not mean that it is possible to find a *formula* for $y = f(x)$. It means its possible to draw a small box around the point (a, b) such that *within that box*, a curve of the equation $F(x, y) = 0$ passing through (a, b) is a function.

Figure 12.1: Left: The unit circle can be partitioned into a pair of functions (equations 12.6 and 12.7) and points ((-1,0) and (1,0)) where the tangent line is vertical. Right: The Folium of Descartes is described implicitly by $x^3 + y^3 = axy$ for any positive constant a. The folium cannot be explicitly solved as a function; in fact it fails the vertical line test in the right-hand plane. But we can described it in terms of the sum of three different functions (represented here by the solid, dotted, and dashed lines), and the two points of vertical tangency. Each of the three arcs are themselves functions. It is these functions that the implicit function theorem tells us exist, and which we are allowed to differentiate.

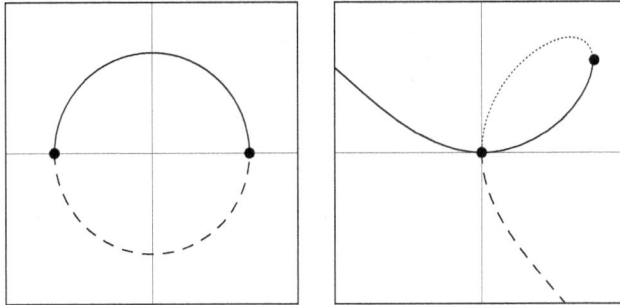

Applying the sum and power rule to the left side of the equation and the product rule to the right,

$$3x^2 + 3y^2 y' = 10xy' + 10yx' \tag{12.8b}$$

On the left hand side, when we calculated $(y^3)'$, we used the chain rule:

$$(y^3)' = 3y^2 y' \tag{12.8c}$$

On the right side, we will use the fact that the derivative $x' = dx/dx = 1$, to give

$$3x^2 + 3y^2 y' = 10xy' + 10y \tag{12.8d}$$

All of our differentiation is finished now. The rest of the problem is algebra.

The next step is to move all terms with y' dependence to the left side of the equation, and all other terms to the right side of the equation.

$$3y^2 y' - 10xy' = 10y - 3x^2 \tag{12.8e}$$

The left hand side of the equation has a common factor of y', which we can factor out.

$$y'(3y^2 - 10x) = 10y - 3x^2 \tag{12.8f}$$

Dividing both sides of the equation by the quantity in brackets gives

$$y' = \frac{10y - 3x^2}{3y^2 - 10x} \tag{12.8g}$$

$\square$

In general when solve for y' using implicit differentiation the result will be a function of both x and y, such as the solution we found in (12.8g). The reason for this can be seen by appealing to figure 12.1. In the right-half plane, there is an interval where, if you pick a value of x an draw a vertical line, it will intersect the curve three times. As we discussed earlier, this is because the function is defined implicitly at each of those points. For the folium of Descartes, there are three different functions in total. Thus we need to know a value of both x and y to be able calculate a value for the slope. The values which we are allowed to plug into the equation for y' correspond to the values of (x, y) that line on the equation for the curve itself, $x^3 + y^3 = 10xy$.

Example 12.2. Find y' when $8\sqrt{x} - 4\sqrt{y} = 2$ using implicit differentiation.

Solution. Differentiating both sides of the equation,

$$(8x^{1/2})' - (4y^{1/2})' = (2)' = 0 \tag{12.9a}$$

$$4x^{-1/2} - 2y^{-1/2}y' = 0 \tag{12.9b}$$

Here we have used the chain rule in the second term above. Bringing the second term to the right side of the equation and rewriting the exponents as roots gives

$$\frac{4}{\sqrt{x}} = \frac{2y'}{\sqrt{y}} \tag{12.9c}$$

Cross multiplying by $\dfrac{\sqrt{y}}{2}$ gives

$$y' = 2\sqrt{\frac{y}{x}} \tag{12.9d}$$

We also note that the original function could be solved for

$$\sqrt{y} = 2\sqrt{x} - \frac{1}{2} \tag{12.9e}$$

so it is possible to remove the y dependence from the solution,

$$y' = \frac{2}{\sqrt{x}}\left(2\sqrt{x} - \frac{1}{2}\right) = 4 - \frac{1}{\sqrt{x}} \tag{12.9f}$$

It is also possible to check our solution by differentiating explicitly, since

$$y' = \frac{d}{dx}\left[\left(2\sqrt{x} - \frac{1}{2}\right)^2\right] = 2\left(2\sqrt{x} - \frac{1}{2}\right)\frac{2}{2\sqrt{x}} = 4 - \frac{1}{\sqrt{x}}. \tag{12.9g}$$

$\square$

Heuristic 12.1. Procedure for Implicit Differentiation

1. Apply d/dx to both sides of the equation and using the power rule (since $\sqrt{x} = x^{1/2}$,
2. Expand $d(LHS)/dx$ and $d(RHS)/dx$ using the rules of differentiation such as the power rule, product rule, quotient rule, and chain rule.
3. Replace every occurrence of dy/dx with y', i.e., set $dy/dx = y'$
4. Replace every occurrence of dx/dx with 1, i.e., set $dx/dx = 1$.
5. Solve for y'. If all the occurrences of y' are alone, not raised to any power, inside any function, and on the numerator, we can usually do this:
   (a) Put all the terms with a y' on the LHS of the equation
   (b) Put all the terms without a y' on the RHS of equation.
   (c) Factor y' out of the LHS.
   (d) Solve for y' algebraically (if possible).

Example 12.3. Find the equation of the tangent line to $x \cos y = y \cos x$ at the point $(\pi/4, \pi/4)$.

Solution. Differentiating implicitly and using the product rule on each side of the equation

$$(x \cos y)' = (y \cos x)' \tag{12.10a}$$

$$(x)(\cos y)' + (x)'(\cos y) = (y)'(\cos x) + (y)(\cos x)' \tag{12.10b}$$

By the chain rule $(\cos y)' = -y' \sin y$, so

$$-xy' \sin y + \cos y = y' \cos x - y \sin x \tag{12.10c}$$

At $(\pi/4, \pi/4)$ we set $y' = m$ to find the slope of the tangent line, and solve for m.

$$-\frac{\pi}{4} m \cdot \left(\sin \frac{\pi}{4} \right) + \cos \frac{\pi}{4} = m \cdot \left(\cos \frac{\pi}{4} \right) - \frac{\pi}{4} \cdot \left(\sin \frac{\pi}{4} \right) \tag{12.10d}$$

$$-\frac{\pi m \sqrt{2}}{8} + \frac{\sqrt{2}}{2} = \frac{m \sqrt{2}}{2} - \frac{\pi \sqrt{2}}{8} \tag{12.10e}$$

$$-\pi m \sqrt{2} + 4\sqrt{2} = 4m\sqrt{2} - \pi\sqrt{2} \tag{12.10f}$$

$$4m\sqrt{2} + \pi y' \sqrt{2} = 4\sqrt{2} + \pi\sqrt{2} \tag{12.10g}$$

$$m(4\sqrt{2} + \pi\sqrt{2}) = 4\sqrt{2} + \pi\sqrt{2} \tag{12.10h}$$

$$m = 1 \tag{12.10i}$$

Using the point-slope formula, the tangent line is

$$y = y_1 + m(x - x_1) = \frac{\pi}{4} + 1 \cdot \left(x - \frac{\pi}{4} \right) = x \tag{12.10j}$$

$\square$

Example 12.4. Find the equation of the tangent line to the hyperbola

$$x^2 + 4xy - 15y^2 + 3x = 0 \tag{12.11}$$

at the point $(1,2/3)$.

Solution. Using implicit differentiation,

$$\frac{d}{dx} \left(x^2 + 4xy - 15y^2 + 3x \right) = 0 \tag{12.12a}$$

Differentiating term by term,

$$(x^2)' + (4xy)' - (15y^2)' + (3x)' = 0 \tag{12.12b}$$

Since $(xy)' = xy' + y$ and $(y^2)' = 2yy'$,

$$2x + 4y + 4xy' - 30yy' + 3 = 0 \tag{12.12c}$$

Substituting $x = 1$ and $y = 2/3$ gives

$$2 \cdot 1 + 4 \cdot \frac{2}{3} + 4 \cdot 1 \cdot y' - 30 \cdot \frac{2}{3} y' + 3 = 0 \tag{12.12d}$$

Simplifying and rearranging gives $y' = \dfrac{23}{48}$. Since the slope at $(1,2/3)$ is $23/48$, the equation of the tangent line is

$$y = \frac{2}{3} + \frac{23}{48}(x - 1) = \frac{23}{48} x + \frac{3}{16} \tag{12.12e}$$

where we have used the point-slope formula at the point $(x_1, y_1) = (1, 2/3)$ $\square$

Example 12.5. Find y' and y'' for $xy + y^4 = 1$.

Solution. Differentiating implicitly and solving for y' gives

$$(xy + y^4)' = (1)' \tag{12.13a}$$

$$xy' + y + 4y^3 y' = 0 \tag{12.13b}$$

$$y'(x + 4y^3) = -y \tag{12.13c}$$

$$y' = -\frac{y}{x + 4y^3} \tag{12.13d}$$

To find the second derivative, implicitly differentiation equation 12.13b.

$$(xy' + y + 4y^3 y')' = 0 \tag{12.13e}$$

$$(xy')' + y' + 4(y^3 y')' = 0 \tag{12.13f}$$

$$x'y' + xy'' + y' + 4(y^3)'y' + 4y^3 y'' = 0 \tag{12.13g}$$

$$y' + xy'' + y' + 12y^2 (y')^2 + 4y^3 y'' = 0 \tag{12.13h}$$

because $(y^3)' = 3y^2 y'$. Solving for y'',

$$xy'' + 4y^3 y'' = -2y' - 12y^2 (y')^2 \tag{12.13i}$$

$$(x + 4y^3)y'' = -2y'(1 + 6y^2 y') \tag{12.13j}$$

Substituting equation 12.13d,

$$(x + 4y^3)y'' = \frac{2y}{x + 4y^3} \left(1 - 6y^2 \cdot \frac{y}{x + 4y^3} \right) = \frac{2y}{(x + 4y^3)^2}(x + 4y^3 - 6y^3) \tag{12.13k}$$

$$y'' = \frac{2y(x - 2y^3)}{(x + 4y^3)^3} \tag{12.13l}$$

$$\square$$

Derivatives of the Inverse Trigonometric Functions

You are most likely familiar with the inverse trigonometric functions form your pre-calculus class or a previous class in trigonometry. The discussion probably went something like this: If $y = \sin x$, then we define the arcsin function by $x = \arcsin y$, and so forth. Sometimes we write this as $\sin^{-1} y$, which is always read as **inverse sin** or **arc-sin** function. Students often confuse the superscript of -1 on the inverse functions: beware – this is never used to indicate an exponent. For this reason we summarize the names of all the inverse trigonometric functions here. Should we ever want to indicate the quantity $1/(\cos(x))$ using an exponent, we would write it as $(\cos(x))^{-1}$. The expressions $\cos^{-1}(x)$ and so forth **always** means inverse function, and **is never used to refer to the reciprocal.**

Definition 12.1. Inverse Trigonometric Functions

Name	Full Name of Inverse	Common Name of Inverse	Fundamental Property
$\sin(x)$	$\arcsin(x)$	$\sin^{-1}(x)$	$y = \sin^{-1}(x)$ means $x = \sin(y)$
$\cos(x)$	$\arccos(x)$	$\cos^{-1}(x)$	$y = \cos^{-1}(x)$ means $x = \cos(y)$
$\tan(x)$	$\arctan(x)$	$\tan^{-1}(x)$	$y = \tan^{-1}(x)$ means $x = \tan(y)$
$\cot(x)$	$\text{arccot}(x)$	$\cot^{-1}(x)$	$y = \cot^{-1}(x)$ means $x = \cot(y)$
$\sec(x)$	$\text{arcsec}(x)$	$\sec^{-1}(x)$	$y = \sec^{-1}(x)$ means $x = \sec(y)$
$\csc(x)$	$\text{arccsc}(x)$	$\csc^{-1}(x)$	$y = \csc^{-1}(x)$ means $x = \csc(y)$

Theorem 12.2. Derivatives of Inverse Trigonometric Functions

$$\frac{d}{dx}\sin^{-1}x = \frac{1}{\sqrt{1-x^2}} \quad (12.14) \qquad \frac{d}{dx}\csc^{-1}x = -\frac{1}{x\sqrt{x^2-1}} \quad (12.15)$$

$$\frac{d}{dx}\cos^{-1}x = -\frac{1}{\sqrt{1-x^2}} \quad (12.16) \qquad \frac{d}{dx}\sec^{-1}x = \frac{1}{x\sqrt{x^2-1}} \quad (12.17)$$

$$\frac{d}{dx}\tan^{-1}x = \frac{1}{x^2+1} \quad (12.18) \qquad \frac{d}{dx}\cot^{-1}x = -\frac{1}{x^2+1} \quad (12.19)$$

Example 12.6. Calculate y' when $y = \sin^{-1}x$.

Solution. Since $y = \sin^{-1}x$ is equivalent to

$$x = \sin y \tag{12.20a}$$

we use implicit differentiation followed by the chain rule,

$$\frac{dx}{dx} = \frac{d}{dx}\sin y \tag{12.20b}$$

$$1 = (\cos y)\frac{dy}{dx} \tag{12.20c}$$

$$\frac{dy}{dx} = \frac{1}{\cos y} \tag{12.20d}$$

Using the trigonometric identity $\sin^2 y + \cos^2 y = 1$ we can solve for $\cos y$:

$$\cos y = \sqrt{1 - \sin^2 y} = \sqrt{1-x^2} \tag{12.20e}$$

where the last step follows from (12.20a). Substituting (12.20e) into (12.20d),

$$\frac{dy}{dx} = \frac{1}{\sqrt{1-x^2}} \tag{12.20f}$$

substituting $y = \sin^{-1}x$ gives us the formula

$$\frac{d}{dx}\sin^{-1}x = \frac{1}{\sqrt{1-x^2}} \tag{12.20g}$$

$$\square$$

In example 12.6 we used the trigonometric identity $\sin^2 y + \cos^2 y = 1$ to to simplify the following result:

$$\frac{dy}{dx} = \frac{1}{\cos y} = \frac{1}{\cos(\sin^{-1}x)} \tag{12.21}$$

An alternative method is to simplify this expression geometrically. We do this by asking the following question: what is the cosine of the angle $y = \sin^{-1}x$. We do this by drawing a picture of a right triangle, and labelling it according to the basic definitions of trigonometric functions. For any angle θ,

$$\sin\theta = \frac{\text{opposite}}{\text{hypotenuse}}, \quad \cos\theta = \frac{\text{adjacent}}{\text{hypotenuse}}, \quad \tan\theta = \frac{\text{opposite}}{\text{adjacent}} \tag{12.22}$$

and so forth. See figure 12.2. We are free to give the sides any lengths we want, so long as they are consistent with the Pythagorean theorem. We choose their lengths so that the calculation of the sines and cosines are easy. We want to calculate $\cos y$ where $y = \sin^{-1}x$.

Figure 12.2: On the left we show a right triangle with one of the acute angles labelled as y. On the right we label the sides so that we can calculate both $\sin y$ and $\cos y$ easily.

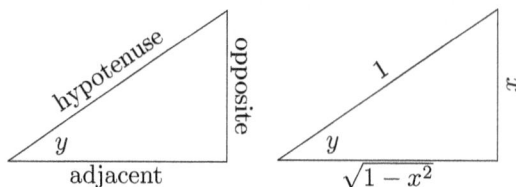

Figure 12.3: Geometry used to simplify equation 12.27c in example 12.7.

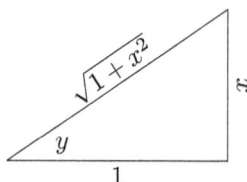

Since $y = \sin^{-1} x$, then by the fundamental property of the arcsin function, $x = \sin y$. So we pick two sides of the triangle to make calculation of $\sin y$ easy. Since

$$x = \sin y = \frac{\text{opposite}}{\text{hypotenuse}} = \frac{x}{1} \tag{12.23}$$

an easy association would be

$$\text{hypotenuse} \rightarrow 1 \tag{12.24}$$

$$\text{opposite} \rightarrow x \tag{12.25}$$

Then since we want to know $\cos y$, we combine the trigonometric definition with the Pythagorean theorem,

$$\cos y = \frac{\text{adjacent}}{\text{hypotenuse}} = \frac{\sqrt{1 - x^2}}{1} = \sqrt{1 - x^2} \tag{12.26}$$

This agrees precisely with equation 12.20e.

Example 12.7. Find $(d/dx)(\tan^{-1}(x))$.

Solution. We use the same procedure as in example 12.6. Let $y = \tan^{-1}(x)$. Then by the definition of the arctangent, this means that

$$x = \tan y \tag{12.27a}$$

Differentiating both sides of the equation,

$$\frac{dx}{dx} = \frac{d}{dx} \tan y = \sec^2 y \frac{dy}{dx} \tag{12.27b}$$

Since $(d/dx)(x) = 1$, we can solve for y'

$$y' = \frac{1}{\sec^2 y} = \frac{1}{1 + \tan^2 y} = \frac{1}{1 + x^2} \tag{12.27c}$$

where we have used the trigonometric identity $\tan^2 y + 1 = \sec^2 y$ in the last step. The geometric calculation is illustrated in figure figure:arctan-geometry-simplification. We use

$$\frac{x}{1} = x = \tan y = \frac{\text{opposite}}{\text{adjacent}} \qquad (12.27\text{d})$$

from equation 12.27a. Thus

$$\sec^2 y = \frac{1}{\cos^2 y} = \left(\frac{\text{hypotenuse}}{\text{adjacent}}\right)^2 = 1 + x^2 \qquad (12.27\text{e})$$

This gives us the denominator in (12.27c).   □

Exercises

Find y' for each of the implicitly defined functions in exercise 1 through 7.

1. $3x^2 - 2y^2 = 1$
2. $x^2 + 2xy + y^2 = 0$
3. $\dfrac{x+y}{x^2+y^2} = x$
4. $5x^3 + x^2y - xy^3 = 4$
5. $3y + x = \sin\left(xy^2\right)$
6. $x + \sqrt{y} = \cos x + \cos y$
7. $\cos x + \cos y = \sin x + \sin y$

In exercises 8 and 9 find the slope of the tangent line to the given curve at the point specified.

8. $3x^2 - 3xy + y^3 = 59$, $(4,-1)$
9. $\dfrac{y}{x-6y} = x^8 + y$, $(1, 8/49)$

In exercises 10 and 11, find the equation of the tangent line at the point given.

10. $xy^3 + xy = 14$, $(7,1)$
11. $x^2 - y^3 = y - 1$, $(3,2)$

In exercises 12 and 13, find y' and y'' using implicit differentiation and simplify.

12. $x^3 + y^3 = 8$
13. $xy = \sin y$

Simplify the expressions in exercises 14 through 19.

14. $\cos(\tan^{-1} x)$
15. $\cos(\csc^{-1} x)$
16. $\tan(\cos^{-1} x)$
17. $\cos(\sin^{-1} x)$
18. $\tan(\cot^{-1} x)$
19. $x\cos(\sec^{-1} x)$

Repeat the same procedure used in examples 12.6 and 12.7 to find general formulas for the derivatives 0f the functions given in exercises 20 through 23.

20. $y = \cos^{-1} x$
21. $y = \sec^{-1} x$
22. $y = \cot^{-1} x$
23. $y = \csc^{-1} x$

Find the derivatives of the functions given in exercises 24 through 29.

24. $y = (\sin^{-1} x)^2$
25. $y = (\sin^{-1}(\tan x))$
26. $y = \tan^{-1}(x^2)$
27. $y = x\sec^{-1} x$
28. $y = \cot^{-1}(x+1)$
29. $y = \dfrac{\sin^{-1} x}{\cos^{-1} x}$

Answers to Selected Exercises

1. $y' = \frac{3x}{2y}$

2. $y' = -1$

3. $\frac{3x^2+y^2-1}{1-2xy}$

4. $\frac{-15x^2+y^3-2xy}{x(x-3y^2)}$

5. $\frac{1-y^2\cos(xy^2)}{2xy\cos(xy^2)-3}$

6. $-\frac{2\sqrt{y}(\sin(x)+1)}{2\sqrt{y}\sin(y)+1}$

7. $y' = \frac{\cos x+\sin x}{\cos y+\sin y}$

8. 3

9. 1/6

10. $y = -\frac{1}{14}x + \frac{3}{2}$

11. $y = \frac{6}{13}x + \frac{8}{13}$

12. $y' = -x^2/y^2$
    $y'' = 16x/y^5$

13. $y' = y/(\cos y - x)$
    $y'' = \frac{xy^2+2\cos y-2x}{(x-\cos y)^3}$

14. $\frac{1}{\sqrt{x^2+1}}$

15. $\sqrt{1-\frac{1}{x^2}}$

16. $\frac{\sqrt{1-x^2}}{x}$

17. $\sqrt{1-x^2}$

18. $1/x$

19. 1

20. $-1/\sqrt{1-x^2}$

21. $\frac{1}{x\sqrt{x^2-1}}$

22. $-1/(1+x^2)$

23. $-\frac{1}{x\sqrt{x^2-1}}$

24. $\frac{2\sin^{-1}(x)}{\sqrt{1-x^2}}$

25. $\frac{\sec^2(x)}{\sqrt{1-\tan^2(x)}}$

26. $\frac{2x}{x^4+1}$

27. $\frac{1}{\sqrt{1-\frac{1}{x^2}}x} + \sec^{-1}(x)$

28. $-\frac{1}{(x+1)^2+1}$

29. $\frac{\pi}{2\sqrt{1-x^2}\cos^{-1}(x)^2}$

Chapter 13

Related Rates

Chapter Summary and Goal

Here we turn to a discussion of problems in which we have some information about the rate of change dy/dt of one variable y and are asked to find the rate of change dz/dt of a different variable z. In these **related rates** problem, the idea is to find some way of relating the superficially unrelated variables y and z together, typically through some sort of **constraint**.

Usually these problems are stated as word problems, and we must extract and translate the necessary information into the appropriate equations. Typically, the relating equation must be solved using implicit differentiation.

For example, R may be the radius of a spherical comet as it passes by the sun. We are told that its radius shrinks at a certain rate, and are asked to find the rate of change of the comet's mass dm/dt at its closest approach to the sun. The two rates will be related by the volume formula for a sphere – $V = 4\pi R^3/3$ – and the density of the comet ρ. Since $m = V\rho$, we can use implicit differentiation to relate $m'(t)$ to $R'(t)$. If we are given dm/dt precisely at the moment of closest approach, then we can plug it into the result of that implicit differentiation and solve that problem. If not, we will need to know more about the trajectory and orbital mechanics of the comet to figure out when closest approach will occur.[1]

Student Learning Objectives

The student will:

1. Identify key words and phrases in related rates problems and translate them to appropriate terms in calculus equations.
2. Translate word problems into calculus problems and solve them.
3. Relate variables to one another via a third variable and use this information to find the derivative.

Relating Variables to One Another

Related rates problems invariably arise as word problems. The first thing you need to do is examine the problem statement thoroughly and identify the exactly (a) **what information is given?** and (b) **what information is requested?**

---

[1] Don't worry, we won't attempt to solve any orbital mechanics problems here!

Start by drawing a diagram (if geometry is involved in the problem), and carefully label each object in the diagram.

Assign a variable to each object in the problem (look for the nouns), and mark it on the diagram.

Look for key words like "change" or "rate," (look for the verbs) because the values of these quantities refer to their derivatives. For example, if a problem says, "water is leaking at a rate of 7 gallons per hour" this means that the volume of water (V) is changing (dV/dt) at a rate of 7 gallons per hour, i.e., we are given that $dV/dt = 7$.

Look for relationships between variables. If there is more than one variable, try to think of how they are related. If you are told that the water is leaking out of an upright cylindrical tank, and are asked to figure out how fast the height h of the water is changing, this means you are looking for dh/dt. You can relate V and h through the formula for the volume of a cylinder, $V = \pi r^2 h$.

Don't look for a formula that relates rates to rates directly. That's where the calculus comes in! By differentiating $V = \pi r^2 h$ you can get the formula you need: $dV/dt = \pi r^2 dh/dt$, which relates dV/dt to to dh/dt.

Heuristic 13.1. Problem Solving Strategy for Related Rates Problems

1. Draw a **diagram** when possible.
2. Assign **symbols** to all quantities.
3. Identify the **given information** symbolically and express rates as derivatives.
4. Identify the **goal** symbolically and express rates as derivatives.
5. Write **equations** that relate the remaining quantities.
6. Use **implicit differentiation**, the **chain rule**, and the **rules of differentiation** to find the derivative in step (5) .
7. Substitute the **given** from (3) into the relationship found in (6).
8. **Solve for the unknown**, i.e., the goal identified in (4).

Example 13.1. A spherical "ice-cube" is melting so that its volume decreases at a rate of 0.25 cm^3/min. How fast is the radius of the "ice-cube" changing when it is 1 cm. in diameter?

Solution. We are given the following information: the volume V of the ice cube is changing at a rate of 0.25 cm^3/min, i.e.,

$$\frac{dV}{dt} = 0.25 \frac{\text{cm}^3}{\text{min}} \tag{13.1a}$$

We are asked to find: the rate of change of the radius, dr/dt, when the diameter d of the ice cube is $d = 1$ cm.

We can relate the volume to the radius by the formula for the volume of a sphere

$$V = \frac{4}{3}\pi r^3 \tag{13.1b}$$

We then relate dr/dt to dV/dt through implicit differentiation and the chain rule,

$$\frac{dV}{dt} = \frac{d}{dt}\left(\frac{4}{3}\pi r^3\right) = 4\pi r^2 \frac{dr}{dt} \tag{13.1c}$$

Solving for dr/dt,

$$\frac{dr}{dt} = \frac{1}{4\pi r^2}\frac{dV}{dt} \tag{13.1d}$$

When $d = 1$ then $r = d/2 = 1/2$ cm. Substituting $dV/dt = 0.25$ cm^3/min,

$$\frac{dr}{dt} = \frac{2^2}{4\pi} \cdot (0.25) = \frac{1}{4\pi} \frac{\text{cm}}{\text{min}}. \tag{13.1e}$$

$\square$

Example 13.2. An upright cylindrical water tank has radius 25cm (=1/4 m). Water is leaking out at a rate of 2500 cm^3 / min. How fast is the height of the water changing?

Solution. Let the height of the water tank be h, and the radius of the tank be r. The volume of a cylinder is given by

$$V = \pi r^2 h \tag{13.2a}$$

We are given the following information:
1. The radius of the tank is $r = \frac{1}{4}$m is **fixed** for all time.
2. The tank is cylindrical, so the volume of water in it is given by

$$V = \pi r^2 h = \pi \left(\frac{1}{4}\right)^2 h = \frac{\pi h}{16} \tag{13.2b}$$

3. We are asked to find dh/dt. Differentiating the formula for V,

$$\frac{dV}{dt} = \frac{d}{dt}\left(\frac{\pi h}{16}\right) = \frac{\pi}{16}\frac{dh}{dt} \tag{13.2c}$$

This can be solved for h:

$$\frac{dh}{dt} = \frac{16}{\pi}\frac{dV}{dt} \tag{13.2d}$$

4. Finally, we are told that water is leaking out at a rate of $dV/dt =$-2500 cm^3/min. The minus sign is there because the volume of water in the tank is decreasing. Since everything else is in meters, we need to make a unit conversion:

$$\frac{dV}{dt} = -2500 \frac{\text{cm}^3}{\text{min}} \cdot \left(\frac{1\text{m}}{100 \text{ cm}}\right)^3 = -0.0025 \frac{\text{m}^3}{\text{min}} \tag{13.2e}$$

Therefore

$$\frac{dh}{dt} = \frac{16}{\pi}\frac{dV}{dt} = \frac{16}{\pi} \cdot (-.0025) = -\frac{.04}{\pi}\frac{\text{m}}{\text{min}} \times \frac{1000\,\text{mm}}{\text{m}} \approx -12.7 \frac{\text{mm}}{\text{min}} \tag{13.2f}$$

$\square$

Example 13.3. The volume of a cube is increasing at a rate of 10 cm^3/minute. How fast is the surface area of the cube changing when each side has length 20 cm?

Solution. Let us define ℓ=the length of a side, and let V and A denote the volume and surface area of the cube. These are given by

$$V = \ell^3 \qquad\qquad \text{volume of a cube} \tag{13.3a}$$

$$A = 6\ell^2 \qquad\qquad \text{surface area of a cube} \tag{13.3b}$$

We are asked to find dA/dt when $\ell = 20$ cm., given that $dV/dt = 10$ cm^3/min. The answer will be in cm.^{2}/min. Differentiating (13.3a) and (13.3b), we find that

$$\frac{dV}{dt} = \frac{d}{dt}\left(\ell^3\right) = 3\ell^2\frac{d\ell}{dt} \tag{13.3c}$$

$$\frac{dA}{dt} = \frac{d}{dt}\left(6\ell^2\right) = 12\ell\frac{d\ell}{dt} \tag{13.3d}$$

Dividing these two equations gives

$$\frac{\frac{dV}{dt}}{\frac{dA}{dt}} = \frac{3\ell^2 \frac{d\ell}{dt}}{12\ell \frac{d\ell}{dt}} = \frac{\ell}{4} \tag{13.3e}$$

Cross-multiplying,

$$\frac{4}{\ell}\frac{dV}{dt} = \frac{dA}{dt} \tag{13.3f}$$

Solving for dA/dt and substituting the known values of $\ell = 20$ and $dV/dt = 10$ gives

$$\frac{dA}{dt} = \frac{4}{\ell}\frac{dV}{dt} = \frac{4(10)}{20} = 2 \text{ cm}^2/\text{min.} \tag{13.3g}$$

□

Example 13.4. A 13 foot long ladder is resting against a vertical wall. If the bottom of the ladder slides away from the wall at a rate of 1/2 foot/sec, how fast is the top of the ladder sliding down the wall when the bottom of the ladder is 5 feet from the wall?

Solution. The first step in a problem with complicated geometry is to draw a picture and use it to define variables (figure 13.1). We are given that $dx/dt = 1/2$ and that the length

Figure 13.1: Geometry and variables used inexample 13.4.

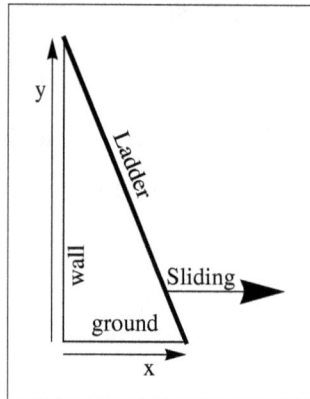

of the ladder ℓ is $\ell = 13$. In terms of the variables illustrated in the figure and the Pythagorean theorem,

$$\ell^2 = x^2 + y^2 \tag{13.4a}$$

Differentiating with respect to time (t)

$$2\ell\frac{d\ell}{dt} = 2x\frac{dx}{dt} + 2y\frac{dy}{dt} \tag{13.4b}$$

Dividing by 2,

$$\ell\frac{d\ell}{dt} = x\frac{dx}{dt} + y\frac{dy}{dt} \tag{13.4c}$$

Since the length of the ladder is fixed $d\ell/dt = 0$,

$$0 = x\frac{dx}{dt} + y\frac{dy}{dt} \tag{13.4d}$$

Solving for dy/dt,

$$\frac{dy}{dt} = -\frac{x}{y}\frac{dx}{dt} \tag{13.4e}$$

We want to find dy/dt when $x = 5$ and $dx/dt = 1/2$. To get this information from (13.4e) we use (13.4a),

$$169 = \ell^2 = x^2 + y^2 = 25 + y^2 \tag{13.4f}$$

Thus when $x = 5$,

$$y^2 = 169 - 25 = 144 \tag{13.4g}$$

or $y = 12$. From (13.4e) we have

$$\frac{dy}{dt} = -\frac{5}{12} \cdot \frac{1}{2} = -\frac{5}{24} \text{ ft/sec} = -2.5 \text{ inches/sec} \tag{13.4h}$$

$\square$

Figure 13.2: Schematic diagram of two resistors in parallel. In example 13.5 the dependence of the total effective resistance R between points A and B when each individual resistor is changing is calculated.

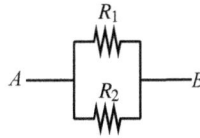

Example 13.5. Resistors. When two resistors are connected together in parallel (see figure 13.2), the total resistance is found from

$$\frac{1}{R} = \frac{1}{R_1} + \frac{1}{R_2} \tag{13.5}$$

Suppose that R_1 are R_2 are increasing at the rates of 1 ohms/sec and 2 ohms/sec, respectively. How fast is R changing when $R_1 = 1000$ ohms and $R_2 = 500$ ohms?

Solution. We are asked to find dR/dt when $R_1 = 1000$ and $R_2 = 500$, given that $dR_1/dt = 1$ and $dR_2/dt = 2$.

Differentiating (13.5) gives

$$-\frac{1}{R^2}\frac{dR}{dt} = -\frac{1}{R_1^2}\frac{dR_1}{dt} - \frac{1}{R_2^2}\frac{dR_2}{dt} \tag{13.6a}$$

Therefore,

$$\frac{dR}{dt} = \frac{R^2}{R_1^2}\frac{dR_1}{dt} + \frac{R^2}{R_2^2}\frac{dR_2}{dt} \tag{13.6b}$$

Before we solve (13.6b) we will make the following useful simplification from (13.5). Putting the right hand side of (13.5) over a common denominator gives

$$\frac{1}{R} = \frac{R_1 + R_2}{R_1 R_2} \tag{13.6c}$$

Taking the reciprocal,

$$R = \frac{R_1 R_2}{R_1 + R_2} \tag{13.6d}$$

Hence

$$\frac{R}{R_1} = \frac{R_2}{R_1 + R_2} \text{ and } \frac{R}{R_2} = \frac{R_1}{R_1 + R_2} \tag{13.6e}$$

Plugging (13.6e) back into (13.6b)

$$\frac{dR}{dt} = \left(\frac{R_2}{R_1 + R_2}\right)^2 \frac{dR_1}{dt} + \left(\frac{R_1}{R_1 + R_2}\right)^2 \frac{dR_2}{dt} \tag{13.6f}$$

$$= \left(\frac{500}{1000 + 500}\right)^2 (1) + \left(\frac{1000}{1000 + 500}\right)^2 (2) \tag{13.6g}$$

$$= \left(\frac{1}{3}\right)^2 + \left(\frac{2}{3}\right)^2 \cdot 2 \tag{13.6h}$$

$$= \frac{1 + 4 \cdot 2}{9} = \frac{7}{9} \approx 0.78 \text{ Ohm/sec} \tag{13.6i}$$

$\square$

Example 13.6. Two sides of a triangle have fixed lengths of 10 cm and 15 cm. The sides are rotating so that the angle between them is increasing at a rate of 5 degrees/minute. How fast is the length of the third side changing when the angle is 45 degrees?

Solution. We are asked to find dz/dt when $\theta = 45$ deg, given that $d\theta/dt = 5$ deg/minute. To get consistent distances, we must convert the rate into radians:

$$\frac{d\theta}{dt} = 5\frac{\text{deg}}{\text{min}} \times \frac{\pi}{180}\frac{\text{radians}}{\text{deg}} = \frac{\pi}{36}\frac{\text{radians}}{\text{min}} \tag{13.7a}$$

Consider the geometry illustrated in figure 13.3. Applying the law of cosines to the

Figure 13.3: Geometry of the triangle with rotating edges from example 13.6.

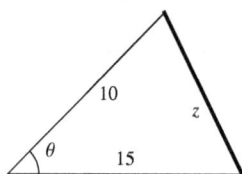

length of the third side, which we will denote by z,

$$z^2 = 15^2 + 10^2 - 2(15)(10)\cos\theta \tag{13.7b}$$

$$= 225 + 100 - 300\cos\theta \tag{13.7c}$$

$$= 325 - 300\cos\theta \tag{13.7d}$$

Differentiating and rearranging

$$2z\frac{dz}{dt} = 300\sin\theta\frac{d\theta}{dt} \tag{13.7e}$$

$$\frac{dz}{dt} = \frac{150\sin\theta}{z}\frac{d\theta}{dt} \tag{13.7f}$$

$$= \frac{150\sin\theta}{\sqrt{325 - 300\cos\theta}}\frac{d\theta}{dt} \tag{13.7g}$$

So when $\theta = 45$ degrees,

$$\frac{dz}{dt} = \frac{150\sin(45)}{\sqrt{325 - 300\cos(45\deg)}} \frac{d\theta}{dt} \tag{13.7h}$$

$$= \frac{150(\sqrt{2}/2)}{\sqrt{325 - 300(\sqrt{2}/2)}} \left(\frac{\pi}{36}\right) \tag{13.7i}$$

$$= \frac{5\pi}{6\sqrt{26 - 12\sqrt{2}}} \approx 0.87\text{meters/Sec} \tag{13.7j}$$

$\square$

Example 13.7. Car A is traveling west at 75 miles/hour and car B is traveling north at 55 miles/hour. Both are headed for the intersection of the two roads. At what rate are the cars approaching each other when car A is 5 mi and car B is 2 mi from the intersection?

Solution.

1. Draw a diagram when possible. See figure 13.4.

Figure 13.4: Two cars are heading toward the same intersection as described in example 13.7.

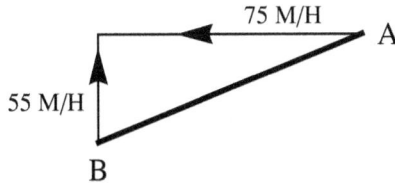

75 M/H

A

55 M/H

B

2. Assign symbols to all quantities. See figure 13.4.
   a is the distance of car A from the origin
   b is the distance of car B from the origin
   r is the distance between car A and car B

3. Identify the **given information** symbolically

$$\frac{da}{dt} = -75 \tag{13.8a}$$

$$\frac{db}{dt} = 55 \tag{13.8b}$$

4. Identify the **goal** symbolically:

$$\left.\frac{dr}{dt}\right|_{a=5,b=-2} \tag{13.8c}$$

5. Write equations that relate the remaining quantities

$$r^2 = a^2 + b^2 \tag{13.8d}$$

6. Use **implicit differentiation**, the **chain rule**, and the rules of differentiation to find dr/dt:

$$2r\frac{dr}{dt} = 2a\frac{da}{dt} + 2b\frac{db}{dt} \tag{13.8e}$$

$$\frac{dr}{dt} = \frac{a}{r} \cdot \frac{da}{dt} + \frac{b}{r} \cdot \frac{db}{dt} \tag{13.8f}$$

$$= \frac{1}{r} \left(a\frac{da}{dt} + b\frac{db}{dt} \right) \tag{13.8g}$$

$$= \frac{1}{\sqrt{a^2 + b^2}} \cdot \left(a\frac{da}{dt} + b\frac{db}{dt} \right) \tag{13.8h}$$

7. Substitute the **given information** into the relationship:

$$\frac{dr}{dt} = \frac{1}{\sqrt{5^2 + (-2)^2}} \cdot (5 \cdot (-75) + (-2) \cdot 55) \tag{13.8i}$$

8. Solve for the unknown, i.e., the goal identified in (4)

$$\frac{dr}{dt} = -\frac{485}{\sqrt{29}} \approx 90.06 \text{ miles/hour} \tag{13.8j}$$

$\square$

Exercises

1. Each side of a cube is increasing at a rate of 10 inches per minute. How fast is the volume of the cube changing when each side has a length of 36 inches?
ans: 38,880 in^3/min

2. Sand is being pulled by a conveyor belt and dumped at a rate of 50 cubic feet per minute. It forms a mound of sand that shaped like a right circular cone whose base and height are always equal to one another. How fast is height of mound growing when it is 25 feet high? (Hint: the volume of a right circular cone with height h and base radius r is $\frac{1}{3}\pi r^2 h$.)
ans: 0.0102 feet/min

3. A snowball is melting so that its diameter is decreasing at a rate of 5 mm/minute. How fast is the volume of the snowball changing when its diameter is 30 cm? (Assume the snowball is spherical.)

4. A 6 foot tall man is walking directly away from a street light at a rate of 5 feet/second. The street light is at the top of a 12 foot tall pole. How fast is the tip of his shadow moving when he is 25 feet way from the base of pole? ans: 10 feet/sec

5. A object is constrained to move along the curve $y = \sqrt{x}$. When is passes through the point $(x, y) = (1, 1)$, its distance from the origin is increasing at a rate of 10 cm/sec. How fast is the y coordinate changing at that moment?
ans: $20\sqrt{2}/3$ cm/sec

6. Water is leaking out of an inverted conical tank at a rate of 15 liters per minute, while at the same time, water is being pumped into the tank at an unknown (but constant) rate of r liters per minute. Suppose the height of the tank is 12 meters and the diameter of the top is 5 meters. What is the value of r if the water level is rising at a rate of 20 cm/minute when the height is 3 meters?

7. At noon the good cruse ship *Calculus Fantasy* is 10 nautical miles due west of the *Royal Derivative*. If the *Calculus Fantasy* is sailing west at 20 knots (1 knot is 1 nautical mile per hour) and the *Royal Derivative* is sailing north at 25 knots, how fast (in knots) is the distance between the ships changing at 4:00 PM?
ans: $430/\sqrt{83}$ knots

Chapter 14

Inverse Functions

Chapter Summary and Goal

It is often useful to think of a function as a machine. You pick some number x – the **input** – from a set called the **domain**, and drop it into the machine. The machine whirs and clanks along for a while (think Willy Wonka) and then spits out a new number – $f(x)$ – the **output** – into a set that we call the **range** of the function (see figure 14.1).

The machine is built in such a way that whenever you put the same thing in, you get the same thing out.

Mathematically, we describe this plotting the (x, y) pairs on a graph and looking at the plot. We describe this "function" property – same thing in/same thing out – by saying that no vertical line ever crosses the graph more than once.

Figure 14.1. Illustration of a function as a machine.[1]

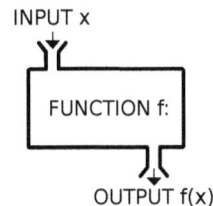

In this chapter we ask a different question: can we reverse the function machine, and if so, when will it work, and when will it break? Thinking Willy-Wonkish, can we just turn figure 14.1 upside down, reverse the current, and drop something into the output spout, and see what drops out the input spout?

Student Learning Objectives

The student will:

1. Understand the difference between functions and their inverses.
2. Determine whether or not a function has an inverse.
3. Be able to calculate inverses for simple functions.
4. Be able to calculate the derivatives of inverse functions.

---

[1]Public domain llustration from Wikimedia Commons http://en.wikipedia.org/wiki/File:Function_machine2.svg

Functions and Their Inverses

A function is a relationship between two variables, typically x (which we might call the **input variable** or **independent variable**) and y (which we typically think of as the **output variable** or **dependent variable**), such that, for any given input, there is only one possible output. We commonly represent functions as machines that produce a specific output for each input (as in figure 14.1).

The plot of a curve is a function if it passes the **vertical line test**: if no vertical line passes through the curve of the function more than once, then the curve is a function (figure 14.2.a).

Figure 14.2: (a) Vertical line test. The curve on the left is a function because any vertical line crosses it only once. The curve on the right is not a function because there is at least one vertical line that crosses it twice. (b) Horizontal line test. The function on the left is not one-to-one because there is at least one horizontal line that crosses it more than once. The function on the right is one-to-one because no horizontal line intersects the curve more than once.

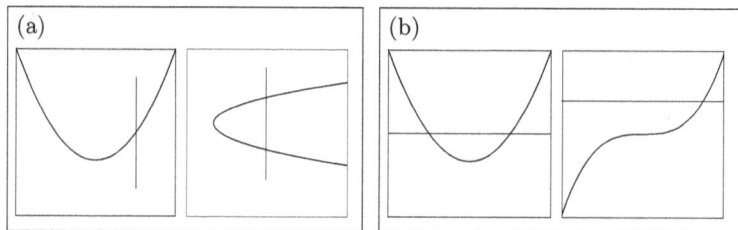

We will be interested in a special class of functions in this chapter, called **one-to-one functions**, because only one-to-one functions are guaranteed to have inverses.

A function is **one-to-one** if it passes the **horizontal line test**: if no horizontal line passes through the curve of the function more than once, the function is one-to-one (fig. 14.2.b).

Definition 14.1. One-to-One (1-1)

A function $y = f(x)$ is said to be **one-to-one** if it never takes on the same y value twice. We can express this symbolically by saying that

$$f(x_1) \neq f(x_2) \text{ whenever } x_1 \neq x_2 \qquad (14.1)$$

When we can list the elements in the domain and range, we sometimes illustrate functions with **dot-and-blob representations** (figure 14.3). In these figures the domain and range are represented by large blobs, and the elements in the domain and range are represented by dots. The functions themselves are represented as arrows – or mappings – that give the relationship between specific input and output values. If any point in the range has more than one arrow incident upon it, the function is not one-to-one.

Figure 14.3: Dot-and-blob representations of functions. A functional mapping is one to one only if there is no more than one arrow incident on any single point in the range

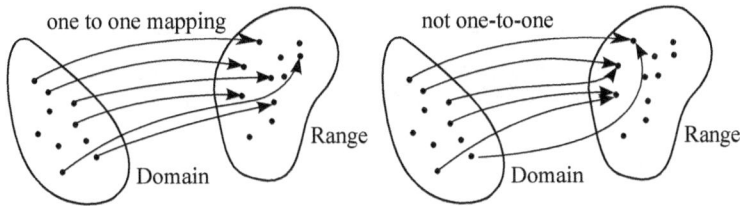

Theorem 14.1. Invertibility

Let $y = f(x)$ be a 1-1 function with domain A and range B. Then $f(x)$ has an inverse $x = f^{-1}(y)$ with domain B and range A.

Furthermore, if $f(x)$ is continuous, then f^{-1} is also continuous.

Figure 14.4: Concept of invertibility.

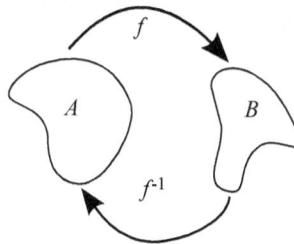

Example 14.1. Find the inverse of $f(x) = x^3$.

Solution. The function $y = x^3$ passes the vertical line test, so it is one-to-one. Because it is one-to-one, it has an inverse. We can find the inverse by solving for x.

$$y = x^3 \tag{14.2a}$$

$$x = y^{1/3} \tag{14.2b}$$

$$f^{-1}(y) = y^{1/3} \tag{14.2c}$$

However, strange as it may seem, we typically write the inverse function as a function of x rather than a function of y, so the last step in the process is to exchange the x and y variables:

$$f^{-1}(x) = x^{1/3} \tag{14.2d}$$

The reason why this is OK is because the function represents a black-box-machine and variables are really dummy arguments, so it doesn't really matter what we call them. Think of the analogy with computer programming - the function is like a subprogram or method and the variables x and y are then local variables, so the outside world shouldn't care what we name them. □

Theorem 14.2. Fundamental Property of Inverses

$$y = f^{-1}(x) \text{ if and only if } x = f(y) \qquad (14.3)$$

Theorem 14.3. Cancellation Property of Inverses

Let $f(x)$ be an invertible function with domain A and range B and suppose that $f^{-1}(x)$ is its inverse. Then

$$f^{-1}(f(x)) = x \text{ for all } x \text{ in } A \qquad (14.4)$$
$$f(f^{-1}(x)) = x \text{ for all } x \text{ in } B \qquad (14.5)$$

Example 14.2. Demonstrate the cancellation property with $f(x) = x^3$.

Solution. In example 14.1 we found that $f(x) = x^3$ had inverse $f^{-1}(x) = x^{1/3}$. Thus

$$f(f^{-1}(x)) = f\left(x^{1/3}\right) = \left(x^{1/3}\right)^3 = x \qquad (14.6a)$$
$$f^{-1}(f(x)) = f^{-1}\left(x^3\right) = \left(x^3\right)^{1/3} = x \qquad (14.6b)$$

as required. □

Heuristic 14.1. Procedure for Finding an Inverse Function

1. Write $y = f(x)$ and verify that $f(x)$ is 1-1.
2. If f is not 1-1 then $f(x)$ is not invertible over its entire domain (although it may be invertible in certain subsets of the domain if the function is 1-1 on those subsets).
3. Solve for x in terms of y.
4. Interchange the variables x and y to write $y = f^{-1}(x)$; more specifically, replace x with $f^{-1}(x)$ and y with x to get the final result.

Example 14.3. Find the inverse of $f(x) = x^3 + 2$.

Solution. Following the procedure outline above, we begin by replacing $f(x)$ with y. This is makes the algebra easier to manipulate.

$$y = x^3 + 2 \qquad (14.7a)$$

Next, we solve for x. We can bring the x^3 to the LHS and take the cube root, which is unique.

$$x^3 = y - 2 \qquad (14.7b)$$
$$x = (y - 2)^{1/3} \qquad (14.7c)$$

Finally, we exchange x and y,

$$y = (x - 2)^{1/3} \qquad (14.7d)$$

Replacing y with $f^{-1}(x)$, we see that the inverse function is $f^{-1}(x) = (x - 2)^{1/3}$. □

The inverse of $y = x^3 + 2$ computed in example 14.3 is plotted in figure 14.5. This figure demonstrates an important property of inverse functions: the inverse of $f(x)$ is the reflection

Figure 14.5: Demonstration of the inverse reflection property (theorem 14.4) for $y = x^3 + 2$.

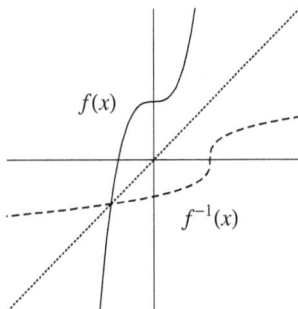

of $f(x)$ across the line $y = x$ (theorem 14.4). We use this property to construct an Oragami-ish exercise to visualize the inverse, as follows.

Theorem 14.4. Reflection Theorem

The graph of $f^{-1}(x)$ is the reflection of $f(x)$ about the line $y = x$.

Visualization Exercise. This gives us a very easy way to visualize the inverse of any function. Start with any square piece of paper, and fold it into quarters (figure 14.6.A,B). The fold lines represent the x and y axes. Then sketch $f(x)$ with a dark marker, such as a Sharpie (C). Fold the paper along the 45-degree diagonal that corresponds to the line $y = x$ (D). Then by tracing through the original curve in a different color we can sketch the inverse function(D, E). When the paper is unfolded, the inverse is seen in the new color (e.g., the dotted line in fig. 14.6).

Figure 14.6: Visualization of the inverse by folding a piece of square paper. See text for details.

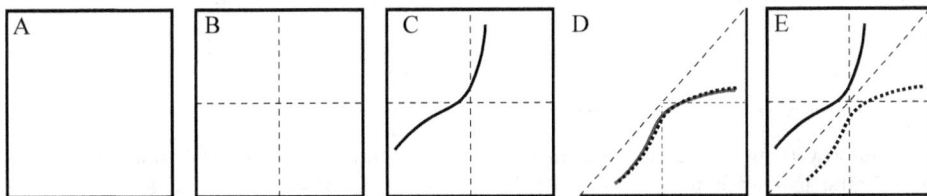

Derivatives of Inverse Functions

Suppose that $f(x)$ is continuous and differentiable. According to the definition of the derivative of $f^{-1}(x)$ at $x = a$,

$$(f^{-1})'(a) = \lim_{x \to a} \frac{f^{-1}(x) - f^{-1}(a)}{x - a} \tag{14.8}$$

where the notation $(f^{-1})'(a)$ means $(f^{-1}(x))'$ evaluated at $x = a$: first calculate the derivative, then plug in $x = a$.

Let $y = f^{-1}(x)$ and $b = f^{-1}(a)$. By the fundamental property of inverses (theorem 14.2),

$$y = f^{-1}(x) \implies x = f(y) \tag{14.9}$$

$$b = f^{-1}(a) \implies a = f(b) \tag{14.10}$$

Therefore

$$\lim_{x \to a} y = \lim_{x \to a} f^{-1}(x) = f^{-1}(a) = b \tag{14.11}$$

Thus as $x \to a$, $y \to b$ so we can change the limit in (14.8) from $x \to a$ to $y \to b$. If we also substitute equations 14.9 and 14.10 into (14.8), we obtain

$$(f^{-1})'(a) = \lim_{y \to b} \frac{\overbrace{f^{-1}(x)}^{y} - \overbrace{f^{-1}(a)}^{b}}{\underbrace{x}_{f(y)} - \underbrace{a}_{f(b)}} \tag{14.12}$$

$$= \lim_{y \to b} \frac{y - b}{f(y) - f(b)} \tag{14.13}$$

$$= \lim_{y \to b} \frac{1}{\dfrac{f(y) - f(b)}{y - b}} \tag{14.14}$$

$$= \frac{1}{\lim\limits_{y \to b} \dfrac{f(y) - f(b)}{y - b}} \tag{14.15}$$

$$= \frac{1}{f'(b)} \tag{14.16}$$

$$= \frac{1}{f'(f^{-1}(a))} \tag{14.17}$$

Theorem 14.5. Inverse Function Theorem

If $f(x)$ is invertible and differentiable at $x = a$ and if $b = f(a)$ then

$$(f^{-1})'(b) = \frac{1}{f'(a)} = \frac{1}{f'(f^{-1}(b))} \tag{14.18}$$

An easier way to prove the formula in the inverse function theorem does not require calculating limits; instead it uses the chain rule. Suppose that $y = f^{-1}(x)$. Then $x = f(y)$ and differentiating by the chain rule,

$$\frac{dx}{dx} = \frac{d}{dx} f(y) \tag{14.19}$$

$$1 = \frac{d}{dy} f(y) \cdot \frac{dy}{dx} = f'(y) \frac{dy}{dx} \tag{14.20}$$

$$\frac{dy}{dx} = \frac{1}{f'(y)} = \frac{1}{\left(\dfrac{dx}{dy} \right)} \tag{14.21}$$

$$\frac{d}{dx} f^{-1}(x) = \frac{1}{f'(f^{-1}(x))} \tag{14.22}$$

Interchanging variables, we got the generalized formula below.

> **Theorem 14.6. General Formula for the Derivative of an Inverse**
>
> If $y = f(x)$ is invertible and differentiable (so that $x = f^{-1}(y)$), then
>
> $$(f^{-1})'(y) = \frac{1}{f'(x)} = \frac{1}{f'(f^{-1}(y))} \qquad (14.23)$$

For most functions, however, it will not be possible to analytically determine a general formula for the inverse and you will be limited to only being able to calculate the inverse at specific points using equation 14.18. This is not a practical limitation for most applications because the complete inverse over any domain of interest can almost always be computed numerically to sufficient accuracy.

Example 14.4. Find $(f^{-1})'(1)$ for $f(x) = x^2$ on the interval [0,2].

Solution. From (14.18) we have

$$(f^{-1})'(1) = \frac{1}{f'(f^{-1}(1))} \qquad (14.24a)$$

To find $f^{-1}(1)$ we let $y = 1$ and solve for x:

$$1 = x^2 \qquad (14.24b)$$

The only solution to this on the interval [0,2] is $x = 1$. Hence $f^{-1}(1) = 1$. Therefore

$$(f^{-1})'(1) = \frac{1}{f'(1)} \qquad (14.24c)$$

But since $f(x) = x^2$ we can easily calculate that $f'(x) = 2x$ so $f'(1) = 2 \cdot 1 = 2$. Thus

$$(f^{-1})'(1) = \frac{1}{2} \qquad (14.24d)$$

$\square$

Example 14.5. Find $(f^{-1})'(1)$ for $f(x) = 2x + \cos x$.

Solution. Again, from (14.18) we have

$$(f^{-1})'(1) = \frac{1}{f'(f^{-1}(1))} \qquad (14.25a)$$

Since $f'(x) = 2 - \sin x$, we can write this as

$$(f^{-1})'(1) = \frac{1}{2 - \sin(f^{-1}(1))} \qquad (14.25b)$$

To find $f^{-1}(1)$ we set $y = 1$ and solve for x:

$$1 = y = 2x + \cos x \qquad (14.25c)$$

By direct substitution we see that $x = 0$ is a solution. The fact that this is the only solution can be verified with a graphing calculator or program. Thus $f^{-1}(1) = 0$. Using this in (14.25b),

$$(f^{-1})'(1) = \frac{1}{2} \qquad (14.25d)$$

$\square$

Procedure for Calculating $(f^{-1})'(a)$

1. Find $f^{-1}(a)$ by setting $y = a$ and solving for x.
2. Find $f'(x)$ by differenting f.
3. Calculate $f'(f^{-1}(a))$ as $f'(x)$ using the x from step 1 and the f' from step 2.
4. The derivative at $y = a$ is $(f^{-1})'(a) = 1/f'(x)$ using $f'(x)$ you from step 2.

Example 14.6. Find $(f^{-1})'(2)$ for $f(x) = 1 + \sqrt{2 + 4x}$.

Solution. (1) <u>Find $f^{-1}(2)$</u>:

$$2 = 1 + \sqrt{2 + 4x} \tag{14.26a}$$

$$1 = \sqrt{2 + 4x} \tag{14.26b}$$

$$1 = 2 + 4x \tag{14.26c}$$

$$-1 = 4x \tag{14.26d}$$

$$f^{-1}(2) = x = -1/4 \tag{14.26e}$$

(2) <u>Find $f'(x)$</u>:

$$f'(x) = \frac{d}{dx}\left(1 + \sqrt{2 + 4x}\right) = \frac{2}{\sqrt{2 + 4x}} \tag{14.26f}$$

(3) <u>Find $f'(f^{-1}(2))$</u>:

$$f'(f^{-1}(2)) = \frac{2}{\sqrt{2 + 4(-1/4)}} = 2 \tag{14.26g}$$

(4) <u>Find $(f^{-1})'(2)$</u>:

$$(f^{-1})'(2) = \frac{1}{f'(f^{-1}(2))} = \frac{1}{2} \tag{14.26h}$$

$\square$

Example 14.7. Find $(f^{-1})'(5)$ for $f(x) = (1 + x)/(1 - x)$.

Solution. (1) <u>Find $f^{-1}(5)$</u>:

$$5 = \frac{1 + x}{1 - x} \tag{14.27a}$$

$$5 - 5x = 5(1 - x) = 1 + x \tag{14.27b}$$

$$5 - 1 = 5x + x \tag{14.27c}$$

$$4 = 6x \tag{14.27d}$$

$$f^{-1}(5) = x = 2/3 \tag{14.27e}$$

(2) <u>Find $f'(x)$</u>:

$$f'(x) = \frac{d}{dx}\frac{1 + x}{1 - x} \tag{14.27f}$$

$$= \frac{(1 - x)(1) - (1 + x)(-1)}{(1 - x)^2} \tag{14.27g}$$

$$= \frac{2}{(1 - x)^2} \tag{14.27h}$$

(3) Find $f'(f^{-1}(5))$:

$$f'(f^{-1}(5)) = \frac{2}{(1 - (f^{-1}(5))^2} \tag{14.27i}$$

$$= \frac{2}{(1 - (2/3))^2} \tag{14.27j}$$

$$= 18 \tag{14.27k}$$

(4) Find $(f^{-1})'(5)$:

$$(f^{-1})'(5) = \frac{1}{f'(f^{-1}(5))} = \frac{1}{18} \tag{14.27l}$$

$\square$

Exercises

Determine whether or not each of the following functions is invertible.

8.

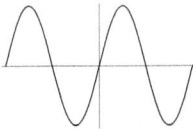

1.

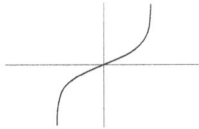

4.

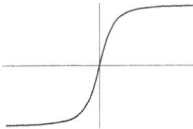

2.

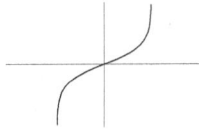

5.

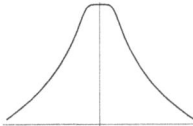

3.

6.

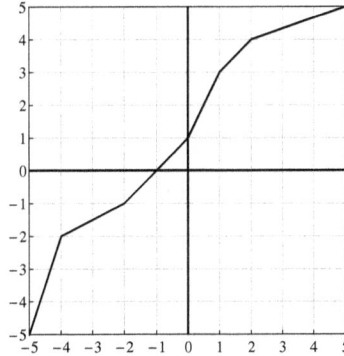

Find the inverse of each of the following functions.

9. $f(x) = 9 + 12x$ ans: $f^{-1}(x) = \dfrac{x - 9}{12}$

10. $f(x) = \dfrac{3x + 5}{7x + 11}$ ans: $f^{-1}(x) = \dfrac{5 - 11x}{7x - 3}$

Find $(f^{-1})'(a)$ for each of the following functions at the given values of a.

11. $f(x) = x^2 + 4x + 5$
    $a = 2$, for $x > 0$.   ans: 1/2 or -1/2

12. $y = \dfrac{x}{1 + x}$ for $a = 3$.   ans: 1/4

13. $f(x) = 5 - 6x$
    $a = 1$ ans: -1/6

14. $f(x) = 12 - x^2$
    $a = 5$ on $x \geq 0$ ans: $-1/2\sqrt{7}$

15. $f(x) = 1/x$
    $a = 3$ ans: -1/9

16. $f(x) = \sqrt{x}$
    $a = 4$ ans: 8

17. $y = \dfrac{\sqrt{7}\,x}{\sqrt{1 - x^2}}$
    $a = 3$ ans: 7/64

18. $y = \dfrac{6}{1 - x}$
    $a = 4$ ans: 3/8

Sketch the inverses of each of the functions illustrated below.

7.

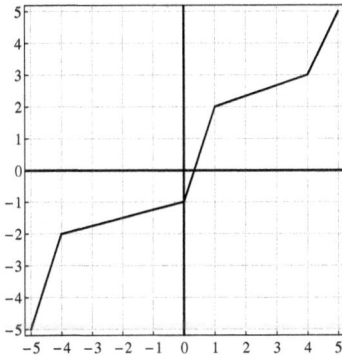

Chapter 15

Exponential and Logarithmic Functions

Chapter Summary and Goal

An **exponential function** is any function of the form $f(x) = a^x$ form some constant x. The corresponding inverse function $f^{-1}(x)$ is called **the logarithm to the base** a. In this chapter we will discuss the properties of exponential and logarithmic functions, the relationships between the two types of functions, and we will derive formulas for their derivatives.

Student Learning Objectives

The student will:

1. Perform basic algebra with exponential and logarithmic functions.
2. Understand the relationships between exponentials and logarithms.
3. Differentiate exponential and logarithmic functions.
4. Understand the distinction between different types of logarithms.
5. Solve problems using logarithmic differentiation.

Exponential Functions

An exponential function is any function of the form $y = a^x$, where a is any fixed constant. An example is $y = 3^x$.

To see that this definition makes sense, let's consider different classes of values in the domain, starting with integers. If x is an integer, it is reasonably easy to conceptualize y:

$$f(1) = a^1 = a \tag{15.1}$$

$$f(2) = a^2 = a \cdot a = a^2 \tag{15.2}$$

$$a(3) = a^3 = a \cdot a \cdot a = a^3 \tag{15.3}$$

$$\vdots$$

$$a(n) = a^n = \underbrace{a \cdot a \cdots a}_{n \text{ times}} \tag{15.4}$$

If x is rational, then there are integers p and q, with $q \neq 0$, such that $x = p/q$ (this is the definition of a rational number). So then we could write

$$f(x) = a^{p/q} = (a^p)^{1/q} = \sqrt[q]{a^p} \tag{15.5}$$

i.e., the qth root of a^p.

If x is irrational, we can use the following reasoning: every irrational number is very close to some rational number (within any small ϵ you might want to choose). Just pick some rational number that is as close as you like. You can extend this in an ϵ-δ fashion, like we did the definition of limits.[1]

Finally, we can extend this definition to include all real numbers. If $x < 0$ then we will use use the properties of exponents: take the reciprocal. For example,

$$f(-8) = a^{-8} = 1/a^8 \tag{15.6}$$

If $f(x) = 2^x$ then $f(-8) = 1/256$. Proving that the function is continuous is a bit more complicated, and you'll have to take my word for it that this is really true.[2]

Theorem 15.1. Continuity of Exponentials

Let $a > 0$ be any positive real number. Then the exponential function $f(x) = a^x$ is a continuous function with domain $-\infty < x < \infty$ and range $0 < x < \infty$.

Figure 15.1: The exponential function $y = a^x$ for $a > 1$ (solid curve) and $0 < a < 1$ (dashed curve).

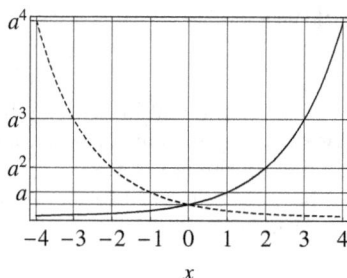

Some of the main properties of the exponential properties can be seen from figure 15.1. In particular, the x axis is always a horizontal asymptote, and $f(0) = 1$ (regardless of the value of a). The exponential function is strictly increasing when $a > 1$ (the solid curve in figure 15.1) and strictly decreasing when $0 < a < 1$ (the dashed curve in figure 15.1.) Furthermore,

$$\lim_{x \to \infty} = \begin{cases} \infty, & \text{if } a > 1 \\ 0, & \text{if } 0 < a < 1 \end{cases} \tag{15.7}$$

and

$$\lim_{x \to -\infty} = \begin{cases} 0, & \text{if } a > 1 \\ \infty, & \text{if } 0 < a < 1 \end{cases} \tag{15.8}$$

Other properties of the exponential function follow from the properties of exponents.

---

[1] But we won't do that, you'll just have take my word for it that it works
[2] Or look it up in a book on advanced calculus.

Theorem 15.2. Properties of Exponential Functions

Let $a > 0$ and $b > 0$ be real numbers. Then

$$a^{x+y} = a^x a^y \quad (15.9) \qquad a^{x-y} = a^x/a^y \quad (15.10)$$
$$(a^x)^y = a^{xy} \quad (15.11) \qquad (ab)^x = a^x b^x \quad (15.12)$$

The Derivative of the Exponential

Let $f(x) = a^x$. Our immediate goal is to find a formula for $f'(x)$. From the Newtonian definition of the derivative of a function (equation 9.1),

$$f'(x) = \frac{d}{dx}(a^x) = \lim_{h \to 0} \frac{a^{x+h} - a^x}{h} = \lim_{h \to 0} \frac{a^x a^h - a^x}{h} \tag{15.13}$$

$$= \lim_{h \to 0} a^x \left(\frac{a^h - 1}{h} \right) = a^x \lim_{h \to 0} \left(\frac{a^h - 1}{h} \right) \tag{15.14}$$

Similarly, from equation 1.13 for the derivative at a point (with $a = 0$ in (1.13)),

$$f'(0) = \lim_{h \to 0} \frac{f(0+h) - f(0)}{h} = \lim_{h \to 0} \frac{a^h - 1}{h} \tag{15.15}$$

Therefore

$$\frac{d}{dx}(a^x) = a^x f'(0) \tag{15.16}$$

where $f(x) = a^x$.

We do not yet know how to calculate $f'(0) = \lim_{h \to 0} \frac{a^h - 1}{h}$ using calculus.

Instead we will attempt to estimate it analytically at $x = 2$ and $x = 3$. Here are the calculations to four decimal places:

h	$(2^h - 1)/h$	h	$(3^h - 1)/h$
0.1	0.7177	0.1	1.612
0.01	0.6956	0.01	1.1047
0.001	0.6934	0.001	1.0992
0.0001	0.6932	0.0001	1.0987

This would suggest the following argument:

$$\text{as } h \to 0 \quad \begin{array}{ccccc} \dfrac{2^h - 1}{h} & < & \dfrac{X^h - 1}{h} & < & \dfrac{3^h - 1}{h} \\ \downarrow & & \downarrow & & \downarrow \\ \approx 0.69 & < & 1.0 & < & \approx 1.1 \end{array} \tag{15.17}$$

where X is some unknown number between 2 and 3. We call this unknown number e. We will define e as the number that satisfies the limit

$$\lim_{h \to 0} \frac{e^h - 1}{h} = 1 \tag{15.18}$$

The number e is irrational and has no repeating sequences, like π. It's approximate value to 50 digits is

$$e \approx 2.71828\ 18284\ 59045\ 23536\ 02874\ 71352\ 66249\ 77572\ 47093\ 69995 \tag{15.19}$$

Other names used for e include the **natural exponential**, **Euler's number**, and **Napier's constant**.

Let $y = e^x$. By equations 15.16 and 15.18,

$$y' = e^x \left[\frac{d}{dx} \left(e^x \right) \right]_{x=0} = e^x \lim_{h \to 0} \frac{e^h - 1}{h} = e^x \cdot 1 = e^x \tag{15.20}$$

Theorem 15.3. Derivative of Natural Exponential

The exponential is its own derivative:

$$\frac{d}{dx} e^x = e^x \tag{15.21}$$

The properties of the **natural exponential** $y = e^x$ are the same as those of the general exponential function $y = a^x$, where a is any positive real number. We repeat them here nevertheless for clarity.

Theorem 15.4. Properties of the Natural Exponential

$$e^{x+y} = e^x e^y \quad (15.22) \qquad e^{x-y} = e^x / e^y \quad (15.23)$$
$$(e^x)^y = e^{xy} \quad (15.24) \qquad e^x > 0 \text{ for all } x \quad (15.25)$$
$$\lim_{x \to -\infty} e^x = \infty \quad (15.26) \qquad \lim_{x \to \infty} e^x = \infty \quad (15.27)$$
$$\lim_{x \to -\infty} e^{-x} = 0 \quad (15.28) \qquad \lim_{x \to \infty} e^{-x} = 0 \quad (15.29)$$

Example 15.1. Find $\lim\limits_{x \to \infty} \dfrac{3e^{3x}}{7e^{3x} + 6}$.

Solution. When we plug ∞ in for x we get ∞/∞, which we can't evaluate.[3] Recall the method we used to find limits of ratios as they became large in chapter 5: we divide the numerator and denominator by the largest quantity in either. In this case, the largest term is e^{3x}. Then

$$\lim_{x \to \infty} \frac{3e^{3x}}{7e^{3x} + 6} = \lim_{x \to \infty} \frac{3e^{3x}}{7e^{3x} + 6} \cdot \frac{1/e^{3x}}{1/e^{3x}} = \lim_{x \to \infty} \frac{3}{7 + 6e^{-3x}} \to \frac{3}{7 + 0} = \frac{3}{7} \tag{15.30a}$$

In the next-to-last step we used the property that $e^{-3x} \to 0$ as $x \to \infty$.   □

Example 15.2. Find $\lim\limits_{x \to \infty} \dfrac{e^{5x} + 8}{7e^{12x} + 6e^{4x}}$.

Solution. This is also ∞/∞, so

$$\lim_{x \to \infty} \frac{e^{5x} + 8}{7e^{12x} + 6e^{4x}} = \lim_{x \to \infty} \frac{e^{5x} + 8}{7e^{12x} + 6e^{4x}} \cdot \frac{1/e^{5x}}{1/e^{5x}} \tag{15.31a}$$

$$= \lim_{x \to \infty} \frac{\dfrac{e^{5x}}{e^{5x}} + \dfrac{8}{e^{5x}}}{7\dfrac{e^{12x}}{e^{5x}} + 6\dfrac{e^{4x}}{e^{5x}}} \tag{15.31b}$$

$$= \lim_{x \to \infty} \frac{1 + 8e^{-5x}}{7e^{7x} + 6e^{-x}} \to \frac{1 + 0}{\infty + 0} \to 0 \tag{15.31c}$$

□

---

[3] We call this an "∞/∞ form."

Example 15.3. Find $f'(x)$ when $f(x) = xe^{-x^2}$.

Solution. By the product rule

$$f'(x) = \frac{d}{dx}\left(xe^{-x^2}\right) = x\frac{d}{dx}e^{-x^2} + e^{-x^2}\frac{dx}{dx}x\frac{d}{dx}e^{-x^2} + e^{-x^2} \qquad (15.32a)$$

Applying the chain rule to the second factor in the first term,

$$f'(x) = xe^{-x^2}\frac{d}{dx}(-x^2) + e^{-x^2} = -2x^2e^{-x^2} + e^{-x^2} = e^{-x^2}(1 - 2x^2) \qquad (15.32b)$$

$\square$

Example 15.4. Find y' if $e^{x/y} = x - y$.

Solution. This equation cannot be solved explicitly for y, so we use implicit differentiation. On the left side of the equation we apply the chain rule:

$$\frac{d}{dx}\left(e^{x/y}\right) = \frac{d}{dx}(x - y) \qquad (15.33a)$$

$$e^{x/y}\frac{d}{dx}\left(\frac{x}{y}\right) = 1 - y' \qquad (15.33b)$$

Expanding the derivative on the left-hand side using the quotient rule gives

$$e^{x/y} \cdot \frac{y \cdot 1 - x \cdot y'}{y^2} = 1 - y' \qquad (15.33c)$$

Multiplying the equation through by y^2 and distributing the $e^{x/y}$ on the left,

$$ye^{x/y} - xy'e^{x/y} = y^2 - y^2y' \qquad (15.33d)$$

Bringing all the terms that contain y' to the left and the terms that do not contain y' to the right side of the equation,

$$y^2y' - xy'e^{x/y} = y^2 - ye^{x/y} \qquad (15.33e)$$

Factoring a y' on the left and dividing by the factor $y^2 - xe^{x/y}$ we can solve for y',

$$y' = \frac{y^2 - ye^{x/y}}{y^2 - xe^{x/y}} \qquad (15.33f)$$

The exponential dependence can actually be removed from this result by recalling that $e^{x/y} = x - y$, so that

$$y' = \frac{y^2 - y(x - y)}{y^2 - x(x - y)} = \frac{2y^2 - xy}{y^2 - x^2 + xy} \qquad (15.33g)$$

$\square$

Definition 15.1. The exp Notation for Natural Exponential

The following are equivalent notations:

$$e^x = \exp(x) \qquad (15.34)$$

Sometimes a special notation is used to represent the natural exponential when there is a large formula formula in the exponent. This exp-notation arose historically because of the difficulty of type-setting and reading long expressions like the formula for the bivariate normal distribution function, which arises frequently in statistics

$$f(x,y) = \frac{1}{2\pi\sigma_x\sigma_y\sqrt{1-\rho_x}} e^{\left(-\frac{1}{2(1-\rho^2)}\left[\frac{(x-\mu_x)^2}{\sigma_x^2} + \frac{(x-\mu_y)^2}{\sigma_y^2} - \frac{2\rho(x-\mu_x)(y-\mu_y)}{\sigma_x\sigma_y}\right]\right)} \tag{15.35}$$

The alternative format is

$$f(x,y) = \frac{1}{2\pi\sigma_x\sigma_y\sqrt{1-\rho_x}} \times$$
$$\exp\left(-\frac{1}{2(1-\rho^2)}\left[\frac{(x-\mu_x)^2}{\sigma_x^2} + \frac{(x-\mu_y)^2}{\sigma_y^2} - \frac{2\rho(x-\mu_x)(y-\mu_y)}{\sigma_x\sigma_y}\right]\right) \tag{15.36}$$

Logarithms

Logarithms are the inverse of exponential functions (figure 15.2). The **natural logarithm** $y = \ln x$ is the inverse of the natural exponential $y = e^x$, and the **base-10 logarithm**, or simply the **logarithm** $y = \log x$, is the inverse of $y = 10^x$. The base-10 logarithm is sometimes called the **common logarithm**. The **base-a logarithm** $y = \log_a x$ is the inverse of $y = a^x$.

> **Definition 15.2. Logarithms**
>
> $$y = \ln x \quad \Longleftrightarrow \quad x = e^y \quad (15.37)$$
> $$y = \log x \quad \Longleftrightarrow \quad x = 10^y \quad (15.38)$$
> $$y = \log_a x \quad \Longleftrightarrow \quad x = a^y \quad (15.39)$$

Logarithms have a number of useful properties that are summarized in theorem 15.5. Among them are the property that multiplication is converted to addition and exponentiation is converted to multiplication. It turns out that these properties can considerably simplify the calculation of numerical quantities; historically, they they were used extensively in engineering and scientific calculations and formed the basis of the slide rule. We will discover later in this section that they are still useful in helping us find derivatives of functions that cannot otherwise be differentiated.

Figure 15.2: The logarithm is the inverse of the exponential function.

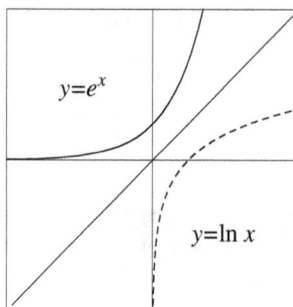

All of these properties follow from the properties of exponents and the definition of the logarithm as the inverse of the exponential. For example, consider the additivity property (15.49). We prove that in the following example.

Example 15.5. Prove that $\ln xy = \ln x + \ln y$.

Solution. From the properties of exponentials

$$e^{p+q} = e^p e^q \tag{15.40a}$$

Define $z = e^{p+q}$, $x = e^p$, and $y = e^q$. Then

$$z = xy \tag{15.40b}$$

Furthermore since $\ln x$ is the inverse of e^x,

$$z = e^{p+q} \implies p + q = \ln z \tag{15.40c}$$
$$x = e^p \implies p = \ln x \tag{15.40d}$$
$$y = e^q \implies q = \ln y \tag{15.40e}$$

Adding,

$$\ln xy = \ln z = p + q = \ln x + \ln y \tag{15.40f}$$

as required. $\qquad\square$

Theorem 15.5. Properties of Logarithms

$$\log_a a = 1 \tag{15.41} \qquad\qquad \ln e = 1 \tag{15.42}$$
$$\log_a 1 = 0 \tag{15.43} \qquad\qquad \ln 1 = 0 \tag{15.44}$$
$$\log_a(a^x) = x \tag{15.45} \qquad\qquad \ln(e^x) = x \tag{15.46}$$
$$a^{\log_a(x)} = x \tag{15.47} \qquad\qquad e^{\ln x} = x \tag{15.48}$$
$$\log_a(xy) = \log_a(x) + \log_a(y) \tag{15.49} \qquad \ln xy = \ln x + \ln y \tag{15.50}$$
$$\log_a\left(\frac{x}{y}\right) = \log_a(x) - \log_a(y) \tag{15.51} \qquad \ln\left(\frac{x}{y}\right) = \ln x - \ln y \tag{15.52}$$
$$\log_a(x^r) = r \log_a x \tag{15.53} \qquad\qquad \ln(x^r) = r \ln x \tag{15.54}$$
$$\lim_{x \to \infty} \log_a(x) = \infty \tag{15.55} \qquad\qquad \lim_{x \to \infty} \ln x = \infty \tag{15.56}$$
$$\lim_{x \to 0^+} \log_a(x) = -\infty \tag{15.57} \qquad\qquad \lim_{x \to 0^+} \ln x = -\infty \tag{15.58}$$

Example 15.6. Find x if $e^{12-4x} = 16$.

Solution. Taking the natural logarithm of both sides of the equation,

$$\ln\left(e^{12-4x}\right) = \ln 16 \tag{15.59a}$$

By the fundamental property of logarithms (see (15.46)), the ln and the exponential cancel on the left side of the equation. Writing $16 = 2^4$ on the right and using property (15.54) the right hand side is also reduced to $4\ln 2$.

$$12 - 4x = \ln(2^4) = 4\ln 2 \tag{15.59b}$$

Rearranging,

$$4x = 12 - 4\ln 2 \tag{15.59c}$$

Dividing by 4 gives $x = 3 - \ln 2$. $\qquad\square$

Change of Base and Generalized Exponential Derivative

The different types of logarithms are related to one another by the **change of base formula**. Suppose we know

$$z = \log_b x \tag{15.60}$$

and we want to find

$$y = \log_a x \tag{15.61}$$

How do we find y in terms of numbers we already know? We can first compute x from (15.61) as

$$x = a^y \tag{15.62}$$

and then taking the logarithm to the known base, b, and applying the properties of logarithms.

$$\log_b x = \log_b(a^y) = y \log_b a \tag{15.63}$$

Solving for y gives us the $y = log_a x$, the logarithm of the number in the new base, in terms of the logarithm of the number in the old base, and the log of the new base in the old base.

$$y = \frac{\log_b x}{\log_b a} \tag{15.64}$$

This formula is useful, for example, because we may have a $\log_b$ key on our calculator, but may need to know the logarithm in another base. Typically calculators only have buttons for log (base-10 logarithm) or ln (the natural logarithm. In some fields of engineering and science, other bases are particularly useful. For example, the base-2 logarithm is a measure of entropy in information theory.

Theorem 15.6. Change of Base Formula

For any positive numbers a and b,

$$\log_a x = \frac{\log_b x}{\log_b a} = \frac{\ln x}{\ln a} \tag{15.65}$$

Finally, we can use the properties of logarithms to derive a formula for the derivative of $y = a^x$. The "trick" is to write $a = e^{\ln a}$.

$$\frac{d}{dx}(a^x) = \frac{d}{dx}\left(\left(e^{\ln a}\right)^x\right) = \frac{d}{dx}\left(e^{x \ln a}\right) \tag{15.66}$$

$$= \left(e^{x \ln a}\right) \cdot \frac{d}{dx}(x \ln a) = \left(e^{\ln a^x}\right) \cdot \ln a \tag{15.67}$$

$$= a^x \ln a \tag{15.68}$$

We observe that when $a = e$ this reduces to theorem 15.3.

Theorem 15.7. Derivative of Generalized Exponential

$$\frac{d}{dx} a^x = a^x \ln a \tag{15.69}$$

Derivatives of Logarithms

To find a formula for the derivative of $y = \ln x$ we start by observing that an equivalent formula is $x = e^y$. Differentiating implicitly,

$$\frac{d}{dx}\left(e^y\right) = \frac{d}{dx}(x) \tag{15.70}$$

$$e^y y' = 1 \tag{15.71}$$

$$y' = \frac{1}{e^y} = \frac{1}{e^{\ln x}} = \frac{1}{x} \tag{15.72}$$

To get a formula for the derivative of the common logarithm (or the log to any other base) we begin with change of base formula (theorem 15.6),

$$\frac{d}{dx}\log_a x = \frac{d}{dx}\frac{\ln x}{\ln a} = \frac{1}{\ln a}\frac{d}{dx}\ln x = \frac{1}{x\ln a} \tag{15.73}$$

There is also a corresponding generalized chain rule for logarithms.

Theorem 15.8. Derivative of the Logarithm

$$\frac{d}{dx}\ln x = \frac{1}{x} \quad (15.74) \qquad \frac{d}{dx}\log_a x = \frac{1}{x\ln a} \quad (15.75)$$

Theorem 15.9. Chain Rule for Logarithms

$$\frac{d}{dx}\ln u = \frac{1}{u}\frac{du}{dx} \quad (15.76) \qquad \frac{d}{dx}\left(\ln g(x)\right) = \frac{g'(x)}{g(x)} \quad (15.77)$$

Example 15.7. Find y' for $y = \ln\sqrt{\tan x}$.

Solution. Using the generalized chain rule for logarithms,

$$y' = \frac{d}{dx}\ln\sqrt{\tan x} = \frac{1}{\sqrt{\tan x}}\frac{d}{dx}\sqrt{\tan x} \tag{15.78a}$$

Using the chain rule,

$$y' = \frac{1}{\sqrt{\tan x}}\cdot\frac{1}{2\sqrt{\tan x}}\frac{d}{dx}\tan x = \frac{\sec^2 x}{2\tan x} \tag{15.78b}$$

Alternatively, one could treat the square root as a power and use (15.54) before differentiating,

$$y' = \frac{d}{dx}\ln\sqrt{\tan x} = \frac{d}{dx}\left(\frac{1}{2}\ln\tan x\right) = \frac{1}{2\tan x}\cdot\sec^2 x \tag{15.78c}$$

As we see, either method leads to the same result.   □

Example 15.8. Find the equation of the line tangent to $y = \dfrac{1}{x^2}\ln\left(\dfrac{x}{3}\right)$ at $x = 3$.

Solution. At $x = 3$, $y = \left(\dfrac{1}{9}\right)\cdot(\ln 1) = 0$, so the line passes through the point $(3,\,0)$. To find the slope we differentiate,

$$y' = (x^{-2})'\ln\left(\frac{x}{3}\right) + x^{-2}\left(\ln\left(\frac{x}{3}\right)\right)' \tag{15.79a}$$

$$= -2x^{-3}\ln\left(\frac{x}{3}\right) + x^{-2}\cdot\frac{1}{x/3}\cdot\frac{1}{3} \tag{15.79b}$$

$$= -\frac{2}{x^3}\ln\left(\frac{x}{3}\right) + \frac{1}{x^3} \tag{15.79c}$$

At $x = 3$, $y' = m$,

$$m = -\frac{2}{3^3}\ln\left(\frac{3}{3}\right) + \frac{1}{3^3} = \frac{1}{27} \tag{15.79d}$$

Using the point-slope equation of a line,

$$y = y_1 + m(x - x_1) = 0 + \frac{1}{27}(x - 3) = \frac{1}{27}x - \frac{1}{9}. \tag{15.79e}$$

$\square$

Example 15.9. Show that $\lim\limits_{x\to 0}(1+x)^{1/x} = e$ and $\lim\limits_{n\to\infty}\left(1+\frac{1}{n}\right)^n = e$.

Solution. To Show the first limit we let $f(x) = \ln x$. Then $f'(x) = 1/x$ and $f'(1) = 1$. Hence from the definition of a derivative,

$$1 = \lim_{h\to 0}\frac{f(1+h) - f(1)}{h} = \lim_{h\to 0}\frac{\ln(1+h) - \ln 1}{h} \tag{15.80a}$$

$$= \lim_{h\to 0}\frac{\ln(1+h)}{h} = \lim_{h\to 0}\left(\frac{1}{h}\ln(1+h)\right) \tag{15.80b}$$

$$= \lim_{h\to 0}\left(\ln(1+h)^{1/h}\right) \tag{15.80c}$$

$$= \ln\left(\lim_{h\to 0}(1+h)^{1/h}\right) \tag{15.80d}$$

Changing the variable h to x and exponentiating both sides of the equation gives the first limit.

$$e = \exp\left(\ln\left(\lim_{x\to 0}(1+x)^{1/x}\right)\right) = \lim_{x\to 0}(1+x)^{1/x} \tag{15.80e}$$

To get the second limit, let $x = 1/n$. When $x \to 0$ then $n \to \infty$ so

$$e = \lim_{n\to\infty}\left(1+\frac{1}{n}\right)^n \tag{15.80f}$$

$\square$

Theorem 15.10. e as a Limit

$$\lim_{x\to 0}(1+x)^{1/x} = e = \lim_{n\to\infty}\left(1+\frac{1}{n}\right)^n \tag{15.81}$$

Logarithmic Differentiation

Sometimes there are functions that we just can't differentiate without using logarithms. We have already seen an example of this with the derivative of the derivative of $y = a^x$ (theorem 15.7). In other situations solving for the derivative will require lengthy and repeated applications of the product and/or quotient rules. By converting to logarithmic space we can convert all of our multiplications and divisions to additions and subtractions, which can sometimes make the calculation of the derivative much less tedious.

Example 15.10. Find $\dfrac{d}{dx} x^x$.

Solution. Let $y = x^x$ and solve for y' using logarithmic differentiation. First, we take natural logarithms of both sides of the equation.

$$\ln y = \ln(x^x) = x \ln x \qquad (15.82a)$$

Then we differentiate both sides of the equation with respect to x. On the left hand side, we use the chain rule; on the right hand side, the product rule.

$$\frac{y'}{y} = x \cdot (\ln x)' + (x)' \cdot \ln x = \frac{x}{x} + \ln x = 1 + \ln x \qquad (15.82b)$$

Multiplying through by y and making a final substitution of $y = x^x$ (from the original step of the problem), we achieve our result.

$$y' = y(1 + \ln x) = x^x(1 + \ln x). \qquad (15.82c)$$

$\square$

Example 15.11. Find y' if $x^y = y^x$.

Solution. Using logarithmic differentiation as before, we begin by taking logarithms of both sides of the equation.

$$\ln(x^y) = \ln(y^x) \qquad (15.83a)$$

First we apply (15.54) to both sides of the equation:

$$y \ln x = x \ln y \qquad (15.83b)$$

and then we differentiate, using the product rule on each side:

$$(y \ln x)' = (x \ln y)' \qquad (15.83c)$$
$$y' \cdot (\ln x) + y \cdot (\ln x)' = (x)' \cdot (\ln y) + x \cdot (\ln y)' \qquad (15.83d)$$

On the right hand side we use the chain rule for $(\ln y)' = y'/y$.

$$y' \ln x + \frac{y}{x} = \ln y + \frac{xy'}{y} \qquad (15.83e)$$

The remainder of the exercise is algebra. Bring all the terms that depend on y' to the left hand side of the equation, and move all the remaining terms to the right:

$$y' \ln x - \frac{xy'}{y} = \ln y - \frac{y}{x} \qquad (15.83f)$$

Factoring out the common factor of y' and dividing by the common $(\ln x - x/y)$ gives

$$y'(\ln x - x/y) = \ln y - y/x \qquad (15.83g)$$
$$y' = \frac{\ln y - y/x}{\ln x - x/y} \qquad (15.83h)$$

□

Logarithms are often useful in practice because they convert multiplication into addition and division into subtraction. In logarithmic differentiation this means we can avoid using the product and quotient rule in favor of the sum and difference rule. In the worst case they turn exponents into multiplication, as we saw in the previous example.

Example 15.12. Find y' for $y = \sqrt{\dfrac{x-1}{x+1}}$.

Solution. Using the properties of logarithms,

$$\ln y = \ln \sqrt{\frac{x-1}{x+1}} = \frac{1}{2} \ln \frac{x-1}{x+1} = \frac{1}{2}\left(\ln(x-1) - \ln(x+1) \right) \tag{15.84a}$$

Differentiating,

$$2 \cdot \frac{y'}{y} = \frac{1}{x-1} - \frac{1}{x+1} = \frac{(x+1) - (x-1)}{(x+1)(x-1)} = \frac{2}{(x+1)(x-1)} \tag{15.84b}$$

$$y' = \frac{y}{(x+1)(x-1)} = \frac{(x-1)^{1/2}}{(x+1)^{1/2}} \cdot \frac{1}{(x+1)(x-1)} \tag{15.84c}$$

$$= \frac{1}{(x+1)^{3/2}(x-1)^{1/2}} \tag{15.84d}$$

□

Example 15.13. Find y' for $y = \dfrac{x^{7/8}\sqrt{\tan^2 x + \cos^2 x}}{(x^8 + 18)^{28}}$.

Solution. The process begins by taking logarithms.

$$\ln y = \ln \left(\frac{x^{7/8}\sqrt{\tan^2 x + \cos^2 x}}{(x^8 + 18)^{28}} \right) \tag{15.85a}$$

$$= \ln \left(x^{7/8} \right) + \ln \left(\sqrt{\tan^2 x + \cos^2 x} \right) - \ln \left((x^8 + 18)^{28} \right) \tag{15.85b}$$

$$= \frac{7}{8} \ln x + \frac{1}{2} \ln \left(\tan^2 x + \cos^2 x \right) - 28 \ln \left(x^8 + 18 \right) \tag{15.85c}$$

Differentiating,

$$\frac{y'}{y} = \frac{7}{8x} + \frac{1}{2} \cdot \frac{2\tan x \sec^2 x - 2 \cos x \sin x}{\tan^2 x + \cos^2 x} - \frac{28(8x^7)}{x^8 + 18} \tag{15.85d}$$

$$y' = y \cdot \left(\frac{7}{8x} + \frac{\tan x \sec^2 x - \cos x \sin x}{\tan^2 x + \cos^2 x} - \frac{224x^7}{x^8 + 18} \right) \tag{15.85e}$$

$$= \left(\frac{x^{7/8}\sqrt{\tan^2 x + \cos^2 x}}{(x^8 + 18)^{28}} \right) \cdot$$

$$\left(\frac{7}{8x} + \frac{\tan x \sec^2 x - \cos x \sin x}{\tan^2 x + \cos^2 x} - \frac{224x^7}{x^8 + 18} \right) \tag{15.85f}$$

□

The Slide Rule

Slide rules were invented in the 17th as the concept of the logarithm slowly disseminated through the scientific community. Although John Napier (1550-1617) is generally credited with the invention of logarithms, the slide rule is most often attributed to his contemporary on Edmund Gunter (1581-1626), who was also credited with the invention of several other nautical calculating devices.

The slide rule is composed two scales like rulers with markings that are measured logarithmically rather than linearly. By sliding the two scales alongside one-another, very rapid multiplications and divisions could be performed. An experienced user can multiply numbers to three significant figures using a foot-long slide rule.

Figure 15.3: Example of slide rule showing how to multiply 2×4. The distance from the 1 to the 8 on the C scale is at a distance $r \log 8$ from the left end of the scale, where r is the total distace (e.g., in millimeters) from the 1 to the 10). Similarly, the distance from the 1 to the 4 on the D scale is a distance $r \log 4$, and the distance from the 1 to the 2 on the C scale is a distance $r \log 2$.

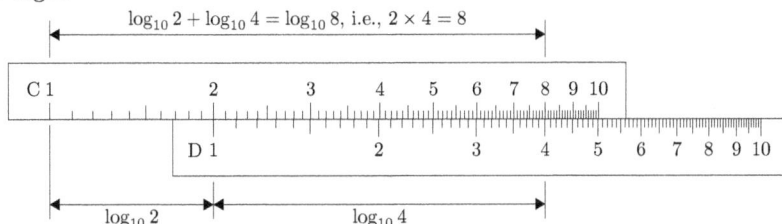

Multiplication proceeds by placing the number 1 on the D scale next to the first operand on the C scale, and reading the answer on the C scale adjacent to the second operand on the D scale; see figure 15.3. Division proceeds in the opposite direction: place the denominator on the D scale adjacent to the numerator on the C scale. The answer is on the C scale, adjacent to the 1 on the D scale.

Most slide rules have been replaced by electronic calculators; up until the early 1970's a slide rule – or *slipstick* as it was known in those days – was part of the standard engineering and science student's repertoire. Why you don't see them much they are still around. They are sometimes still kept as backups on board ships in the event of power loss, or in places where high radiation exposure may damage key electronics.

Hyperbolic Functions

The hyperbolic functions are to a hyperbolic central angle what the circular functions are to a circular central angle (fig. 15.5). While the circular functions are identical to the trigonometric functions (e.g., the circular $\sin \theta$ and the trigonometric $\sin \theta$ are identical) the hyperbolic functions are not identical to their trigonometric counterparts, and thus they have the letter "h" appended to their abbreviation.

Figure 15.4: Detail of a Log Log Duplex Decitrig slide rule model N4081-3, manufactured by Keuffel & Esser in 1948. This was one of the most popular engineering models sold in North America and various versions of it were manufactured from 1937 to 1972.

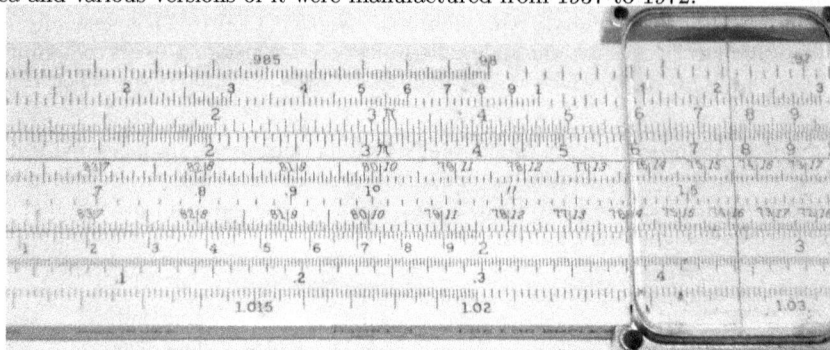

Definition 15.3. Hyperbolic Functions

$$\cosh x = \frac{1}{2}\left(e^x + e^{-x}\right) \quad (15.86) \qquad \sinh x = \frac{1}{2}\left(e^x - e^{-x}\right) \quad (15.87)$$

$$\operatorname{sech} x = \frac{1}{\cosh x} \quad (15.88) \qquad \operatorname{csch} x = \frac{1}{\sinh x} \quad (15.89)$$

$$\tanh x = \frac{\sinh x}{\cosh x} \quad (15.90) \qquad \coth x = \frac{\cosh x}{\sinh x} \quad (15.91)$$

Figure 15.5: Definition of hyperbolic angle on the unit hyperbola $x^2 - y^2 = 1$.

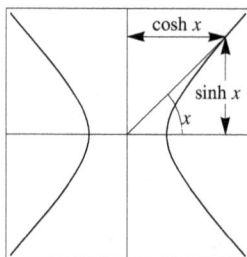

The analogous Pythagorean identities for hyperbolic functions are given in theorem 15.11.

Theorem 15.11. Hyperbolic Pythagorean Identities

$$\cosh^2 x - \sinh^2 x = 1 \tag{15.92}$$

$$\operatorname{sech}^2 x = 1 - \tanh^2 x \tag{15.93}$$

$$\operatorname{csch}^2 x = \coth^2 x - 1 \tag{15.94}$$

The derivatives of the hyperbolic functions can be found by applying the rules for the derivatives of $y = e^u$ and are given in theorem 15.12.

Theorem 15.12. Derivatives of Hyperbolic Functions

$$\frac{d}{dx}\sinh x = \cosh x \qquad (15.95)$$
$$\frac{d}{dx}\cosh x = \sinh x \qquad (15.96)$$
$$\frac{d}{dx}\tanh x = \operatorname{sech}^2 x \qquad (15.97)$$
$$\frac{d}{dx}\coth x = -\operatorname{csch}^2 x \qquad (15.98)$$
$$\frac{d}{dx}\operatorname{sech} x = -\tanh x \operatorname{sech} x \qquad (15.99)$$
$$\frac{d}{dx}\operatorname{csch} x = -\coth x \operatorname{csch} x \qquad (15.100)$$

Example 15.14. Find the derivative of $y = \tanh(x^3)$.

Solution. Using the chain rule,

$$\frac{d}{dx}\tanh(x^3) = \operatorname{sech}^2(x^3)\frac{d}{dx}x^3 = 3x^2\operatorname{sech}^2(x^3) \qquad (15.101a)$$

$\square$

The inverse hyperbolic functions have the word "arc" in front of them (e.g., arccosh, arcsinh, etc.) or use the inverse symbol ($\cosh^{-1} x$, $\sinh^{-1} x$, etc.). The derivatives are shown in theorem 15.13.

Example 15.15. Find y' for $y = \cosh^{-1} x$.

Solution. The method is similar to the method we used to find the derivative of the inverse trigonometric functions. If $y = \cosh^{-1} x$ then $x = \cosh y$. Thus

$$(\cosh y)' = x' \qquad (15.102a)$$
$$(\sinh y)y' = 1 \qquad (15.102b)$$
$$y' = \frac{1}{\sinh y} \qquad (15.102c)$$

Using the identity $\cosh^2 y - \sinh^2 y = 1$ (equation 15.92), along with $x = \cosh y$ gives

$$x^2 - \sinh^2 y = 1 \qquad (15.102d)$$

Solving for $\sinh y$,

$$\sinh y = \sqrt{x^2 - 1} \qquad (15.102e)$$

gives

$$\frac{d}{dx}\cosh^{-1} x = y' = \frac{1}{\sinh y} = \frac{1}{\sqrt{x^2 - 1}} \qquad (15.102f)$$

which agrees with the formula given in theorem 15.13. $\square$

Theorem 15.13. Derivatives of Inverse Hyperbolic Functions

$$\frac{d}{dx}\sinh^{-1} x = \frac{1}{\sqrt{x^2 + 1}} \qquad (15.103)$$
$$\frac{d}{dx}\cosh^{-1} x = \frac{1}{\sqrt{x^2 - 1}} \qquad (15.104)$$
$$\frac{d}{dx}\tanh^{-1} x = \frac{1}{1 - x^2} \qquad (15.105)$$
$$\frac{d}{dx}\coth^{-1} x = -\frac{1}{1 - x^2} \qquad (15.106)$$
$$\frac{d}{dx}\operatorname{sech}^{-1} x = -\frac{1}{x\sqrt{1 - x^2}} \qquad (15.107)$$
$$\frac{d}{dx}\operatorname{csch}^{-1} x = -\frac{1}{|x|\sqrt{1 + x^2}} \qquad (15.108)$$

Exercises

Solve for x in exercises 1 through 5.

1. $\ln x + \ln(x+2) = 12$

2. $2^{x+4} = 6$

3. $3^{5x+7} = 9$

4. $e^{2x} + 4e^x + 4 = 0$

5. $e^{e^x} = 9$

Find the inverse of each function in exercises 6 through 9.

6. $y = \ln(x+5)$

7. $y = e^{x^2}$

8. $\dfrac{e^x}{1-3e^x}$

9. $y = (\ln x)^3$

Find y' for each of the functions in exercises 10 through 39.

10. $y = -3e^{x\sin x}$

11. $y = x^2 e^x$

12. $y = 5\sqrt{x}\ln x$

13. $y = 8^{-3x^2}$

14. $y = x^{4\sin(5x)}$

15. $y = 3\sin^{-1}(x^4)$

16. $y = \ln(\ln x)$

17. $y = x^4 \tan^{-1} 4x$

18. $y = \cosh\sqrt{x}$

19. $y = \tanh(\sin x)$

20. $y = \sin^{-1}(e^{7x})$

21. $y = \dfrac{x^3(x-9)^7}{(x^2+2)^2}$

22. $y = \ln\sqrt{\dfrac{8x+4}{3x-7}}$

23. $y = \cot^{-1}\sqrt{\dfrac{2-x}{2+x}}$

24. $y = (\tan^{-1}x)^2$

25. $y = \tan^{-1}(x^2)$

26. $y = \ln(x + e^x)$

27. $y = \ln(x + 7x^2)$

28. $y = \dfrac{x + \ln x}{x^2 + 1}$

29. $y = \sqrt{1 + \ln x}$

30. $y = x^2 + \ln(3x)$

31. $y = \dfrac{\ln x}{x^2}$

32. $y = \ln(\sec x + \tan x)$

33. $y = (2 + \sin^{-1}x)^{1/4}$

34. $y = x^{\sin x}$

35. $y = x^{(x^x)}$

36. $y = (x^x)^x$

37. $y = (\tan x)^x$

38. $y = (x^2 + x)^x$

39. $y = x^{1/x}$

In exercises 40 through 42 find the slope of the given functions at the specified point.

40. $f'(3)$ for $f(x) = 8\log_2 x$

41. $f'(2)$ for $f(x) = 2x^{2x}$

42. $f'(1)$ for $f(x) = 5(\sin x)^x$

In exercises 43 through 47 find the equation of the tangent line to the given curve at the specified point.

43. $f(x) = xe^x$ at $(2, f(2))$

44. $xe^y + ye^x = 1$ at $(0,1)$

45. $y = \dfrac{e^{2x}}{8x-5}$ at $(2, f(2))$

46. $\ln(x+y) = x^2 - y - 55$ at $(7,-6)$

47. $y = \ln(2x^3 - 53)$ at $(3,0)$

Verify the derivatives of each of the hyperbolic functions in theorem 15.12.

48. $y = \cosh x$

49. $y = \sinh x$

50. $y = \tanh x$

51. $y = \operatorname{sech} x$

52. $y = \operatorname{csch} x$

53. $y = \coth x$

Verify the derivatives of each of the remaining inverse hyperbolic functions in theorem 15.13.

54. $y = \sinh^{-1} x$

55. $y = \tanh^{-1} x$

56. $y = \operatorname{sech}^{-1} x$

57. $y = \operatorname{csch}^{-1} x$

58. $= \coth^{-1} x$

59. Verify the hyperbolic Pythagorean identity $\cosh^x - \sinh^2 x = 1$

60. Show that $\sinh^{-1} x = \ln\left(x + \sqrt{x^2 + 1}\right)$. Hint: Let $y = \sinh^{-1}x$. Solve for x. Find a quadratic equation in terms of e^y and solve it for e^y. Then take the natural logarithm.

61. Show that $\cosh^{-1} x = \ln\left(x + \sqrt{x^2 - 1}\right)$ (for $x \geq 1$). (See in the hint in exercise 60.

62. Show that $\tanh^{-1} x = \dfrac{1}{2}\ln\left(\dfrac{1+x}{1-x}\right)$ (for $-1 < x < 1$). See the hint in exercise 60.

63. Prove the power rule: if n is any real number then $(y^n)' = nx^{n-1}$. Hint: Let $x^n = e^{n\ln x}$.

Chapter 16

Indeterminate Limits

Chapter Summary and Goal

An **indeterminate form** is a limit that appears to approach $\frac{0}{0}$, $\frac{\infty}{\infty}$, $\infty - \infty$, $\infty \cdot 0$, ∞^0 or 1^∞ when substitution is used. One example is the ratio of two functions that are both clearly approaching zero as $x \to 0$, such as

$$\lim_{x \to 0} \frac{2^x - 1}{x} \tag{16.1}$$

(see, for example, equation 15.14). The substitution rule (theorem 3.9) does not actually apply in this case, because the denominator is zero at $x = 0$, but the ratio certainly *appears* to approach a value of $0/0$ and hence we refer to it as a '0/0 form.' Other examples include the products

$$\lim_{x \to 0^+} x^x \quad \text{and} \quad \lim_{x \to \infty} x^2 e^{-x} \tag{16.2}$$

These limits appear to be approaching 0^0 (in the first case) and ∞^0 (in the second). It is not at all obvious how to interpret these concepts; in fact, these are not the correct answers at all, but because the limits *appear* to approaching these values we call them '0^0 forms' and '∞^0' forms respectively.

In this chapter we will learn a valuable tool that allows us to solve indeterminate limits of the '0/0' and '∞/∞' forms, known as **L'Hôpital's Rule**. We will then be able to combine this tool with logarithmic transformation[1] to solve other types of indeterminate limits.

Student Learning Objectives

The student will:

1. Recognize when L'Hôpital's Rule can be applied.
2. Use L'Hôpital's Rule to solve $\frac{0}{0}$ and $\frac{\infty}{\infty}$ forms.
3. Solve indeterminate limits using logarithmic techniques.
4. Solve indeterminate limits of the $\infty \cdot 0$, ∞^0, 1^∞ and $(\infty - \infty)$ forms.

---

[1] i.e., a transformation of the equation into logarithmic space by taking the ln of both sides of the equation.

Indeterminate Ratios: L'Hôpital's Rule

L'Hôpital's[2] Rule[3] uses derivatives to simplify the evaluation of indeterminate limits when direct substitution indicates they approach either $\frac{0}{0}$ or $\frac{\infty}{\infty}$. To see why the rule works in the $\frac{0}{0}$ case, suppose that $f(x)$ and $g(x)$ are continuously differentiable (this means that both the functions and their derivatives are continuous) in some neighborhood of $x = a$ where $f(a) = g(a) = 0$ and that $g'(a) \neq 0$. Because $f(a) = g(a) = 0$, then by substitution

$$\lim_{x \to a} \frac{f(x)}{g(x)} \to \frac{f(a)}{g(a)} \to \frac{0}{0} = \text{indeterminate} \qquad (16.3)$$

However,

$$\lim_{x \to a} \frac{f(x)}{g(x)} = \lim_{x \to a} \frac{f(x) - 0}{g(x) - 0} \qquad (16.4)$$

$$= \lim_{x \to a} \frac{f(x) - f(a)}{g(x) - g(a)} \qquad \text{because } f(a) = g(a) = 0 \qquad (16.5)$$

$$= \lim_{x \to a} \frac{f(x) - f(a)}{g(x) - g(a)} \cdot \frac{\frac{1}{x-a}}{\frac{1}{x-a}} \qquad \text{multiply by 1} \qquad (16.6)$$

$$= \lim_{x \to a} \frac{\left(\frac{f(x)-f(a)}{x-a}\right)}{\left(\frac{g(x)-g(a)}{x-a}\right)} \qquad (16.7)$$

$$= \frac{\lim_{x \to a} \left(\frac{f(x)-f(a)}{x-a}\right)}{\lim_{x \to a} \left(\frac{g(x)-g(a)}{x-a}\right)} \qquad \text{theorem 3.6} \qquad (16.8)$$

$$= \frac{f'(a)}{g'(a)} \qquad \text{definition of } f'(a), g'(a); g'(a) \neq 0 \qquad (16.9)$$

$$= \lim_{x \to a} \frac{f'(x)}{g'(x)}. \qquad \text{continuity of } f'(x), g'(x) \text{ at } x = a \qquad (16.10)$$

A similar result can be derived for infinite limits, but the proof is more difficult.

Theorem 16.1. L'Hôpital's Rule

Suppose that $f(x)$ and $g(x)$ are continuously differentiable in an interval about $x = a$, and that $g'(a) \neq 0$. If either of the following conditions hold:

1. $\lim_{x \to a} f(x) = 0$ and $\lim_{x \to a} g(x) = 0$; or
2. $\lim_{x \to a} f(x) = \pm\infty$ and $\lim_{x \to a} g(x) = \pm\infty$

then

$$\lim_{x \to a} \frac{f(x)}{g(x)} = \lim_{x \to a} \frac{f'(x)}{g'(x)} \qquad (16.11)$$

---

[2] The spelling of L'Hôpital's name (for Guillaume Francois Antoine, the marquis de l'Hôpital (1661 - 1704)) is often confusing and erroneous when printed in English language literature. During L'Hôpital's life, it was spelled L'Hospital. In 1740 the Académie Française introduced the circumflex (ô) to indicate that a silent letter had been removed. While including the s and omitting the circumflex conforms to the original spelling, this tends to induce English speakers to incorrectly call L'Hôpital's Rule the "hospital" rule. Here we conform to the modern French.

[3] Originally developed by Johann Bernoulli (1667-1748). In exchange for an annual payment of 300 Francs, Bernoulli agreed in 1694 to provide all of his discoveries to the wealthier marquis de l'Hôpital rather than publish them elsewhere. The rule was published by l'Hôpital in the first calculus textbook, *Analyse des Infiniment Petits pour l'Intelligence des Lignes Courbes* (Infinitesimal calculus with applications to curved lines) in 1696.

We will use the symbol $\overset{H}{=}$ in place of the usual equal sign to indicate that we have made the substitution

$$\lim_{x \to a} \frac{f(x)}{g(x)} \overset{H}{=} \lim_{x \to a} \frac{f'(x)}{g'(x)} \tag{16.12}$$

Example 16.1. Find $\lim\limits_{x \to \infty} \dfrac{\ln x}{x^2}$.

Solution. Since $\lim\limits_{x \to \infty} \dfrac{\ln x}{x^2} \to \dfrac{\infty}{\infty}$ we can use L'Hôpital's Rule. Therefore

$$\lim_{x \to \infty} \frac{\ln x}{x^2} \overset{H}{=} \lim_{x \to \infty} \frac{1/x}{2x} = \lim_{x \to \infty} \frac{1}{2x^2} \to 0. \tag{16.13a}$$

$\square$

Example 16.2. Find $\lim\limits_{x \to 1} \dfrac{\ln x}{x - 1}$.

Solution. Since $\lim\limits_{x \to 1} \dfrac{\ln x}{x - 1} \to \dfrac{0}{0}$ we can use L'Hôpital's Rule. Therefore

$$\lim_{x \to 1} \frac{\ln x}{x - 1} \overset{H}{=} \lim_{x \to 1} \frac{1/x}{1} \to 1. \tag{16.14a}$$

$\square$

Example 16.3. Find $\lim\limits_{x \to \infty} \dfrac{x^2}{3x^2 + 1}$

Solution. Although we already know from theorem 5.1 that the limit should be 1/3, we will nevertheless verify it using L'Hôpital's Rule. This is an $\frac{\infty}{\infty}$ form and so we are allowed to use the rule.

$$\lim_{x \to \infty} \frac{x^2}{3x^2 + 1} \overset{H}{=} \lim_{x \to \infty} \frac{2x}{6x} = \lim_{x \to \infty} \frac{1}{3} \to \frac{1}{3} \tag{16.15a}$$

as expected. $\square$

Example 16.4. Find $\lim\limits_{x \to 0} \dfrac{e^x - 1}{x}$.

Solution. This is a $\frac{0}{0}$ form. Thus

$$\lim_{x \to 0} \frac{e^x - 1}{x} \overset{H}{=} \lim_{x \to 0} \frac{e^x}{1} \to e^0 = 1. \tag{16.16a}$$

$\square$

Example 16.5. Find $\lim\limits_{x \to \infty} \dfrac{e^x}{x}$

Solution. This is an $\frac{\infty}{\infty}$ form. Thus

$$\lim_{x \to \infty} \frac{e^x}{x} \overset{H}{=} \lim_{x \to \infty} \frac{e^x}{1} \to \infty. \tag{16.17a}$$

$\square$

Example 16.6. Find $\lim\limits_{x \to 1} \dfrac{x \ln x}{x^2 - 1}$.

Solution. This is a $\frac{0}{0}$ form. Using L'Hôpital's Rule,

$$\lim_{x \to 1} \frac{x \ln x}{x^2 - 1} \overset{H}{=} \lim_{x \to 1} \frac{1 + \ln x}{2x} \to \frac{1 + \ln 1}{2 \cdot 1} = \frac{1}{2}. \tag{16.18a}$$

$\square$

Example 16.7. Find $\lim\limits_{x\to\infty} \dfrac{x^3}{e^{x^2}}$.

Solution. Since $\dfrac{x^3}{e^{x^2}} \to \dfrac{\infty}{\infty}$ we can use L'Hôpital's Rule,

$$\lim_{x\to\infty} \frac{x^3}{e^{x^2}} \stackrel{H}{=} \lim_{x\to\infty} \frac{3x^2}{2xe^{x^2}} = \lim_{x\to\infty} \frac{3x}{2e^{x^2}} \to \frac{\infty}{\infty}. \tag{16.19a}$$

We can use L'Hôpital's Rule a second time,

$$\lim_{x\to\infty} \frac{x^3}{e^{x^2}} \stackrel{H}{=} \lim_{x\to\infty} \frac{3}{4xe^{x^2}} \to \frac{3}{\infty} = 0. \tag{16.19b}$$

$\square$

Indeterminate Products

Typical **indeterminate products** are limits of the form $0 \cdot \infty$, such as

$$\lim_{x\to 0^+} x\ln x \tag{16.20}$$

This limit is an indeterminate product because $x \to 0$ and $\ln x \to -\infty$. The trick is to turn it into an indeterminate ratio of the form $\frac{0}{0}$ or $\frac{\infty}{\infty}$ by taking the reciprocal of one of the factors, i.e., use a rewrite rule such as

$$\lim_{x\to a} f(x) \cdot g(x) \to \lim_{x\to a} \frac{f(x)}{1/g(x)} \tag{16.21}$$

Example 16.8. Find $\lim\limits_{x\to 0^+} x\ln x$.

Solution. Since this is a $0 \cdot \infty$ form, it can be changed to an $\frac{\infty}{\infty}$ form by replacing the 0 factor with $1/$(its reciprocal), as follows:

$$\lim_{x\to 0^+} x\ln x = \lim_{x\to 0^+} \frac{\ln x}{1/x} \to \frac{-\infty}{\infty} \tag{16.22a}$$

$$\stackrel{H}{=} \lim_{x\to 0^+} \frac{1/x}{-1/x^2} = \lim_{x\to 0^+} (-x) = 0. \tag{16.22b}$$

$\square$

Example 16.9. Find $\lim\limits_{x\to \pi/2} (\pi/2 - x)\sec x$.

Solution. This is a $0 \cdot \infty$ form. We can write $\sec x = 1/(cos x)$ to make it a $\frac{0}{0}$ form.

$$\lim_{x\to \pi/2} = \lim_{x\to \pi/2} \frac{\pi/2 - x}{\cos x} \to \frac{0}{0} \tag{16.23a}$$

$$\stackrel{H}{=} \lim_{x\to \pi/2} \frac{-1}{-\sin x} = 1. \tag{16.23b}$$

$\square$

Indeterminate Powers

Typical **indeterminate** powers are 0^0, ∞^0, and 1^∞ forms. These can all be handled using the natural logarithm, sometimes following by L'Hôpital's Rule, and are illustrated in the following examples.

Example 16.10. Find $\lim_{x \to 0^+} x^x$.

Solution. This is a 0^0 form. If we let

$$L = \lim_{x \to 0^+} x^x \tag{16.24a}$$

and then take the natural logarithm of both sides of the equation,

$$\ln L = \ln \lim_{x \to 0^+} x^x = \lim_{x \to 0^+} (\ln x^x) = \lim_{x \to 0^+} (x \ln x) = 0 \tag{16.24b}$$

where the last step follows from example 16.8. Thus since $\ln L = 0$, $L = e^0 = 1$, i.e., $\lim_{x \to 0^+} x^x = 1$. □

Example 16.11. Find $\lim_{x \to \infty} (\ln x)^{1/x}$.

Solution. This is an ∞^0 form so we first define

$$L = \lim_{x \to \infty} \left((\ln x)^{1/x} \right) \tag{16.25a}$$

and then take natural logarithms of both sides of the equation.

$$\ln L = \ln \lim_{x \to \infty} \left((\ln x)^{1/x} \right) = \lim_{x \to \infty} \ln \left((\ln x)^{1/x} \right) \tag{16.25b}$$

$$= \lim_{x \to \infty} \frac{\ln(\ln x)}{x} \to \frac{\infty}{\infty} \tag{16.25c}$$

$$\overset{H}{=} \lim_{x \to \infty} \frac{\frac{1}{\ln x} \cdot \frac{1}{x}}{1} = 0 \tag{16.25d}$$

Exponentiating both sides of the equation,

$$L = e^{\ln L} = e^0 = 1 \tag{16.25e}$$

Hence $\lim_{x \to \infty} \left((\ln x)^{1/x} \right) = 1$. □

Indeterminate Differences

Indeterminate Differences of the $\infty - \infty$ form typically have to be handled on a case-by-case basis depending on the nature of the problem. Usually they require some sort of algebraic manipulation to reduce them to one of the other forms.

Example 16.12. Find $\lim_{x \to \infty} (x^2 - \ln x)$.

Solution. Since both $x^2 \to \infty$ and $\ln x \to \infty$, this is an $\infty - \infty$ form. We can turn it into a product by factoring the the x^2:

$$\lim_{x \to \infty} (x^2 - \ln x) = \lim_{x \to \infty} x^2 \left(1 - \frac{\ln x}{x^2} \right) \tag{16.26a}$$

From example 16.1, $((\ln x)/x^2) \to 0$ and thus the term in parenthesis approaches 1 as $x \to \infty$. Thus the limit is ∞. □

Example 16.13. Find $\lim\limits_{x \to 1} \left(\dfrac{x}{x-1} - \dfrac{1}{\ln x} \right)$.

Solution. This has the form

$$\lim_{x \to 1} \left(\frac{x}{x-1} - \frac{1}{\ln x} \right) \to \left(\frac{1}{1-1} - \frac{1}{0} \right) \to (\infty - \infty) \qquad (16.27\mathrm{a})$$

so it is an indeterminate difference. We begin by putting everything over a common denominator and then find we can apply L'Hôpital's rule twice.

$$\lim_{x \to 1} \left(\frac{x}{x-1} - \frac{1}{\ln x} \right) = \lim_{x \to 1} \frac{x \ln x - x + 1}{(x-1)\ln x} \to \frac{0}{0} \qquad (16.27\mathrm{b})$$

$$\overset{H}{=} \lim_{x \to 1} \frac{x \cdot (1/x) + \ln x - 1}{(x-1) \cdot \frac{1}{x} + \ln x} \qquad (16.27\mathrm{c})$$

$$= \lim_{x \to 1} \frac{\ln x}{1 + \ln x - (1/x)} \to \frac{0}{0} \qquad (16.27\mathrm{d})$$

$$\overset{H}{=} \lim_{x \to 1} \frac{1/x}{(1/x) - (-1/x^2)} = \frac{1}{2} \qquad (16.27\mathrm{e})$$

$\square$

Exercises

Show that each of the following limits is indeterminate and then find the limit.

1. $\lim\limits_{x \to 1} \dfrac{x^3 - 1}{x^9 - 1}$ Ans: 1/3

2. $\lim\limits_{x \to 0} \dfrac{\sin^2 x}{x}$ Ans: 0

3. $\lim\limits_{x \to \infty} \dfrac{e^{3x}}{x^2}$ Ans: ∞

4. $\lim\limits_{x \to \infty} \dfrac{\ln(2x + 5)}{3x + 4}$ Ans: 0

5. $\lim\limits_{x \to 0} \dfrac{e^x - 1}{\sin(12x)}$ Ans: 1/12

6. $\lim\limits_{x \to 0} \dfrac{\tan 4x}{\sin 7x}$ Ans: 4/7

7. $\lim\limits_{x \to \infty} \dfrac{\ln x}{\sqrt{x}}$ Ans: 0

8. $\lim\limits_{x \to 0} \dfrac{\sin 9x}{x^3}$ Ans: ∞

9. $\lim\limits_{x \to 0} \dfrac{\sin^2 x}{\sin x^2}$ Ans: 1

10. $\lim\limits_{x \to 0} (1 - 3x)^{4/x}$ Ans: e^{12}

11. $\lim\limits_{x \to 0^+} (5x)^{7x}$ Ans: 1

12. $\lim\limits_{x \to 0} \left(\dfrac{3}{x^4} - \dfrac{4}{x^2} \right)$ Ans: ∞

13. $\lim\limits_{x \to 0} \dfrac{\cos x - 1}{x}$ Ans: 0

14. $\lim\limits_{x \to 0^+} (1 - 2x)^{1/x}$ Ans: e^{-2}

15. $\lim\limits_{x \to 0^+} \sqrt{x} \ln(12x)$ Ans: 0

16. $\lim\limits_{x \to \infty} (6x e^{1/x} - 6x)$ Ans: ∞

17. $\lim\limits_{x \to 0^+} (\tan x)^x$ Ans: 1

18. $\lim\limits_{x \to 0^+} (1 - e^x)^{1/x}$ Ans: 0

19. $\lim\limits_{x \to \infty} \dfrac{4x + 3}{\ln(3 + 5e^x)}$ Ans: 4

20. $\lim\limits_{x \to \infty} \left(1 + \dfrac{12}{x} \right)^{x/6}$ Ans: e^2

21. $\lim\limits_{x \to (\pi/2)^+} (\sec x - \tan x)$ Ans: 0

22. $\lim\limits_{x \to 0} \dfrac{\tan x - x}{x^3}$ Ans: 1/3

23. $\lim\limits_{x \to 0} (1 - \cos x)^{\tan x}$ Ans: 1

24. $\lim\limits_{x \to \infty} (1 + x)^{1/x}$ Ans: 1

Chapter 17

Extrema: The Maximum and Minimum Value of a Function

Chapter Summary and Goal

Every continuous function on a closed interval must have both a largest (maximum) value and a smallest (minimum) value on that interval. These existence of these values **extrema** are guaranteed by the following theorem:

> **Theorem 17.1. Extreme Value Theorem (Short Form)**
>
> A continuous function on a closed interval has both a maximum and a minimum.

In this chapter we will learn to identify the locations of these **maxima** and **minima**.

Student Learning Objectives

The student will:

1. Understand the difference between absolute and local extrema.
2. Understand the implications of the Extreme Value Theorem.
3. Understand Fermat's theorem.
4. Identify and calculate critical points.
5. Be able to identify local and global extrema on a closed interval.

Maximum and Minimum Values of a Function

The largest or smallest value of a function on its domain is called the **absolute** (or **global**) **maximum** (or **minimum**). Taken together, the absolute maximum and absolute minimum, if they exist, are called the **extreme values** of $f(x)$.

> **Definition 17.1. Absolute (or Global) Maximum**
>
> $f(c)$ is an **absolute (or global) maximum** of f if $f(c) \geq f(x)$ for all x in the domain of $f(x)$.

> **Definition 17.2. Absolute (or Global) Minimum**
>
> $f(c)$ is an **absolute (or global) minimum** of f if $f(c) \leq f(x)$ for all x in the domain of $f(x)$.

There is no guarantee that a function will have extreme values. For example, the line

$$y = mx + b \tag{17.1}$$

has neither an absolute minimum nor an absolute maximum. The parabola

$$y = (x - a)^2 \tag{17.2}$$

has an absolute minimum at $(a, 0)$ but no absolute maximum. Functions with vertical asymptotes in their domains also do not have extrema; for example,

$$y = \frac{x - 3}{x + 5} \tag{17.3}$$

has neither an absolute maximum nor an absolute minimum.

Continuous functions on closed intervals, however, are guaranteed to have both a maximum value and a minimum value. The function given in equation 17.3 does have both a maximum and minimum value on any closed interval $[a, b]$ that does not include the point $x = -5$ as part of its domain. Thus (17.3) has extrema on [17, 29] and on [-19,-11]. This is guaranteed by the **Extreme Value Theorem**. Any interval that contains $x = -5$, however, such as [-8,-3] does not have either a maximum or a minimum, because the function has a vertical asymptote there (figure 17.1).

Figure 17.1: Plot of equation 17.3. On the closed interval [-19,-11], the absolute maximum is at $(-11, f(-11))$, and the absolute minimum is at $(-19, f(-19))$. Because of the vertical asymptote at $x = -5$, there is neither an absolute maximum nor an absolute minimum on the interval [-8,-3]. Asymptotes are indicated as dashed lines.

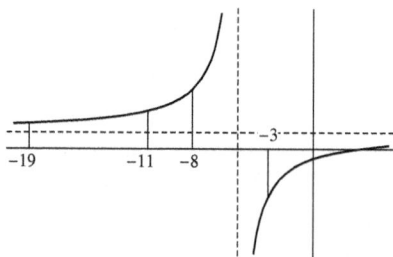

> **Theorem 17.2. Extreme Value Theorem.**
>
> If f is continuous on a closed interval $[a, b]$ then f takes on both an absolute maximum value $f(c)$ at some point $c \in [a, b]$, and an absolute minimum value $f(d)$ at some point $d \in [a, b]$.

While the Extreme Value Theorem guarantees the existence of extrema for continuous functions on closed intervals, it does not rule out the existence of extreme values of functions on infinite intervals. For example, the function

$$y = \frac{1}{1 + x^2} \tag{17.4}$$

has an absolute maximum at (0,1) and the function

$$\frac{-x^2 - 4x + 3}{(x^2 + 1)(x^2 - 4x + 5)} \tag{17.5}$$

has both an absolute maximum and an absolute minimum (figure 17.2).

Figure 17.2: Equations (17.4) (left) has an absolute maximum; (17.5) (right) has both an absolute maximum and an absolute minimum.

In addition to absolute or global extrema, we can define extreme values in a local neighborhood. An easy analogy is to consider the function $f(x)$ as representing local topography. When you are standing on the top of a hill you can see other, higher hills around you, and multiple valleys beneath you. The hill you are standing on, as well as the other hills you see around you, all correspond to **local** (or **relative**) **maxima**, because the point where you are standing is higher than every point nearby – if you take a step in any direction *nearby*, you will go downhill.

Similarly, the valleys correspond to **local** (or **relative**) **minima**. If you are standing at the bottom of any valley, you are at a locally low point, because if you take a single step in any direction, you will move upwards.

These concepts are illustrated in figure 17.3. We can formalize these definitions mathematically in the following boxes.

Definition 17.3. Local (or Relative) Maximum

$f(c)$ is a **local** (or **relative**) **maximum** if the $f(c) \geq f(x)$ for all x close to c.

Definition 17.4. Local (or Relative) Minimum

$f(c)$ is a **local** (or **relative**) **minimum** if the $f(c) \leq f(x)$ for all x close to c.

Notice the difference in the difference between the definitions of absolute and local extrema. Absolute extrema must be extreme **over the entire domain of** $f(x)$. Local extrema only need to be **extrema over all values near** c.

When $f(x)$ is only defined on a closed interval, t*his distinction rules out the existence of local extrema at end points* because because points on both sides of the extrema are not included in the function domain. Thus *end points may considered absolute extrema but are not considered local extrema.*

The extreme value theorem only applies on a closed interval, not an open interval, and **the function must be continuous**. An example of what happens when the functions has a jump discontinuity is given in figure 17.4. On the closed interval $[0, 2.25]$, $f(x)$ has no maximum value, even though it is defined for all x in the interval. This is because $f(x)$ is not continuous on $[0,2.25]$. Similarly, The function does not have a maximum on the half-open interval $(0, 2.25]$, even though it is defined on the entire interval. On the other hand, it does have a maximum on the interval $[1.5, 2.25]$, and on the interval $[0, 1]$.

Figure 17.3: Illustration of extreme values. D is the absolute maximum; E is the absolute minimum; C is a relative minimum; B is a relative maximum; and A and F are endpoints.

Point	Description
A	End Point
B	Local Maximum
C	Local Minimum
D	Absolute and Local Maximum
E	Absolute and Local Minumum
F	End Point

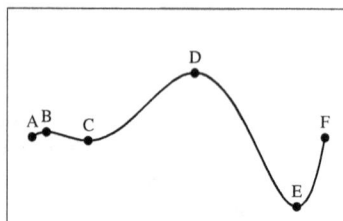

Figure 17.4: A function defined on a closed interval with no maximum value.

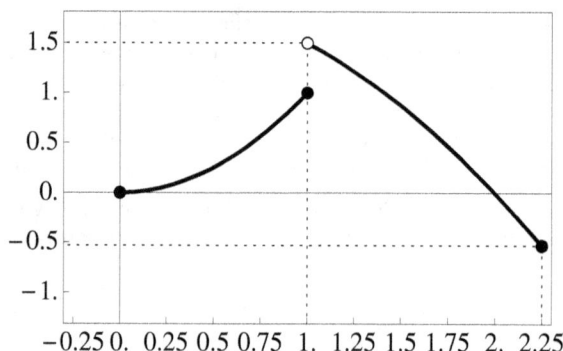

Finding Extreme Values

The key observation about extrema of differentiable functions is that their tangent lines are horizontal. Since the slope of a horizontal line is zero, this means that if the function is differentiable at an extreme value, the derivative is zero at that point. This result is known as Fermat's theorem.

Theorem 17.3. Fermat's Theorem

If f has a local maximum or minimum at c, and if $f'(c)$ exists, then $f'(c) = 0$.

Some words of caution about Fermat's Theorem:

1. $f(x)$ will not necessarily be differentiable at all extrema, so Fermat's theorem will not apply. An example is $y = |x|$, which has a minimum at $x = 0$.

2. The converse of the theorem does not apply; it is possible to have $f'(c) = 0$ at points that **are not extrema**. An example is $f(x) = x^3$ at $x = 0$.

What Fermat's theorem does give us are candidate locations for the extrema. These candidates are known as critical numbers.

> **Definition 17.5. Critical Numbers**
>
> A **critical number** of $f(x)$ is a number c where
>
> 1. $f'(c) = 0$, or
> 2. $f'(c)$ is undefined
>
> The point $(c, f(c))$ is called a **critical point**.

Example 17.1. Find the critical numbers of $f(t) = 3t^4 + 4t^3 - 6t^2$.

Solution. Since the domain is not given, we must infer it from the function, and since a polynomial is defined on all x, the domain is $(-\infty, \infty)$. Hence there are no endpoints.

Since a polynomial is everywhere differentiable, there are no points where the derivative is discontinuous (i.e., no corners).

The remaining critical points occur when $f'(x) = 0$. So we solve

$$0 = f'(x) = 12t^3 + 12t^2 - 12t = 12t(t^2 + t - 1) \tag{17.6a}$$

There are two roots r_1 and r_2 of this quadratic, given by

$$r_{1,2} = \frac{-b \pm \sqrt{b^2 - 4ac}}{2a} = \frac{-2 \pm \sqrt{5}}{2} = -1 \pm \frac{\sqrt{5}}{2} \tag{17.6b}$$

Therefore the critical points occur at

$$x = 0, -1 - \frac{\sqrt{5}}{2}, -1 + \frac{\sqrt{5}}{2} \tag{17.6c}$$

$\square$

> **Theorem 17.4. Corollary of Fermat's Theorem**
>
> Local extrema of f – maxima and minima – occur at critical numbers.

Fermat's theorem tells us that **local extrema** occur at critical numbers. What Fermat's theorem does not tell us is anything about global extrema. It also doesn't tell us anything about end points. So we also need to add them to the list of candidate points.

> **Heuristic 17.1. Finding Extrema: The Closed Interval Method**
>
> If $f(x)$ is defined on a closed interval $[a, b]$:
>
> 1. Find all the critical numbers c of $f(x)$ in (a, b).
> 2. Find the values of $f(c)$ at the critical numbers of in (a, b).
> 3. Find the values $f(a)$ and $f(b)$ at the end points.
> 4. The **largest** of the values from steps 2 and 3 is **the absolute maximum value**; the **smallest** of these values is the **absolute minimum value**.

If the function is defined on an infinite interval, we omit the calculation at the end points, but the remainder of the method remains the same.

Example 17.2. Find the absolute maximum and minimum values of

$$f(x) = x^3 - 6x^2 + 9x + 2 \tag{17.7}$$

on the interval $[-1, 4]$.

Solution. At the endpoints,

$$f(-1) = (-1)^3 - 6(-1)^2 + 9(-1) + 2 = -14 \tag{17.8a}$$

$$f(4) = 4^3 - 6(4)^2 + 9(4) + 2 = 64 - 96 + 36 + 2 = 6 \tag{17.8b}$$

There are no points of discontinuity or corners, because $f(x)$ is a polynomial, which is differentiable everywhere. The remaining critical points occur where $f'(x) = 0$:

$$0 = f'(x) = 3x^2 - 12x + 9 = 3(x^2 - 4x + 3) = 3(x - 1)(x - 3) \tag{17.8c}$$

Hence $x = 1$ and $x = 3$ are critical numbers. The function values at these points are

$$f(1) = 1^3 - 6(1) + 9(1) + 2 = 6 \tag{17.8d}$$

$$f(3) = 3^3 - 6(3^2) + 9(3) + 2 = 27 - 54 + 27 + 2 = 2 \tag{17.8e}$$

The critical points are summarized in the following table:

x	$f(x)$	type of critical point
-1	-14	left end point
1	6	$f'(x) = 0$
3	2	$f'(x) = 0$
4	6	right endpoint

Thus $x = 1$ and $x = 4$ tie for absolute maximum; and the absolute minimum occurs at $x = -1$. The absolute maximum value of the function on $[-1, 4]$ is 6, and the absolute minimum value of the function on $[-1, 4]$ is -14 (see figure 17.5). □

Figure 17.5: Functions $f(x)$ used in examples 17.2 (left) and 17.3 (right).

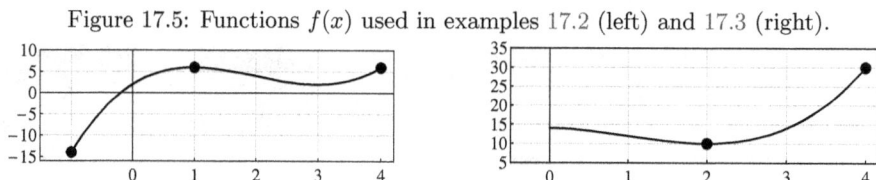

Example 17.3. Find the absolute maximum and minimum of $f(x) = x^3 - 3x^2 + 14$ on $[-\frac{1}{2}, 4]$.

Solution. Differentiating, $f'(x) = 3x^2 - 6x$. Setting this equal to zero gives the critical points on the interior of the interval.

$$0 = 3x^2 - 6x = 3x(x - 2) \tag{17.9a}$$

Hence $x = 0$ or $x = 2$. Both critical points occur in the given interval $[-\frac{1}{2}, 4]$. So we calculate the values of $f(x)$ at each candidate location.

$$f(0) = 14 \qquad\qquad \text{critical point} \tag{17.9b}$$

$$f(2) = 10 \qquad\qquad \text{critical point} \tag{17.9c}$$

$$f(-1/2) = 13.125 \qquad\qquad \text{end point} \tag{17.9d}$$

$$f(4) = 30 \qquad\qquad \text{end point} \tag{17.9e}$$

The absolute maximum and minimum are found by comparing values. The absolute maximum is at $(4,30)$ and the absolute minimum at $(2,10)$ (see figure 17.5). □

Example 17.4. Find the absolute maximum and absolute minimum of $f(x) = x + \dfrac{1}{x}$ on $[0.2, 4]$.

Solution. Interior critical points occur when $f'(x) = 0$. Differentiating and setting $f'(x) = 0$

$$f'(x) = 1 - \frac{1}{x^2} = 0 \tag{17.10a}$$

Solving this for x we find two roots, $x = \pm 1$. Since only one of the critical points falls within the given interval $[.2, 4]$, we only need consider the critical point at $x = 1$. The

Figure 17.6: $f(x)$ used in example 17.4.

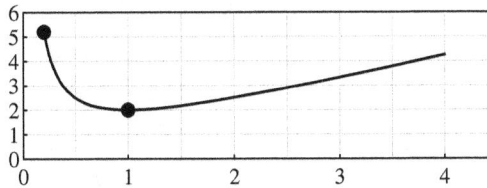

critical numbers are $\{0.2, 1, 4\}$. We evaluate the function at each critical number.

$$f(1) = 1 + \frac{1}{1} = 2 \tag{17.10b}$$

$$f(.2) = 0.2 + \frac{1}{.2} = 5.2 \tag{17.10c}$$

$$f(4) = 4 + \frac{1}{4} = 4.25 \tag{17.10d}$$

The absolute maximum is at $(.2, 5.2)$ and the absolute minimum is at $(1,2)$, as illustrated in figure 17.6. There is nothing "special" about the right endpoint at $x = 4$.   □

Example 17.5. Find the absolute maximum and minimum of $f(x) = \dfrac{x}{x^2 - x + 1}$ on $[0, 3]$.

Differentiating,

$$f'(x) = \frac{(x^2 - x + 1)(1) - x(2x - 1)}{(x^2 - x + 1)^2} \tag{17.11}$$

$$= \frac{x^2 - x + 1 - 2x^2 + x}{(x^2 - x + 1)^2} \tag{17.12}$$

$$= \frac{1 - x^2}{(x^2 - x + 1)^2} \tag{17.13}$$

The derivative can only be zero when the numerator is zero; setting the numerator zero gives $x = \pm 1$. Since $x = -1$ lies outside the interval $[0, 3]$, there is only one interior critical number, at $x = 1$. Thus the complete set of critical numbers is $\{0, 1, 3\}$. Evaluating the function at each of these,

$$f(0) = 0 \tag{17.14}$$

$$f(3) = \frac{3}{7} \tag{17.15}$$

$$f(1) = 1 \tag{17.16}$$

Thus the absolute maximum is $(1,1)$ and the absolute minimum is $(0, 0)$.   □

Exercises

In exercises 1 through 11 find the critical numbers of the given function.

1. $y = 3x^4 - 4x^3 - 12x^2 + 17$
   Ans: 0, -1, 2

2. $y = x^3\sqrt{x^2 - 9}$
   Ans: $0, \pm\frac{3\sqrt{3}}{2}$

3. $y = 10 + \cos(x/2) + 2\sin(x/2)$
   Ans: $2\arctan(2)$

4. $y = 3x^4 + 8x^3 - 90x^2 + 5$
   Ans: 0, 3, -5

5. $y = x^{3/2}(3x - 5)$
   Ans: 0, 1

6. $y = x + \ln x$
   Ans: -1

7. $y = \dfrac{e^{9x}}{x - 2}$, for $x > 2$.
   Ans: 19/9

8. $f(x) = 11x - 3\ln x$, for $x > 0$.
   Ans: 3/11

9. $y = 8\ln(x + 1) - \ln(12x)$
   Ans: 1/7

10. $y = \dfrac{e^x - x}{e^x + x}$
    ans: 1

11. $y = xe^{-x^2}$
    ans: $\pm 1/\sqrt{2}$

In exercises 12 through 24, use the closed interval method (Heuristic 17.1) to find the absolute maximum and absolute minimum of the given function on the specified interval.

12. $y = x^3 - 3x^2 + 1$ on $[-1/2, 4]$
    Ans: (2, -3); (4, 17)

13. $y = x^2 - \ln x$ on $[1/4, 1]$
    Ans: $(\frac{1}{\sqrt{2}}, \frac{1}{2} + \ln\sqrt{2})$; $(\frac{1}{4}, \frac{1}{16} + \ln 4)$

14. $y = 5 + 54x - 2x^3$ on $[0, 4]$
    Ans: (0, 5); (3, 113)

15. $y = 3x^4 - 4x^3 - 12x^2 + 1$ on $[-2, 3]$
    Ans: (2, -31); (-2, 33)

16. $f(x) = \begin{cases} 0, & x = 0 \\ x(1 - \ln x), & x > 0 \end{cases}$ on $[0, 2]$
    Ans: (0, 0); (1, 1)

17. $y = x\sqrt{9 - x^2}$ on $[-1, 2]$
    Ans: $(-1, -2\sqrt{2})$; $(2, 2\sqrt{5})$

18. $y = 2\cos x + \sin(2x)$ on $[0, \frac{\pi}{2}]$
    Ans: $(\frac{\pi}{2}, 0)$; (0, 2)

19. $y = e^{-(x-2)^2}$ on $[1, 4]$
    Ans: $(1, e^{-9})$; (4, 1)

20. $y = x - e^x$ on $[-2, 2]$
    Ans: $(-2, -2 - 1/e^2)$; (0,-1)

21. $y = xe^{-x}$ on $[-1, 2]$
    Ans: $(-1, -e)$; (1, 1/e)

22. $y = x^2 e^{-x^2}$ on $[-3, 3]$
    Ans: (0,0); (-1, 1/e); (1, 1/e)

23. $y = \dfrac{(x - 4)^2}{(x + 5)^2}$ on $[2, 6]$
    Ans: (4,0); (2, 4/49)

24. $y = \dfrac{1}{x}\ln x$ on $[0.1, 5]$
    Ans: (.1, -10\ln 10); (e, 1/e)

25. Prove that $\pi^e < e^\pi$, as follows.

    (a) Let $y = x^{1/x}$. Use implicit differentiation to find y' and y''.

    (b) Show that the maximum of y occurs when $x = e$.

    (c) Use the result of part 25b to show that $x^{1/x} < e^{1/e}$ for all positive $x \neq e$.

    (d) Substitute $x = \pi$ into the result of part 25c and derive the conclusion $\pi^e < e^\pi$.

Chapter 18

The Mean Value Theorem

Chapter Summary and Goal

This chapter will present two foundational theorems of differential calculus: **Rolle's Theorem** and the **Mean Value Theorem**.

Rolle's Theorem tells us that any continuously differentiable function with $f(a) = f(b)$ has a point at which the tangent line is horizontal: you can't get to the other side of the hill (in one dimension) without going over the top (figure 18.1).

The **Mean Value Theorem** (MVT) generalizes Rolle's Theorem to the case where $f(a) \neq f(b)$. In this case the secant line (the line joining the end points) is not horizontal, but has some slope m. The MVT tells us that there is some point c in the interval (a, b) where the derivative $f'(c) = m$ (figure 18.2).

Student Learning Objectives

The student will:

1. Learn the conditions under which Rolle's Theorem applies.
2. Learn the conditions under which the Mean Value Theorem applies.
3. Be able to locate points at which Rolle's Theorem applies.
4. Be able to locate points at which the Mean Value Theorem applies.
5. Understand the proofs of Rolle's Theorem and the Mean Value Theorem.
6. Use Rolle's theorem and the Intermediate Value Theorem to prove the existence and uniqueness of roots in an interval.
7. Learn about proof by contradiction.
8. Learn about proof by cases.

Rolle's Theorem

Rolle's Theorem is illustrated in figure 18.1. The basic idea is this: if the two end points of the curve lie on the same horizontal line, there is some spot on the curve connecting point A to point B where the tangent line is also horizontal. The only requirements for this to hold is that f be continuous on the entire interval $a \leq x \leq b$, and differentiable on the interior of the interval, $a < x < b$.

Theorem 18.1. Rolle's Theorem

Let f be function that is continuous[a] on $[a, b]$ and differentiable on (a, b), such that $f(a) = f(b)$. Then there is a number $c \in (a, b)$ such that $f'(c) = 0$.

[a]Since differentiability implies continuity (theorem 7.13)) the first condition is really not needed.

Figure 18.1: Illustration of Rolle's Theorem. Points $(a, f(a))$ and $(b, f(b))$ of the differentiable function $f(x)$ lie on the same horizontal line. Rolle's theorem says that there is some point c between a and b where the tangent line at $(c, f(c))$ is horizontal.

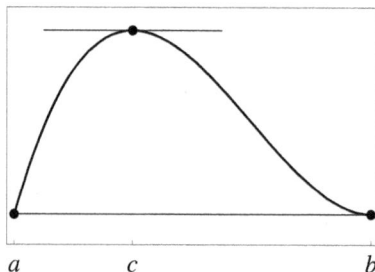

Proof. (Rolle's Theorem). To see why Rolle's theorem works we will consider three cases. At least one of these conditions must always be true:

    a) $f(x)$ is a constant

    b) There is some number x in (a, b) such that $f(x) > f(a)$

    c) There is some number x in (a, b) such that $f(x) < f(a)$

Case (a): If f is a constant, say $f(x) = K$ for some constant K, then every point $x \in (a, b)$ satisfies $f'(x) = 0$, because the derivative of a constant is zero. Pick any point in the interval (a, b) and call it c. Then the theorem holds for this point, and we are done with the proof (in this case).

Case (b): There is some number x in the interval (a, b) such that $f(x) > f(a)$.

Since $f(x)$ is continuous, we know by the extreme value theorem (theorem 17.2) that f has an absolute maximum at some point $x = c$ on $[a, b]$.

Because we are told (in case (b)) that there is some point x with $f(x) > f(a)$, then the absolute maximum $f(c)$ is at least as big as $f(x)$, i.e.,

$$f(c) \geq f(x) > f(a) \tag{18.1}$$

Since $f(x) > f(a)$ we also know that x is distinct from a hence there is a point in the interior of (a, b) where the function is larger than $f(a)$. Hence the maximum does not occur at a, i.e, $c \neq a$.

Since $f(a) = f(b)$, we can apply the same argument to $f(b)$ and conclude that $c \neq b$. Therefore

$$a < c < b \tag{18.2}$$

i.e., $c \neq a$ and $c \neq b$. By Fermat's theorem (theorem 17.3), $f'(c) = 0$. But since $f'(c) = 0$, we have shown that there is a point c in (a, b) where $f'(c) = 0$, therefore completing the proof (for case (b)).

Case (c): The proof for case (c) is identical to the proof for case (b), with the "greater-than" sign ($>$) replaced by a "less-than" sign ($<$) everywhere.   □

Example 18.1. Show that $f(x) = x^3 - 4x^2 + 4x$ satisfies the conditions of Rolle's Theorem on the interval $[0,2]$ and find a point c at at which it applies.

Solution. $f(x)$ is a polynomial which is continuous and differentiable on all x, so the first two conditions apply. To see that the value of the function at the endpoints apply, we calculate the value of $f(x)$:

$$f(0) = 0 \tag{18.3a}$$

$$f(2) = 2^3 - 4(2^2) + 4(2) = 8 - 16 + 8 = 0 \tag{18.3b}$$

Since $f(0) = f(2)$, the condition $f(a) = f(b)$ also applies. Thus there is some point c in $(0,2)$ where the derivative is zero. To find this point we calculate the derivative and set it equal to zero.

$$0 = f'(x) = 3x^2 - 8x + 4 = (3x - 2)(x - 2) \tag{18.3c}$$

Hence $x = 2$ or $x = 2/3$. Since $x = 2$ is an endpoint, we conclude that the only solution that satisfies the theorem is $x = 2/3$.   □

Example 18.2. Show that $f(x) = x^3 + x$ has exactly one real root on the interval $[-1, 1]$.

Solution. To complete this exercise we have to do two things: (a) prove **existence** of a root, i.e., show that the function has a root; and (b) prove its **uniqueness**, i.e., show that $f(x)$ has exactly one root.

To show existence, we will need to use the intermediate value theorem (IVT). We will use Rolle's theorem to show uniqueness.

<u>Existence</u>. (Compare with example 7.16.) At the end points we have

$$f(-1) = (-1)^3 - 1 = -2 \tag{18.4a}$$

$$f(1)(1)^3 + 1 = 2 \tag{18.4b}$$

which have opposite signs. Since f is continuous on $[-1, 1]$, the intermediate value theorem tells us that there is some point c, where $-1 < c < 1$, such that

$$f(c) = 0 \tag{18.4c}$$

Thus there is (at least) one root between -1 and 1. This proves the existence part of the exercise.

We will use the method of contradiction to prove that there cannot be any additional roots. In this method, we assume that there at least two roots, and show that this assumption leads to a nonsensical conclusion.

<u>Uniqueness</u>. Suppose that there are two roots at $x = p$ and $x = q$. Since both p and q are roots, then

$$f(p) = f(q) = 0 \tag{18.4d}$$

Since both p and q are inside the interval $[-1,1]$, then we have $-1 < p < q < 1$. Since $f(x)$ is a polynomial it is continuous and differentiable everywhere, so, in particular, it is continuous on the interval $[p, q]$, and it is differentiable on the interval (p, q).

Thus by Rolle's Therem, there is some point r such that

$$-1 < p < r < q < 1 \tag{18.4e}$$

where

$$f'(r) = 0 \tag{18.4f}$$

But

$$f'(x) = 3x^2 + 1 > 0 \tag{18.4g}$$

In other words, it is impossible to have $f'(r) = 0$ *at any point*.

This is our nonsensical conclusion, which we know is false. When this happens, it means we made an incorrect assumption. In this case, the only assumption we throw away is that there are multiple roots. Hence the root is unique, completing the second part of the proof.   □

The method of proof use in example 18.2 is know as the **proof by contradiction** or **reductio ad absurdum**. This method is often used to prove that an object is unique: we assume that there are two such objects, and show that such an assumption leads to nonsense. The "nonsense" result is known as a logical contradiction, because it is really a contradiction of a previous statement in our proof, or a reduction of our original assumption (that two objects exist) to absurdity.

Typical *logical contradictions* that one might obtain in such proofs might be results that disagree with earlier steps in the proof, or statements of mathematical nonsense, such as "$0 > 1$"' or "$2 = 1$." In the last example we obtained two different, and contradictory statements: $f'(r) = 0$ and $f'(r) > 0$, which cannot simultaneously be true.

Proof by contradiction (*reductio ad absurdium*)

To prove Q is true, assume that Q is false, and prove that the logical consequence of that assumption is a logical contradiction.

Proving Uniqueness

To prove that some object z is unique, assume that there are two of z, and show that this leads to nonsense using *reductio ad absurdium*.

The Mean Value Theorem

The **Mean Value Theorem** (MVT) (see figure 18.2) is a generalization of Rolle's Theorem. Now the end points are no longer required to be fixed on the same vertical; instead they can be canted at any angle. We consider a secant line connecting the endpoints $(a, f(a))$ to $(b, f(b))$. Then this powerful theorem tells us that $f(x)$ has a tangent line that is parallel to this secant line, and that the point of tangency lies somewhere between a and b.

Figure 18.2: The Mean Value Theorem (MVT). If f is continuous on $[a, b]$ and differentiable on (a, b) then there is some point c in the interval (a, b) where $f'(c)$ is equal to the slope of the secant line joining the points $(a, f(a)$ and $(b, f(b))$.

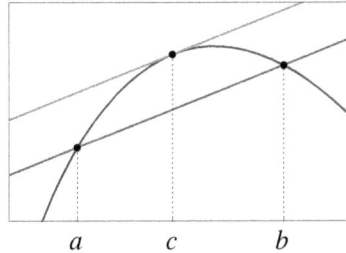

Theorem 18.2. Mean Value Theorem.

Suppose that $f(x)$ is continuous on $[a, b]$ and differentiable on (a, b). Then there is some number $c \in (a, b)$ such that

$$f'(c) = \frac{f(b) - f(a)}{b - a} \tag{18.5}$$

or equivalently

$$f(b) - f(a) = f'(c)(b - a) \tag{18.6}$$

Proof. The slope the secant line between $(a, f(a))$ and $(b, f(b)$ is

$$m = \frac{f(b) - f(a)}{b - a}. \tag{18.7a}$$

Starting with the point-slope form of the equation of a line

$$y = y_1 + m(x - x_1) \tag{18.7b}$$

Using m from equation 18.7a and $(x_1, y_1) = (a, f(a))$ as the point, this becomes

$$y = f(a) + \frac{f(b) - f(a)}{b - a}(x - a) \tag{18.7c}$$

Now define a new function $g(x)$ by the following difference:

$$g(x) = f(x) - y = f(x) - f(a) - \frac{f(b) - f(a)}{b - a}(x - a) \tag{18.7d}$$

The function $g(x)$ measures the vertical distance between $f(x)$ and the secant line at any x value. By substitution at the left end point $x = a$ we see that this distance is

$$g(a) = \overbrace{f(a) - f(a)}^{=0} - \frac{f(b) - f(a)}{b - a} \overbrace{(a - a)}^{=0} = 0 \tag{18.7e}$$

At the right end point $x = b$,

$$g(b) = f(b) - f(a) - \frac{f(b) - f(a)}{b - a}(b - a) \tag{18.7f}$$

$$= f(b) - f(a) - f(b) + f(a) = 0 \tag{18.7g}$$

Doing Calculus

Hence (1) $g(x)$ is continuous on $[a, b]$; (2) differentiable on (a, b); and (3) satisfies $g(a) = g(b)$. These are the three requirements for Rolle's Theorem to hold.

Hence (by Rolle's Theorem) there is a point c, where $a < c < b$, such that $g'(c) = 0$. Differentiating (18.7d)

$$g'(x) = f'(x) - \frac{f(b) - f(a)}{b - a} \tag{18.7h}$$

To find the point at which Rolle's Theorem holds, we set $x = c$ and solve $g'(c) = 0$:

$$0 = g'(c) = f'(c) - \frac{f(b) - f(a)}{b - a} \tag{18.7i}$$

Rearranging gives

$$f'(c) = \frac{f(b) - f(a)}{b - a} \tag{18.7j}$$

which is the equation we were trying to prove. □

Example 18.3. Show that the function $y = x^3 + 3x$ satisfies the conditions of the Mean Value Theorem on the interval $[2,5]$, and find a point c that corresponds to the conditions of the theorem.

Solution.   Since $f(x) = x^3 + 3x$ is a polynomial it is continuous and differentiable for all values of x. Thus it it is continuous on $[2,5]$ and differentiable on $(2,5)$. The value of the function at the endpoints is

$$f(2) = 2^3 + 3(2) = 8 + 6 = 14 \tag{18.8a}$$
$$f(5) = 5^3 + 3(5) = 125 + 15 = 140 \tag{18.8b}$$

The MVT then says that there is some point c, $2 < c < 5$, where $f'(c)$ is the slope of the secant line between $(2, 14)$ and $(5, 140)$. This slope is

$$m = \frac{f(5) - f(2)}{5 - 2} = \frac{140 - 14}{3} = \frac{126}{3} = 42 \tag{18.8c}$$

Hence $f'(c) = 42$. But since $f'(x) = 3x^2 + 3$,

$$42 = 3c^2 + 3 \implies 39 = 3c^2 \implies c = \sqrt{13} \approx 3.6056. \tag{18.8d}$$

□

Example 18.4. Suppose all that we know about $f(x)$ is (a) it is continuous and differentiable; (b) that $f(0) = 0$ and (c) that $f'(x) \leq 10$. How large can $f(5)$ possibly be?

Solution. To solve this problem we apply the mean value theorem on the interval

$$[a, b] = [0, 5] \tag{18.9a}$$

Since f is continuous and differentiable, the MVT applies. Thus there is some number c in $(0, 5)$ such that

$$f(5) - f(0) = f'(c)(5 - 0) = 5f'(c) \leq 5 \times 10 = 50 \tag{18.9b}$$

where the last step follows because $f'(c) \leq 10$. Since we are also told that $f(0) = 0$,

$$f(5) \leq f(0) + 50 = 50 \tag{18.9c}$$

Thus the largest possible value that $f(5)$ can possibly have is 50. □

Example 18.5. Verify the mean value theorem for $f(x) = \dfrac{x}{x-7}$ on $[1,3]$ and find all numbers c that satisfy the conclusion of the Mean Value Theorem.

Solution. Since $f(x)$ is a rational function, it is continuous and differentiable everywhere except where the denominator is zero, which is at $x = 7$. Since $x = 7$ does not fall into the domain of the problem, the conditions of the mean value theorem apply with $a = 1$ and $b = 3$.

Thus there is a point c between 1 and 3 such that

$$f'(c) = \frac{f(3) - f(1)}{3 - 1} = \frac{-3/4 - (-1/6)}{3 - 1} = -\frac{7}{24} \tag{18.10a}$$

By the quotient rule,

$$f'(x) = \frac{(x-7)(1) - (x)(1)}{(x-7)^2} = -\frac{7}{(x-7)^2} \tag{18.10b}$$

To find the point that satisfies $f'(c) = -7/24$, we solve

$$-\frac{7}{24} = -\frac{7}{(x-7)^2} \tag{18.10c}$$

Cross multiplying and simplifying,

$$(x-7)^2 = 24 \tag{18.10d}$$

Thus $x = 7 \pm \sqrt{24}$. Of these two values, only the one with the minus sign, $x = 7 - 2\sqrt{6} \approx 2.101$, is in the given domain of (1,3). $\qquad\square$

Constant Functions

We have frequently made use of the fact that if a function is constant then its derivative is zero. One application of the mean value theorem is a proof of the converse.

> **Theorem 18.3. Constant Function Theorem**
>
> If $f'(x) = 0$ for all x in (a, b) then f is constant on (a, b).

Proof. Let u, v be any two numbers such that

$$a < u < v < b \tag{18.11a}$$

Since $f(x)$ is differentiable on (a, b), then it is differential on (u, v). Similarly, since $f(x)$ is continuous on $[a, b]$, it is also continuous on (u, v).

Hence $f(x)$ meets both conditions that are required by the mean value theorem on $[u, v]$.

This tells us that there is some number c, where $u < c < v$, and

$$f(u) - f(v) = f'(c)(u - v) \tag{18.11b}$$

But since we are told that $f'(x) = 0$ for all x, then it must be true that $f'(c) = 0$. Hence the right hand side of (18.11b) is zero. Substituting this tells us that

$$f(u) = f(v) \tag{18.11c}$$

But when we picked u and v, they were completely arbitrary numbers chosen anywhere in the interval (a, b) subject to the condition (18.11a). In other words, we have shown that $f(u) = f(v)$ for *any pair of numbers* with $u < v$ in (a, b). Thus $f(x)$ is constant on (a, b), as stated. $\square$

An immediate consequence of this is that if two functions have the same derivative, then they differ by at most a constant. We say "at most" because it is possible for the constant to be zero.

Theorem 18.4. Equal Derivatives

If $f'(x) = g'(x)$ for all x in (a, b) then $f(x) - g(x)$ is a constant on (a, b), i.e.,

$$f(x) = g(x) + C \qquad (18.12)$$

for some constant C.

Proof. Define

$$h(x) = f(x) - g(x) \qquad (18.13a)$$

Then since $f'(x) = g'(x)$ for all x

$$h'(x) = f'(x) - g'(x) = 0 \qquad (18.13b)$$

for all x in (a, b). Hence by the previous theorem $h(x) = C$ for some constant C, and substituting $h(x) = f(x) - g(x)$ gives

$$f(x) - g(x) = C \qquad (18.13c)$$

Therefore

$$f(x) = g(x) + C \qquad (18.13d)$$

as required. $\square$

Exercises

In exercise 1 through 6 verify that the given functions satisfy Rolle's theorem and find all numbers c that satisfy the conclusion of the theorem on the intervals specified.

1. $3 - x^2$ on [2,-2]

2. $x^2 - 10x + 28$ on [4,6]

3. $2x^3 - 15x^2 + 36$ on [3/2, 27] (ans: 2)

4. $y = \sqrt{x} + 1/\sqrt{x}$ on [1,4]

5. $y = x\sqrt{10 - x}$ on [0, 10]

6. $y = \cos(\sqrt{\pi x})$ on $[4\pi, 16\pi]$ (ans: 9π)

Use Rolle's Theorem and the Intermediate Value Theorem to show that each of the functions in exercises 7 through 10 has at most one root in the specified interval.

7. $y = x^3 - 15x$ on [-2,2]

8. $y = x^3 + 2x - 2$ on [1/2, 1]

9. $y = x - 3\cos x$ on $[0, \pi]$

10. $y = x^2 - 1/x^2$

Verify that the functions given in exercises 11 through 13 satisfy the mean value theorem on the specified interval, and find all numbers c that satisfy the conclusion of the MVT for that interval.

11. $y = x^3 - 3x + 2$ on [-2, 2]

12. $y = \dfrac{1}{x}$ on [1,3]

13. $y = x + \sin x$ on $[0,2\pi]$

Robin Hood takes aim in Sherwood Forest.

Chapter 19

The Shape of a Curve: The First and Second Derivative Tests

Chapter Summary and Goal

In this chapter we use the first and second derivatives to tell us about the shape of the plot of a curve. We will present two general methods for finding the locations of local extrema. These methods, the **first derivative test** and **the second derivative test**, taken together with what we know about any asymptotes and intercepts that the function may have, will allow us to sketch a rather complete picture of the curve.

Student Learning Objectives

The student will:

1. To find the local and absolute extrema of a function using the first and second derivative tests.
2. To locate and identify the intervals where a function is increasing and decreasing.
3. To locate and identify the intervals of concavity of a function and its inflection points.
4. To sketch a rough plot of a function showing all intervals of increase/decrease, concavity, inflection, local extrema, and intercepts and asymptotes, if they exist.

Intervals of Increase and Decrease

One of the more important applications of the mean value theorem is that it helps us to interpret the meaning of the sign of the derivative. Suppose that $f(x)$ is continuous and differentiable on a closed interval $[a, b]$; then $f(x)$ satisfies the conditions of mean value theorem on $[a, b]$. If we pick any two points x_1 and x_2 inside this interval, such that

$$a < x_1 < x_2 < b \tag{19.1}$$

then $f(x)$ also satisfies the mean value theorem on $[x_1, x_2]$. So by (18.6) we conclude that there must be some number c, where $x_1 < c < x_2$, such that

$$f(x_2) - f(x_1) = f'(c) \overbrace{(x_2 - x_1)}^{>0} \tag{19.2}$$

The second factor must be positive because $x_2 > x_1$ (by (19.1)). Therefore

$$f(x_2) - f(x_1) > 0, \text{ if } f'(c) > 0 \qquad\qquad (19.3)$$
$$f(x_2) - f(x_1) < 0, \text{ if } f'(c) < 0 \qquad\qquad (19.4)$$

Stating this in another way,

$$f'(c) > 0 \implies f(x_2) > f(x_1) \qquad\qquad (19.5)$$
$$f'(c) < 0 \implies f(x_2) < f(x_1) \qquad\qquad (19.6)$$

The result applies to any pair of numbers $x_1 < x_2$ in the interval $[a, b]$, so they may be arbitrarily close. Thus **when the derivative is positive, the value of the function increases** as you pass through $x = c$ moving from the left to the right (the direction of increasing x); and **when the derivative is negative, the value of the function decreases** as you pass through $x = c$, moving from the left to the right. We summarize this in the following box.

Theorem 19.1. Increasing/Decreasing Test

If f is differentiable at any point $x = c$,
   a) If $f'(c) > 0$, f is increasing at c.
   b) If $f'(c) < 0$, f is decreasing at c.
   c) If $f'(c) = 0$, f is horizontal at c.

The points where f is horizontal are a subset of the critical points, and are often (though not necessarily) among the local extrema of a function. It is important to remember, however, that there are situations where the tangent line is horizontal that are not local extrema. Usually in these situations the tangent line will cross the the curve of the function in one direction, but not the other (as with the the x axis and the curve $y = x^3$ at the origin).

By extension of the above ideas, we define the intervals of increase and decrease as those intervals where $f'(x)$ is increasing or decreasing throughout the interval.

Definition 19.1. Intervals of Increase and Decrease

Let $f(x)$ be differentiable on any interval (a, b). If
   a) $f'(x) > 0$ for all x in (a, b) we say that f is increasing on (a, b) and call (a, b) and **interval of increase** of $f(x)$.
   b) $f'(x) < 0$ for all x in (a, b) we say that f is decreasing on (a, b) and call (a, b) an **interval of decrease** of $f(x)$.

A function may have two adjacent intervals of increase or decrease, and students are often tempted to combine them together into a single interval; however, this is not correct. For example, the function $f(x) = x^3$ has two intervals of increase: $(-\infty, 0)$ and $(0, \infty)$. It would not be correct to combine them together into a single interval $(-\infty, \infty)$ because $f'(0) = 0$, i.e., f is not increase at $x = 0$, so $x = 0$ cannot be considered part of the interval of increase.

Example 19.1. Sketch a plot of $y = 4x^3 - 12x^2 + 5$

Solution. The function is a polynomial on an unbounded interval so it is everywhere continuous and differentiable and there are not asymptotes. The only critical points are when $f'(x) = 0$. In addition, the ends of the intervals of increase and decrease will always occur when $f'(x) = 0$. Differentiating,

$$y' = 12x^3 - 24x \qquad\qquad (19.7a)$$

We are interested in where y' is positive and where it is negative. At the transition points between intervals we have

$$0 = 12x^3 - 24x = 12x(x - 2) \tag{19.7b}$$

The transition points are at $x = 0$ and $x = 2$. So we have three intervals to consider: $(-\infty, 0), (0, 2), (2, \infty)$. We can determine what is going on in each interval by considering a point inside each interval. We will pick a point inside each interval **completely arbitrarily**. Because the derivative cannot change signs except at an endpoint we are free to pick any point we like inside each interval - it is in our best interest to pick a point at which it is easy to evaluate the sign of $f'(x)$.

Interval	Point (x)	$f'(x) = 12x(x - 2)$	Slope
$(-\infty, 0)$	-1	12(-1)(-1-2)=36	increasing
$(0, 2)$	1	12(1)(1-2)=-12	decreasing
$(2, \infty)$	3	12(3)(3-2)=36	increasing

So we can begin to make a sketch. All we really know so far is the direction of increase or decrease, so we just draw a slanted line segment in each interval (figure 19.1). The

Figure 19.1: Slanted lines indicate direction of increase or decrease in example 19.1.

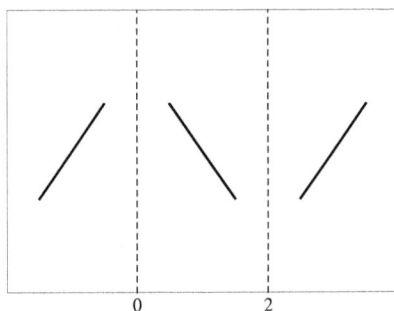

diagonal bars tell us the general direction of increase or decrease. But we have not yet looked at enough information to figure out where the values lie. To do that, we want to find the extrema.

The critical points occur at $x = 0$ and $x = 2$. By definition of a local maximum, since $f(x)$ is increasing to the left of $x = 0$ and decreasing to right of $x = 0$, a local maximum must occur at $x = 0$. The value of this local maximum is

$$f(0) = 4(0^3) - 12(0^2) + 5 = 5 \tag{19.7c}$$

so the local maximum is at $(0,5)$. Similarly, by definition of a local minimum, since $f(x)$ is decreasing to the left of $x = 2$ and increasing to the right of $x = 2$, a local minimum must occor at $x = 2$. The value of this local minimum is

$$f(2) = 4(2^3) - 12(2^2) + 5 = -11 \tag{19.7d}$$

so the local minimum is at $(2, -11)$. There is no absolute maximum or minimum because the function goes to both ∞ and $-\infty$ in the limit as $x \to \pm\infty$. Plotting the additional points leads to a sketch such as the one in figure 19.2. $\qquad\square$

Combining the Increasing/Decreasing Test for an extemum (theorem 19.1) with the definition of a critical number (definition 17.5) and concept of Intervals of Increase and Decrease (definition 19.1) (definitions 17.4 and 17.3) gives us the following rule, which we have already demonstrated

Figure 19.2: Revised sketch for example 19.1.

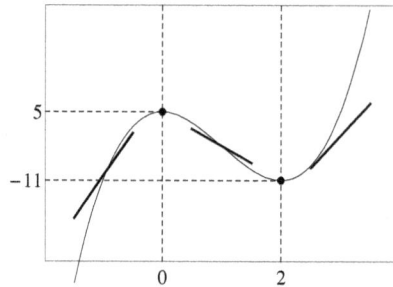

in the previous example, for finding the local maxima and minima of a function. Since it depends entirely on the first derivative of $f(x)$ it is called the **first derivative test**.

Theorem 19.2. The First Derivative Test

Let c be a critical number of $f(x)$ with $f'(c) = 0$. If $f'(x)$ changes from

a) **positive to negative** at $x = c$, then $f(c)$ is a **local maximum** of $f(x)$.

b) **negative to positive** at $x = c$ then $f(c)$ is a **local minimum** of $f(x)$.

If $f'(x)$ does not change sign at $x = c$, then $f(c)$ is neither a maximum nor a minimum.

Concavity and the Second Derivative

The curvature of a function can be described qualitatively by its **concavity**. Think of a bowl: if it is right-side up, then it will hold its contents; if it is upside-down, then its contents will spill out (figure 19.3). We can carry over this idea to the shape of a curve. We will call a curve **concave up** if holds its contents, and **concave down** if it does not. *Formally*, a function $f(x)$ is said to be concave up in an interval (a, b) if the curve of $f(x)$ lies above all of its tangent lines; and concave down in (a, b) if it lies below all of its tangent lines.

Definition 19.2. Intervals of Concavity

Let $f(x)$ be twice differentiable on (a, b). We define the **intervals of concavity** by

a) If $f''(x) > 0$ on (a, b) then $f(x)$ is **concave up** on (a, b).

b) If $f''(x) < 0$ on (a, b) then $f(x)$ is **concave down** on (a, b).

A point where the concavity changes - either from concave up $(f'' > 0)$ to concave down $(f'' < 0)$ or concave down to concave up, is called an **inflection point**. At an inflection point $f''(c) = 0$ (figure 19.3). One should be cautioned that $f''(c) = 0$ does not ensure that $x = c$ is an inflection point (example 19.4).

Figure 19.3: Left and center images: Concave up curves resemble bowls that hold their contents, and concave down functions resemble curves whose contents will spill out. Right: Inflection points occur where the concavity changes.

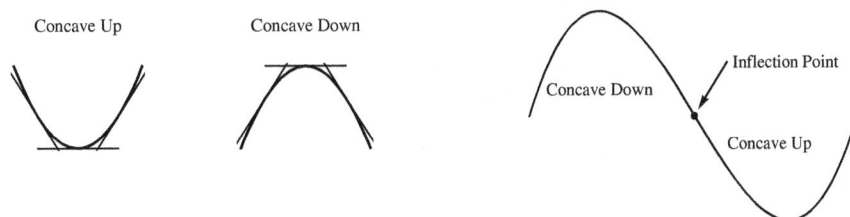

Definition 19.3. Inflection Point

An **inflection point** is a point where the concavity changes, either **from concave up to concave down**, or **from concave down to concave up**.

At an inflection point $f''(x) = 0$, although the converse is not necessarily true.

Example 19.2. Find the intervals of concavity of $y = 4x^3 - 12x^2 + 5$ from example 19.1.

Solution. The first and second derivatives are

$$y' = 12x^2 - 24x \tag{19.8a}$$
$$y'' = 24x - 24 \tag{19.8b}$$

The function is concave up when $y'' > 0$, i.e., when

$$24x - 24 > 0 \tag{19.8c}$$

Thus it is concave up when $x > 1$ and concave down when $x < 1$. Thus the intervals of concavity can be summarized as follows:

Intervals of Concavity	
$(-\infty, 1)$	Concave Down
$(1, \infty)$	Concave Up

The point where the concavity changes, $x = 1$, is an inflection point.   □

Critical Numbers and the Second Derivative Test

We can make two observations that lead us to formulate the second derivative test.

1. Suppose a function has a critical number c in a region where it is concave up (figure 19.4). Then there is some region around $x = c$ where $f''(x) > 0$, and the tangent line to $f(x)$ at c is horizontal. Since the function is concave up, then by definition of concave up, the tangent line must lie *beneath the plot of the function*. Thus the critical point is a local minimum. *Thus critical points where $f'(c) = 0$ and $f''(x) > 0$ are local minima.*

2. Similarly, if the function has a critical point c in a region where it is concave down, there is some region around $x = c$ where $f''(x) < 0$. The tangent line at c is still horizontal, but

now since the function is concave down, then by definition of concave down, the tangent line must lie *above the plot of the function.* Thus the critical point is a local maximum. Thus the critical point is a local minimum. *Hence critical points where $f'(c) = 0$ and $f''(x) < 0$ are local maxima.*

Figure 19.4: The second derivative test. A function is concave up ($f'' > 0$) at a local minimum and concaved down ($f'' < 0$) at a local minimum.

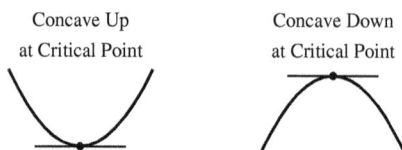

Concave Up
at Critical Point

Concave Down
at Critical Point

Theorem 19.3. Second Derivative Test

Suppose that $f(x)$ has a critical number at $x = c$ with $f'(c) = 0$. Then

a) If $f''(c) > 0$ then f has a local minimum at $x = c$.

b) If $f''(c) < 0$ then f has a local maximum at $x = c$.

If $f''(c) = 0$, then the second derivative test is inconclusive. In this case we can use use the first derivative test to determine the nature of the critical point at $x = c$.

First and Second Derivative Tests

	On an interval		
	Negative	Zero	Positive
$f'(x)$	Decreasing	Critical Point	Increasing
$f''(x)$	Concave Down	Possible inflection point	Concave Up

	At a critical point $x = c$ where $f'(c) = 0$		
	Negative	Zero	Positive
$f''(c)$	Minimum	Inconclusive Possible inflection point	Maximum

Example 19.3. Use the second derivative to classify the critical points of the function $y = 4x^3 - 12x^2 + 5$ studied in examples 19.1 and 19.5.

Solution. As before we find

$$y' = 12x^2 - 24x = 12x(x - 2) \tag{19.9a}$$
$$y'' = 24x - 24 = 24(x - 1) \tag{19.9b}$$

Critical numbers occur when $y' = 0$; thus they are $x = 0$ and $x = 2$. Since the function is everywhere differentiable, these are the only critical numbers.

Since $y''(0) = -24 < 0$, there is a local maximum at $x = 0$. Since $f(0) = 5$, the local maximum occurs at $(0, 5)$.

Since $y''(2) = 24 > 0$, there is a local minimum at $x = 2$. Since $f(2) = -11$, the local minimum occurs at $(2, -11)$. $\quad\square$

Example 19.4. Show that $y = \dfrac{3}{2}x^2 - x^3 + \dfrac{1}{4}x^4$ satisfies $f''(x) = 0$ at $x = 1$ but there is not an inflection point at $x = 1$.

Solution. Differentiating

$$y' = 3x - 3x^2 + x^3 \tag{19.10a}$$

$$y'' = 3 - 6x + 3x^2 = 3(x - 1)^2 \tag{19.10b}$$

Setting $y'' = 0$ gives $x = 1$, so $f''(1) = 0$. But since y'' is a perfect square, the second derivative is non-negative. Thus the function is never concave down. It has two intervals of concavity, both of them concave up: $(-\infty, 1)$ and $(1, \infty)$. The function has a single local and global minimum at $(0, 0)$. $\quad\square$

Example 19.5. Sketch the curve $y = x^4 - 4x^3$.

Solution. Calculating the first two derivatives we have

$$y' = 4x^3 - 12x^2 = 4x^2(x - 3) \tag{19.11a}$$

$$y'' = 12x^2 - 24x = 12x(x - 2) \tag{19.11b}$$

From the first equation we find critical points at $x = 0$ and $x = 3$. The second derivative test tells us nothing about the critical point at $x = 0$, since

$$f''(0) = 0 \tag{19.11c}$$

At $x = 3$ we have

$$f''(3) = 12(3)(3 - 2) > 0 \tag{19.11d}$$

which means that f is concave up at $x = 3$, so $f(3) = 3^4 - 4 \times (3^3) = -27$ is a local minimum.

The critical points divide the domain into three regions:

$$(-\infty, 0), \ (0, 3), \ (3\infty) \tag{19.11e}$$

These give the *intervals of increase and decrease*. To classify each interval, it is convenient to list them in a table. We pick a point (completely arbitrarily) in each interval, and calculate whether $y' > 0$ or $y' < 0$, using the formula $y' = 4x^2(x - 3)$ in each interval. It is best to pick points that are easy to perform calculations with. We choose the points $x = -1, x = 1, x = 4$. Then

Intervals of Increase and Decrease			
interval	point	$f'(x)$	Classification
$(-\infty, 0)$	-1	$(4)(-1)^2(-1 - 3) < 0$	decreasing
$(0, 3)$	1	$4(1)^2(1 - 3) < 0$	decreasing
$(3, \infty)$	4	$4(4)^2(4 - 3) > 0$	increasing

Since $f'(x)$ is decreasing on both sides of the critical point $c = 0$, we now know that the the critical point at $c = 0$ is neither a maximum nor a minimum, and can discard it from further consideration among the extrema.

critical point	classification
$(0, 0)$	Not an extremum
$(3, -27)$	Local minimum

Next we examine the intervals of convexity. These have endpoints when $y'' = 0$, so we have three intervals (from 19.11b)

$$(-\infty, 0), \ (0, 2), \ (2, \infty) \tag{19.11f}$$

We already know from having tested $x = 3$ that $(2, \infty)$ is concave down. Using $y'' = 12x(x - 2)$ and an arbitrary point in each interval, we build up the following table of the remaining intervals of convexity.

Intervals of Concavity			
interval	point	y''	Concavity
$(-\infty, 0)$	-1	$12(-1)(-3) > 0$	Up
$(0, 2)$	1	$12(1)(-1) < 0$	Down
$(2, \infty)$	3	$12(3)(1) > 0$	Up

Since the concavity changes at $x = 0$ and $x = 2$, we conclude that there are also inflection points at $(0,0)$ and $(2,-16)$. The curve is sketched in figure 19.5. □

Figure 19.5: Sketch of $y = x^4 - 4x^3$ (example 19.5).

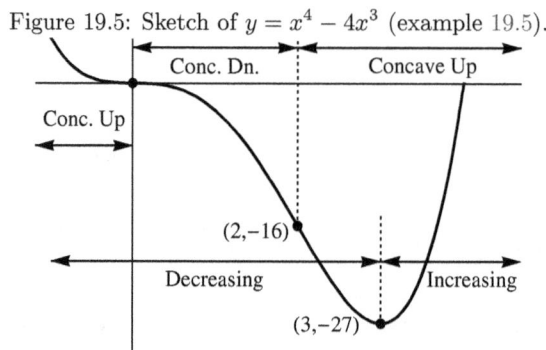

Example 19.6. Sketch $y = \dfrac{x}{(x - 2)(x + 2)}$.

Solution. The natural domain of $f(x)$ is all x except $x = \pm 2$, and these points are both vertical asymptotes of $f(x)$.

The only intercepts of either axis occur at the origin $(0, 0)$.

Both the positive and negative x axis are horizontal asymptotes because $\lim\limits_{x \to \infty} y = 0$ and $\lim\limits_{x \to -\infty} y = 0$. There are also vertical asymptotes at $x = \pm 2$.

We next observe that $f(x)$ is an odd function (see def. 25.2) since

$$f(-x) = \frac{-x}{x^2 - 4} = -f(x) \tag{19.12a}$$

so its plot will be reflected through the origin.

To see how the curves will approach the vertical asymptote we examine the sign of the function in each interval interrupted by a possible change in sign (x axis crossing or vertical axis). The function is positive when

$$x \cdot \frac{1}{x - 2} \cdot \frac{1}{x + 2} > 0 \tag{19.12b}$$

The vertical asymptotes and the x intercept divide the x axis into four possible such intervals, and we can use 19.12b to classify them.

interval	test point	Sign of f
$(-\infty, -2)$	-3	$(-)/(-)^2 < 0$
$(-2, 0)$	-1	$(-)/(-)(+) > 0$
$(0, 2)$	1	$(+)/(-)(+) < 0$
$(2, \infty)$	3	$(+)/(+)(+) > 0$

Thus the function approaches $-\infty$ from negative y, and approaches ∞ from positive y. Furthermore, this tells us that

$$\lim_{x \to -2^-} y = -\infty \qquad\qquad x < 0 \text{ and -2 is a vert. asymp.} \qquad (19.12c)$$

$$\lim_{x \to -2^+} y = \infty \qquad\qquad x > 0 \text{ and -2 is a vert. asymp.} \qquad (19.12d)$$

$$\lim_{x \to 2^-} y = -\infty \qquad\qquad x < 0 \text{ and 2 is a vert. asymp.} \qquad (19.12e)$$

$$\lim_{x \to 2^+} y = \infty \qquad\qquad x > 0 \text{ and 2 is a vert. asymp.} \qquad (19.12f)$$

To find the intervals of increase and decrease we differentiate,

$$y' = \frac{(x^2 - 4)(1) - x(2x)}{(x^2 - 4)^2} = -\frac{x^2 + 4}{(x^2 - 4)^2} < 0 \qquad (19.12g)$$

Since $y' < 0$ for all x in its domain, the function is always decreasing. There are no critical points and no local maxima or minima. The second derivative is

$$y'' = -\frac{(x^2 - 4)^2(2x) - (x^2 + 4)(2)(x^2 - 4)(2x)}{(x^2 - 4)^4} \qquad (19.12h)$$

$$= -\frac{2x(x^2 - 4)}{(x^2 - 4)^4}\left[(x^2 - 4) - 2(x^2 + 4)\right] = \frac{2x(x^2 + 12)}{(x^2 - 4)^3} \qquad (19.12i)$$

This has a single zero at $x = 0$, and gives the following intervals of concavity.

Figure 19.6: Sketch of $y = \dfrac{x}{(x - 2)(x + 2)}$ (example 19.6).

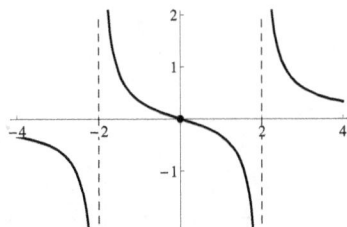

$(-\infty, -2)$	$y'' < 0$	Down
$(-2, 0)$	$y'' > 0$	Up
$(0, 2)$	$y'' < 0$	Down
$(2, \infty)$	$y'' > 0$	Up

There is an inflection point at $x = 0$, but not at the other places where $f(x)$ changes concavity because they are asymptotes. Having compiled all the necessary information, the function is sketched in figure 19.6.   □

Example 19.7. Classify all the extrema of $f(x) = x^2 e^{-x}$.

Solution. Differentiating and simplifying,

$$f'(x) = x^2 \frac{d}{dx} e^{-x} + e^{-x} \frac{d}{dx} x^2 \tag{19.13a}$$

$$= x^2 \cdot (e^{-x}) \cdot (-1) + e^{-x} \cdot (2x) \tag{19.13b}$$

$$= e^{-x} (2x - x^2) \tag{19.13c}$$

$$= e^{-x} x (2 - x) \tag{19.13d}$$

Critical points occur when $f'(x) = 0$; since $e^{-x} > 0$ for all x, the only critical points are $x = 0$ and $x = 2$. There are no end points because the domain of $f(x)$ is all real numbers.

The second derivative is

$$f''(x) = \frac{d}{dx} \left(e^{-x} (2x - x^2) \right) \tag{19.13e}$$

$$= e^{-x} \frac{d}{dx} (2x - x^2) + (2x - x^2) \frac{d}{dx} e^{-x} \tag{19.13f}$$

$$= e^{-x} \cdot (2 - 2x) + (2x - x^2) \cdot e^{-x} \cdot (-1) \tag{19.13g}$$

$$= e^{-x} \left(2 - 4x + x^2 \right) \tag{19.13h}$$

Therefore

$$f''(0) = e^{-0} \left(2 - 0 + 0 \right) = 2 > 0 \tag{19.13i}$$

$$f''(2) = e^{-2} \left(2 - 8 + 4 \right) = -2e^{-2} < 0 \tag{19.13j}$$

Furthermore,

$$f(0) = 0 \text{ and } f(2) = 2^2 e^{-2} \approx 0.541 \tag{19.13k}$$

At $x = 0$, $f(x)$ is concave up, and so $(0, 0)$ is a local minimum.

At $x = 2$, $f(x)$ is concave down, and so $(0, .541)$ is a local maximum. Since $f(1) = (-1)^2 e(-(-1)) = e \approx 2.7 > f(2)$, we conclude by counter-example that $(0, .541)$ is not a global maximum.

$\square$

Figure 19.7: Plot of $y = x \ln x$ on $(0, \infty)$ showing the absolute minimum at $x = 1/e$.

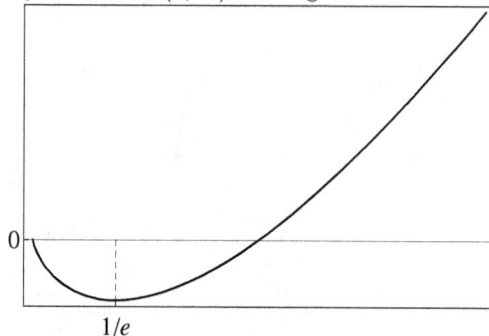

Example 19.8. Find the absolute minimum of $y = x \ln x$ on $(0, \infty)$.

Solution. Differentiating,

$$y' = (x \ln x)' = x(\ln x)' + (x)'(\ln x) = x \cdot \frac{1}{x} + 1 \cdot \ln x = 1 + \ln x \qquad (19.14a)$$

Critical points occur when $y' = 0$ or $\ln x = -1$. Exponentiating, this means $x = e^{-1} = 1/e \approx 0.368$. To verify that this is a minimum we find the second derivative,

$$y'' = 1/x > 0 \text{ for all } x > 0 \qquad (19.14b)$$

Since the second derivative is positive, the function is concave up, and thus the critical point is a minimum (figure 19.7). □

Curve Sketching Procedure

1. Determine the **natural domain** of the function.
   Exclude division by zero and square roots of negative numbers.

2. Find all the **intercepts** in the natural domain, if they exist.
   Set $x = 0$ and solve for y; set $y = 0$ and solve for x.

3. Look for **symmetry**.
   Is $f(x) = f(-x)$ (even function)? $f(x) = -f(-x)$ (odd)? (defs. 25.1, 25.2.)

4. Look for **asymptotes**.
   Horizontal: $\lim\limits_{x \to \infty} f(x) = L$ or $\lim\limits_{x \to -\infty} = M$
   Vertical: any of $\lim\limits_{x \to a^\pm} f(x) = \pm \infty$

5. Determine points where $f' = 0$ and $f'' = 0$.

6. Find the value of $f''(c)$ where $f'(c) = 0$.
   Use the second derivative test to classify these points as local maxima or minima.

7. Find any **other critical points** where f' is undefined.

8. Identify regions of **increase** and **decrease** from sign of f'.
   Pick an arbitrary $x = p$ in each interval between each pair of critical points and determine the sign of $f'(p)$.

9. Identify regions of **concavity** and **inflection** points from f''.
   Pick an arbitrary $x = q$ in each interval between each pair of points where $f''(x) = 0$ and determine the sign of $f''(q)$.
   Inflection points only occur when $f''(q) = 0$ and the the sign of $f''(x)$ is different on either side of q.

10. Use the first derivative test to **identify any extrema** where the second derivative test failed or did not apply because the $f'(c)$ does not exist at the critical point.

11. Sketch!

Exercises

Sketch each of the following curves. Label the intervals of increase, the intervals of decrease, the local and global maxima and minima, the inflection points, and the regions of concavity.

1. $y = 6x^3 + 27x^2 - 324 - 3$

2. $y = x^5 (x+6)^2$

3. $y = 12x^5 + 60x^4 - 100x^3 + 4$

4. $y = 4(x-2)^{2/3}$

5. $y = \dfrac{3x+8}{6x+3}$

6. $y = -2x^3 + 30x^2 - 96x + 10$

7. $y = 3x + \frac{3}{x}$

8. $y = x^{2/3}(6-x)^{1/3}$

9. $f(x) = 4x^3 + 3x^2 - 6x + 1$

10. $f(x) = \sin x + \cos x$ on $[0, 2\pi]$

11. $f(x) = \dfrac{x}{x^2 + 1}$

12. $h(x) = (x+1)^5 - 5x - 2$

13. $h(x) = 5x^3 - 3x^5$

14. $y = x^2 e^{3x}$

15. $y = \dfrac{e^{2x}}{8x - 5}$

Chapter 20

Optimization

Chapter Summary and Goal

In an optimization problem we try to find the value of some variable, say x, that best meets some collection of conditions. We call these conditions **constraints**. We are given optimization problems in words, and the difficult part is translating these word problems into the appropriate equations. Sometimes there are several different equations related to one another, similar to related rates problems. Once we are able to perform this translation, we are able to find a appropriate maximum or minimum value in the appropriate domain using the first and/or second derivative tests.

Student Learning Objectives

The student will:
1. Identify key words and phrases in word problems and convert them to variables.
2. Identify the variable to be optimized in the problem.
3. Translate optimization problems to symbolic (mathematical) format.
4. Solve optimization problems by finding extrema.

Solving Optimization Problems

The hardest part of an optimization problem is translating it from a word problem to its mathematical representation. When reading the problem try to identify the following:
1. The **quantity** to be **optimized**.
2. The **constraints** in the problem.
3. The **variables** that **relate** the quantity to be optimized to the constraints in the problem.
4. The **domain** of the variables in the problem

Once the quantity to be optimized is identified and the equation in item 1 is known, it needs to be simplified to an equation in a single variable. You can usually do this by making substitutions from the constraint equations in item 2. Once you have completed making your substitutions you should have an equation in a single variable. Using the first and second derivative tests, you can use the techniques you have learned to find the appropriate maximum or minimum (remembering, of course, to check for endpoints of the domain when appropriate).

Figure 20.1: Geometry used in example 20.1.

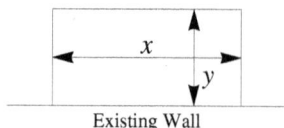

Existing Wall

Example 20.1. Suppose you just bought a new house with a large walled back yard. You want to fence in a small rectangular area for your pets to play in. You have only 150 feet of fence. What are the dimensions of the largest area you can fence in, assuming that one edge of the play area is against an existing wall and that you will use up all of your fence?

Solution. The geometry is illustrated in figure 20.1. Define variables as follows:

x length of fence parallel to existing wall
y length of fence perpendicular to existing wall
A area of enclosed region
P perimeter of enclosed region

We are given that we have 150 feet of fence; thus the perimeter $P = 150$ is fixed if we want to use up all of our fence. We do this by using some of it for the x side and some of it for the two y sides, so that

$$x + 2y = P = 150 \tag{20.1a}$$

We want to enclose the largest possible area, so we need a formula for area. The area of a rectangle is

$$A = xy \tag{20.1b}$$

To get the largest possible area we need to maximize A with respect to either x or y. We do this by eliminating one of the variables:

1. Solve the P equation either for x or for y (your choice);
2. Substitute the result into the A equation;
3. Differentiate the resulting A equation with respect to the remaining variable to find the critical points ($A' = 0$);
4. Find the appropriate critical point and verify that it is an maximum by checking A''.

We can proceed by first solving the P equation for $x = 150 - 2y$. Substituting into the A equation gives

$$A = xy = y(150 - 2y) = 150y - 2y^2 \tag{20.1c}$$

Differentiating with respect to y,

$$A' = 150 - 4y \tag{20.1d}$$

The critical point occurs when $y = 150/4 = 37.5$ We know that this is a maximum because $A'' = -4 < 0$. To get the remaining dimension, use the P equation, $x = 150 - 2y = 150 - 2(37.5) = 75$. Thus the largest region that can be encloses is 37.5×75 feet. $\square$

Example 20.2. Find two positive numbers whose product is 676 and whose sum is a minimum.

Solution. We want to find x and y such that their product $P = xy = 676$ and their sum $S = x + y$ is minimized. The number to be optimized, then, is S, since it must be minimized.

We can solve the P equation for $y = P/x = 676/x$ so that we can eliminate y from the S equation

$$S = x + y = x + \frac{676}{x} \tag{20.2a}$$

Differentiating with respect to x,

$$S' = 1 - \frac{676}{x^2} \tag{20.2b}$$

Critical numbers occur at

$$x^2 = 676 \tag{20.2c}$$

or at

$$x = \pm\sqrt{676} = \pm 26 \tag{20.2d}$$

We rule out the negative value because the problem statement specifically says to find two positive numbers. To verify that the sum is a minimum we use the second derivative test:

$$S'' = \frac{3 \cdot 676}{x^3} > 0 \tag{20.2e}$$

Since $S''(x) > 0$ for all positive x, $S(26)$ is a minimum. The corresponding y value is $y = 676/x = 676/26 = 26$. The sum is $S = 26 + 26 = 52$. □

Example 20.3. Suppose you are a manufacturer of juice boxes and want to make a rectangular box to hold 750 ml with a square bottom that uses the least amount of material. Find the dimensions of the box.

Solution. The box needs to have four sides, which are all the same size; and a top and a bottom, which are the same size as each other. Let the dimensions of the box be $a \times a \times h$ where h is the height and the square cross sections is $a \times a$.

The top and the bottom will each have area a^2.

The sides will each have area ah.

Then the total area will be

$$A = 4ah + 2a^2 \tag{20.3a}$$

The volume will be

$$V = a^2h = 750 \text{ cm}^3 \tag{20.3b}$$

Solving the volume equation for $h = 750/a^2$, we substitute it into the area equation:

$$A = 4a\left(\frac{750}{a^2}\right) + 2a^2 = \frac{3000}{a} + 2a^2 \tag{20.3c}$$

Differentiating,

$$\frac{dA}{da} = -\frac{3000}{a^2} + 4a \tag{20.3d}$$

The only critical point occurs when

$$a^3 = 750 \tag{20.3e}$$

or $a = (750)^{1/3} \approx 9.0856$ cm. The corresponding height is $h = 750/a^2 = 750/(750^{2/3}) = 750^{1/3} = a$. Thus the optimal solution is a cube. It is a minimum because $d^2A/da^2 = 6000/a^2 > 0$. □

Example 20.4. Find the point on the line $y = 23x - 17$ that is closest to the origin, and find the distance.

Solution. The distance from the origin to a point (x, y) is

$$s = \sqrt{x^2 + y^2} \tag{20.4a}$$

Letting the point (x, y) be on the line $y = 23x - 17$ then

$$s = \sqrt{x^2 + (23x - 17)^2} \tag{20.4b}$$

Differentiating,

$$\frac{ds}{dx} = \frac{1}{2\sqrt{x^2 + (23x - 17)^2}} \cdot [2x + 2(23x - 17) \cdot 23] \tag{20.4c}$$

Critical points occur when the numerator is zero,

$$2x + 2(23x - 17) \cdot 23 = 0 \tag{20.4d}$$

Solving for x gives $x = 391/530 \approx 0.73774$. Hence

$$y = 23x - 17 = 23\left(\frac{391}{530}\right) - 17 = \frac{8993}{530} - 17 = -\frac{17}{530} \approx -0.0321 \tag{20.4e}$$

The total distance from the origin is

$$s = \sqrt{\left(\frac{391}{530}\right)^2 + \left(-\frac{17}{530}\right)^2} = \frac{17}{\sqrt{530}} \approx .7384 \tag{20.4f}$$

□

Remark. In example 20.4, it would have been easier to calculate the derivative if we had known in advance that the square root in the denominator was going to cancel out (going from (20.4c) to (20.4d)). By squaring the distance, we would not have gotten the square root in the denominator in the first place, and, as it turns out, the distance in minimized whenever the square of the distance is minimized.

Definition 20.1. Distance Formula

The distance between two points $P = (x_1, y_1)$ and $Q = (x_2, y_2)$ is

$$d(P, Q) = \sqrt{(x_1 - x_2)^2 + (y_1 - y_2)^2} \tag{20.5}$$

Definition 20.2. Squared Distance Formula

The square of the distance between two points $P = (x_1, y_1)$ and $Q = (x_2, y_2)$ is

$$d(P, Q)^2 = (x_1 - x_2)^2 + (y_1 - y_2)^2 \tag{20.6}$$

Remark 20.1. Minimizing Distances

The distance is minimized whenever the square of the distance is minimized.

Example 20.5. Find the nearest point on the line $y = x - 4$ to the parabola $y = x^2$.

Solution. A point on the line is given by

$$P = (x, y) = (x, x - 4) \tag{20.7a}$$

and a point on the parabola is given by

$$Q = (x, y) = (x, x^2) \tag{20.7b}$$

The square of the distance between the points P and Q is given by

$$f(P, Q) = (x - x)^2 + ((x - 4) - x^2)^2 = ((x - 4) - x^2)^2 \tag{20.7c}$$

This is minimized when $f'(x) = 0$,

$$0 = 2(x - 4 - x^2)(1 - 2x) \tag{20.7d}$$

Since there is no real solution to $x - 4 - x^2 = 0$ the only solution occurs when $1 - 2x = 0$ or at $x = 1/2$. The nearest point on the line to the parabola occurs at $(1/2, -7/2)$. □

Example 20.6. A rectangle is inscribed with its base on the x-axis and its upper corners on the parabola $y = 12 - x^2$. What are the dimensions of such a rectangle with the greatest possible area?

Solution.   The geometry is illustrated in figure 20.2. Let the rectangle have a width of $2x$ and height y in such a way that a point on the parabola, given by (x, y), is on the top right hand corner of the rectangle. We want to maximize the area, A, of the rectangle

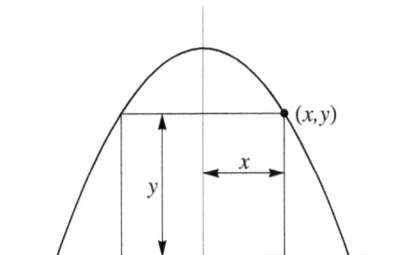

Figure 20.2: Geometry for example 20.6.

$$A = 2xy \tag{20.8a}$$

subject to the coordinates (x, y) lying on the parabola:

$$y = 12 - x^2 \tag{20.8b}$$

Substituting for y into the A equation gives

$$A = 2x(12 - x^2) = 24x - 2x^3 \tag{20.8c}$$

Differentiating twice,

$$A' = 24 - 6x^2 \tag{20.8d}$$

$$A'' = -12x \tag{20.8e}$$

The critical points occur when

$$0 = A' = 24 - 6x^2 \tag{20.8f}$$

Hence $x = \pm 2$. We choose the positive square root $x = 2$ to ensure that $A''(x) < 0$ (hence the function is concave down and at a local maximum). The y-coordinate is

$$y = 12 - x^2 = 12 - 4 = 8 \tag{20.8g}$$

Therefore the upper right hand corner is at (2,8). The dimensions of the rectangle are 4×8 and its area is 32. □

Exercises

1. Find the point on the line $3x + 3y + 5 = 0$ that is closest to the point (-3,-2).

2. Find the point on the line $\dfrac{x}{4} + \dfrac{y}{2} = 1$ that is closest to the point (5,1).

3. A rectangle is inscribed with its base on the x axis and its upper corners on the parabola $y = 20 - x^2$. What are the dimensions of the rectangle with the greatest possible area?

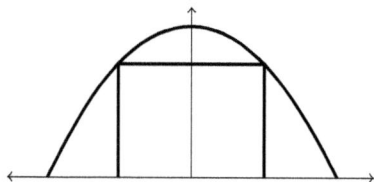

4. You have one square meter of cardboard for a school project. What are the dimensions of the largest rectangular box you can construct from this material if the box does not require a top, i.e., it will only have four sides and a bottom?

5. Find the dimensions of a 300 ml. cylindrical beer can that uses the minimum amount of material.

6. A box is to be made out of a 12 inch by 24 inch piece of cardboard. Squares of equal size will be cut out of each corner, and then the ends and sides will be folded up to form a box with an open top. Find the dimensions of the resulting box that has the largest volume.

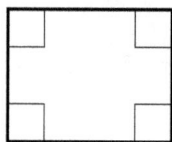

7. A cylinder is inscribed in a right circular cone of height 12 cm. and radius at the base equal to 10 cm. What are the dimensions of the cylinder of maximum volume?

8. A car rental agency rents 150 cars per day at a rate of 75 dollars per day. For each 1 dollar increase in the daily rate, 3 fewer cars are rented. At what rate should the cars be rented to produce the maximum income, and what is the maximum income?

9. A piece of wire 36 inches long is cut into two pieces. One piece is bent into a square, and the other is bent into an equilateral triangle. How should the wire be cut so that the total area is (a) maximized? (b) minimized?

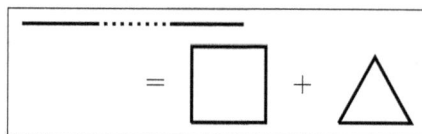

10. Find the positive number that minimizes the sum of itself plus 17 times its reciprocal.

11. Suppose you want to choose a rectangular region of area 4000 square feet in the following way. You will surround it with fence, and you also want to divide the region in half with fence. The fence that divides the region in half will be parallel to one of the sides. What is the minimum total amount of fence in linear feet needed to do this?

Chapter 21

Finding Roots: Newton's Method

Chapter Summary and Goal

The value of x where an equation in x intersects the x axis is called a **root**. More precisely, we say that a root is solution of an equation of the form $f(x) = 0$. We are already familiar with finding roots when we factor equations such as quadratics. For example, the roots of $x^2 - 6x + 8 = 0$ are $x = 2$ and $x = 4$.

Sometimes, however, we are faced with equations whose curves intersect the x axis but we do not know how to find a formula for the value of the roots. For example, the intermediate value theorem (theorem 7.11) tells us that the function $f(x) = \sin(\cos x^2) - x^3$ has a root between 0.8 and 1.2, because the function is continuous, $f(0.8) \approx 0.207 > 0$, and $f(1.2) \approx -1.598 < 0$, but it does not give us a any way to solve for the value of the root.

In this chapter we will study a procedure, **Newton's Method**, that tells us how to find the root of any differentiable function.

Student Learning Objectives

The student will:

1. Be able to explain in words what a root is.
2. Understand the concept of iteration.
3. Understand the derivation of Newton's method.
4. Be able to calculate roots using Newton's method.

Finding the Root

Newton's method is an **iterative technique**(figure 21.1) to find the root of an equation. That means we start with a first guess (call it p_0), and then use a formula to make a better guess (call it p_1). If we are happy with p_1, we stop there; if not, we use the formula again to get another guess (p_2), and so forth. We keep repeating, inserting guess p_n into our iteration formula to get estimate p_{n+1}, stopping with p_{n+1} if we are happy with it; but continuing to iterate if we are not happy.

Newton's Method

Newton's Method is an **iterative technique** used solve an equation $f(x) = 0$ for the value of x. This number x is called the **root** of the equation.

Figure 21.1: Iteration process. An initial guess is fed into an iteration formula which produces a new guess. If we are happy with the guess, we stop. If not, we feed the guess back into the formula.

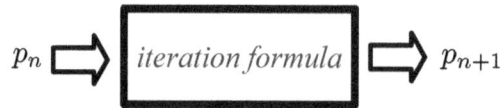

$$p_n \quad \Rightarrow \quad \boxed{iteration\,formula} \quad \Rightarrow \quad p_{n+1}$$

Control of when we stop is determined by an input **error tolerance** ϵ. This is the error in the result that we will be happy with. The most common stopping condition is when

$$|p_{n+1} - p_n| < \epsilon \tag{21.1}$$

where the user (you) specify the value ϵ. In other words, suppose you want to find the root to an accuracy of $\epsilon = 0.001$. This means you keep repeating the iteration process until the difference between two successive solutions is less than 0.001 in absolute value.

Figure 21.2: Newton's method follows the slope of the tangent line to the x axis to find the next guess.

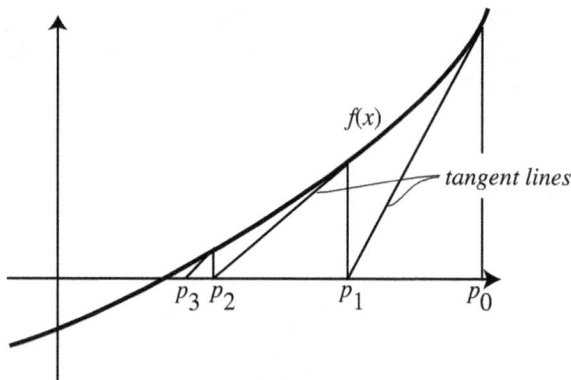

The idea behind Newton's method is to follow the slope of the tangent line. We start by labelling the location of our guess p_0 on the x axis and then project up to $f(x)$ (figure 21.2). At the point $(p_0, f(p_0))$ we construct a tangent line back down from $f(x)$ to the x axis. At some point $(p_1, 0)$, the tangent line intersects the x axis. This intersection should give us a **better guess**, as illustrated in fig. 21.2.

At p_1 we project up to the curve of $f(x)$, construct a new tangent line, follow it to the x axis, and find p_2, and **even better guess** than p_1.

We repeat the process at p_2, p_3, etc. As we keep repeating this process we get closer and closer to the root.

To get a formula for p_1 as a function of p_0, we observe that the slope of the tangent line at

$(p_0, f(p_0))$ if $f'(p_0)$, and therefore the equation of the tangent line is

$$y = f(p_0) + f'(p_0)(x - p_0) \tag{21.2}$$

The value of p_1 is the value of x when this line intersects the x axis. So we set $x = p_1$ and $y = 0$, to give us

$$0 = f(p_0) + f'(p_0)(p_1 - p_0) \tag{21.3}$$

$$-f'(p_0)(p_1 - p_0) = f(p_0) \tag{21.4}$$

$$p_1 - p_0 = -\frac{f(p_0)}{f'(p_0)} \tag{21.5}$$

$$p_1 = p_0 - \frac{f(p_0)}{f'(p_0)} \tag{21.6}$$

Repeating the process at each successive step

$$p_2 = p_1 - \frac{f(p_1)}{f'(p_1)} \tag{21.7}$$

$$p_3 = p_2 - \frac{f(p_2)}{f'(p_2)} \tag{21.8}$$

$$p_4 = \cdots \tag{21.9}$$

Theorem 21.1. Newton's Method

To find the root of a differentiable function $f(x) = 0$, iterate on

$$p_{n+1} = p_n - \frac{f(p_n)}{f'(p_n)} \tag{21.10}$$

Example 21.1. Find $\sqrt{2}$ using Newton's method to 3 decimal places starting with $p_0 = 2$.

Solution. Since $\sqrt{2}$ is a root of $x^2 = 2$, we can use $f(x) = x^2 - 2 = 0$ in Newton's method. Since $f'(x) = 2x$, the iteration formula is

$$p_{n+1} = p_n - \frac{f(p_n)}{f'(p_n)} \tag{21.11a}$$

$$= p_n - \frac{p_n^2 - 2}{2p_n} \tag{21.11b}$$

$$= \frac{2p_n^2 - p_n^2 + 2}{2p_n} \tag{21.11c}$$

$$= \frac{p_n^2 + 2}{2p_n} \tag{21.11d}$$

$$= \frac{p_n}{2} + \frac{1}{p_n} \tag{21.11e}$$

The first several iterations are then

$$p_1 = \frac{2}{2} + \frac{1}{2} = \frac{3}{2} \approx 1.5 \tag{21.11f}$$

$$p_2 = \frac{1.5}{2} + \frac{1}{1.5} \approx 1.41666 \tag{21.11g}$$

$$p_3 = \frac{1.4166}{2} + \frac{1}{1.4166} \approx 1.4142 \tag{21.11h}$$

$$p_4 = \frac{1.4142}{2} + \frac{1}{1.4142} \approx 1.4142 \tag{21.11i}$$

Since the last two estimates are the same to 4 decimal places, the answer is 1.4142.   □.
□

Example 21.2. Find the value of x satisfies $x = \cos x$.

Solution. To find the root of $f(x) = x - \cos x$ we calculate $f'(x) = 1 + \sin x$ and form the Newtons' method formula, which gives

$$p_{n+1}(x) = p_n - \frac{f(p_n)}{f'(p_n)} \tag{21.12a}$$

$$= p_n - \frac{p_n - \cos p_n}{1 + \sin p_n} \tag{21.12b}$$

$$= \frac{\cos p_n + p_n \sin p_n}{1 + \sin p_n} \tag{21.12c}$$

Since we don't have any idea of where the root is, any starting guess is as good as any other. Say we start with $p_0 = 1$. Then

$$p_1 = \frac{\cos 1 + (1) \sin 1}{1 + \sin 1} = \frac{0.540302 + .841471}{1 + .841471} = 0.750364 \tag{21.12d}$$

$$p_2 = \frac{\cos .750364 + (.750364) \sin .750364}{1 + \sin .750364} = .739113 \tag{21.12e}$$

$$p_3 = \frac{\cos .759113 + (.759113) \sin .739113}{1 + \sin .739113} = .739085 \tag{21.12f}$$

$$p_4 = \frac{\cos .739085 + (.739085) \sin .739085}{1 + \sin .739085} = .739085 \tag{21.12g}$$

So to three decimals $x = .739085$ radians.   □

Example 21.3. Find the root of $f(x) = x^3 - x$ using Newton's method starting with $x = 2$, to 2 decimal places.

Solution. We first observe that we could calculate the root exactly, by factoring:

$$0 = f(x) = x^3 - x = x(x^2 - 1) = x(x - 1)(x + 1) \tag{21.13a}$$

So there are really three roots, at $x = -1, 0, 1$.

Let's see what Newton's method tells us. We are told to start with a first guess of $p_0 = 2$. Since

$$f'(x) = 3x^2 - 1, \tag{21.13b}$$

Newton's Iteration Formula gives

$$p_{n+1} = p_n - \frac{x^3 - x}{3x^2 - 1} \tag{21.13c}$$

Starting with $p_0 = 2$, the next several iterations give us

$$p_1 = 2 - \frac{2^3 - 2}{3(2^2) - 1} = 2 - \frac{6}{11} = \frac{16}{11} \approx 1.45455 \tag{21.13d}$$

$$p_2 = 1.45 - \frac{(1.45)^3 - (1.45)}{3(1.45)^2 - 1} \approx 1.14 \tag{21.13e}$$

$$p_0 = 1.14 - \frac{1.14^3 - (1.14)}{3(1.14)^2 - 1} \approx 1.02 \tag{21.13f}$$

$$p_4 = 1.02 - \frac{1.02^3 - (1.02)}{3(1.02)^2 - 1} \approx 1.00 \tag{21.13g}$$

$$p_5 = 1.00 - \frac{1.00^3 - (1.00)}{3(1.00)^2 - 1} = 1 \tag{21.13h}$$

In this example, if we had kept more decimals, our first five iterations would be

$$2.0, 1.45455, 1.15105, 1.02533, 1.00091, 1.000001, \ldots . \tag{21.13i}$$

If we had started with a different p_0, say $p_0 = 2.1$, we would get

$$2.1, 1.51447, 1.18133, 1.03470, 1.01671, 1.000004, \ldots \tag{21.13j}$$

which still converges to 1.0. These sequences demonstrate how

$$p_k \to p \text{ as } k \to \infty \tag{21.13k}$$

If we wanted to find the other roots we'd have to start with different values of p_0; we will discuss this further below. If we start at, say, $p_0 = 0.3$, then we would find

$$0.3, -0.07397, 0.000823, -1.1151 \times 10^{-9}, 0, \ldots \tag{21.13l}$$

which converges to the root at $x = 0$; and if we were to start at $p_0 = -1.5$,

$$-1.5, -1.17391, -1.032307, -1.00146, -1.000003, -1.0, \ldots \tag{21.13m}$$

which converges to the root at $x = -1$. This show that to find all the roots we have to start with different guesses (see figure 21.3). Unfortunately, since we don't, in general, know how many roots there are, we don't know how many times to keep making new guesses. Furthermore, we won't, in general, know where the roots are, so we will not have a good idea of where to place our starting guess. $\qquad\square$

Figure 21.3: Illustration of convergence domains for Newton's method in example 21.3. The triangles show initial guesses at $x = -1.5$, $x = 3$, and $x = 2$, and the arrows illustrate that the successive guesses will eventually converge to the roots at $x = -1$, $x = 0$, and $x = 1$.

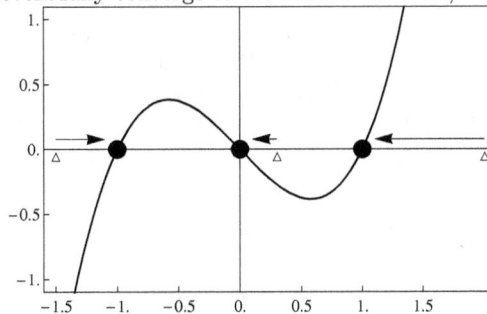

This example reminds us that a single iteration process with Newton's method is only guaranteed to find a single root. To find other roots, we have to start with other initial guesses. Unfortunately, there is no foolproof method to ensure that we will find every root.

Convergence

If we define

$$g(x) = x - \frac{f(x)}{f'(x)} \tag{21.14}$$

for any continuously differentiable function $f(x)$, then the list of numbers $p_0, p_1, p_2, p_3, \ldots,$ where

$$\left.\begin{array}{l} p_1 = g(p_0) \\ p_2 = g(p_1) \\ p_3 = g(p_2) \\ \vdots \end{array}\right\} \tag{21.15}$$

is a **converging sequence**. This sequence converges to a root of $f(x)$. We will have more to say about convergent sequences in chapter 38.

Theorem 21.2. Convergence of Newton's Method

The sequence of numbers 21.15 defined by equation 21.14,

$$p_0, p_1, p_2, p_3, \ldots \tag{21.16}$$

converges to a root of $f(x)$ as $n \to \infty$, i.e., $\lim\limits_{n \to \infty} p_n = p$ where p is a root of $f(x)$, for some p_0 in the domain of $f(x)$.

Example 21.4. Use Newton's method to estimate the root of $x^4 = 1 + x$ to three decimal places.

Solution. We first write the function as

$$g(x) = x^4 - 1 - \frac{x}{2} = 0 \tag{21.17a}$$

To estimate where the roots are we find extrema:

$$g'(x) = 4x^3 - \frac{1}{2} \tag{21.17b}$$

Critical points occur when $g' = 0$, or when $4x^3 = 1/2$. This occurs when $x = 1/2$.

Since $f(x)$ is a polynomial, there are no asymptotes, and the function is everywhere differentiable. Hence we can safely use the second derivative to test the local extrema. Since $g''(x) = 12x^2 > 0$ (everywhere) the function everywhere concave up. Thus a local minimum occurs at $x = 1/2$. Since $f(x)$ is a quartic, this is also an absolute minimum.

At the minimum we find that

$$g(1/2) = (1/2)^4 - 1 - (1/2)/2 = \frac{1}{16} - 1 - \frac{1}{4} \tag{21.17c}$$

$$= \frac{1 - 16 - 4}{16} = \frac{-19}{16} < 0. \tag{21.17d}$$

Since the minimum is below the x axis, it opens upward, and $f(x) \to \infty$ as $x \to \pm\infty$, we know that $f(x)$ must cross the x axis twice, once on either side of the minimum. Thus there are two roots. One root is in the interval $(-\infty, 1/2)$ and the other is in the interval $(1/2, \infty)$.

Next, to find the roots, we write down the formula for Newton's method.

$$n(x) = x - \frac{f(x)}{f'(x)} \tag{21.17e}$$

$$= x - \frac{x^4 - 1 - x/2}{4x^3 - 1/2} \tag{21.17f}$$

$$= x - \frac{2x^4 - x - 2}{8x^3 - 1} \tag{21.17g}$$

$$= \frac{2(3x^4 + 1)}{8x^3 - 1} \tag{21.17h}$$

To find the root, a good rule-of-thumb is to start with a guess in each interval where a root is known to occur (this is not always enough; sometimes we have to start even closer to the root). If we start with $p_0 = 1$, then are successive guesses are

$$p_1 = \frac{2(3(1) + 1)}{8(1) - 1} = \frac{8}{7} \approx 1.14286 \tag{21.17i}$$

$$p_2 = \frac{2(3(1.14286^4) + 1)}{8(1.14286^3) - 1} \approx 1.11827 \tag{21.17j}$$

$$p_3 = \frac{2(3(1.11827^4) + 1)}{8(1.11827^3) - 1} \approx 1.11735 \tag{21.17k}$$

$$p_4 = \frac{2(3(1.11735^4) + 1)}{8(1.11735^3) - 1} \approx 1.11735 \tag{21.17l}$$

Since the last two estimates agree to 5 decimal places, the first root root is 1.11735 to five decimal places, or 1.117 to three decimal places.

To find the other root we can try $p_0 = -1$. The first four guess give us

$$p_1 = -0.888889 \tag{21.17m}$$

$$p_2 = -0.868117 \tag{21.17n}$$

$$p_3 = -0.867471 \tag{21.17o}$$

$$p_4 = -0.867471 \tag{21.17p}$$

so to three decimal places the second root is at -0.867. □

Exploring Newton's Method in a Spreadsheet

Newton's Method is extremely easy to implement numerically, once you have formulas for $f(x)$ and $f'(x)$. Example ?? illustrates this in LibreOffice Calc.

Example 21.5. Find a root of $y = x^2 - 7x + 3$ using Newton's Method.

Solution Differentiating, we find that $y' = 2x - 7$. We will solve this numerically by implementing the formula

$$x_{n+1} = x_n - f(x_n)/f'(x_n) \tag{21.18}$$

in a spreadsheet. The idea is to use each column for a separate variable. For convenience, we will use the first column as an iteration counter, and the second column for the value of the variable x.

Say, for example, that x is in cell **F1**. Then the syntax in any spreadsheet for the polynomial $x^2 - 7x + 3$ in any spreadsheet is

```
=F5^2-7*F5+3
```

Thus if we place the text =F5^2-7*F5+3 in any cell of the spreadsheet, it will know to look into cell F5 and substitute the value from that cell in place of the string F5 in the formula.

For convenience, we place this text in cell G5, just to the right of our first value for x.

	E	F	G	H	I
4	n	x	f(x)	f'(x)	x-f(x)/f'(x)
5	1	1	=F5^2-7*F5+3		
6					
7					
8					
9					

Similarly, the formula for the derivative is =2*F5-7, so we place that in cell H5.

	E	F	G	H	I
4	n	x	f(x)	f'(x)	x-f(x)/f'(x)
5	1	1	=F5^2-7*F5+3	=2*F5-7	
6					
7					
8					
9					

The formula for $x - f(x)/f'(x)$ is then =F5-G5/H5 We can type that into cell I5.

	E	F	G	H	I
4	n	x	f(x)	f'(x)	x-f(x)/f'(x)
5	1	1	=F5^2-7*F5+3	=2*F5-7	=F5-G5/H5
6					
7					
8					
9					

To number the iterations, we type =1+E5 in cell E6. Here's a first trick: we don't have to type the cell name. Whenever we want a cell name like E5, just click on the cell. Then we copy the formula =1+E5 (using the cut and paste menu) into cells E5 through E9. The spreadsheet magically changes the number, knowing that you want the relative formula to the cell just above you, not a direct reference to cell E5 in each case.

	E	F	G	H	I
4	n	x	f(x)	f'(x)	x-f(x)/f'(x)
5	1	1	=F5^2-7*F5+3	=2*F5-7	=F5-G5/H5
6	=1+E5				
7	=1+E6				
8	=1+E7				
9	=1+E8				

To copy the first update into the next update, we type =I5 into cell F6 and then copy it down to the bottom of the table. This takes the result of the last calculation and copies it over to the start of the next calculation

Figure 21.4: Screen grab of spreadsheet used in example 21.5. Observe how the equation is displayed for the selected cell in the equation bar. The entire calculation could be repeated with a different starting value merely by changing the number in cell F5.

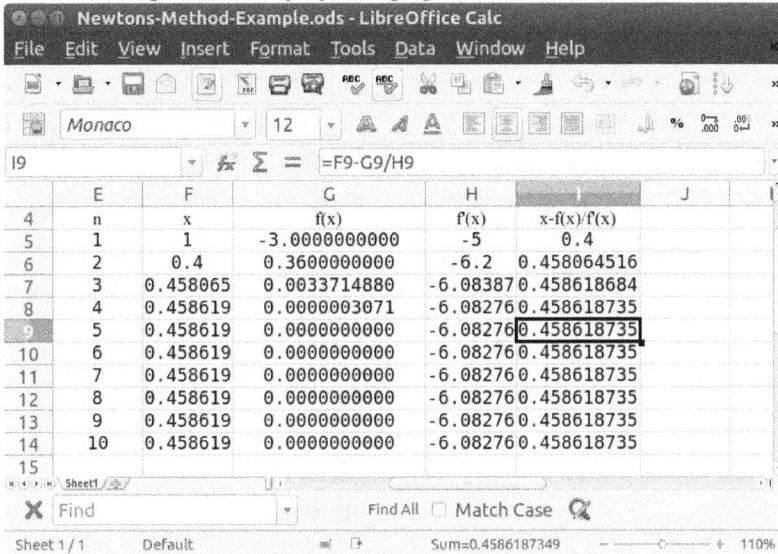

Repeat the copying with columns G, H, and I.

	E	F	G	H	I
4	n	x	f(x)	f'(x)	x-f(x)/f'(x)
5	1	1	=F5^2-7*F5+3	=2*F5-7	=F5-G5/H5
6	=1+E5	=I5	=F6^2-7*F6+3	=2*F6-7	=F6-G6/H6
7	=1+E6	=I6	=F7^2-7*F7+3	=2*F7-7	=F7-G7/H7
8	=1+E7	=I7	=F8^2-7*F8+3	=2*F8-7	=F8-G8/H8
9	=1+E8	=I8	=F9^2-7*F9+3	=2*F9-7	=F9-G9/H9

Note that there is NO CODING except in the first line! Everything was cut and paste, and the spreadsheet does everything else, including changing the variable names. The spreadsheet should automatically display the values rather than the formulas. Normally you just see the formulas in the formula bar at the top of the spreadsheet, and the numbers in the grid. You can display the formulas in the grid by using the menu option Tools / Options / Libre Office Calc / View / Display Formulas. To see the numbers, un-check the box labeled Formulas. You can also toggle back and forth by pressing the CTRL and back-quote key at the same time (this will also work in Excel. The calculation is illustrated in figure 21.4

Exercises

Use Newton's method in these exercises.

1. Estimate a root of $x^2 - 5 = 0$ to 6 significant figures, using a first guess of $x_0 = 5$. Repeat the problem, this time using $x_0 = -2.5$. What happens? Can you explain? What should the exact answer be? Estimate your error.

2. Find a root of $x^2 + x - 6 = 0$, starting with a first guess of $x_0 = 1$. Find two additional approximations x_1 and x_2.
   Ans: 7/3, 103/51

3. Estimate $\sqrt{2}$ by solving $x^2 - 2 = 0$, to five figures to the right of decimal point.
   Ans: $1, \frac{3}{2}, \frac{17}{12}, \frac{577}{408}, \frac{665857}{470832}$ or 1, 1.5, 1.4147, 1.41422, 1.41421

4. Find $\sqrt[3]{59}$ to 3 decimal places.
   Ans: Using $x^3 - 59 = 0$ and $x_0 = 4$ gives $4, 3.895, 3.893, 3.893$

5. Find the root of $\cos(x^2 + 5) = x^3$ starting with $x_0 = 1$. Find the next two iterations.
   Ans: 0.983684, 0.983151

6. Solve $x = \cos x$ in radians to 3 significant figures. Use whatever starting point you like.
   Ans: 0.739085

7. Find a solution of $x^3 - 6x + 10 = 0$ to 3 decimal places.   Ans: -3.047

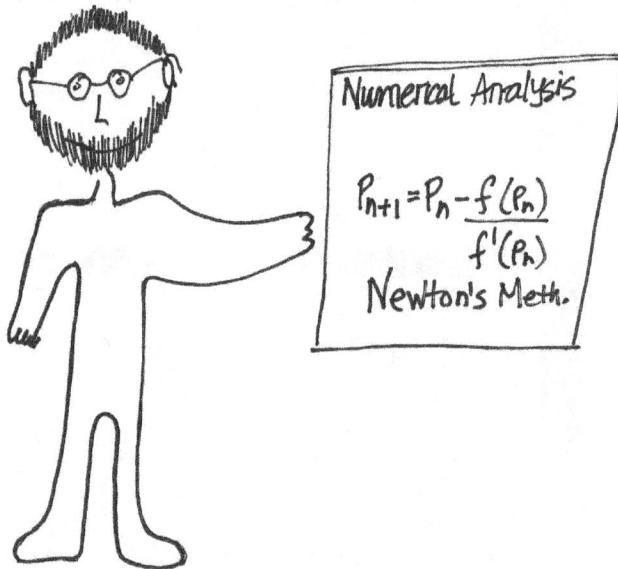

Numerical Analysis

$$P_{n+1} = P_n - \frac{f(P_n)}{f'(P_n)}$$

Newton's Meth.

Chapter 22

Antiderivatives

Chapter Summary and Goal

Reversing the process of differentiation is called **antidifferentiation**, as illustrated in figure 22.1.

When we calculate an antiderivative we are answering the following questions: what function $F(x)$ has a derivative equal to $f(x)$.

If $f'(x) = F(x)$, then we call $F(x)$ an **antiderivative** of $f(x)$. Antiderivatives are also called **indefinite integrals**.

Figure 22.1. Schema of the Differentiation and antidifferentiation processes.

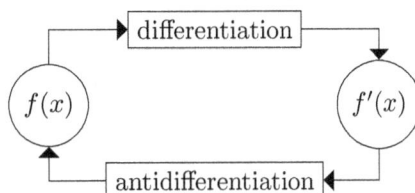

Student Learning Objectives

The student will:

1. Understand the difference between a derivative and an antiderivative.
2. Understand the basic properties of antierivatives.
3. Understand the relationship between antiderivatives and indefinite integrals.
4. Calculate basic antiderivatives from the table of derivatives.

Antiderivatives: Inverting the Derivative

If the derivative of a function $f(x)$ is $f'(x)$, it is natural to ask a similar question: what function (or functions) $F(x)$ has a derivative $F'(x)$ given by $f(x)$? If such a function exists it is called an antiderivative.

> **Definition 22.1. Antiderivative**
>
> $F(x)$ is called an **antiderivative** of a function $f(x)$ on an interval $[a, b]$ if $F'(x) = f(x)$ for all x in (a, b).

As it turns out we had to be careful in asking the question above not just to say "what function" but what "functions" because the antiderivative is not unique. Recall from theorem 18.4 that

if two functions share the same derivative than they must differ by at most constant.[1] Thus
if any two functions $F(x)$ and $G(x)$ are both antiderivatives of f, then since they both share
the same derivative, they differ by at most a constant. Expressed differently, if $F(x)$ is any
antiderivative of $f(x)$, then the most general form of antiderivative of $f(x)$ is $F(x) + C$ where
C represents a general constant that is allowed to run over all real numbers.

For example, $F(x) = x^2$ is an antiderivative of $f(x) = 2x$ because $(d/dx)(x^2) = 2x$. However,
$G(x) = x^2 + 7$ and $H(x) = x^2 - 103$ are also antiderivatives of $f(x) = 2x$. The most general
antiderivative is $F(x) = x^2 + C$.

Theorem 22.1. Most General Antiderivative

If $F(x)$ is any antiderivative of $f(x)$ on an interval $[a, b]$ then the most general antiderivative of $f(x)$ on $[a, b]$ is

$$F(x) + C \qquad\qquad (22.1)$$

where C is an arbitrary constant.

To avoid any suspense or confusion we introduce the following notation now, without any
mathematical justification. It will be justified when we prove the Fundamental Theorem of
Calculus in chapter 24.[2] The symbol $\int$ is refered to as the **integral** symbol, and the expression
$\int f(x)dx$ is read as "the integral of $f(x)$." (See figure 22.2.)

Figure 22.2: The antiderivative is represented using the integral sign.

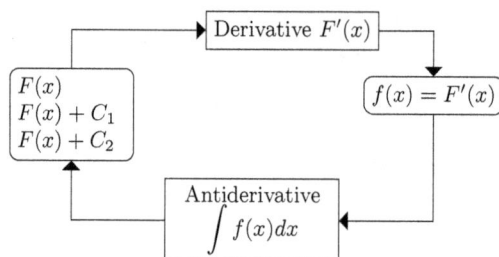

Representing Antiderivatives by Indefinite Integrals

If $F(x)$ is any antiderivative of $f(x) = F'(x)$ then we write

$$\int f(x)dx = \int F'(x)dx = F(x) + C \qquad\qquad (22.2)$$

The arbitrary constant C is called a **constant of integration**. The function $f(x)$,
when it is written in the form $\int f(x)\,dx$ is called an **integrand**.

---

[1]This was a consequence of the Mean Value Theorem.

[2]Many authors prefer a more formal development where the integral symbol is completely avoided in the
discussion of antiderivatives. However, I have found that this only confuses students, as we allow it later,
and many are already familiar with integrals. Since these students are going to refer to antiderivatives as
integrals for the rest of they professional lives anyway, it makes ore sense to introduce the notation they are
going to use sooner, rather than later, and avoid long expressions "the antiderivative of foo is foobar" every
other line.

Example 22.1. Find the most general antiderivative of $f(x) = \sin x$.

Solution. Let $F(x) = -\cos x$. Then $F'(x) = f(x)$. Hence the most general form is

$$F(x) = -\cos x + C. \tag{22.3a}$$

In terms of the integral notation

$$\int \sin x dx = -\cos x + C. \tag{22.3b}$$

$\square$

The linearity of antiderivatives follows from the linearity of derivatives (theorem 9.8).

Theorem 22.2. Linearity of Antiderivatives

The antiderivative is linear: for functions f and g, and constants a and b,[a]

$$\int (af(x) \pm bg(x))dx = a \int f(x)dx \pm b \int g(x)dx \tag{22.4}$$

[a]The most general antiderivative in (22.4) is still subject to the addition of an arbitrary constant.

Example 22.2. Find the most general antiderivative of $f(x) = x^n$ for $n \neq -1$.

Solution. Since

$$\frac{d}{x}(x^p) = px^{p-1} \tag{22.5a}$$

then

$$\frac{1}{p} \cdot \frac{d}{dx}(x^p) = x^{p-1} \tag{22.5b}$$

Let $n = p - 1$. Then $p = n + 1$, and

$$x^n = \frac{1}{n+1} \cdot \frac{d}{dx}(x^{n+1}) = \frac{d}{dx}\left(\frac{x^{n+1}}{n+1}\right) \tag{22.5c}$$

By equation 22.2

$$\int x^n dx = \int \frac{d}{dx}\left(\frac{x^{n+1}}{n+1}\right)dx = \frac{x^{n+1}}{n+1} + C \tag{22.5d}$$

The restriction $n \neq -1$ was needed was needed because otherwise the last result would not be valid. $\square$

Theorem 22.3. Power Law for Integrals

$$\int x^n \, dx = \frac{x^{n+1}}{n+1} + C \tag{22.6}$$

Example 22.3. Find the most general antiderivative of $f(x) = x - 3$.

Solution. We combine example 22.2 and theorem 22.2, to find a function whose derivative is $x - 3$.

$$\int (x-3)dx = \int xdx - \int 3dx \qquad \text{by theorem 22.2} \tag{22.7a}$$

$$= \frac{1}{2}x^2 - 3x + C \qquad \text{by example 22.2} \tag{22.7b}$$

We can differentiate to verify that we have a correct result. $\square$

Example 22.4. Find the most general antiderivative of $f(x) = x^{-3/2}$.

Solution. By theorem 22.3,

$$\int x^{-3/2} dx = \frac{x^{-3/2+1}}{-3/2+1} + C = \frac{x^{-1/2}}{-1/2} + C = -\frac{2}{\sqrt{x}} + C \qquad (22.8a)$$

for all $x > 0$. $\qquad\qquad\qquad\qquad\qquad\qquad\qquad\qquad\qquad\qquad\qquad\qquad\qquad\square$

Example 22.5. Find the most general antiderivative of $f(x) = 10/x^9$.

Solution. We write $f(x) = 10x^{-9}$ and apply theorem 22.3:

$$F(x) = \int 10x^{-9} dx = \frac{10x^{-9+1}}{-9+1} + C = -\frac{10x^{-8}}{8} + C = -\frac{5x^{-8}}{4} + C \qquad (22.9a)$$

As before, however, the function is not defined at $x = 0$. Unlike the earlier example, where the domain only included positive values of x, this function includes both positive and negative values of x. Since there is a disconnect between the two regions where the function is not defined, we must allow for the possibility of different constants of integration in each domain. The most general solution is

$$F(x) = \begin{cases} -(5/4)x^{-8} + C_1, & x > 0 \\ -(5/4)x^{-8} + C_2, & x < 0 \end{cases} \qquad (22.9b)$$

where C_1 and C_2 are any two general unrestricted real constants. $\qquad\qquad\square$

Example 22.6. Find y such that $y' = x^2 + \sin x$.

Solution. Applying table 22.1 term by term (which is allowed, by the sum rule),

$$y = \frac{1}{3}x^3 - \cos x + C \qquad (22.34a)$$

is the most general antiderivative. $\qquad\qquad\qquad\qquad\qquad\qquad\qquad\qquad\qquad\square$

Example 22.7. Find y such that $y' = x^3 - 4x + \sin x$ and $y(0) = 0$.

Solution. From table 22.1,

$$y = \frac{1}{4}x^4 - 2x^2 - \cos x + C \qquad (22.35a)$$

gives the most general form for y. To get a value of C, we substitute $x = 0$ and $y = 0$,

$$0 = 0 - 0 - \cos 0 + C \qquad (22.35b)$$

Hence $C = 1$ and therefore

$$y = \frac{1}{4}x^4 - 2x^2 - \cos x + 1 \qquad (22.35c)$$

satisfies all of the required conditions. $\qquad\qquad\qquad\qquad\qquad\qquad\qquad\qquad\square$

Example 22.8. Find $f(x)$ if $f'(x) = x\sqrt{x}$ and $f(1) = 2$.

Solution. Since $y' = x\sqrt{x} = x^{3/2}$, table 22.1 gives us the general form

$$y = \frac{2}{5}x^{5/2} + C \qquad (22.36a)$$

Setting $y = 2$ when $x = 1$ gives

$$2 = \frac{2}{5} \cdot 1 + C \qquad (22.36b)$$

So $C = 8/5$. Substituting back into (22.36a) gives

$$y = \frac{2}{5}x^{5/2} + \frac{8}{5}. \qquad (22.36c)$$

$$\square$$

Table 22.1: Common antiderivatives based on the derivatives derived in the preceding chapters.

Formulas for Antiderivatives*

$$\int C f(x)\,dx = C \int f(x)\,dx \qquad (22.10)$$

$$\int \frac{1}{x}\,dx = \ln|x| \qquad (22.11)$$

$$\int (f \pm g)\,dx = \int f\,dx \pm \int g\,dx \qquad (22.12)$$

$$\int e^x\,dx = e^x \qquad (22.13)$$

$$\int x^n\,dx = \frac{x^{n+1}}{n+1}, n \neq -1 \qquad (22.14)$$

$$\int a^x\,dx = \frac{a^x}{\ln a} \qquad (22.15)$$

$$\int \sin x\,dx = -\cos x \qquad (22.16)$$

$$\int \sinh x\,dx = \cosh x \qquad (22.17)$$

$$\int \cos x\,dx = \sin x \qquad (22.18)$$

$$\int \cosh x\,dx = \sinh x \qquad (22.19)$$

$$\int \sec^2 x\,dx = \tan x \qquad (22.20)$$

$$\int \operatorname{sech}^2 x\,dx = \tanh x \qquad (22.21)$$

$$\int \csc^2 x\,dx = -\cot x \qquad (22.22)$$

$$\int \operatorname{csch}^2 x\,dx = -\coth x \qquad (22.23)$$

$$\int \sec x \tan x\,dx = \sec x \qquad (22.24)$$

$$\int \operatorname{sech} x \tanh x\,dx = -\operatorname{sech} x \qquad (22.25)$$

$$\int \csc x \cot x\,dx = -\csc x \qquad (22.26)$$

$$\int \operatorname{csch} x \coth x\,dx = -\operatorname{csch} x \qquad (22.27)$$

$$\int \frac{dx}{\sqrt{1-x^2}} = \sin^{-1} x \qquad (22.28)$$

$$\int \frac{dx}{\sqrt{x^2+1}} = \sinh^{-1} x \qquad (22.29)$$

$$\int \frac{dx}{1+x^2} = \tan^{-1} x \qquad (22.30)$$

$$\int \frac{dx}{\sqrt{x^2-1}} = \cosh^{-1} x \qquad (22.31)$$

$$\int \frac{dx}{x\sqrt{x^2-1}} = \sec^{-1} \qquad (22.32)$$

$$\int \frac{dx}{1-x^2} = \tanh^{-1} x \qquad (22.33)$$

*Additive constants suppressed to save space.

Exercises

In exercise 1 through 12, find the most general antiderivative of the given function.

1. $f(x) = 20x^3 - 18x^2 + 4x - 2$

2. $y = \sqrt[4]{x^3} + \sqrt[5]{x^4}$

3. $\int \sqrt{x^3}\,dx$

4. $\int \dfrac{1}{\sqrt[3]{x^3}}\,dx$

5. $\int \dfrac{x^3 - 4x^2 + 3}{x^2}\,dx$

6. $\int \left(3x^4 + \dfrac{1}{x^3}\right)\,dx$

7. $\int (x-2)(x+3)\,dx$

8. $\int (x^{2/3} - x^{3/2})\,dx$

9. $\int (x^2 - 3\sin x + 4)\,dx$

10. $\int \dfrac{\sqrt{x} + x^3}{x^2}\,dx$

11. $\int (\sec^2 x - 1)\,dx$

12. $\int (1+x)\sqrt{x}\,dx$

In exercises 13 through 15 find the antiderivative of given function that satisfies the specified condition.

13. $f(x) = 3\sec^2 x - 9x^2$ with $F(0) = 0$

14. $f(x) = \dfrac{3}{x^3} - \dfrac{9}{x^2}$ with $F(1) = 0$

15. $f(x) = \dfrac{5 - 4x^3 + 2x^6}{x^2}$ if $F(1) = 0$

16. Find y such that $y' = x^2 + x$ and $y(0) = 1$.

17. Find y such that $y' = \cos y$ and $(\pi/2) = 1$.

18. Find y such that $y' = x^3 + x^2$ and $y(1) = 1$.

19. Find y such that $y'' = 2\sin x - \cos x$, $y(0) = 1$, and $y'(0) = 1$.

20. Suppose $y'' = 6x+3$, $y'(0) = 4$ and $y(0) = -4$. Find $f'(x)$ and $f(2)$.

21. If $f''(x) = 8x + 10\sin x$, $f(0) = 2$ and $f'(0) = 2$, find $f(x)$.

"Integrals"

Chapter 23

Area Under a Curve

Chapter Summary and Goal

The expression **area under the curve** refers to the area of the region beneath the plot of $f(x)$, above the x-axis, and between two specified vertical lines $x = a$ and $x = b$. More precisely, we call **this the area under the curve on the interval** $[a, b]$. In this chapter we will learn how to calculate this area by approximating the region by very small rectangles, and adding up the area of each of the rectangles. The area under the curve then becomes exact in the limit as the size of each rectangle becomes very narrow and the number of rectangles approaches (in the limit) infinity. In chapter 24 we will relate this calculation to the antiderivative of the function.

Figure 23.1. The region described by **area under the curve** of $f(x)$ is shaded.

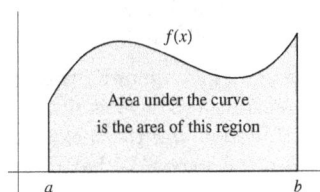

Student Learning Objectives

The student will:

1. Learn to estimate an area by rectangles.
2. Learn to distinguish the different ways of approximating an area by rectangles.
3. Understand the limiting process when the area calculation becomes exact.
4. Be able to calculate exact areas.
5. Understand the summation (big-sigma) notation.

Areas

In geometry area is calculate by by partitioning larger shapes into smaller regions whose areas we already know. The simplest shapes are polygons, where the edges are composed of straight lines. These we can easily partition into triangles. The of a triangle is

$$A_{\text{triangle}} = \frac{1}{2}(\text{base}) \times (\text{height}) \tag{23.1}$$

If we partition a polygon into n triangles (figure 23.2) and label each triangle as T_i, then the area of a polygon is

$$A = A(T_1) + A(T_2) + \cdots + A(T_n) \tag{23.2}$$

213

where by $A(T_i)$ we mean the area of triangle i.

Figure 23.2: Left: The area of a convex polygon can be determined by placing a point at any random location inside the polygon and then summing the areas of the triangles formed by that point with each pair adjacent boundary vertices. Right: approximation of smooth curve by triangles.

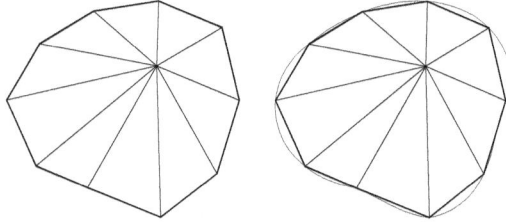

We could, in principle, extend this concept to find the area of any blob shaped region (see the right-hand side of figure 23.2). If the curve is convex, we repeat the same process – just pick any point in the interior. If the curve is not convex, we may have to pick more than one central point, and decompose the triangles recursively. Then we slowly increase the number of triangles – while at the same time we decrease the size of their bases. As this happens, our approximation becomes more accurate. In the limit as $n \to \infty$, the approximation becomes exact. We call the limit of the sum of the areas the exact area of the blob.

We will, in fact, quantify this approximation of blobs by triangles more precisely in 37. Now we will address a somewhat simpler question: find the area of the region between a curve described by a function and the x-axis. Even defining this region is not so simple. What if the function crosses the x-axis? What do we do then? Do we distinguish between regions that are above the axis and regions below the axis? (See figure 23.3.) Furthermore, if the function is entirely above the x-axis, then the right and left edges will be defined by vertical lines. We will define this area to be a positive number if $f(x) > 0$ and a negative number if $f(x) < 0$ (see figure 23.3).

Figure 23.3:   The area under the curve on the left is entirely positive; the area under the curve shown on the right has components that are both positive and negative.

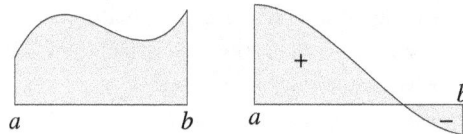

Here our approach is similar to the approach we described above. We partition the region we want to find the area of into smaller regions, but this time, instead of triangles, we will use rectangles. The area of a rectangle is

$$A_{\text{rectangle}} = (\text{base}) \times (\text{height}) \tag{23.3}$$

We will draw the rectangles so that one corner of the rectangle falls on the curve of $f(x)$. There are two ways to do this, as we see in figure 23.4. If all of the upper left hand corners of the rectangles lie on the curve of $f(x)$. then this technique is called the **upper left hand corner method**. If all of the upper right hand corners lie on the curve, the technique is called the **upper right hand corner method**. A third variation (not shown) is to have the midpoint of

Figure 23.4: Ways of approximating the area under a curve by rectangles. Left: upper left hand corner approximation. Right: Upper right hand corner approximation.

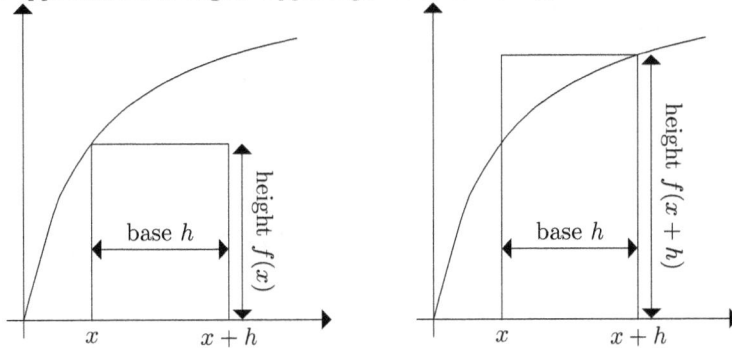

the rectangle exactly lie on the curve. This last variation is called the **midpoint method**. It actually doesn't matter which method we use. Some methods will overestimate the error, and some methods will underestimate it, and the choice of which method does better will depend on the curvature of the individual function.

Figure 23.5: Using left and right endpoints of rectangles to fill up the curve.

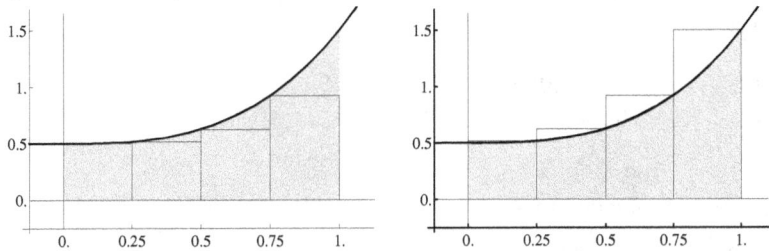

Example 23.1. Estimate the area under the curve of $y = x^3 + 1/2$ on $[0, 1]$ with 4 rectangles, using both the upper left hand method and the upper right hand method, and compare the results.

Solution. The approximations for each method are illustrated in figure 23.5. Inspecting the figure, we expect the left endpoint method to underestimate the area and the right endpoint method to overestimate the area.

Dividing the interval [0,1] into four subintervals tells us that the base of each rectangle is $h = 1/4$. The left edges occur at $x = 0$, $x = 1/4$, $x = 1/2$ and $x = 3/4$. The corresponding heights of each rectangle are then $f(0)$, $f(1/4)$, $f(1/2)$, and $f(3/4)$. The area approximation is

$$A_{\text{Left}} = \frac{1}{4}f(0) + \frac{1}{4}f\left(\frac{1}{4}\right) + \frac{1}{4}f\left(\frac{1}{2}\right) + \frac{1}{4}f\left(\frac{3}{4}\right) \tag{23.4a}$$

$$= \frac{1}{4}\left(\frac{1}{2} + 0\right) + \frac{1}{4}\left(\frac{1}{2} + \frac{1}{4^3}\right) + \frac{1}{4}\left(\frac{1}{2} + \frac{1}{2^3}\right) + \frac{1}{4}\left(\frac{1}{2} + \frac{3^3}{4^3}\right) \tag{23.4b}$$

$$= \frac{1}{4}\left[\left(\frac{32}{64}\right) + \left(\frac{32}{64} + \frac{1}{64}\right) + \left(\frac{32}{64} + \frac{8}{64}\right) + \left(\frac{32}{64} + \frac{27}{64}\right)\right] \tag{23.4c}$$

Doing Calculus

$$= \frac{32 + 32 + 1 + 32 + 8 + 32 + 27}{256} = \frac{164}{256} = \frac{41}{64} = 0.640625 \qquad (23.4\text{d})$$

The right edges occur at $x = 1/4$, $x = 1/2$, $x = 3/4$ and $x = 1$; the corresponding heights are then $f(1/4)$, $f(1/2)$, $f(3/4)$ and $f(1)$. The upper right corner approximation becomes

$$A_{\text{Right}} = \frac{1}{4} f\left(\frac{1}{4}\right) + \frac{1}{4} f\left(\frac{1}{2}\right) + \frac{1}{4} f\left(\frac{3}{4}\right) + \frac{1}{4} f(1) \qquad (23.4\text{e})$$

$$= \frac{1}{4}\left(\frac{1}{2} + \frac{1}{4^3}\right) + \frac{1}{4}\left(\frac{1}{2} + \frac{1}{2^3}\right) + \frac{1}{4}\left(\frac{1}{2} + \frac{3^3}{4^3}\right) + \frac{1}{4}\left(\frac{1}{2} + 1\right) \qquad (23.4\text{f})$$

$$= \frac{1}{4}\left[\left(\frac{32}{64} + \frac{1}{64}\right) + \left(\frac{32}{64} + \frac{8}{64}\right) + \left(\frac{32}{64} + \frac{27}{64}\right) + \left(\frac{32}{64} + \frac{64}{64}\right)\right] \qquad (23.4\text{g})$$

$$= \frac{32 + 1 + 32 + 8 + 32 + 27 + 32 + 64}{256} = \frac{228}{256} = \frac{57}{64} = 0.890625 \qquad (23.4\text{h})$$

Since we believe (from looking at the picture) that

$$A - \text{left} < A < A_{\text{right}} \qquad (23.4\text{i})$$

we conclude that

$$0.640625 < A < 0.890625 \qquad (23.4\text{j})$$

Taking the average and the range of values we might estimate that $A \approx 0.76 \pm 0.13$ (rounded to two decimal places). $\qquad \square$

Example 23.2. Repeat example 23.1 using 10 rectangles.

Solution. The upper left hand corner method gives

$$A_{\text{Left}} = 0.10(f(0) + f(0.1) + \cdots + f(0.9)) \qquad (23.5\text{a})$$

$$\approx 0.10(0.5 + 0.501 + 0.508 + 0.527 + 0.564 + 0.625 +$$

$$0.716 + 0.843 + 1.012 + 1.229) \approx 0.702 \qquad (23.5\text{b})$$

where we have rounded to 3 decimal places. and the upper right hand corners give

$$A_{\text{Right}} = 0.10[f(0.1) + f(0.2) + \cdots + f(1.0)] \qquad (23.5\text{c})$$

$$\approx 0.10(0.501 + 0.508 + 0.527 + 0.564 + 0.625 + 0.716 +$$

$$0.843 + 1.012 + 1.229 + 1.5) \approx 0.803 \qquad (23.5\text{d})$$

This gives us a tighter bounds

$$.702 < A < .803 \qquad (23.5\text{e})$$

Now taking the average and interval we get $A \approx 0.75 \pm 0.5$. $\qquad \square$

Examples 23.1 and 23.2 suggest that by increasing the number of intervals, the size of the bounds on our approximation will decrease. We can treat this bounds as a kind of approximation error:

$$\epsilon \approx \frac{|A_{\text{right}} - A_{\text{left}}|}{2} \qquad (23.6)$$

When $n = 4$, we calculated $\epsilon \approx .13$; for $n = 10$ we calculated $\epsilon \approx 0.05$. The trend suggested by these two calculations does, in general, continue: the larger the number of rectangles, the smaller the approximation error.

If we needed to know the area to some specific precision δ, we *could* attempt to keep increasing n larger and larger until our calculation gave us $\epsilon < \delta$. But perhaps a better way to look at this would be to turn the question around and instead ask the following question: how big does n have to be to ensure that the $\epsilon < \delta$ (ϵ=calculation error; δ=accuracy requirement). This type of question occupies much of the field of numerical analysis.

Example 23.3. Find a general formula for the area under the curve of $y = x^3 + 1/2$ on $[0,1]$ using n rectangles and the upper left hand corner method.

Solution. If we use n rectangles than the rectangles will have width

$$h = 1/n \tag{23.7a}$$

and the ith rectangle, counting from the left, will have its left edge at

$$x = ih = \frac{i}{n} \tag{23.7b}$$

where $i = 0, 1, \ldots n - 1$. The height of the ith rectangle is

$$f(x) = f\left(\frac{i}{n}\right) = \frac{1}{2} + \frac{i^3}{n^3} \tag{23.7c}$$

The area of the ith rectangle is

$$A_i = (i^{th} \text{ width}) \times (i^{th} \text{ height}) = \frac{1}{n}\left(\frac{1}{2} + \frac{i^3}{n^3}\right) \tag{23.7d}$$

Summing over all the rectangles,

$$A = A_1 + A_2 + \cdots + A_n \tag{23.7e}$$

$$= \frac{1}{n}\left(\frac{1}{2} + \frac{1^3}{n^3}\right) + \frac{1}{n}\left(\frac{1}{2} + \frac{2^3}{n^3}\right) + \frac{1}{n}\left(\frac{1}{2} + \frac{3^3}{n^3}\right) + \cdots + \frac{1}{n}\left(\frac{1}{2} + \frac{(n-1)^3}{n^3}\right) \tag{23.7f}$$

$$= \frac{1}{n}\overbrace{\left(\frac{1}{2} + \cdots + \frac{1}{2}\right)}^{n-1 \text{ terms}} + \frac{1}{n}\left(\frac{1^3}{n^3} + \frac{2^3}{n^3} + \cdots + \frac{(n-1)^3}{n^3}\right) \tag{23.7g}$$

$$= \frac{n-1}{2n} + \frac{1}{n^4}(1^3 + 2^3 + 3^3 + \cdots + (n-1)^3) \tag{23.7h}$$

From the sum of cubes formula (equation A.40)

$$1^3 + 2^3 + 3^3 + \cdots + n^3 = \frac{n^2(n+1)^2}{4} \tag{23.7i}$$

Since our sum ends at $n - 1$ instead of n, we replace n with $n - 1$ everywhere,

$$1^3 + 2^3 + 3^3 + \cdots + (n-1)^3 = \frac{(n-1)^2 n^2}{4} \tag{23.7j}$$

Using (23.7j) in (23.7h) gives

$$A = \frac{n-1}{2n} + \frac{(n-1)^2 n^2}{4n^4} \tag{23.7k}$$

If we let $n \to \infty$, we get a formula for the exact area:

$$\text{Area} = \lim_{n\to\infty}\left(\frac{n-1}{2n} + \frac{(n-1)^2 n^2}{4n^4}\right) = \frac{1}{2} + \frac{1}{4} = \frac{3}{4}. \tag{23.7l}$$

$\square$

Statement of General Area Problem

Suppose that in general we want to find the area under the curve of $f(x)$ on $[a, b]$. We begin by dividing $[a, b]$ into n sub-intervals of width h. These subintervals form the bases of each rectangle. Then

$$h = \frac{b - a}{n} \tag{23.8}$$

Then if the corners are at $x_0, x_1, \ldots, x_n$, the left rectangles measure their heights at $x_0, \ldots, x_{n-1}$, while the right rectangles measure their heights at $x_1, \ldots, x_n$. Consequently

$$A_{\text{Left}} \approx h[f(x_0) + f(x_1) + \cdots + f(x_{n-1})] = \sum_{i=0}^{n-1} hf(x_i) \tag{23.9}$$

$$A_{\text{Right}} \approx h[f(x_1) + f(x_1) + \cdots + f(x_n)] = \sum_{i=1}^{n} hf(x_i) \tag{23.10}$$

In each sum there are n terms.

Big-Sigma Notation

If $f(i)$ is any function that depends on an integer variable i, then

$$\sum_{i=1}^{n} f(i) = f(1) + f(2) + \cdots + f(n) \tag{23.11}$$

Here we have continued to refer to the width of the rectangle as h. Sometimes we will also refer to the width of the rectangle as Δx; we will often use these terms interchangeably, as they are both infinitesimals (see chapter 8). However, it is more common notation to use the symbol h when we are calculating a numerical value, and the symbol Δx when we are referring to an infinitesimal change in x. Since we are not specifically talking about differentials (yet) there is no particular distinction between the two. However, after we prove the fundamental theorem of calculus in chapter 24 we will make an association between h and dx, and because of the similarity in appearance (between Δx and dx) it is common to use the notation Δx, rather than h, in this type of expression. Then we can just wave our hands in the next chapter and say that $\Delta x \to dx$ as $n \to \infty$ and if you don't like that you should read a book on really advanced analysis.[1]

Here we have introduced the **Big-Sigma** symbol $\sum$ to indicate a sum: the expression

$$\sum_{i=0}^{n-1} f(x_i)\Delta x \qquad \left(\text{or } \sum_{i=0}^{n-1} hf(x_i) \text{ if you prefer} \right) \tag{23.12}$$

is read as "the sum of eff of x delta x as i goes from zero to $n - 1$." The big sigma equation means the same thing as

$$\Delta x f(x_0) + \Delta x f(x_1) + \Delta x f(x_2) + \cdots \Delta x f(x_{n-1}) \tag{23.13}$$

The symbol $\sum$ is really just a large upper-case greek letter sigma (Σ in normal-size font). To remember what it means, think of Σ as a secret geek speak symbol for the letter "S" as in "Sum."

---

[1] The fact of the matter is, we leave it there, in the integrals, to remind ourselves that it measures the width of the rectangle, while the integrand measures the height.

We don't really have to choose the heights of the rectangle to be the endpoints of the interval. In fact, we can choose a height anywhere in the interval. Suppose instead of the endpoints we arbitrarily choose **any point in each interval**:

$$x_1 \text{ anywhere in } [a, h]$$
$$x_2 \text{ anywhere in } [a + h, a + 2h]$$
$$\vdots$$
$$x_n \text{ anywhere in } [a + (n - 1)h, a + nh]$$

There does not need to any pattern, so long as each point x_i is in the ith interval. The left-endpoints correspond to choosing

$$x_i = a + (i - 1)h \tag{23.14}$$

and the right endpoints correspond to choosing

$$x_i = a + ih \tag{23.15}$$

In either case we still have

$$(i - 1)h \leq x_i \leq ih \tag{23.16}$$

Instead of using the left or right endpoints to determine the height of the rectangles, we will use the randomly chosen point in each interval to determine where the height is measured. All we know about the height of the rectangle is that it is somewhere between the minimum height and the maximum height determined by the range of values taken on by the function in that interval. When we take the limit as $n \to \infty$, the width of the rectangle still goes to zero, and the maximum and the minimum get all squooshed up together, so it doesn't really matter, as they both go to the same limit. So the area approximation becomes

$$A \approx \sum_{i=1}^{n} f(x_i)h = \sum_{i=1}^{n} f(x_i)\frac{b - a}{n} \tag{23.17}$$

where x_i is anywhere in the ith interval, and the **exact area** is

$$A = \lim_{n \to \infty} \sum_{i=1}^{n} f(x_i)h = \lim_{n \to \infty} \sum_{i=1}^{n} f(x_i)\frac{b - a}{n} \tag{23.18}$$

This limit converges for all reasonable functions (we will talk more about this later).

Exercises

In exercises 1 through 8, estimate the area under each curve using (a) the left endpoints, (b) the right endpoints, and (c) the average of the two methods, first using $n = 4$ points, then using $n = 8$ points.

1. $\cos x$ on $[0, \pi]$
2. $\cos x$ on $[0, \pi/2]$
3. $\sin x$ on $[0, \pi]$
4. $\cos x$ on $[0, \pi/2]$
5. x on $[-1,1]$
6. x on $[0,1]$
7. $\sqrt{x}$ on $[1, 2]$
8. $\sqrt{1 - x^2}$ on $[-1, 1]$

For each function in exercise 9 through 16,

(a) Compute $g(x) = \displaystyle\int f(x)\, dx$

(b) Find $g(b) - g(a)$

(c) Compare with the corresponding result in exercises 1 through 9.

9. $\cos x$ on $[0, \pi]$
10. $\cos x$ on $[0, \pi/2]$
11. $\sin x$ on $[0, \pi]$
12. $\cos x$ on $[0, \pi/2]$
13. x on $[-1,1]$
14. x on $[0,1]$
15. $\sqrt{x}$ on $[1, 2]$
16. $\sqrt{1 - x^2}$ on $[-1, 1]$

The problem of finding an area is similar to the problem of calculating distance travelled. If one travels at a fixed velocity for a specific period of time, then the distance travelled is

distance traveled

$$= (\text{velocity}) \times (\text{time})   \qquad (23.19)$$

Here we examine what happens if the velocity changes with time, and is given by a function $v = f(t)$

17. Suppose you travel for a total duration T. Divide the time interval T into fixed time increments Δt and show that the total distance travelled is

$$d = \sum_{i=1}^{n} f(t_i)\Delta t   \qquad (23.20)$$

What is the restriction on each t_i?

18. Suppose that a black-box in your car measures the speed that you are driving in feet per second, and gives the following results.

t(s)	v (ft/s)
0	10
10	15
20	25
30	22
40	17
50	16
60	15

Using the result of exercise 17 and a left-endpoint calculation, estimate the total distance travelled after one minute.

19. Repeat the previous problem using the right-endpoint calculation.

Instead of using right or left endpoints, another way to calculate the area is to use the midpoint of the rectangle to define its height.

20. Letting

$$x_i = \frac{x_{i-1} + x_i}{2}, i = 1, \ldots, n   \qquad (23.21)$$

derive a formula for the midpoint rule.

21. Estimate the area under each of the following curves using the midterm rule using $n = 4$ points.

(a) $\cos x$ on $[0, \pi]$
(b) $\cos x$ on $[0, \pi/2]$
(c) $\sin x$ on $[0, \pi]$
(d) $\cos x$ on $[0, \pi/2]$
(e) x on $[-1,1]$
(f) x on $[0,1]$
(g) $\sqrt{x}$ on $[1, 2]$
(h) $\sqrt{1 - x^2}, [-1, 1]$

Chapter 24

The Definite Integral

Chapter Summary and Goal

The **definite integral** gives us a formula for the exact area under a curve. In chapter 23 we studied how to calculate this area approximately and ended with a discussion of what happens as the number of rectangles we use in our mosaic to fill up the area approaches infinity.

The most important results of this chapter (theorems 24.10 through 24.14) are collectively know as the **Fundamental Theorems of Calculus**. These relate the area under the curve of $f(x)$ to the antiderivative of $f(x)$:

$$\text{Area under the curve} = F(b) - F(a) \tag{24.1}$$

This fundamental relationship ties integral and differential calculus together, and justifies our using the integral symbol for both area and antidifferentiation.

Student Learning Objectives

The student will:

1. Understand the distinction between definite and indefinite integrals.
2. Understand the proof of the Fundamental Theorem of Calculus.
3. Be able to calculate areas using definite integrals.
4. Be able to manipulate and simplify definite integrals using their basic properties such as symmetry, linearity, and area addition.
5. Be able to differentiate integrals with respect to the upper or lower bounds, and to apply the chain rule for integrals.
6. Set up and solve simple differential equations using the net change theorem.

The Definite Integral

In chapter 23 we divided the interval $[a, b]$ into n rectangles, each of fixed width

$$\Delta x = x_{i+1} - x_i = \frac{b - a}{n} \tag{24.2}$$

where

$$a = x_0 < x_a < x_x < \cdots < x_n = b \tag{24.3}$$

and randomly selected points x_i^* each interval such that

$$x_i \leq x_i^* \leq x_{i+1}. \tag{24.4}$$

We found (eq 23.18) that the area under the curve of $f(x)$ on $[a, b]$ is given by the following limit (if it exists), known as the **Riemann Sum**:

$$A = \lim_{n \to \infty} \sum_{i=1}^{n} \Delta x f(x_i^*) = \lim_{n \to \infty} \sum_{i=1}^{n} \frac{b-a}{n} f(x_i^*) \tag{24.5}$$

We take (24.5) as the **definition** of a **definite integral**: in other words, **we define the definite integral as the area under the curve**, completely out of the blue.

Definition 24.1. Definite Integral

The **definite integral from a to b** is defined by

$$A = \int_a^b f(x)dx = \lim_{n \to \infty} \sum_{i=1}^{n} \Delta x f(x_i^*) \tag{24.6}$$

If the limit exists we say that f is **integrable** on $[a, b]$, and A gives the **area under the curve** of $f(x)$ on $[a, b]$

The numbers a and b are called the **upper bound** or **upper limit**(b) and **lower bound** or **lower limit** (a), respectively

---

The definite integral exists whenever the limit in equation 24.6 exists. This limit will exist whenever $f(x)$ is continuous, or has at most only a finite number of jump discontinuities in the interval $[a, b]$. This is because we can always fill up the curve with infinitesimally small rectangles in these situations.

Theorem 24.1. Existence of the Definite Integral

If $f(x)$ is continuous on $[a, b]$ except possibly at a finite number of jump discontinuities, then $f(x)$ is integrable on $[a, b]$.

Remark 24.1. Hey! That Looks Like the Antiderivative!

Even though we are introducing a symbol $\int_a^b f(x)dx$ that closely **resembles** the symbol $\int f(x)dx$ that we used previously for the antiderivative, we have not provided any justification for using the same symbol.

In fact, a student seeing this for the first time should be very confused and wonder why we are doing something so silly.

There is no reason at all (at this point) to believe at that there is any relationship whatsoever between the antiderivative and the definite integral.

Everything will become clear when we present the Fundamental Theorems of Calculus.

Properties of the Definite Integral

Suppose that we exchange the the numbers a and b in equation 24.2; then the sign of Δx changes. Thus if we exchange the upper bound with lower bound on a definite integral, the sign of Δx will change. This causes the sign of resulting Riemann sum, and hence of the corresponding integral, to also change.

Theorem 24.2. Reversal of Bounds of Definite Integrals

$$\int_a^b f(x)dx = -\int_b^a f(x)dx \qquad (24.7)$$

Looking again at equation 24.2, if $a = b$, then Δx is zero, so the sum is zero, so the limit is zero, so the integral is zero. Geometrically, this is because we are attempting to find the area under a point. Not only have we shrunk the width of all the rectangles down to zero, we have shrunk the area of the entire interval down to zero.

Theorem 24.3. Area Under a Point

$$\int_a^a f(x)dx = 0 \qquad (24.8)$$

If $f(x) = C$ (a fixed constant) then the function does not change between a and b and we can calculate the area immediately: it is the area of a rectangle of width $b - a$ and height C. Therefore

Theorem 24.4. Area of a Rectangle

If C is any constant then

$$\int_a^b Cdx = C(b - a) \qquad (24.9)$$

From the properties of limits, the limit of the sum (difference) is the sum (or difference) of the limits, if the limits exist. Similarly, we can pull a constant out of the limit. Thus the same result holds for the definite integral, and therefore the integral is linear (see definition 9.2).

Theorem 24.5. Linearity of the Definite Integral

Let A and B be any fixed constants and f and g be any integrable functions. Then

$$\int_a^b (Af(x) \pm Bg(x))\ dx = A\int_a^b f(x)\ dx \pm B\int_a^b g(x)\ dx \qquad (24.10)$$

Finally, if c is any point (it doesn't actually have to be between a and b, but it helps to visualize it that way) then the areas to the left of c and the right of c must add up to equal the area from a to b (see figure 24.1).

Figure 24.1: Additivity of area: the area under the curve, given by $\int_a^b f(x)dx$ on the left can be divided into two parts and is equal to the sum of the two areas $A_1 = \int_a^c f(x)dx$ and $A_2 = \int_c^b f(x)dx$ illustrated on the right (theorem 24.6)

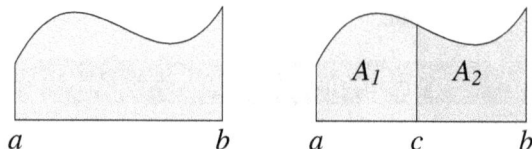

Theorem 24.6. Additivity of Area

$$\int_a^b f(x)dx = \int_a^c f(x)dx + \int_c^b f(x)dx \qquad (24.11)$$

Example 24.1. Find $\int_0^5 (1+3x)dx$.

Solution. From linearity (theorem 24.5),

$$\int_0^5 (1+3x)dx = 1 \times \int_0^5 dx + 3 \int_0^5 xdx \qquad (24.12a)$$

The first term is $1(5-0) = 5$ by theorem 24.4, and the second term is area of a triangle of base 5 and height 5. Thus

$$\int_0^5 (1+3x)dx = 5 + 3 \times \frac{1}{2}(5 \times 5) = 42.5 \qquad (24.12b)$$

□

Theorem 24.7. Positivity of Area

If $f(x) \geq 0$ for $a \leq x \leq b$ then $\int_a^b f(x)dx \geq 0$.

Note, however, that the integral can be negative, if the function is negative! If the function is positive, then the area is positive. If the function is negative, the area is negative.

The following theorem is true because the area under the smaller function is smaller than the area under the bigger function (figure 24.2).

Theorem 24.8. Area Comparison Property

If $f(x) \geq g(x) \geq 0$ for $a \leq x \leq b$ then

$$\int_a^b f(x)dx \geq \int_a^b g(x)dx \qquad (24.13)$$

Figure 24.2: Area Comparison Property (theorem 24.8). If $f(x) \geq g(x)$ on $[a, b]$ then $A_1 \geq A_2$.

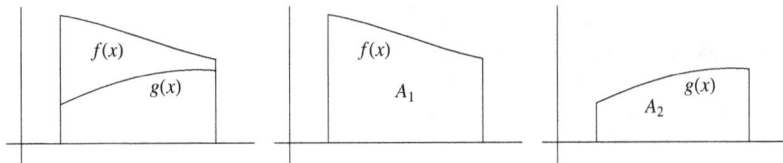

Finally, suppose that a function is always positive, and that we can bound it by two rectangles: a big rectangle with height $M = \max(f(x))$ and a smaller rectangle $m = \min(f(x))$. Think of using a rectangular approximation to find the area with only a single rectangle, but instead of using a corner point we use a tangent point (or a just touching point) of the curve.

Then the area of small rectangle will always underestimate the area under the curve, and the area of the big rectangle will always overestimate the area (figure 24.3).

Theorem 24.9. Upper and Lower Bounds on Definite Integral

If $m \leq f(x) \leq M$ for $a \leq x \leq b$ then

$$m(b - a) \leq \int_a^b f(x)dx \leq M(b - a) \tag{24.14}$$

Figure 24.3: Illustration of the upper and lower bounds comparison theorem (theorem 24.9). The top of the rectangle on the left is the minimum of $f(x)$ on $[a, b]$, and has area smaller than $\int_a^b f(x)dx$ (the figure on the right). The top of the rectangle in the center is the maximum of $f(x)$ on $[a, b]$ and has area larger than $\int_a^b f(x)dx$.

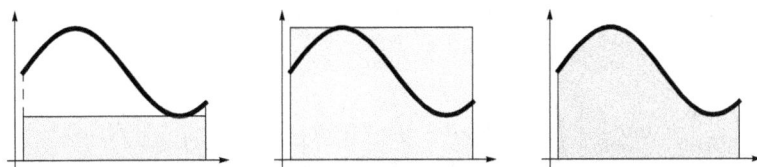

Example 24.2. Find upper and Lower Bounds of $\int_1^4 \sqrt{x}\,dx$ without actually solving the integral.

Solution.   Letting $f(x) = \sqrt{x}$. On the interval $[1, 4]$, the maximum of $f(x)$ is 2 and the minimum of $f(x)$ is 1. Hence

$$3 = 1(4 - 1) \leq \int_1^4 \sqrt{x}\,dx \leq 2(4 - 1) = 6 \tag{24.15a}$$

so the integral is somewhere between 3 and 6.   □

The Fundamental Theorem of Calculus

Let $f(x)$ be any integrable function on an interval $[a,b]$. Suppose we define an area function $A(x)$ that returns the area illustrated in figure 24.3. This function, $A(x)$ will give us the area under the curve of $f(x)$ between a and x for any value of x. By definition 24.1,

Figure 24.3. The function $A(x)$ gives the area under the curve of $f(x)$ between a and x.

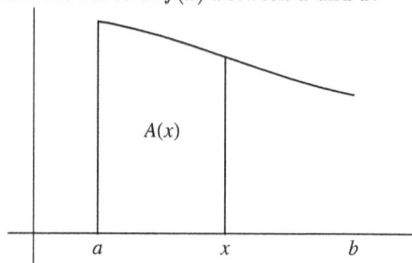

$$A(x) = \int_a^x f(t)dt \qquad (24.16)$$

be the area under the curve of $f(x)$ and above the x axis between a and x, where where $a \leq x \leq b$.

Let's find $dA(x)/dx$ using the definition of the derivative. This is

$$\frac{dA}{dx} = \lim_{h \to 0} \frac{A(x+h) - A(x)}{h} = \lim_{h \to 0} \frac{1}{h} \left[\int_a^{x+h} f(t)dt - \int_a^x f(t)dt \right] \qquad (24.17)$$

From additivity of area (equation (24.11))

$$\int_a^b f(t)dt = \int_a^c f(t)dt + \int_c^b f(t)dt \qquad (24.18)$$

Setting $b = x + h$ and $c = x$ gives

$$\int_a^{x+h} f(t)dt = \int_a^x f(t)dt + \int_x^{x+h} f(t)dt \qquad (24.19)$$

hence

$$\int_a^{x+h} f(t)dt - \int_a^x f(t)dt = \int_x^{x+h} f(t)dt \qquad (24.20)$$

and therefore

$$\frac{dA}{dx} = \lim_{h \to 0} \frac{1}{h} \left[\int_a^{x+h} f(t)dt - \int_a^x f(t)dt \right] = \lim_{h \to 0} \frac{1}{h} \int_x^{x+h} f(t)dt \qquad (24.21)$$

Let m$= \min_{x \leq t \leq x+h} f(t)$ and $M = \max_{x \leq t \leq x+h} f(t)$. Then by theorem 24.9

$$mh \leq \int_x^{x+h} f(t)dt \leq Mh \qquad (24.22)$$

Dividing by h,

$$m \leq \frac{1}{h} \int_x^{x+h} f(t)dt \leq M \qquad (24.23)$$

In the limit as $h \to 0$,

$$\lim_{h \to 0} m = \lim_{h \to 0} \left(\min_{x \leq t \leq x+h} f(t) \right) = \lim_{h \to 0} f(x) = f(x) \qquad (24.24)$$

$$\lim_{h \to 0} M = \lim_{h \to 0} \left(\max_{x \leq t \leq x+h} f(t) \right) = \lim_{h \to 0} f(x) = f(x) \qquad (24.25)$$

Therefore by the squeeze theorem (theorem 3.11),

$$\lim_{h \to 0} m \quad \leq \quad \lim_{h \to 0} \frac{1}{h} \int_x^{x+h} f(t)dt \quad \leq \quad \lim_{h \to 0} M \tag{24.26}$$
$$\downarrow \qquad\qquad\qquad \downarrow \qquad\qquad\qquad \downarrow$$
$$f(x) \qquad\qquad\qquad L \qquad\qquad\qquad f(x)$$

The only possible value for L is $f(x)$ and therefore

$$\frac{dA}{dx} = f(x) \tag{24.27}$$

This means that **the area function $A(x)$ is an antiderivative of** $f(x)$. We call this result the (first) fundamental theorem of calculus.

> **Theorem 24.10. First Fundamental Theorem of Calculus**
>
> Let $f(x)$ be continuous on $[a, b]$, and let
>
> $$A(x) = \int_a^x f(t)dt, \ a \leq t \leq b \tag{24.28}$$
>
> be the area under the curve of f on $[a, x]$. Then $A(x)$ is continuous on $[a, b]$ and
>
> $$A'(x) = f(x) \tag{24.29}$$
>
> i.e., $A(x)$ is an antiderivative of f.

> **Remark 24.2. Antiderivatives and Integrals**
>
> This is the first place in the book that we actually know that the definite integral (area) is somehow related to the antiderivative.

We have derived that $A(x)$ is one particular antiderivative of $f(x)$; therefore the most general antiderivative of $f(x)$ is

$$F(x) = A(x) + C \tag{24.30}$$

for any constant C. Substituting $x = a$ gives

$$F(a) - C = \int_a^a f(t)dt = 0 \tag{24.31}$$

because the area under a point is zero (theorem 24.3). Thus $C = F(a)$. Substituting this into equation 24.30 (and replacing $A(x)$ with its definition from equation 24.16),

$$F(x) - F(a) = \int_a^x f(x)dx \tag{24.32}$$

Letting $x = b$ then gives

$$F(b) - F(a) = \int_a^b f(x)dx \tag{24.33}$$

This is a more common way of stating the Fundamental Theorem of Calculus. We will refer to equation 24.33 as the second Fundamental Theorem of Calculus.

Theorem 24.11. Second Fundamental Theorem of Calculus

Let $f(x)$ be continuous on $[a, b]$ and let $F(x)$ be any anti-derivative of f. Then

$$\int_a^b f(x)dx = F(b) - F(a) \tag{24.34}$$

Solving Definite Integrals

The distinction between *definite* integrals – integrals with endpoints – and *indefinite* integrals is important. The term *indefinite integral* is used interchangeably with antiderivative and refers to a family of functions.

The term *definite integral* refers to an area and *always has a numerical value.*

You can tell the two types of integrals apart visually because definite integrals have bounds (limits) and indefinite integrals do not.

Indefinite Integral

An Indefinite Integral is always a function.

Definite Integral

A Definite Integral is always a number.

The second FTIC[1] gives us an algorithm for solving definite integrals (or finding areas by evaluating antiderivatives. This algorithm is demonstrated in the following example.

Example 24.3. Find $\displaystyle\int_{-2}^3 x^3 dx$.

Solution. This is a definite integral, but we evaluate it by finding the antiderivative. By the power rule,

$$\int x^3 \, dx = \frac{1}{4}x^4 + C \tag{24.35a}$$

The antiderivative is

$$F(x) = \frac{1}{4}x^4 + C \tag{24.35b}$$

Therefore the definite integral

$$\int_{-2}^3 x^3 \, dx = F(3) - F(-2) \tag{24.35c}$$

Observe that when we calculate the difference $F(3) - F(-2)$, the constant C cancels out, so when evaluating *definite* integrals, we can forget about C. We have a special notation for the difference $F(3) - F(-2)$ that we write as follows:

$$F(3) - F(-2) = F(x)\Big|_{-2}^3 \tag{24.35d}$$

---

[1]FTIC = Fundamental Theorem of Calculus

This vertical line notation means the following: (a) first substitute the upper number (3) into $F(x)$; then (b) substitute the lower number into $F(x)$; then (c) subtract. Thus the answer to example is as follows.

$$\int_{-2}^{3} x^3 dx = \frac{x^4}{4}\bigg|_{-2}^{3} = \frac{(3)^4}{4} - \frac{(-2)^4}{4} = \frac{81}{4} - \frac{16}{4} = \frac{65}{4} \qquad (24.35e)$$

□

Notation

$$g(x)\bigg|_{a}^{b} = g(b) - g(a) \qquad (24.36)$$

Heuristic 24.1. Solving a Definite Integral

To evaluate $\displaystyle\int_{a}^{b} f(x)\,dx$:[a]

1. Find any antiderivative $F(x) = \displaystyle\int f(x)dx$, e.g., from an integral table.

2. Find $F(b)$ and $F(b)$.

3. The definite integral (or the area under the curve) is $\displaystyle\int_{b}^{a} f(x)\,dx = F(b) - F(a)$.

---
[a]or to find the area under the curve of $f(x)$ on $[a, b]$

Example 24.4. Find $\displaystyle\int_{0}^{1} x^{4/5} dx$.

Solution. Using the power rule,

$$\int_{0}^{1} x^{4/5} = \frac{5}{9}x^{9/5}\bigg|_{0}^{1} = \frac{5}{9}(1 - 0) = \frac{5}{9}. \qquad (24.37a)$$

□

Example 24.5. Evaluate $\displaystyle\int_{1}^{4} (1 + 3y - y^2)dy$.

Solution. FRom the linearity of integrals and the power rule,

$$\int_{1}^{4} (1 + 3y - y^2)dx = \left(y + \frac{3}{2}y^2 - \frac{1}{3}y^3\right)\bigg|_{1}^{4} \qquad (24.38a)$$

$$= \left(4 + \frac{3}{2}4^2 - \frac{1}{3}4^3\right) - \left(1 + \frac{3}{2}1^2 - \frac{1}{3}1^3\right) \qquad (24.38b)$$

$$= \left(4 + 24 - \frac{64}{3}\right) - \left(1 + \frac{3}{2} - \frac{1}{3}\right) = \frac{9}{2} \qquad (24.38c)$$

□

Example 24.6. Find the area under the parabola $y = x^2$ from $x = 3$ to $x = 4$.

Solution. The area under the curve is the same as the definite integral. From the power rule,

$$\int_{3}^{4} x^2 dx = \frac{x^3}{3}\bigg|_{3}^{4} = \frac{4^3 - 3^3}{3} = \frac{64 - 27}{3} = \frac{37}{3}. \qquad (24.39a)$$

□

Example 24.7. Find the area under the curve of $y = \sqrt[3]{x}$ from $x = 0$ to $x = 27$.

Solution. From the power rule,

$$\int_0^{27} x^{1/3} dx = \frac{x^{4/3}}{4/3} \bigg|_0^{27} = \frac{3 \cdot 27^{4/3}}{4} = \frac{243}{4}. \tag{24.40a}$$

□

Example 24.8. Find the area under the curve $y = \sin x$, from $x = 0$ to $x = \pi$.

Solution. Using the formula for the antiderivative of the sine function,

$$\int_0^{\pi} \sin x\, dx = -\cos x \bigg|_0^{\pi} = -\cos \pi + \cos 0 = -(-1) + 1 = 2. \tag{24.41a}$$

□

Differentiating Definite Integrals With Respect to a Bound

Let $F(x)$ be any anti-derivative of $f(x)$. From the second fundamental theorem of calculus (theorem 24.11),

$$\int_a^x f(t)dt = F(x) - F(a) \tag{24.42}$$

Differentiating both sides of the equation,

$$\frac{d}{dx} \int_a^x f(t)dt = \frac{d}{dx}\left(F(x) - F(a)\right) = \frac{d}{dx}F(x) - \frac{d}{dx}F(a) = F'(x) - 0 = f(x) \tag{24.43}$$

where we have used the facts that $(F(a))' = 0$ (derivative of a constant); and $F'(x) = f(x)$ (definition of an antiderivative) in the last two steps. Thus **the derivative of a definite integral with respect to its upper bound is the integrand evaluated at that upper limit.** We will call this result the third FTIC.

Theorem 24.12. Third Fundamental Theorem of Calculus

$$\frac{d}{dx} \int_a^x f(t)dt = f(x) \tag{24.44}$$

Example 24.9. Find $\dfrac{d}{dx} \displaystyle\int_1^x \frac{1}{t^3 + 1}dt$.

Solution. Using theorem 24.12, since we are differentiating with respect to x, and there is an x in the upper bound, we replace the t in the integrand everywhere with x.

$$\frac{d}{dx} \int_1^x \frac{1}{t^3 + 1}dt = \frac{1}{x^2 + 1} \tag{24.45a}$$

□

Example 24.10. Find $\dfrac{d}{dx}\displaystyle\int_x^1 \cos\sqrt{t}\,dt$.

Solution. Here we are differentiating with respect to the lower bound instead of the upper bound. To apply theorem 24.12 we need to reverse the direction of integration (theorem 24.2). This introduces a change in the sign of the integral.

$$\frac{d}{dx}\int_x^1 \cos\sqrt{t}\,dt = -\frac{d}{dx}\int_1^x \cos\sqrt{t}\,dt = -\cos\sqrt{x} \qquad (24.46a)$$

$\square$

What if the upper bounds of the integral is a function, rather than just a variable. For example, how do we find

$$\frac{d}{dx}\int_0^{g(x)} f(t)dt? \qquad (24.47)$$

This sounds like a job for the chain rule. We begin by making the substitution

$$u = g(x) \qquad (24.48)$$

Then we write

$$\frac{d}{dx}\int_0^{g(x)} f(t)dt = \frac{d}{dx}\int_0^u f(t)dt \qquad (24.49)$$

where $u = g(x)$. To make sense of this we can first define the function

$$I(u) = \int_0^u f(t)dt \qquad (24.50)$$

Then

$$\frac{d}{dx}\int_0^{g(x)} f(t)dt = \frac{d}{dx}I(g(x)) = I'(g(x)))g'(x) \qquad (24.51)$$

or in the Liebniz notation

$$\frac{d}{dx}\int_0^{g(x)} f(t)dt = \left(\frac{d}{du}\int_0^u f(t)dt\right)\Bigg|_{u=g(x)} \times \left(\frac{dg(x)}{dx}\right) \qquad (24.52)$$

Theorem 24.13. Chain Rule - Fundamental Theorem of Calculus

$$\frac{d}{dx}\int_0^{g(x)} f(t)dt = \left(\frac{d}{du}\int_0^u f(t)dt\right)\Bigg|_{u=g(x)} \times \left(\frac{d}{dx}g(x)\right) \qquad (24.53)$$

Example 24.11. Find $\dfrac{d}{dx}\displaystyle\int_0^{1/x} \sin^4 t\,dt$.

Solution. Define $u = g(x) = 1/x$. Then $g'(x) = -1/x^2$, so that

$$\frac{d}{dx}\int_0^{1/x} \sin^4 t\,dt = -\frac{1}{x^2}\left(\frac{d}{du}\int_0^u \sin^4 t\,dt\right)\Bigg|_{u=g(x)=1/x} \qquad (24.54a)$$

$$= -\frac{1}{x^2}\sin^4\frac{1}{x} \qquad (24.54b)$$

$\square$

Example 24.12. Find $\dfrac{d}{dx}\displaystyle\int_x^{x^2}\sqrt{1+t^2}\ dt$.

Solution. Now both bounds depend on x; to use theorem 24.13 we must have the function on the upper bound only. So we have to do two things: (1) split the integral into two parts, with an arbitrary point in the middle; then (2) reverse the order of integration on the integral that has the function on the wrong bounds. It is only after performing these that we can apply theorem 24.13. First we pick any point c in the domain of integration – we don't actually care what the value is. By theorem 24.6,

$$\frac{d}{dx}\int_x^{x^2}\sqrt{1+t^2}\ dt = \frac{d}{dx}\int_x^{c}\sqrt{1+t^2}\ dt + \frac{d}{dx}\int_c^{x^2}\sqrt{1+t^2}\ dt \qquad (24.55\text{a})$$

We can reverse the order of integration on the first integral (remembering to include a minus sign) and then apply the chain rule to each integral.

$$\frac{d}{dx}\int_x^{x^2}\sqrt{1+t^2}\ dt = -\frac{d}{dx}\int_c^{x}\sqrt{1+t^2}\ dt + \frac{d}{dx}\int_c^{x^2}\sqrt{1+t^2}\ dt \qquad (24.55\text{b})$$

$$= -\sqrt{1+x^2} + \left(\frac{d}{du}\int_c^{u}\sqrt{1+t^2}\ dt\right)\Bigg|_{u=x^2} \times \left(\frac{d}{dx}x^2\right) \qquad (24.55\text{c})$$

$$= -\sqrt{1+x^2} + \left(\sqrt{1+u^2}\right)\Big|_{u=x^2} \times (2x) \qquad (24.55\text{d})$$

$$= -\sqrt{1+x^2} + 2x\sqrt{1+x^4} \qquad (24.55\text{e})$$

$\square$

Example 24.13. Find a continuous function $f(x)$ such that

$$\int_5^{x} f(t)\sin t\ dt = \frac{1-x}{x} + \int_{\pi}^{x} f(t)\cos t\ dt. \qquad (24.56)$$

Solution. We can solve this problem by using implicit differentiation to eliminate the integrals. Differentiating both sides of the equation gives us

$$\frac{d}{dx}\int_5^{x} f(t)\sin t\ dt = \frac{d}{dx}\frac{1-x}{x} + \frac{d}{dx}\int_{\pi}^{x} f(t)\cos t\ dt \qquad (24.57\text{a})$$

We can differentiate each integral using theorem 24.12:

$$f(x)\sin x = \frac{d}{dx}\frac{1-x}{x} + f(x)\cos x \qquad (24.57\text{b})$$

Since $\dfrac{1-x}{x} = \dfrac{1}{x} - 1$, the derivative of the first term on the right hand side of the equation is $-1/x^2$, and thus

$$f(x)\sin x = -\frac{1}{x^2} + f(x)\cos x \qquad (24.57\text{c})$$

We solve for $f(x)$ by collecting all terms that have $f(x)$ on one side and solving for it as an algebraic quantity:

$$f(x)\sin x - f(x)\cos x = -\frac{1}{x^2} \qquad (24.57\text{d})$$

$$f(x)(\sin x - \cos x) = -\frac{1}{x^2} \qquad (24.57\text{e})$$

$$f(x) = \frac{1}{x^2(\cos x - \sin x)}. \qquad (24.57\text{f})$$

$\square$

The Net Change Theorem

If F is an antiderivative of f then the fundamental theorem of calculus tells us that

$$\int_a^b f(t)dt = F(b) - F(a) \tag{24.58}$$

But since F is an antiderivative, $F'(t) = f(t)$. We can substitute this on the left hand side of the equation:

$$\int_a^b f(t)dt = \int_a^b F'(t)dt \tag{24.59}$$

Setting the last two results equal gives us our final FTIC: The Net Change Theorem.

Theorem 24.14. Net Change Theorem
Fourth Fundamental Theorem of Calculus

$$\int_a^b \frac{dF(t)}{dt}dt = F(b) - F(a) \tag{24.60}$$

or more succinctly,

$$\int_{t=a}^{t=b} dF(t) = F(b) - F(a) \tag{24.61}$$

Example 24.14. Suppose an object is moving on a trajectory described by the equation $v(t) = -t + 3t^2$, where v is the object's velocity in meters per second. How far has the object moved away from its original position in 5 seconds? 10 seconds? between $t = 3$ and $t = 7$?

Solution. The definition of velocity is $v = ds/dt$, where s is the position. According to the net change theorem, the distance moved in the first 5 seconds is

$$s(5) - s(0) = \int_0^5 v(t)dt = \int_0^5 (-t + 3t^2)dt \tag{24.62a}$$

$$= -\frac{t^2}{2} + \frac{3t^3}{3}\Big|_0^5 = -\frac{25}{2} + 125 = \frac{225}{2} = 112.5 \tag{24.62b}$$

In the first 10 seconds, the corresponding displacement is

$$s(10) - s(0) = -\frac{t^2}{2} + \frac{3t^3}{3}\Big|_0^{10} = -\frac{100}{2} + 1000 = 950 \tag{24.62c}$$

The displacement between $t = 3$ and $t = 7$ is

$$s(7) - s(3) = \int_3^7 v(t)dt = -\frac{t^2}{2} + \frac{3t^3}{3}\Big|_3^7 \tag{24.62d}$$

$$= \left(-\frac{49}{2} + 343\right) - \left(-\frac{9}{2} + 27\right) \tag{24.62e}$$

$$= \frac{-49 + 686 + 9 - 54}{2} = \frac{592}{2} = 296 \tag{24.62f}$$

$\square$

Example 24.15. Suppose a baseball is thrown upwards at 20 feet/second from the top of 500 foot tall building. (a) How high does the ball get before it reaches the top of its trajectory, assuming there is no drag, and that the acceleration of gravity is 32 feet/sec²? (b) When does the ball hit the ground at the base of the building? (c) How fast is the ball moving when it hits the ground?

Solution. We are given the following initial conditions:

$$y(0) = 500 \qquad \text{the height of the ball at } t = 0 \qquad (24.63\text{a})$$
$$y'(0) = 20 \qquad \text{the speed of the ball at } t = 0 \qquad (24.63\text{b})$$

Since the only force on the ball is gravity,

$$\frac{dv}{dt} = -32 \qquad (24.63\text{c})$$

Integrating,

$$\int dv = -32 \int dt \qquad (24.63\text{d})$$
$$v = 32t + C \qquad (24.63\text{e})$$

Substituting $v(0) = 20$ gives $C = 20$, hence

$$\frac{dy}{dt} = v(t) = -32t + 20 \qquad (24.63\text{f})$$

Integrating again,

$$y = \int dy = \int (-32t + 20)dt = -16t^2 + 20t + C \qquad (24.63\text{g})$$

Substituting $y(0) = 500$ gives $C = 500$, so

$$y(t) = -16t^2 + 20t + 500 \qquad (24.63\text{h})$$

(a) The ball reaches its peak when the velocity is zero, which occurs when

$$-32t + 20 = 0 \qquad (24.63\text{i})$$

or $t = 20/32 = 5/8$. The maximum height is then $y(5/8) = 2025/4 = 506.25$ feet.

(b) The ball hits the ground when

$$0 = y = -16t^2 + 20t + 500. \qquad (24.63\text{j})$$

The only positive root is $t = 25/4 = 6.25$ seconds.

(c) The speed at $t = 6.25$ seconds is $v(6.25) = -32(6.25) + 20 = -200$ ft/sec. The negative sign indicates that the ball is moving downwards when it hits the ground. $\square$

Summary: Fundamental Theorems of Calculus

1) If $f(x)$ is continuous on (a, b) and we define $A(x)$ as the area under the curve from a to x, where $a < x < b$, then the function $A(x)$ is an antiderivative of $f(x)$.

2) If $F(x)$ is any antiderivative of $f(x)$ then $\displaystyle\int_a^b f(x)dx = F(b) - F(a)$.

3) $\displaystyle\frac{d}{dx} \int_a^x f(t)dt = f(x)$ or $\displaystyle\int_0^{g(x)} f(t)dt = g'(x) \left[\int_a^u f(t)dt \right]_{u=g(x)}$

4) $\displaystyle\int_a^b \frac{dF(t)}{dt} dt = F(b) - F(a)$ (Net Change Theorem)

Exercises

Evaluate the definite integral in exercises 1 through 6.

1. $\displaystyle\int_{-2}^3 (x^2 - 3)dx$

2. $\displaystyle\int_{-1}^1 t(1 - t)^2\, dt$

3. $\displaystyle\int_{\pi/4}^{\pi/3} \csc^2 \theta\, d\theta$

4. $\displaystyle\int_1^4 \sqrt{\frac{5}{x}}\, dx$

5. $\displaystyle\int_{-2}^1 (1 + x^2)x\, dx$

6. $\displaystyle\int_1^3 \left(\frac{1}{x^2} - \frac{1}{x^3} \right) dx$

Find $f'(x)$ for each of the functions in exercises 7 through 13.

7. $f(x) = \displaystyle\int_3^x (t + t^2)dt$

8. $f(x) = \displaystyle\int_7^x \frac{1}{u}\, du$

9. $f(x) = \displaystyle\int_0^{\sin x} t^2\, dt$

10. $f(x) = \displaystyle\int_{x^2}^{x^3} t \sin t\, dt$

11. $f(x) = \displaystyle\int_1^{x^2} \sec t\, dt$

12. $f(x) = \displaystyle\int_x^{\pi/2} \frac{\sin \sqrt{t}}{\sqrt{t}}\, dt$

13. $f(x) = \displaystyle\int_3^{x^2+x} \frac{1}{1 + t^4}\, dt$

In exercises 14 through 17 find an explicit formula for a continuous function $f(t)$ that satisfies each of the following equations.

14. $x^2 = \displaystyle\int_0^{x^2} \frac{1 - t^2}{f(t)}\, dt$

15. $\displaystyle\int_0^x t^4 f(t)^2\, dt = \int_0^x 2t f(t)dt$

16. $\displaystyle\int_x^1 \frac{f(t)}{1 - \cos t}\, dt = x + \int_0^x f(t)dt$

17. $\displaystyle\int_\pi^x t f(t)dt = x^2 \cos x + \int_0^x \frac{t f(t)}{1 - t^2}\, dt$

18. Just as the velocity is the derivative of the position, the position is the integral of the derivative. Suppose an object is moving for a time interval $t_1 < t < t_2$. Then we can define both a **displacement** and a **total distance travelled** by

Displacement $=$
$$\int_{t_1}^{t_2} v(t)dt = \int_{t_1}^{t_2} s'(t)dt = s(t_2) - s(t_1)$$

Distance Traveled $= \displaystyle\int_{t_1}^{t_2} |v(t)|dt$

Suppose that the velocity of a particle is given by $v(t) = t^2 - 8t + 15$.

(a) Determine when the particle is moving forward and when it is moving backwards.

(b) Determine the total distance travelled

(c) Determine the total displacement

Chapter 25

Substitution Methods for Integration

Chapter Summary and Goal

Most integrals cannot be determined by simply looking at a table of derivatives. Instead, we will have to make one (or more) substitutions. The purpose of these substitutions is to change a complicated-looking integral (that we cannot solve) into something simple, that we already know how to solve. We will learn how to make basic "u-substitutions" in this chapter.

Student Learning Objectives

The student will:

1. Recognize how and when u-substitutions will work.
2. Learn to solve integrals by making u-substitutions.
3. Apply symmetry to simplify definite integrals.

Method of u-substitution

In this method we define a new variable as a function of variables that already exist in the integral. For example, if an integral $\int f(x)dx$ depends on x, we define u as some function of x, and then by using algebraic techniques, replace all references to the variable x with references to the new variable u in both the integrand and dx. We demonstrate via an example.

Example 25.1. Solve $\displaystyle\int 2x\sqrt{1+x^2}\ dx$ by u-substitution.

Solution. In many u-substitution problems there are multiple ways to proceed, and there is rarely a single "correct" method to solve the problem. We will use the substitution

$$u = 1 + x^2 \tag{25.1a}$$

Then

$$du = 2x\ dx \tag{25.1b}$$

This implies that

$$dx = \frac{du}{2x} \tag{25.1c}$$

237

Substituting back into the original integral,

$$\int 2x\sqrt{1+x^2}\ dx = \int 2x\sqrt{u}\left(\frac{du}{2x}\right) = \int \sqrt{u}\ du = \int u^{1/2}du \qquad (25.1\text{d})$$

$$= \frac{u^{3/2}}{3/2} + C = \frac{2}{3}(1+x^2)^{3/2} + C \qquad (25.1\text{e})$$

In the last step we replaced u with its original definition in terms of x, thereby eliminating it from the final answer. We will always do this when we solve indefinite integrals via u-substitution: *the answer must always depend solely on the original variable, and the new variable can never be in the solution.*   □

Why this worked. We were able to find a substitution that turned our integral into something like this:

$$\int f(x)\ dx = \int F(g(x))\frac{dx}{du}du = \int F(u)\frac{dx}{du}du \qquad (25.2)$$

for some function $F(u)$ in such a way that x disappeared from the equation.

To develop a heuristic, we will try to write the integral as $\int F'(g(x))g'(x)dx$. If we are successful at doing this, then by the chain rule

$$\int F'(g(x))g'(x)dx = F(g(x)) + C \qquad (25.3)$$

This works because by the chain rule

$$\frac{d}{dx}F(g(x)) = F'(g(x))g'(x) \qquad (25.4)$$

so that

$$\int F'(g(x))g'(x)\ dx = \int \frac{d}{dx}[F(g(x))]\ dx = F(g(x)) + C \qquad (25.5)$$

If we define

$$f(x) = F'(x) \qquad (25.6)$$

then $F(x)$ is an anti-derivative of f, so

$$\int f(g(x))g'(x)\ dx = \int F'(g(x))g'(x)]dx \qquad (25.7)$$

$$= F(g(x)) + C \qquad (25.8)$$

$$= \int F'(u)\ du \qquad (25.9)$$

$$= \int f(u)\ du \qquad (25.10)$$

Heuristic 25.1. Substitution Rule

If $u = g(x)$ is differentiable then

$$\int f(g(x))g'(x)\ dx = \int f(u)\ du. \qquad (25.11)$$

Example 25.2. Find $\int x^5 \cos(x^6 + 7)dx$.

Solution. If we make the substitution $u = x^6 + 7$ then $du = 6x^5 dx$. By rearrangement,

$$x^5 dx = du/6 \tag{25.12a}$$

Hence

$$\int x^5 \cos(x^6 + 7)\ dx = \int \cos(x^6 + 7) \times (x^5\ dx) \tag{25.12b}$$

$$= \int (\cos u) \times (du/6) \tag{25.12c}$$

$$= \frac{1}{6} \int \cos u\ du = \frac{1}{6} \sin u + C \tag{25.12d}$$

$$= \frac{1}{6} \sin(x^6 + 7) + C \tag{25.12e}$$

$\square$

Example 25.3. Solve $\int \dfrac{\sin \sqrt{x}}{\sqrt{x}}\ dx$.

Solution. We let $u = \sqrt{x}$. Then $du = \dfrac{dx}{2\sqrt{x}}$ or $dx = 2\sqrt{x}\,du$, so that

$$\int \frac{\sin \sqrt{x}}{\sqrt{x}}\ dx = \int \frac{\sin u}{\sqrt{x}} \cdot 2\sqrt{x}\,du = 2 \int \sin u\ du \tag{25.13a}$$

$$= -2\cos u + C = -2\cos \sqrt{x} + C \tag{25.13b}$$

$\square$

Example 25.4. Find $\int \sec^2 x \tan x\,dx$.

Solution. Since $(\tan x)' = \sec^2 x$, we use the substitution $u = \tan x$. Then $du = \sec^2 x\,dx$, and the integral becomes

$$\int \sec^2 x \tan x\,dx = \int u\,du = \frac{u^2}{2} + C = \frac{1}{2} \tan^2 x + C \tag{25.14a}$$

$\square$

Example 25.5. Find $\int \tan x\,dx$.

Solution. The definition of the the tangent is $\tan x = \dfrac{\sin x}{\cos x}$, and since

$$(\cos x)' = -\sin x \tag{25.15a}$$

then

$$\int \tan x\,dx = \int \frac{\sin x}{\cos x}\,dx \tag{25.15b}$$

If we try $u = \cos x$. Then $du = -\sin x\,dx$, so that

$$\int \tan x\,dx = -\int \frac{du}{u} = -\ln|u| = -\ln|\cos x| = \ln|\sec x|. \tag{25.15c}$$

$\square$

Example 25.6. Find $\displaystyle\int \sqrt{4t-1}\; dt$.

Solution. Here we have two choices. We can either use $u = 4t - 1$ or $u = \sqrt{4t-1}$.

Method 1: Let $u = 4t - 1$. Then $du = 4\, dt$.

$$\int \sqrt{4t-1}\; dt = \int \sqrt{u} \times \frac{dt}{4} = \frac{1}{4}\int u^{1/2}\; du \qquad (25.16a)$$

$$= \frac{1}{4(3/2)} u^{3/2} + C = \frac{1}{6}(4t-1)^{3/2} + C \qquad (25.16b)$$

Method 2: Let $u = \sqrt{4t-1}$. Then by the chain rule

$$du = \frac{1}{2\sqrt{4t-1}} \times 4dt = \frac{2dt}{\sqrt{4t-1}} = \frac{2dt}{u} \qquad (25.16c)$$

Hence $dt = udu$ and therefore

$$\int \sqrt{4t-1}\; dt = \int u \times \frac{u\; du}{2} = \frac{1}{2}u^2\; du = \frac{1}{6}u^3 + C = \frac{1}{6}(4t-1)^{3/2} + C \qquad (25.16d)$$

□

The procedure is similar for definite integrals, only we must remember that the bounds apply to the old variable. When we change variables, we must also change the values of the bounds to apply to the new variable.

Example 25.7. Find $\displaystyle\int_0^2 (2x-1)^{25} dx$.

Solution. Let $u = 2x - 1$. Then $du = 2dx$.

When $x = 0$, $u = 2 \times 0 - 1 = -1$; and when $x = 1$, $u = 2 \times 1 - 1 = 1$.

$$\int_0^1 (2x-1)^{25} dx = \int_{-1}^1 u^{25} du = \left.\frac{u^{26}}{26}\right|_{-1}^1 = \frac{1}{26}(1-1) = 0. \qquad (25.17a)$$

□

Example 25.8. Find $\displaystyle\int_0^{\sqrt{\pi/2}} x\cos(x^2) dx$.

Solution. Let $u = x^2$; then $du = 2x\; dx$ or $x\; dx = du/2$.

When $x = 0$, $u = 0^2 = 0$; and when $x = \sqrt{\pi/2}$, $u = (\sqrt{\pi/2})^2 = \pi/2$. Thus

$$\int_0^{\sqrt{\pi/2}} x\cos(x^2)\; dx = \frac{1}{2}\int_0^{\pi/2} \cos u\; du = \left.\frac{1}{2}\sin u\right|_0^{\pi/2} \qquad (25.18a)$$

$$= \frac{1}{2}\sin\frac{\pi}{2} - \frac{1}{2}\sin 0 = \frac{1}{2}. \qquad (25.18b)$$

□

Symmetry and Definite Integrals

Recall the definitions of **odd** and **even** functions from algebra. A function is odd if it can be reflected through the origin, and even if it can be reflected about the y axis (figure 25.1).

Definition 25.1. Even Functions

A function $f(x)$ is called an **even function** if $f(x) = f(-x)$ for all x.

Definition 25.2. Odd Functions

A function $f(x)$ is called an **odd function** if $f(x) = -f(-x)$ for all x.

Figure 25.1: An odd function (left) and an even function (right).

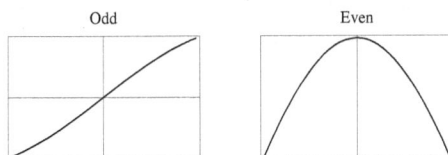

If $f(x)$ is even then if you fold a plot of the curve along the y axis, the curve on the right-hand half plane will overlay the curve on the left-hand half plane. The area under the curve on the left – between $-a$ and 0 – is identical to the area under the curve on the right – between 0 and a.

$$\int_{-a}^{0} f(x)dx = \int_{0}^{a} f(x)dx \tag{25.19}$$

Consequently

$$\int_{-a}^{a} f(x)dx = \int_{-a}^{0} f(x)dx + \int_{0}^{a} f(x)dx = 2\int_{0}^{a} f(x)dx \tag{25.20}$$

Theorem 25.1. Symmetry Property of Even Functions

If $f(x)$ is even, then

$$\int_{-a}^{a} f(x)dx = 2\int_{0}^{a} f(x)dx \tag{25.21}$$

Example 25.9. Find $\displaystyle\int_{-\pi/2}^{\pi/2} \cos x dx$.

Solution. The integrand is even and the integration is over an interval from $-\pi/2$ to $\pi/2$. Hence we can use the symmetry property.

$$\int_{-\pi/2}^{\pi/2} \cos x dx = 2\int_{0}^{\pi/2} \cos x dx = 2\left(\sin x|_{0}^{\pi/2}\right) = 2\left(\sin\frac{\pi}{2} - \sin 0\right) = 2. \tag{25.22a}$$

$\square$

If $f(x)$ is an odd function, then the area on the left hand side of the origin is the same magnitude as the area on the right side of the origin, but it is opposite in sign. Thus they cancel out.

$$\int_{-a}^{0} f(x)dx = -\int_{0}^{a} f(x)dx \tag{25.23}$$

Theorem 25.2. Area Cancellation of Odd Functions

If $f(x)$ is odd, then

$$\int_{-a}^{a} f(x)dx = 0 \tag{25.24}$$

Example 25.10. Find $\displaystyle\int_{-5}^{5} e^{x^2} \sin x\, dx$.

Solution. The integrand is an odd function and we are integrating over an interval from -5 to 5. Hence the area on the left cancels the area on the right, and

$$\int_{-5}^{5} e^{x^2} \sin x\, dx = 0. \tag{25.25a}$$

$\square$

Exercises

Find an appropriate substitution and solve each of the indefinite integrals in exercisess 1 through 19.

1. $\displaystyle\int \sqrt{2x-1}\ dx$

2. $\displaystyle\int (3x+7)^3 dx$

3. $\displaystyle\int 3x\sqrt{1-4x^2}\ dx$

4. $\displaystyle\int \frac{\cos x}{\sin x}\ dx$

5. $\displaystyle\int \frac{1}{x\ln x}\ dx$

6. $\displaystyle\int \frac{x^3\ dx}{\sqrt[5]{x4+6}}$

7. $\displaystyle\int 2x\sec^2(x^2-1)dx$

8. $\displaystyle\int \sqrt{x}\sin(x^{3/2})dx$

9. $\displaystyle\int x^2\sin(x^3+3)dx$

10. $\displaystyle\int x^2(x^3-8)^{48}dx$

11. $\displaystyle\int 9\sin^6 x\cos x\, dx$

12. $\displaystyle\int \sin^2 x\cos x\, dx$

13. $\displaystyle\int x\cos(x^2)dx$

14. $\displaystyle\int x^5\sqrt{3+x^6}\, dx$

15. $\displaystyle\int \frac{2x-2}{(2x^2-4x+6)^3}\, dx$

16. $\displaystyle\int \frac{\sec^2(9/x)}{x^2}\, dx$

17. $\displaystyle\int \frac{\sqrt{1+\sqrt{x}}}{\sqrt{x}}\, dx$

18. $\displaystyle\int \sec(x+\pi)\tan(x+\pi)\ dx$

19. $\displaystyle\int \cos x(\sin^3 x + 3\sin^2 x + 12)dx$

Find an appropriate substitution and use it solve each of the definite integrals in exercises 20 through 26.

20. $\displaystyle\int_{0}^{\pi/4} x^3\sin(x^4)dx$ Ans: $\frac{1}{2}\sin^2\left(\frac{\pi^4}{512.}\right)$

21. $\displaystyle\int_{0}^{\sqrt{81\pi}} x\cos\left(\frac{x^2}{18}\right)\, dx$ Ans: 9

22. $\displaystyle\int_{0}^{\pi/3} \sin(3t)dt$ Ans: 2/3

23. $\displaystyle\int_{1}^{2} \frac{2x\ dx}{x^2-16}$

24. $\displaystyle\int_{0}^{3} \frac{1}{\sqrt{1+x}}\, dx$

25. $\displaystyle\int_{0}^{1} xe^{x^2}\ dx$

26. $\displaystyle\int_{-1}^{1} \sin(x^3)e^{x^2}\ dx$

Chapter 26

Area Between Curves

Chapter Summary and Goal

In this chapter we will learn to calculate the area between two curves using definite integrals. In short, the area is simply the area under the top curve minus the area under the bottom curve.

Student Learning Objectives

The student will:

1. Be able to calculate areas of regions between two non-intersecting curves.
2. Be able to calculate areas of regions between intersecting curves.
3. Be able to justify the use of definite integrals for calculating the areas of regions between curves.

Finding the Area Between Two Curves

We can use the method of filling up any area with small vertically oriented rectangles and summing their area (the Riemann Sum) to find the area between two curves. We begin by observing that if we draw a horizontal line across the middle of any rectangle we have two new rectangles, and the area of the total rectangle is equal to the area of top rectangle plus the area of the bottom rectangle:

$$\text{Area of a Rectangle} = (\text{Area of Top Rectangle}) + (\text{Area of Bottom Rectangle}) \qquad (26.1)$$

It doesn't matter where we draw the horizontal line; this result will always be true. Now fill up the top curve $f(x)$ with rectangles, and divide the ith rectangle in half by drawing a horizontal line across it in such a way that it crosses the bottom curve $g(x)$ at $x = x_i$ (see figure 26.1). Let the top rectangles have areas T_i, the bottom rectangles have areas B_i, and the big rectangles have area A_i. Since

$$A_i = T_i + B_i \qquad (26.2)$$

then

$$\sum_{i=1}^{n} A_i = \sum_{i=1}^{n} T_i + \sum_{i=1}^{n} B_i \qquad (26.3)$$

Figure 26.1: Divide the area under each curve into Riemann Sums as before, but separate the rectangles into their top parts and their bottom parts as illustrated on the left. The area between the two curves is approximated by the rectangles on the top, as shown on the right.

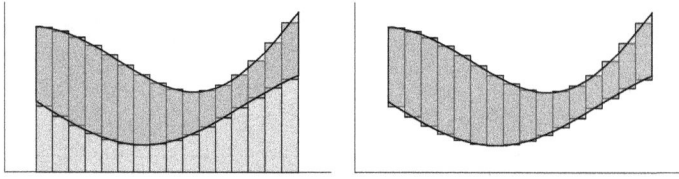

or

$$\sum_{i=1}^{n} T_i = \sum_{i=1}^{n} A_i - \sum_{i=1}^{n} B_i \tag{26.4}$$

Let the top curve be $f(x)$ and the bottom curve be $g(x)$. Then the area of rectangle A_i (the rectangle from the top curve to the x-axis at x_i) is given by $A_i = f(x_i)\Delta x$, and the area of rectangle B_i (the rectangle from the bottom curve to the x-axis at x_i) is $B_i = g(x_i)\Delta x$. Thus

$$\sum_{i=1}^{n} T_i = \sum_{i=1}^{n} f(x_i)\Delta x - \sum_{i=1}^{n} g(x_i)\Delta x = \sum_{i=1}^{n} (f(x_i) - g(x_i))\Delta x \tag{26.5}$$

Taking the limit as $n \to \infty$ (while $\Delta x \to 0$) we get a Riemann Sum of the top rectangles which gives the area between the two curves (see figure 26.2).

Theorem 26.1. Area Between Two Curves

If $f(x) \geq g(x)$ on $[a, b]$ and f and g are integrable on $[a, b]$, then the area of region bounded by $f(x)$, $g(x)$ and the lines $x = a$ and $x = b$ is

$$\text{Area} = \int_a^b (f(x) - g(x))dx \tag{26.6}$$

Figure 26.2: We can calculate the area between two curves by subtracting the area beneath the lower curve from the area beneath the top curve. This is illustrated in the following "equation."

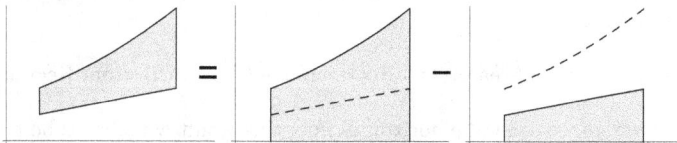

Example 26.1. Find the area bounded by $y = 3 - x^2$, $y = x$, $x = 0$ and $x = 1$.

Solution. The area to be found is illustrated in figure 26.3.

The top curve is $y = 3 - x^2$ and the bottom curve is $y = x$. The area is

$$A = \int_0^1 ((3 - x^2) - x)dx \tag{26.7a}$$

$$= \left(3x - \frac{1}{3}x^3 - \frac{1}{2}x^2\right)\Big|_0^1 \tag{26.7b}$$

$$= 3 - \frac{1}{3} - \frac{1}{2} = \frac{18 - 2 - 3}{6} = \frac{13}{6}. \tag{26.7c}$$

$\square$

Figure 26.3: Area found in examples 26.1 (figure on left) and 26.2 (right).

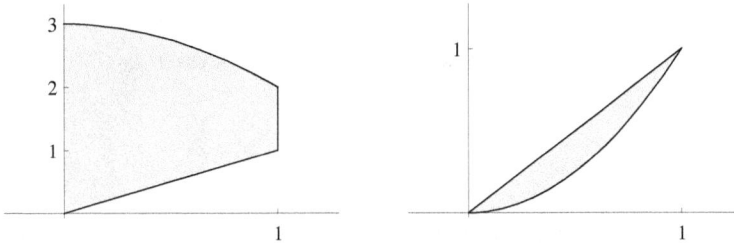

Example 26.2. Find the area between the curves $y = x^2$ and $y = x$.

Solution. The two curves intersect when they are equal, i.e., when $x^2 = x$. Rearranging gives

$$x(x - 1) = 0 \tag{26.8a}$$

The only points of intersection are $x = 0$ and $x = 1$. In this region, $x^2 < x$, so $y = x^2$ is the lower curve and $y = x$ is the upper curve (see figure 26.3). The area is

$$A = \int_0^1 (x - x^2)dx = \left(\frac{1}{2}x^2 - \frac{1}{3}x^3\right)\Big|_0^1 = \frac{1}{2} - \frac{1}{3} = \frac{1}{6}. \tag{26.8b}$$

$\square$

When the two curves intersect it is important to locate the intersection points so that we can identify the regions where $f(x)$ is the top curve and the regions where $g(x)$ is the top curve.

Figure 26.4: The shaded area illustrates the region between $\sin x$ and $\cos x$ on $[0, \pi/3]$. To the left of the intersection point at $[\pi/4, \sqrt{2}/2]$, $\sin x < \cos x$; to the right of the intersection point, $\sin x > \cos x$.

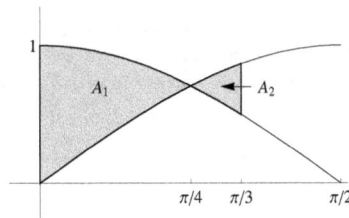

Example 26.3. Find the area between $y = \sin x$, $y = \cos x$, $x = 0$, and $x = \pi/3$.

Solution. The two curves intersect when

$$\sin x = \cos x \tag{26.9a}$$

This occurs when $\tan x = 1$ or $x = \pi/4$.

Denote the area on $[0, \pi/4]$ by A_1 and the area on $[\pi/4, \pi/3]$ by A_2 (figure 26.4). Then the total area is

$$A = A_1 + A_2 \qquad (26.9\text{b})$$

On $[0, \pi/4]$, the top curve is $y = \cos x$ and the bottom curve is $y = \sin x$. Thus

$$A_1 = \int_0^{\pi/4} (\cos x - \sin x)\; dx \qquad (26.9\text{c})$$

$$= (\sin x + \cos x)|_0^{\pi/4} \qquad (26.9\text{d})$$

$$= \sin \frac{\pi}{4} + \cos \frac{\pi}{4} - \sin 0 - \cos 0 \qquad (26.9\text{e})$$

$$= \frac{\sqrt{2}}{2} + \frac{\sqrt{2}}{2} - 0 - 1 \qquad (26.9\text{f})$$

$$= \sqrt{2} - 1 \qquad (26.9\text{g})$$

On the interval $[\pi/4, \pi/3]$, the top curve is $y = \sin x$ and the bottom curve is $y = \cos x$. So

$$A_2 = \int_{\pi/4}^{\pi/3} (\sin x - \cos x)\; dx \qquad (26.9\text{h})$$

$$= (-\cos x - \sin x)|_{\pi/4}^{\pi/3} \qquad (26.9\text{i})$$

$$= -\cos \frac{\pi}{3} - \sin \frac{\pi}{3} + \cos \frac{\pi}{4} + \cos \frac{\pi}{4} \qquad (26.9\text{j})$$

$$= -\frac{1}{2} - \frac{\sqrt{3}}{2} + \frac{\sqrt{2}}{2} + \frac{\sqrt{2}}{2} \qquad (26.9\text{k})$$

$$= \sqrt{2} - \frac{1 + \sqrt{3}}{2} \qquad (26.9\text{l})$$

and the total area is

$$A = A_1 + A_2 = \sqrt{2} - 1 + \sqrt{2} - \frac{1 + \sqrt{3}}{2} = -\frac{3 + \sqrt{3} - 4\sqrt{2}}{2}. \qquad (26.9\text{m})$$

$\square$

Figure 26.5: The area between the plots of the parabola $y^2 = 12 - 2x$ and the line $y = x - 2$ is shaded below. The two curves intersect at $x = -2$ and $x = 4$. The figure on the right illustrates how the integration must be broken up into two parts to reflect the fact that the top curve changes at $x = 4$.

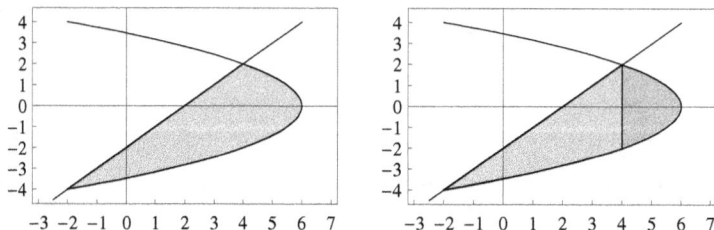

Example 26.4. Find the area between the curves $y'' = 12 - 2x$ and $y = x - 2$ (see figure 26.5).

Solution. The line and the parabola intersect when they both have the same y value. This occurs when

$$12 - 2x = (x - 2)^2 \tag{26.10a}$$

$$12 - 2x = x^2 - 4x + 4 \tag{26.10b}$$

$$0 = x^2 - 2x - 8 \tag{26.10c}$$

$$0 = (x + 2)(x - 4) \tag{26.10d}$$

Hence $x = -2$ and $x = 4$. Substituting into $y = x - 2$ we get the intersection points of $(-2, -4)$ and $(4, 2)$. For $x < 4$, the line $y = x - 2$ is the top curve, but for $x > 4$, the upper branch of the parabola $y = \sqrt{12 - 2x}$ is the top branch of the curve. In both instances, the lower branch of the parabola $y = -\sqrt{12 - 2x}$ is the bottom curve.

The area between the two functions is

$$A = \int_{-2}^{4} (x - 2 - (-\sqrt{12 - 2x}))\,dx +$$

$$\int_{4}^{6} (\sqrt{12 - 2x} - (-\sqrt{12 - 2x}))\,dx \tag{26.10e}$$

$$= \int_{-2}^{4} (x - 2 + \sqrt{12 - 2x})\,dx + 2\int_{4}^{6} \sqrt{12 - 2x}\,dx \tag{26.10f}$$

$$= \left(\frac{1}{2}x^2 - 2x + \frac{1}{3}(12 - 2x)^{3/2} \right)\Big|_{-2}^{4} - \frac{2}{3}(12 - 2x)^{3/2}\Big|_{4}^{6} \tag{26.10g}$$

$$= \frac{38}{3} + \frac{16}{3} = \frac{54}{3} = 18 \tag{26.10h}$$

$\square$

Figure 26.6: Comparison of integration over x, as done in example 26.4 (shown on the left), and integration of y (shown on the right). Individual rectangles in the Riemann Sum are hilighted.

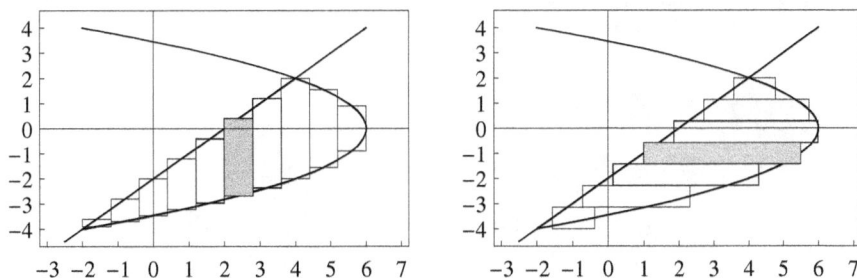

Example 26.5. Repeat example 26.4, this time by integrating along the y axis, as illustrated on the right-half of figure 26.6.

Solution. We want to find the area between the line $y = x - 2$ and the parabola $y^2 = 12 - 2x$. Figure 26.5 a horizontal line test tells us that these curves are functions of y. Since we are integrating along the y axis we need to solve each of these curves for x as a function of y. The line $y = x - 2$ gives us

$$x = 2 + y \tag{26.11a}$$

From the parabola $y^2 = 12 - 2x$ we can rearrange to find

$$x = 6 - \frac{1}{2}y^2 \tag{26.11b}$$

Observing figure 26.5, the parabola is on the right and the line is on the left for the entire region.

The left endpoint, which we found in example 26.4 at $x = -2$, is also the bottom endpoint. This corresponds (using the equation of the line) to $y = x - 2 = -2 - (-2) = -4$.

The intersection, which we found in example 26.4 at $x = 4$, is the top endpoint. Also using the line, this occurs at at $y = 4 - 2 = 2$.

Thus the range of integration over y is $-4 < y < 2$, and the area is

$$\int_{-4}^{2} \left(\left(6 - \frac{1}{2}y^2 \right) - (2 + y) \right) dy = \int_{-4}^{2} \left(4 - \frac{1}{2}y^2 - y \right) dy \tag{26.11c}$$

$$= \left(4y - \frac{1}{2 \cdot 3}y^3 - \frac{1}{2}y^2 \right) \Big|_{-4}^{2} \tag{26.11d}$$

$$= \left(4 \cdot 2 - \frac{8}{6} - \frac{4}{2} \right) - \left(4 \cdot (-4) - \frac{-64}{6} - \frac{16}{2} \right) \tag{26.11e}$$

$$= 8 - \frac{4}{3} - 2 + 16 - \frac{32}{3} + 8 \tag{26.11f}$$

$$= 30 - \frac{36}{3} = 30 - 12 = 18 \tag{26.11g}$$

□

Exercises

Find the area between the specified curves. Sketch the region first to determine if you should integrate over x or over y. Use the indicated interval if specified, otherwise determine the endpoints from the intersection points.

1. $f(x) = 0.5x^2 + 7$ and $g(x) = x$ on $[-3, 5]$
2. $f(x) = 3x$ and $g(x) = 5x^2$

3. $y = x$ and $y = 9 - x^2$
4. $x = y^2 - t$ and $x = 6 - 2y^2$
5. $x + y^2 = 56$ and $x + y = 0$
6. $x = 2y^2$, $x = 4 + y^2$
7. $y = 4|x|$ and $y = x^2 - 5$
8. $y = (x - 3)^2$ and $y = x$
9. $y = \cos x$, $y = 8 \cos x$, and $x = 0$

Chapter 27

Volumes of Rotation

Chapter Summary and Goal

In this chapter we extend the concept of the Riemann, that was used to define a the definite integral in chapter 24, in a way that allows us to find volumes of solids of rotation. A **solid of rotation** is formed by rotating a finite section of a curve about a straight line. We call the straight line the **axis of rotation**.

The methods we will describe can be visualized by the following **salami metaphor**.[1] We think of the axis of symmetry of the salami as the axis of rotation. If we cut the salami by any plane that contains the axis of rotation, the edge of the salami with intersect the plane in a shape that is identical to the shape of the curve of the original plot of $f(x)$. But normally we will cut the salami in slices, oriented orthogonally to the axis of rotation. The total volume of the salami is equal to the sum of volumes of the individual slices, but each slice is a flat, circular disk. We know how to find the volume of a flat, circular disk. It is just the area of the disk (πr^2) times the thickness of the disk. We call this disk a **volume element**. Then we add up the volume of all the volume elements to get the total volume.

Student Learning Objectives

The student will:

1. Be able to explain how Riemann Sums are applied to find volumes of solids of rotation.

2. Be able to define volume elements for the methods of disks, washers, and shells.

3. Be able to set up integrals for volumes of rotations using the methods of disks, washers, and shells.

4. Be able to calculate the volumes of solids of rotation about the coordinate axes.

5. Be able to calculate the volumes of solids of rotation about lines parallel to the coordinate axes.

6. Be able to calculate the the volume of solids formed by rotating the region between two curves about the coordinate axis or a line parallel to one of the coordinate axes.

7. Understand the difference between an analogy, a metaphor, and a simile.

---

[1] Technically, we are making an analogy here, and not a metaphor. But when we rotate a curve and call the volume a salami, and the volume elements slices of salami, then we are using a metaphor. So we will call it a metaphor and forget about any possible grammatical misrepresentation. If we say its *like a salami*, then it becomes a simile, which is a type of metaphor.

Finding Volumes Using Riemann Sums

To find a volume we will generalize the idea of the Riemann Sum to three dimensions. In this and the next chapter we will only consider solids of rotation. These are solid objects found by rotating a curve about a straight line.

Imagine that you cut the solid object into thin slabs orthogonally to the axis of rotation like a stack of coins. The total volume is then the sum of the volumes of individual slabs (figure 27.1). If we designate the volume of the kth slab by ΔV_k, then the total volume is

$$V = \sum_{k=1}^{n} \Delta V_k \approx \sum_{k=1}^{n} A(u_k)\Delta u \qquad (27.1)$$

where $A(u_k)$ is the cross-sectional area at $u = u_k$ and the slab thickness is Δu. Since the cross-section of each slab is a circle, its cross-sectional area is

$$A(u_k) = \pi r(u_k)^2 \qquad (27.2)$$

where u is measured along the axis of rotation and $r(u)$ is the radius of the disk perpendicular to the u-axis. If we take the limit as $n \to \infty$ while $\Delta u \to u$, the the volume between the points

Figure 27.1: Illustration of a rotation about the y-axis. After the curve is rotated (thick red line), we decompose the resulting solid into a stack of coin-shaped objects.

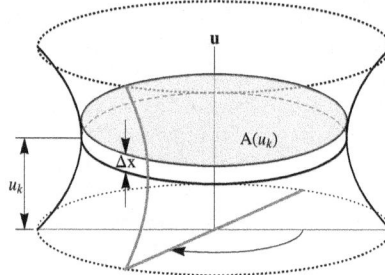

$u = a$ and $u = b$ is

$$V = \lim_{n \to \infty} \sum_{k=1}^{n} \pi r(u_k)^2 \Delta u = \int_a^b \pi r(u)^2 du \qquad (27.3)$$

Rotations Around the x-axis

The calculation is slightly simpler to illustrate when we are rotating about the x-axis, because then the radius at x is simply $f(x)$, as we demonstrate in the next two examples.

Theorem 27.1. Volume of Rotation About the x-axis

Suppose that the function $f(x)$ is integrable on $[a,b]$, Then the volume of the solid formed when the curve of $f(x)$, $a \le x \le b$ is rotated about the x-axis, is given by

$$V = \int_a^b \pi f(x)^2 dx \qquad (27.4)$$

Example 27.1. Find the volume of the object formed by rotating the curve $y = \sqrt{x}$ about the x axis, between the origin and $x = 4$.

Solution.   The curve of $y = \sqrt{x}$ is a parabola centered on the $x-$axis (figure 27.2). The coin-shaped slab at x has radius $f(x)$. So the volume element dv is given by

Figure 27.2: Rotation of the function $y = \sqrt{x}$ about the x-axis.

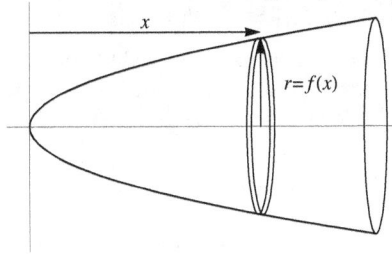

$$dV = \pi f(x)^2 dx = \pi(\sqrt{x})^2 dx = \pi x dx \tag{27.5a}$$

We are asked for the volume of rotation between the x axis and $x = 4$, so we need to vary x (the arrow in figure 27.2) from $x = 0$ to $x = 4$. The net volume is then given by the integral

$$V = \int_0^4 \pi x dx = \left. \frac{\pi}{2} x^2 \right|_0^4 = 8\pi. \tag{27.5b}$$

$\square$

Example 27.2. Find the volume of the object obtained by rotating the following area about the x-axis: the region bounded by the curves $y = 1/x$, $x = 1$, $x = 5$, and $y = 0$.

Solution.   The region to be rotated is illustrated on the left hand side of figure 27.3). Proceeding as before, we are the volume is

$$V = \int_1^5 \pi f(x)^2 dx = \int_1^5 \frac{\pi}{x^2} dx = \left. \frac{-\pi}{x} \right|_1^5 = -\frac{\pi}{5} + \frac{\pi}{1} = \frac{4\pi}{5} \tag{27.6a}$$

$\square$

Figure 27.3: The function $y = 1/x$ (left) is rotated about the x axis on the right.

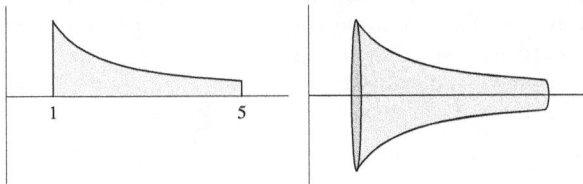

Rotations about the y-axis

When the curve is rotated about the y axis, the function itself does not directly give the radius. This is because the slices of salami are centered on the y axis and have radii measured parallel to the x axis. To determine this radius, we need to do some sort of calculation: we know $y = f(x)$, but we need to know $x = f^{-1}(y)$, i.e., when we can solve for x as some function of y.

Theorem 27.2.   Volume of Rotation About the y-axis

Suppose that the function $y = f(x)$ is to be rotated about the y axis between the points $y = A$ and $y = B$. If the equation $y = f(x)$ can be solved to find $x = x(y)$ for some function $x(y)$ then the volume of the solid is given by

$$V = \int_A^B \pi x(y)^2 dy \qquad (27.7)$$

Figure 27.4: Plot of the the region to be rotated (left) and the solid generated (center) in example 27.3.The figure on the right illustrates how the Riemann-sum might be generated for the region on the left by exteding the rectangle shown on the left through the region.

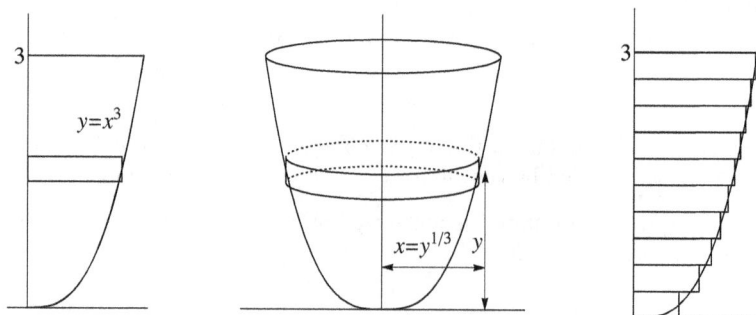

Example 27.3. Find the volume of the solid generated by rotating the region bounded by curves $y = x^3$, the y-axis, and the line $y = 3$ about the y-axis.

Solution.   As we see from figure 27.4, the slices of salami are now parallel to the x axis and perpendicular to the y axis. The *variable of integration* is now y, and the radius is the distance, measured parallel to the x axis, between the y axis and the function. Since $y = x^3$, we determine this distance by solving for x, and find that $x = y^{1/3}$. (This is the *inverse function*: if $y = f(x)$, then $x = f^{-1}(y)$ is the inverse function, read as "f-inverse of x.")

To see this, imagine rotating the rectangle on the left-hand side illustration in figure 27.4 about the y-axis. If we rotate this rectangle, it forms a disk, shown in the figure on the right. The radius of this disk is $y^{1/3}$, and the distance of the disk from the x axis is y. We can then image the range of values that we need to move this rectangle through to fill up the entire region on the left. It would range from $y = 0$ to $y = 3$ (see the illustration of the Riemann sum on the right illustration of figure 27.4). This gives the range of integration.

Summarizing:

$$\text{we integrate in } y \text{ from } 0 \text{ to } 3$$

$$\text{radius} = x(y) = y^{1/3}$$

Thus the volume element is

$$dV = \pi(\text{radius})^2\,dy = \pi y^{2/3}\,dy \tag{27.8a}$$

and the volume is

$$V = \int_0^3 \pi y^{2/3}\,dy = \pi \left(\frac{y^{5/3}}{5/3} \right)\Big|_0^3 = \frac{3\pi \cdot 3^{5/3}}{5} = \frac{3^{8/3}}{5}\pi. \tag{27.8b}$$

$\square$

Figure 27.5: The quarter circle $x^2 + y^2 = x^2$ in the first quadrant (left) is rotated about the y-axis to form the half-sphere shown on the right.

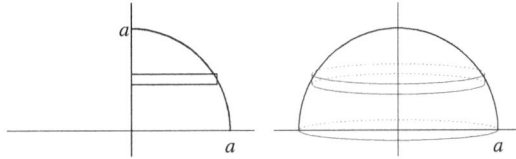

Example 27.4. Find the volume of solid formed by rotating the region bounded in first quadrant by $x^2 + y^2 = a^2$ about the y-axis.

Solution. The function $y = \sqrt{a^2 - x^2}$, for $0 \le x \le a$, is the first 90-degrees of a circle, as illustrated in figure 27.5. When we rotate this quarter-circle about the y-axis we expect to get a half-sphere. We already know the answer to this: the volume of a sphere is $\frac{4}{3}\pi a^3$ so the volume of the half-sphere should be $\frac{2}{3}\pi a^3$. Let's see if we get the correct answer by integration.

We appeal to the Riemann sum concept to determine the limits: the rectangle on the left-hand side in figure 27.5 is imagined to be rotated about the y-axis to form the disks, then moved from the bottom to the top of the figure. The thickness of the rectangle is dy, and the width of the rectangle is the radius. Its range of movement is the range of the integral.

Since the rectangle can move from the bottom to the top of the circle, we can imagine the value of y changing during this move from $y = 0$ to $y = a$. These are the limits of integration.

The width of the rectangle is the distance from the y-axis to the curve, or the value of x. Solving $x^2 + y^2 = a^2$ for x gives

$$x(y) = \sqrt{a^2 - y^2} \tag{27.9a}$$

Hence the volume is

$$V = \int_0^a \pi \left(\sqrt{a^2 - y^2} \right)^2 dy = \pi \int_0^a (a^2 - y^2)\,dy \tag{27.9b}$$

$$= \pi \left(a^2 y - \frac{1}{3}y^3 \right)\Big|_0^a = \frac{2\pi a^3}{3} \tag{27.9c}$$

as expected!

$\square$

Figure 27.6: Rotation of $y = x^3$ on $[0,1]$ about $x = 1$.

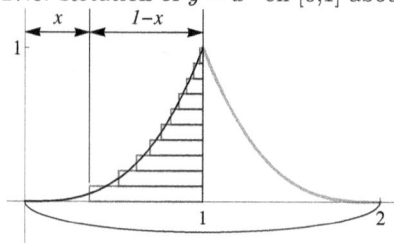

Example 27.5. Find the volume of solid obtained as follows: the region bounded by the curve $y = x^3$, $x = 1$, and the x axis, when it is rotated about the line $x = 1$.

Solution. Examining figure 27.6, we see that the radius at a given point (x, y) is

$$\text{radius} = 1 - x = 1 - y^{1/3} \qquad (27.10a)$$

since the inverse of $y = x^3$ is $x = y^{1/3}$. Hence the volume element is

$$dV = \pi(1 - y^{1/3})^2 dy \qquad (27.10b)$$

and the total volume is

$$V = \int_0^1 \pi(1 - y^{1/3})^2 dy \qquad (27.10c)$$

$$= \pi \int_0^1 (1 - 2y^{1/3} + y^{2/3}) dy \qquad (27.10d)$$

$$= \pi \left(y - \frac{2}{4/3}y^{4/3} + \frac{1}{5/3}y^{5/3} \right) \Big|_0^1 \qquad (27.10e)$$

$$= \pi \left(1 - \frac{3}{2} + \frac{3}{5} \right) \qquad (27.10f)$$

$$= \pi \left(\frac{10 - 15 + 6}{10} \right) = \frac{\pi}{10} \qquad (27.10g)$$

Thus the total volume of the solid of rotation is $V = \pi/10$ □

Rotations of Regions Bounded by Two Curves

Let the region to be rotated be bounded above by $f(x)$, below by $g(x)$, on the left by $x = a$, and on the right by $x = b$; and that we want to rotate this region about the x axis. We know that the area of this region to be rotated is

$$A = \int_a^b (f(x) - g(x)) dx \qquad (27.11)$$

We can now think of the the upper and lower functions (f and g) as forming and outer and inner radii of a washer. When the rectangles in the Riemann Sum are rotated about the x-axis each one will form into the shape of an **annulus**, or two concentric circles, in cross-section. We will call these disks with holes in the center **washers** (see figure 27.7).

The area of an annulus (or the area of the flat surface of a washer) is the area of the outer disk minus the area of the inner disk.

$$\text{Area of Annulus} = \text{Area of Outer Disk} - \text{Area of Inner Disk} \qquad (27.12)$$

Figure 27.7: When the region bounded by two curves is rotated about the x-axis, the rectangles in the Riemann Sum become washers. The circles illustrate the washer formed by a single Rieman rectangle.

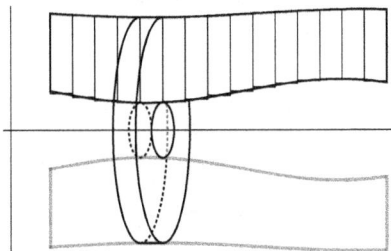

$$= \pi f(x)^2 - \pi g(x)^2 \tag{27.13}$$

$$= \pi \left(f(x)^2 - g(x)^2 \right) \tag{27.14}$$

and therefore the volume of the washer, our volume element, is

$$dV = \pi \left(f(x)^2 - g(x)^2 \right) dx \tag{27.15}$$

Figure 27.8: The region bounded by the parabolas $y = (x-6)^2 + 4$ and $y = (x-5)^2 + 4$ betwen $x = 1$ and $x = 3$ is rotated about the x-axis. A single volume element dV is illustrated on the left; the right shows how the stack of volume elemnents line up next to one-another.

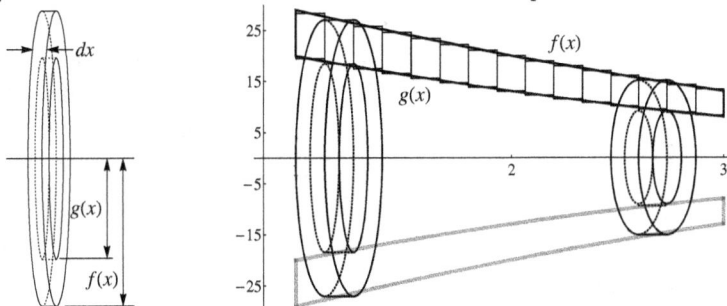

Example 27.6. Find the volume of the solid obtained when the region bounded by the parabolas $y = (x - 6)^2 + 4$, $y = (x - 5)^2 + 4$, and the lines $x = 1$ and $x = 3$ is rotated about the x-axis.

Solution. The region to be rotated and sample area elements (rectangles, before rotation) are illustrated in figure 27.8, along with the footprint of the rotated region in lower-half plane.

The volume element is given by the area of a single washer at x,

$$dV = \pi \left(f(x)^2 - g(x)^2 \right) dx \tag{27.16a}$$

$$= \pi \left([(x-6)^2 + 4]^2 - [(x-5)^2 + 4]^2 \right) dx \tag{27.16b}$$

$$= \pi \left((x^2 - 12x + 36 + 4)^2 - (x^2 - 10x + 25 + 4)^2 \right) dx \tag{27.16c}$$

$$= \pi \left((x^2 - 12x + 40)^2 - (x^2 - 10x + 29)^2 \right) dx \tag{27.16d}$$

Doing Calculus

$$= \pi \left((x^4 - 24x^3 + 224x^2 - 960x + 1600) - \right.$$
$$\left. (x^4 - 20x^3 + 158x^2 - 580x + 841)\right) dx \qquad (27.16\text{e})$$
$$= \pi \left(-4x^3 + 66x^2 - 380x + 759\right) dx \qquad\qquad (27.16\text{f})$$

The volume of the solid object is then

$$V = \int_1^3 \pi \left(-4x^3 + 66x^2 - 380x + 759\right) dx \qquad (27.16\text{g})$$

$$= \pi \left(-x^4 + 22x^3 - 190x^2 + 759x\right)\Big|_1^3 \qquad (27.16\text{h})$$

$$= \pi \left((-3^4 + 22(3^3) - 190(3^2) + 759(3))\right.$$
$$\left. -(-1 + 22 - 190 + 759)\right) \qquad (27.16\text{i})$$

$$= \pi(-81 + 594 - 1710 + 2277 - 590) = 490\pi \qquad (27.16\text{j})$$

□

Theorem 27.3. Method of Washers

If f and g are integrable on $[a, b]$ then the volume of the solid formed when the region between $f(x)$ and $g(x)$, where $f(x) \geq g(x) \geq 0$ for all x in $[a, b]$, is rotated about the x axis is

$$V = \int \pi \left(f(x)^2 - g(x)^2\right) dx \qquad (27.17)$$

Example 27.7. Find the volume of the solid that results when the region bounded by $y = x^2$ and $y = x$ is rotated about the y-axis.

Solution. When the solid is rotated it forms a funnel-shaped object, with a conical interior and paraboloidal exterior (figure 27.9). We calculate the inner and outer radii as follows.

	Outer Radius	Inner Radius
$y = y(x)$	$y = x^2$	$y = x$
$x = x(y)$	$x = \sqrt{y}$	$x = y$

In each case, the second line was obtained by solving the first line for x as a function of y. Thus the volume is

$$V = \pi \int_0^1 \left((\sqrt{y})^2 - (y)^2\right) dy = \pi \int_0^1 \left(y - y^2\right) dy = \pi \left(\frac{1}{2}y^2 - \frac{1}{3}y^3\right)\Big|_0^1 = \frac{\pi}{6}. \qquad (27.18\text{a})$$

□

Example 27.8. Find the volume of the solid when the region rotated in example 27.7 is rotated about the line $x = 2$.

Solution. The inner and outer radii are found from figure 27.10 and summarized as follows.

	Outer Radius	Inner Radius
Description	Distance from $y = 2$ to $y = x$	Distance from $y = 2$ to $y = x^2$
$x = x(y)$	$2 - y$	$2 - \sqrt{y}$

The outer radius is $r = 2 - y$ and the inner radius as $r = 2 - \sqrt{y}$ Therefore the volume element is given by

$$dV = \pi \left((2 - y)^2 - (2 - \sqrt{y})^2\right) dy \qquad (27.19\text{a})$$

Figure 27.9: When the area bounded by $y = x^2$ and $y = x$ is rotated about the y axis it forms a funnel-shaped object with a conical interior as illustrated here. A single volume element is illustrated on the left, and the stack of volume elements on the right.

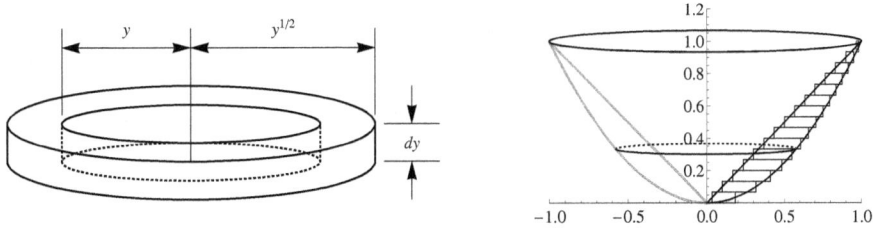

Figure 27.10: The shape bounded by $y = x$ and $y = x^2$ is rotated about the line $x = 2$ giving something of a volcano shaped - a mesa whose center is hollow. Left: A single area element, along the footprint of the rotated region in the xy plane. Right: Schematic of the volume element formed by rotating a single rectangle about the line $x = 2$.

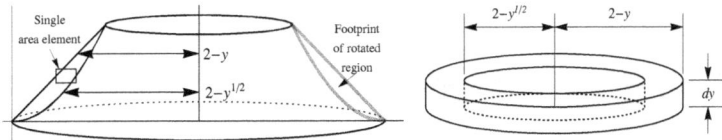

$$= \pi \left(4 - 4y + y^2 \right) - (4 - 4\sqrt{y} + y) \right) dy \qquad (27.19b)$$

$$= \pi \left(-5y + y^2 + 4\sqrt{y} \right) dy \qquad (27.19c)$$

The volume is given by the integral from 0 to 1,

$$V = \int_0^1 \pi \left(-5y + y^2 + 4\sqrt{y} \right) dy \qquad (27.19d)$$

$$= \pi \left(-\frac{5}{2}y^2 + \frac{1}{3}y^3 + \frac{8y^{3/2}}{3} \right) \Bigg|_0^1 \qquad (27.19e)$$

$$= \pi \left(-\frac{5}{2} + \frac{1}{3} + \frac{8}{3} \right) = \frac{\pi}{2}. \qquad (27.19f)$$

$\square$

The Method of Shells

The method of shells uses a different kind of volume element. Instead of slicing the solid like a salami it breaks it down into concentric cylinders. Unlike washers, these cylinders have a finite (and not infinitesimal) height, and the difference between the inner and outer radii is infinitesimal (and not finite).

We have already used the formula for the volume of a cylinder:

$$\text{Volume of a Cylinder} = \pi(\text{radius})^2 \times \text{height} = \pi r^2 h \qquad (27.20)$$

If we drill a hole of radius r_{in} down the central axis of a cylinder of radius r_{out}, we will be removing a cylinder of volume $\pi r_{in}^2 h$. Thus the total volume of the shell that remains is

$$dV = (\text{Total Volume of Cylinder}) - (\text{Volume Removed}) \qquad (27.21)$$

$$= \pi r_{out}^2 h - \pi r_{in}^2 h \tag{27.22}$$

$$= \pi h (r_{out}^2 - r_{in}^2) \tag{27.23}$$

$$= \pi h (r_{out} - r_{in})(r_{out} + r_{in}) \tag{27.24}$$

Let

$$\Delta r = r_{out} - r_{in} = \text{infinitesimal shell thickness} \tag{27.25}$$

$$r = \frac{r_{in} + r_{out}}{2} = \text{ average shell radius} \tag{27.26}$$

Then

$$dV = 2\pi r h \Delta r \tag{27.27}$$

If the height of the cylinder h is some function of the radius, $h = f(r)$, and the radius varies over some domain $a \le r \le b$, we can divide the interval $[a, b]$ into n points of width $\Delta r = (b - a)/n$. To form the Riemann sum for the volume, we pick any point r_i in each interval $[a, b]$ so that $a_i \le r_i \le b_i$, and compute the sum:

$$V = \lim_{n \to \infty} \sum_{i=1}^{n} 2\pi r_i f(r_i) \Delta r = 2\pi \int_a^b r f(r) dr \tag{27.28}$$

Theorem 27.4. Method of Shells

If $f(x) > 0$ is integrable on $0 \le x \le b$, then the volume solid formed by rotating the region bounded by $f(x)$, the lines $x = a$, $x = b$ and the x axis about the y is

$$V = 2\pi \int_a^b x f(x) dx \tag{27.29}$$

Figure 27.11: Left: Region to be rotated, $y = 1 - x^2$, in example 27.9. Middle and right: Comparison of volume elements when the function $y = 1 - x^2$ is rotated around the y-axis using the "cut-like-a-salami" (on the left) method and the "method of shells" (on the right).

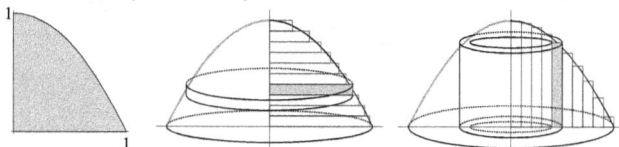

Example 27.9. Find the volume of the region bounded by the function $y = 1 - x^2$ and the coordinate axes using the method of shells. and then compare the answer with the result you would get using the method of disks to break up the volume element.

Solution. The region to be rotated is illustrated on the far left of figure 27.11; the function intersects the coordinate axes at $(0, 1)$ and $(1, 0)$. Schematics of typical volume elements have been drawn in the middle and right of figure 27.11.

Examining the picture on the right-hand side of figure 27.11 we see that the volume by the method of shells is given by

$$V = \int_0^1 2\pi x (1 - x^2) dx = \int_0^1 2\pi (x - x^3) dx = 2\pi \left(\frac{x^2}{2} - \frac{x^4}{4} \right) \Big|_0^1 = \frac{\pi}{2}, \tag{27.30a}$$

Using the method discussed earlier in the chapter, using the fact that the inverse of $y = 1 - x^2$ is $x = \sqrt{1-y}$, the volume element would have been (refer to the middle of figure 27.11)

$$dV = \pi \left(\sqrt{1-y} \right)^2 dy = \pi(1-y)dy \qquad (27.30b)$$

and therefore the volume becomes

$$V = \int_0^1 \pi(1-y)dy = \pi \left(y - \frac{y^2}{2} \right) \Big|_0^1 = \frac{\pi}{2}. \qquad (27.30c)$$

This demonstrates that both methods of integration give the same result.   □

Figure 27.12: Rotation of $y = x^2$ about the y axis using the method of shells.

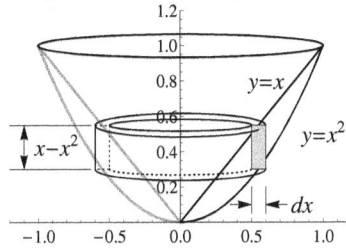

Example 27.10. Repeat example 27.7 using the method of shells.

Solution. The geometry is illustrated in figure 27.12. The shells have a height of $x - x^2$, so the volume element is

$$dV = 2\pi \times (\text{radius}) \times (\text{height}) \times dx = 2\pi x(x - x^2)dx \qquad (27.31a)$$

so the total volume of the rotated solid is

$$V = \int_0^1 2\pi x(x - x^2)dx = 2\pi \left(\frac{x^3}{3} - \frac{x^4}{4} \right) \Big|_0^1 = 2\pi \left(\frac{1}{3} - \frac{1}{4} \right) = \frac{\pi}{6}. \qquad (27.31b)$$

as in example 27.7 (see eq. 27.18a).   □

Example 27.11. Repeat example 27.8 using the method of shells.

Solution. The height of each shell is

$$\text{height} = x - x^2 \qquad (27.32a)$$

and the radius is

$$\text{radius} = 2 - x \qquad (27.32b)$$

Thus the volume element is

$$dV = 2\pi \times (\text{radius}) \times (\text{height})dx \qquad (27.32c)$$

$$= 2\pi(2 - x)(x - x^2)dx \qquad (27.32d)$$

$$= 2\pi(2x - 3x^2 + x^3)dx \qquad (27.32e)$$

Even though the fully rotated object stretches from $x = 0$ to $x = 4$, we only need to integrate from $x = 0$ to $x = 1$ because this will encompass the entire shell. The total volume is thus

$$V = \int_0^1 2\pi(2x - 3x^2 + x^3)dx = 2\pi \left(\frac{2x^2}{2} - \frac{3x^3}{3} + \frac{x^4}{4} \right) \Big|_0^1 = \frac{\pi}{4} \qquad (27.32f)$$

as before (see equation 27.19f).   □

Summary of Rotations

Method of Disks (Salami)	$V = \displaystyle\int_a^b \pi f(u)^2\, du$ $u = x$ or $u = y$ is along the axis of rotation
Method of Washers	$V = \displaystyle\int_a^b \pi \left(f^2_{\text{bigger}}(u) - f^2_{\text{smaller}}(u) \right) du$ $u = x$ or $u = y$ is along the axis of rotation
Method of Shells	$V = \displaystyle\int_a^b 2\pi r f(r)\, dr$ r is measured normal to the axis of rotation.

Washers and and Shells and Disks! Oh My!

Evil Confusion

It is natural for a student to be confused: when faced with a rotation, which method should you use?

In general, there is no right answer to this question. But if you are able to set up and then evaluate the integral correctly by more than one method, they will give you the same answer.

In many cases, one of the methods will be more natural and will lead to simpler integrals than the others. Similarly, you will often find that there are situations where at least one of the methods leads to very difficult (or intractable) integrals.

With time and practice you will develop an intuitive feel for the appropriate method to use in any given situation. Until then, here is a list of steps that may help you get things started.

Suppose that you are attempting to find the volume of some region R that is being rotated about the $\vec{u}$ axis, as schematized in figure 27.13. Let let $r_{big}(u)$ and $r_{small}(u)$ be functions the describe the edges of the region furthest (r_{big}) and nearest to the $\vec{u}$ axis, expressed as a function of the radius r, and let height(r) denote the extent of the region to be rotated as a function of u.

1. Always start by sketching the region to be rotated.

2. On the same figure, sketch the axis of rotation.

3. Sketch the infinitesimal rectangle dA that will be rotated to create the infinitesimal volume element dV.

   - For the **methods of disks or washers**
     - the long edge of the rectangle will be normal to the axis of rotation
     - the short axis of the rectangle will have length du
     - the variable of integration is u

   - For the **method of shells**

— the long edge of the rectangle will be parallel to the axis of rotation

— the short axis of rotation will have length dr, where r is measured along a perpendicular dropped from the rectangle to the axis of rotation.

— the variable of integration is r

4. Write down equations for the area element (the small rectangle you just drew in the previous step).

   • **for washers**, you want to find a formula for the endpoints of the rectangle as a function of r. You will need to calculate $r_{big}^2 - r_{small}^2$. This is because the area of a washers is the difference of two circles, so that $dV = \pi(r_{big}^2 - r_{small}^2)dr$.

   • **for disks**, you will do exactly the same thing, only one of the endpoints is zero, so you end up with $dV = \pi r^2 dr$.

   • **for shells** you want to find the height of the area element as function of r. Then since $dA = \text{height}(r) \times dr$, the volume element is $dV = 2\pi r dA = 2\pi r \times \text{height}(r) \times dr$.

5. Determine the limits of integration and set up the integral.

6. Solve the integral.

Figure 27.13: Schematics of the different methods of rotations.

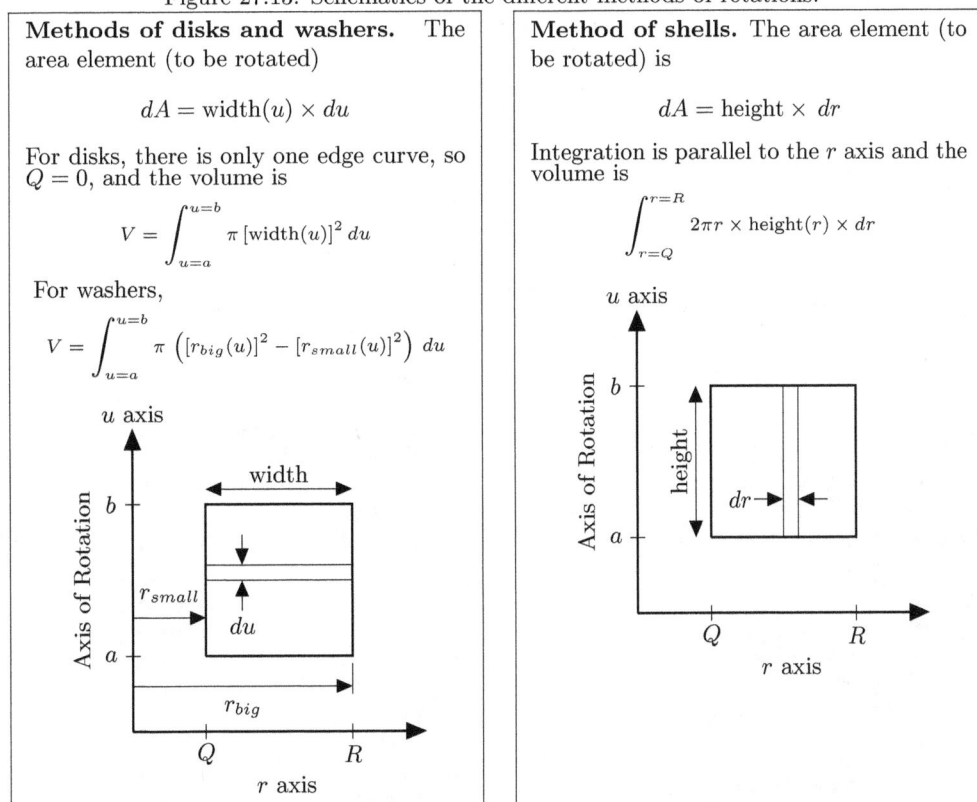

Methods of disks and washers.   The area element (to be rotated)

$$dA = \text{width}(u) \times du$$

For disks, there is only one edge curve, so $Q = 0$, and the volume is

$$V = \int_{u=a}^{u=b} \pi \left[\text{width}(u)\right]^2 du$$

For washers,

$$V = \int_{u=a}^{u=b} \pi \left([r_{big}(u)]^2 - [r_{small}(u)]^2\right) du$$

Method of shells. The area element (to be rotated) is

$$dA = \text{height} \times dr$$

Integration is parallel to the r axis and the volume is

$$\int_{r=Q}^{r=R} 2\pi r \times \text{height}(r) \times dr$$

Exercises

In exercises 1 through 7, find the volumes of the solids formed when the regions bounded by the given curves are rotated about the specified axis.

1. $y = \dfrac{1}{2}x$ on $[0, 3]$ about the x axis.

2. $y = x^2 + 6x + 0$ and $y = 0$
   about the x axis.

3. $x = 8y$ and $x = y$ for $y \geq 0$
   about the y axis.

4. $y = x^6$ and $y = 1$
   about $y = 5$.

5. $y = \sin x$ and the x axis on $[0, \pi]$
   about the x axis

6. $y = \sin x$ and the x axis on $[0, \pi]$
   about $x = \dfrac{\pi}{2}$

7. $y = \sin x$ and the x axis on $[0, \pi]$
   about $x = \pi$

In exercises 8 through 15, use the method of shells to find the volumes of the solids formed when the regions bounded by the given curves are rotated about the specified axis.

8. $y = 10x - 2x^2$ and $y = 0$
   about the y axis.

9. $x + y = 1$ and $x = 2 - (y - 1)^2$
   about the x axis.

10. $y = x^3$, $y = 0$, $x = 1$, and $x = 2$
    about the y-axis.

11. $xy = 1$, $x = 0$, $y = 0$, and $y = 3$
    about the x-axis.

12. $y = 6x^2$ and $y = 1 + 5x^2$
    about $x = 1$

13. $x = 3y - y^2$
    about $y = -8$

14. $y = \sqrt{x}$, $y = 0$ and $x = 1$
    about $x = -1$.

15. $y = x^3$, $y = 0$, and $x = 1$
    about $y = 1$.

Chapter 28

Average Values

Chapter Summary and Goal

One useful application of the definite integral is the calculation of the average values of continuous function. The reason why this works follows directly from the definition of the integral as a Riemann Sum. We will describe this method of calculating averages here.

Student Learning Objectives

The student will:

1. Understand the relationship between averages and integrals.
2. Understand the derivation of the formula for the average value of a function.
3. Be able to calculate the average value of an integrable function.
4. Understand and be able to apply the mean value theorem for integrals.

Calculating Average Values

Suppose we want to calculate the average value of a function $y = f(x)$ on some interval $[a, b]$. Lets recall how we calculate the average value of a list of numbers; the average of a list of numbers $\{q_1, q_2, \ldots, q_n\}$ is given by

$$\bar{q} = \frac{q_1 + q_2 + \cdots + q_n}{n} = \frac{1}{n} \sum_{i=1}^{n} q_i \tag{28.1}$$

So what we would like to do is pick an collection of points (x_i, y_i) on the function, and average the values of the y_i (see figure 28.1).

One way to determine how which values to pick is the same way we set up the Riemann Sum. Divide the interval $[a, b]$ into n sub-intervals and number them $i = 1, 2, .., n$, with end points $[x_i, x_{i+1}]$

$$x_i = a + (i-1)\frac{b-a}{n} = a + (i-1)\Delta x \tag{28.2}$$

where

$$\Delta x = \frac{b-a}{n}. \tag{28.3}$$

Then pick some point x_i^* inside each interval such that

$$x_{i-1} \leq x_i^* \leq x_i \tag{28.4}$$

Define the points on the curve as

$$y_i^* = f(x_i^*), \ i = 1, \ldots, n \tag{28.5}$$

We will define the average of the function as the limit, as $n \to \infty$, of the $\{y_1^*, y_2^*, \ldots, y_n^*\}$:

$$\overline{f(x)} = \lim_{n\to\infty} \overline{y^*} \tag{28.6}$$

$$= \lim_{n\to\infty} \frac{1}{n} \sum_{i=1}^{n} y_i^* \tag{28.7}$$

$$= \lim_{n\to\infty} \frac{1}{n} \sum_{i=1}^{n} f(x_i^*) \tag{28.8}$$

$$= \lim_{n\to\infty} \frac{\Delta x}{b-a} \sum_{i=1}^{n} f(x_i^*) \tag{28.9}$$

$$= \frac{1}{b-a} \lim_{n\to\infty} \sum_{i=1}^{n} f(x_i^*)\Delta x \tag{28.10}$$

$$= \frac{1}{b-a} \int_a^b f(x)dx \tag{28.11}$$

where the last line follows from equation 24.6 (the definition of the Riemann Sum).

Definition 28.1. Average Value of a Function

The **average value of** $f(x)$ on the interval $[a, b]$ is the number

$$\overline{f} = \frac{1}{b-a} \int_a^b f(x) \ dx \tag{28.12}$$

Figure 28.1: Calculation of the average of $y = \sin x$ on $[0, \pi]$ using eight points. One point is selected inside each interval.

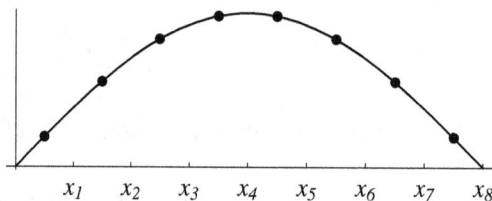

Example 28.1. Find the average value of $f(x) = \sin x$ on the interval $[0, \pi]$

Solution.

$$\overline{f} = \frac{1}{\pi - 0} \int_0^\pi \sin x \, dx \tag{28.13a}$$

$$= \frac{1}{\pi}(-\cos\pi + \cos 0) = \frac{2}{\pi} \tag{28.13b}$$

□

Example 28.2. Demonstrate the numerical convergence of the average in example 28.1 by calculating the average at (a) 4 equally spaced points; (b) 8 equally spaced points

Solution. Using 4 points.

i	x_i	$\sin x_i$
1	$\frac{\pi}{8}$	0.3827
2	$\frac{3\pi}{8}$	0.9239
3	$\frac{5\pi}{8}$	0.9239
4	$\frac{7\pi}{8}$	0.3827
sum		2.6132
$\overline{y}$		0.6533

Using 8 points

i	x_i	$\sin x_i$
1	$\frac{\pi}{16}$	0.1951
2	$\frac{3\pi}{16}$	0.5556
3	$\frac{5\pi}{16}$	0.8315
4	$\frac{7\pi}{16}$	0.9808
5	$\frac{9\pi}{16}$	0.9808
6	$\frac{11\pi}{16}$	0.8315
7	$\frac{13\pi}{16}$	0.5556
8	$\frac{15\pi}{16}$	0.1951
sum		5.126
$\overline{y}$		0.6408

By comparison the integrated average is $\frac{2}{\pi} \approx = 0.6366$.   □

Mean Value Theorem for Integrals

Let $f(x)$ be a continuous function on $[a, b]$ and define the function

$$g(x) = \int_a^x f(t)dt \tag{28.15}$$

Then $g(x)$ is an anti-derivative of $f(x)$ with $g'(x) = f(x)$ (by the third fundamental theorem of calculus 24.12). By the mean value theorem (theorem 18.2), there is some number c, where $a \le c \le b$, such that

$$g'(c)(b - a) = g(b) - g(a) \tag{28.16}$$

Substituting $g'(c) = f(c)$ into (28.16) gives

$$f(c)(b - a) = g(b) - g(a) \tag{28.17}$$

By the first fundamental theorem of calculus,

$$\int_a^b f(t)dt = g(b) - g(a) = f(c)(b - a) \tag{28.18}$$

This proves the mean value theorem for integrals.

Theorem 28.1. Mean Value Theorem for Integrals

If $f(x)$ is continuous on $[a, b]$, theen there is some number c, $a \le c \le b$, such that

$$\int_a^b f(t)dt = g(b) - g(a) = f(c)(b - a) \qquad (28.19)$$

or equivalently,

$$\overline{f} = f(c) = \frac{1}{b - a} \int_a^b f(x)dx \qquad (28.20)$$

Theorem 28.2. Mean Value Theorem for Integrals, Second Form

A continuous function on $[a, b]$ takes on its average value at some point c in $[a, b]$, where $f(c) = \overline{f}$

Figure 28.2: At the point c, the average of the function $y = \cos x$ on the interval $[\pi/2, \pi]$ is equal to the $f[c]$, as determined by example 28.3.

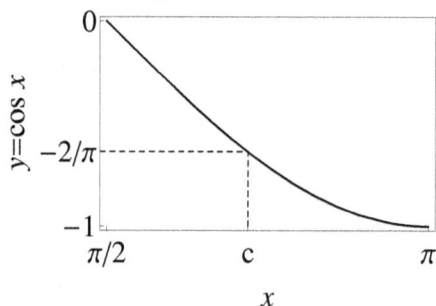

Example 28.3. Find the value of c for which the mean value theorem for integrals says that the function is equal to its average for the function $y = \cos x$ on $[\pi/2, \pi]$

Solution. According to equation 28.19, we need to find the value of c for which

$$\int_a^b f(t)dt = f(c)(b - a) \qquad (28.21a)$$

where $a = \frac{\pi}{2}$, $b = \pi$ and $f(x) = \cos x$. Making these substitutions gives

$$\int_{\pi/2}^{\pi} \cos t \; dt = (\cos c)\left(\pi - \frac{\pi}{2}\right) = \frac{\pi \cos c}{2} \qquad (28.21b)$$

$$\sin \pi - \sin \frac{\pi}{2} = \frac{\pi \cos c}{2} \qquad (28.21c)$$

$$0 - 1 = \frac{\pi \cos c}{2} \qquad (28.21d)$$

$$\cos c = -\frac{2}{\pi} \qquad (28.21e)$$

$$c = \cos^{-1}\left(-\frac{2}{\pi}\right) \qquad (28.21f)$$

The function is illustrated in figure 28.2. $\square$

Weighted Averages

A weighted average is defined by assigning a weight to each object. For example, if we have a bag of marbles and want to find the average mass of a marble we add up the mass off all the marbles and divide by the total number of marbles. If we have two types of marbles, 10 of which are plastic, 5 grams each, and 20 of which are glass and are 25 grams each, then the average mass would be:

$$\overline{M} = \frac{\overbrace{5 + 5 + \cdots 5}^{10 \text{ repeats}} + \overbrace{25 + 25 + \cdots + 25}^{20 \text{ repeats}}}{10 + 20} = \frac{10 \times 5 + 20 \times 25}{10 + 20} \tag{28.22}$$

We can look at this way: we are averaging the two numbers, 5 and 25, but we are putting more emphasis on the 25 then on the 5 because there are more marbles with this mass. The number in front of the 25 is called a **weight**. In general, the weighted average of the numbers $\{q_1, q_2, \ldots, q_n\}$ subject to weights $\{w_1, w_2, \ldots, w_n\}$ is given by

$$\overline{Q} = \frac{w_1 q_1 + w_2 q_2 + \cdots + w_n q_n}{w_1 + w_2 + \cdots + w_n} \tag{28.23}$$

This reduces to the usual average when all the weights are 1. We can define the weighted average of a function $f(x)$ over the interval $[a, b]$, with weight function $w(x)$, in a similar manner.

Definition 28.2. Weighted Average of a Function

The **weighted average** of $f(x)$ on $[a, b]$ with **weight function** $w(x)$ is

$$\overline{f(x)} = \frac{\int_a^b w(x) f(x) dx}{\int_a^b w(x) dx} \tag{28.24}$$

If $f(x) = x$ and $w(x)$ is a density distribution, then this gives the center of mass in one dimension. If $f(x) = x$ and $w(x)$ is a probability distribution, then this gives the expectation value (mean value) of x subject to that probability distribution. The idea can be extending to higher dimensions as well; the weighted averages of x^2, y^2, x^2, xy, xz, and yz give the unique non-zero elements of the three-dimensional inertial matrix when $w(x)$ is a density, for example.

Theorem 28.3. Second Mean Value Theorem For Integrals

Let $f(x)$ be monotonic and integrable on $[a, b]$. Then for any integrable weight function $w(x)$ there is some number c such that

$$\int_a^b w(x) f(x) dx = f(a) \int_a^c w(x) + f(b) \int_c^b w(x) dx \tag{28.25}$$

where c is between a and b.

Example 28.4. Show that $f(x) = x$ satisfies the second mean value theorem for integrals, and find the value of c that applies with a weight function of $w(c)$ on $[5, 10]$.

Solution. Since $f(x) = x$ is a polynomial it is integrable; since $f'(x) = 1 > 0$ for all x it is monotonic. Hence it satisfies the requirements of the theorem. Therefore there is some number c such that

$$\int_5^{10} w(x) f(x) dx = f(5) \int_5^c w(x) dx + f(10) \int_c^{10} w(x) dx \tag{28.26a}$$

where $f(x) = x$ and $w(x) = x$. Since $f(5) = 5$ and $f(10) = 10$,

$$\int_5^{10} x^2 dx = 5 \int_5^c x dx + 10 \int_c^{10} x dx \tag{28.26b}$$

Integrating both sides of the equation,

$$\frac{1}{3}x^3 \Big|_5^{10} = 5 \times \frac{1}{2}x^2 \Big|_5^c + 10 \times \frac{1}{2}x^2 \Big|_c^{10} \tag{28.26c}$$

$$\frac{875}{3} = 5\left(\frac{c^2}{2} - \frac{25}{2}\right) + 10\left(50 - \frac{c^2}{2}\right) = \frac{875}{2} - \frac{5c^2}{2} \tag{28.26d}$$

Multiplying both sides of the equation by 2, and substituting the bounds on both sides,

$$\frac{2}{3}\left(10^3 - 5^3\right) = 5(c^2 - 5^2) + 10(10^2 - c^2) \tag{28.26e}$$

$$\frac{2}{3}(1000 - 125) = 5c^2 - 125 + 1000 - 10c^2 \tag{28.26f}$$

$$\frac{1750}{3} = -5c^2 + 875 \tag{28.26g}$$

$$5c^2 = 875 - \frac{1750}{3} = \frac{875}{3} \tag{28.26h}$$

$$c^2 = \frac{175}{3} \tag{28.26i}$$

$$c = \sqrt{\frac{175}{3}} = 5\sqrt{\frac{7}{3}} \approx 7.63763 \tag{28.26j}$$

which is between 5 and 10 as expected.   □

Exercises

In exercises 1 through 10, find the average values of the given functions on the specified interval.

1. $y = x^2$ on $[0, 10]$

2. $y = \cos x$ on $[0, \pi/2]$

3. $y = \tan x$ on $[0, \pi/3]$

4. $y = xe^{-x^2}$ on $[0, 2]$

5. $y = \dfrac{\ln x}{x}$ on $[1, e]$

6. $y = \cosh x$ on $[-\ln 100, \ln 100]$

7. $y = x^4$ on $[1, 5]$ Ans: $781/5$

8. $y = \sqrt{x} + 1$ on $[4, 6]$ Ans: $-10/6 + 2\sqrt{6}$

9. $y = 3\sin x + 2\cos x$ on $[0, 5\pi/3]$
   Ans: $(9 - 6\sqrt{3})/(10\pi)$

10. $y = 25 - x^2$ on $[0, 1]$ Ans: $74/3$

In exercises 11 through 14 find the value of c for which the mean value theorem for integrals says that the given function is equal to its average.

11. $y = x^4$ on $[1, 5]$

12. $y = \sqrt{x} + 1$ on $[4, 6]$

13. $y = 3\sin x + 2\cos x$ on $[0, 5\pi/3]$

14. $y = 25 - x^2$ on $[0, 1]$

15. In an alternating current (AC) circuit the voltage is $V(t) = V_{\max} \cos(\omega t)$, where $V_{\max}$ and ω are fixed constants. In the US, $\omega = 2\pi f$ where $f = 60$ cycles/second. According to Ohm's law, the current is $P(t) = V(t)/R$, where R is the load resistance, another fixed constant, and the power through the circuit is $P(t) = I(t)V(t)$.

    (a) Show that $\overline{I(t)} = 0$ over the period $[0, 2\pi/\omega]$.
    (b) Show that $\overline{I(t)} = 0$ over the period $[0, 2\pi/\omega]$.
    (c) Show that $P(t) = V^2(t)/R$ and hence the average power is $\overline{P(t)} = \dfrac{1}{R}\overline{V^2(t)}$, where the average is taken over a single cycle $[0, 2\pi/\omega]$.

    (d) Find a formula for $\overline{P(t)}$ in terms of ω, R, and $V_{\max}$.
    (e) Explain whey the average power is positive even though though the average power and voltage are both zero.

16. In probability theorem the expected value $E[x]$ of a variable x is the integral

$$E(x) = \int_a^b x f(x)\,dx$$

where $f(x)$ is a probability density function and $[a, b]$ is the domain of $f(x)$. Relate $E(x)$ to the average of value of the function $xf(x)$.

17. Suppose that $f(x) = \dfrac{1}{\sigma\sqrt{2\pi}} e^{(x-\mu)^2}/2\sigma^2$
    where μ and σ are constants. Find $E(x)$.

18. Using the same $f(x)$ as in exercise 17, find $E(x^2)$, where

$$E(p(x)) = \int_a^b p(x) f(x)\,dx$$

for any function $p(x)$.

19. From the definition of E in the previous exercise, show that

$$E((x-\mu)^2) = E(x^2) - E(x)^2$$

This number is sometimes called the variance of the probability distribution.

20. Repeat exercises 17 and 18 for the exponential distribution

$$f(x) = \begin{cases} \lambda e^{-\lambda x} & , x \geq 0 \\ 0 & , x \leq 0 \end{cases}$$

21. Repeat exercises 17 and 18 for a uniform distribution on $[a, b]$, $f(x) = 1/(b - a)$.

Pink Floyd (Not)

Chapter 29

Integration by Parts

Chapter Summary and Goal

Integration by parts reverses the product rule for differentiation and gives us an equivalent formula for integration. However, it does not take the form of an integral of the product of two functions, but rather, the integral of the product of one function with the derivative of the second function. The basic formula in its most compact form is

$$\int u \, dv = uv - \int v \, du \qquad (29.1)$$

We will learn to derive this formula, and why it is useful, in this chapter.

Student Learning Objectives

The student will:

1. Understand the derivation of integration by parts.
2. Be able to assign the function u and differential dv in the integration by parts formula.
3. Be able to solve both indefinite and definite integrals using integration by parts.

The Method of Integration by Parts

Recall from the product rule (theorem 10.1) that

$$(uv)' = uv' + u'v \qquad (29.2)$$

With a slight rearrangement,

$$uv' = (uv)' - u'v \qquad (29.3)$$

Writing this in the Leibniz notation

$$u\frac{dv}{dx} = \frac{d(uv)}{dx} - v\frac{du}{dx} \qquad (29.4)$$

Multiplying through by dx gives

$$udv = d(uv) - vdu \qquad (29.5)$$

When we integrate both sides of the equation, we get

$$\int u\,dv = \int d(uv) - \int v\,du \qquad (29.6)$$

We can integrate the first term on the right by theorem 22.3 (with $n = 0$):

$$\int d(uv) = uv \qquad (29.7)$$

The result is given by theorem 29.1.

Theorem 29.1. Integration by Parts

$$\int u\,dv = uv - \int v\,du \qquad (29.8)$$

The most difficult part of applying theorem 29.1 is deciding which part of the integral to substitute as u and which part to substitute as dv. There are no general rules that will make this easier. It is merely a matter of practice and experience.

Example 29.1. Find $\displaystyle\int x \sin x\, dx$.

Solution. There are four possible choices that we can assign u and dv.

u	dv	v	du	$v\,du$
1	$x \sin x\, dx$	$\int x \sin x\, dx$		
x	$\sin x\, dx$	$\int \sin x\, dx$	dx	$-\cos x\, dx$
$\sin x$	$x\, dx$	$\int x\, dx$	$\cos x\, dx$	$\frac{1}{2}x^2 \cos x\, dx$
$x \sin x$	dx	$\int dx$	$(x\cos x + \sin x)dx$	$(x^2 \cos x + x \sin x)dx$

In the third column above we've listed the integral we would have to solve to find v. The integral in the first row is the original integral, so that doesn't help us, and we throw out that idea. That leaves use with the remaining three choices.

Keeping in mind that we will ultimately need to find $\int v\,du$, we write du in the fourth column. Since the idea is to simplify the integration, we look for a product of results in the third and fourth column that will give a *simpler* integral than the original integral.

This table tells us that the only substitution that simplifies things is the one in the second row. (In time, you will develop an instinct about whether or not the $v\,du$ column will be more complicated or less complicated than the original integral without actually having to fill in all the values.) Thus we will make the following substitutions.

$$\int \underbrace{x}_{u} \underbrace{\sin x\, dx}_{dv} \qquad (29.9a)$$

Then since $u = x$, $du = dx$, and since $dv = \sin x\, dx$,

$$v = \int \sin x\, dx = -\cos x \qquad (29.9b)$$

Thus by (29.8)

$$\int x \sin x\, dx = uv - \int v\,du \qquad (29.9c)$$

$$= x \cdot (-\cos x) - \int (-\cos x) \cdot dx \qquad (29.9d)$$

$$= -x \cos x + \sin x + C. \qquad (29.9e)$$

$\square$

Example 29.2. Solve $\int xe^x dx$.

Solution. Let's start by making a table of the possibilities like we did in example 29.1.

u	dv	v	du	vdu
1	$xe^x dx$	$\int xe^x dx$	(no help)	
x	$e^x dx$	$\int e^x dx$	dx	$e^x dx$
e^x	$x dx$	$\int x dx$	$e^x dx$	$\frac{1}{2}x^2 e^x dx$
xe^x	dx	$\int dx$	$(x+1)e^x dx$	$x(x+1)e^x dx$

As before, the only simplifying option appears to be the one in the second row. So we try $u = x$ and $dv = e^x dx$. Then $du = dx$ and $v = e^x$. Thus

$$\int xe^x dx = uv - vdu = x \cdot e^x - \int e^x \cdot dx = xe^x - e^x + C \qquad (29.10a)$$

$\square$ The last two examples were integrals of the form $\int xf(x)dx$ where we could easily integrate $\int f(x)dx$. In the first example, $f(x) = \sin x$, and the second $f(x) = e^x$. This suggests the following heuristic to us.

Heuristic 29.1. Integration by Parts

To solve $\int xf(x)dx$:

  IF You know how to solve $\int f(x)dx$

  THEN Try $u = x$ and $dv = f(x)dx$

Example 29.3. Solve $\int \ln x \, dx$ using integration by parts.

Solution. There are even fewer options in this integral, so we can still list them all out.

u	dv	v	du	vdu
1	$\ln x \, dx$	$\int \ln x \, dx$	(no help)	
$\ln x$	dx	$\int dx$	dx/x	dx

As usual, the first option returns the original integral and is no help, so really, there is only one way to do this using integration by parts. We let $u = \ln x$, and $dv = dx$. Then $du = dx/x$ and $dv = dx$, as shown in the second row. Integration by parts gives

$$\int \underbrace{\ln x}_{u} \, \underbrace{dx}_{dv} = uv - \int vdu = x \ln x - \int x \cdot \frac{dx}{x} = x \ln x - x + C \qquad (29.11a)$$

This result gives rise to the trivial heuristic (29.2) for integration by parts, which rarely, though on occasion, works.   □

Heuristic 29.2. Trivial Heuristic for Integration by Parts

To solve $\int f(x)dx$ by integration by parts:

    TRY $u = f(x)$ and $dv = dx$

Example 29.4. Solve $\int x \ln x \, dx$.

Solution. We can begin our analysis by writing a table of the possibilities. The integrand is a product of two functions so there are only four possibilities. As usual, the first one returns the original integral, and is not helpful.

u	dv	v	du	vdu
1	$x \ln x \, dx$	$\int x \ln x \, dx$	(not helpful)	
x	$\ln x \, dx$	$x(\ln x - 1)^a$	dx	$x(\ln x - 1)dx$
$\ln x$	$x dx$	$\frac{1}{2}x^2$	dx/x	$\frac{1}{2}x \, dx$
$x \ln x$	dx	x	$(1 + \ln x)dx$	$x(1 + \ln x)dx$

    ^{a}From example 29.3.

The second and fourth row return the original integral, which is not helpful at all, and so we are left with the third row. This makes the original logarithm go away, which is very helpful, because it turns everything into powers of x. So we let $u = \ln x$ and $dv = x dx$. Then $du = dx/x$ and $v = x^2/2$. Integrating by parts,

$$\int x \ln x \, dx = \int \underbrace{\ln x}_{u} \underbrace{x \, dx}_{dv} = uv - v \, du = \frac{1}{2}x^2 \ln x - \int \frac{1}{2}x^2 \cdot \frac{dx}{x} \tag{29.12a}$$

$$= \frac{1}{2}x^2 \ln x - \frac{1}{2}\int x dx = \frac{1}{2}x^2 \ln x - \frac{1}{4}x^2 + C. \tag{29.12b}$$

    □

Sometimes integration by parts will return the original function to be integrated, but reversed in sign, or multiplied by a constant. When this happens, one can usually write an algebraic equation for the result, as illustrated in the following example.

Example 29.5. Find $\int e^x \sin x \, dx/$

Solution. Let $u = e^x$ and $dv = \sin x \, dx$. Then $du = e^x dx$ and $v = -\cos x$. Using integration by parts,

$$\int \underbrace{e^x}_{u} \underbrace{\sin x \, dx}_{dv} = uv - vdu = \underbrace{e^x}_{u} \cdot \underbrace{(-\cos x)}_{v} - \int \underbrace{(-\cos x)}_{v} \cdot \underbrace{e^x dx}_{du} \tag{29.13a}$$

$$= -e^x \cos x + \int e^x \cos x dx \tag{29.13b}$$

Now we integrate by parts a second time, using similar substitutions: $u = e^x$ and $dv = \cos x \, dx$. Then $du = e^x dx$ and $v = \sin x$.

$$\int e^x \sin x \, dx = -e^x \cos x + uv - \int v du \qquad (29.13c)$$

$$= -e^x \cos x + \underbrace{e^x}_{u} \underbrace{\sin x}_{v} - \int \underbrace{e^x}_{v} \underbrace{\sin x \, dx}_{du} \qquad (29.13d)$$

At first glance it looks like we are back to where we started, but look again. However, the sign in front of the term

$$\int e^x \sin x \, dx \qquad (29.13e)$$

is different on each side of the equation. Thus we can bring the entire integral over to the left hand hand side:

$$2 \int e^x \sin x \, dx = e^x \sin x - e^x \cos x \qquad (29.13f)$$

Dividing by 2 and factoring the e^x we get the following:

$$\int e^x \sin x \, dx = \frac{e^x}{2} \left[\sin x - \cos x \right] + C. \qquad (29.13g)$$

$\square$

Sometimes it is necessary to combine the substitution method with integration by parts, as illustrated in the following example.

Example 29.6. Find $\displaystyle\int \cos \sqrt{x} \, dx$.

Solution. If we let $z = \sqrt{x}$, then $dz = \dfrac{dx}{2\sqrt{x}} = \dfrac{dx}{2z}$. Thus $dx = 2z dz$, so that the integral becomes

$$\int \cos \sqrt{x} \, dx = \int (\cos z) \cdot 2z dz = 2 \int z \cos z \, dz \qquad (29.14a)$$

We already know how to solve the integral on the right using integration by parts (see exercise 1).

$$\int \cos \sqrt{x} \, dx = 2 \left(\cos z + z \sin z \right) \qquad (29.14b)$$

$$= 2 \cos \sqrt{x} + 2\sqrt{x} \sin \sqrt{x} + C. \qquad (29.14c)$$

$\square$

Exercises

Use integration by parts to solve the integrals in exercises 1 through 12.

In exercises 29.2 through 16, make a substitution and then use integration by parts to solve the given integral.

1. $\int x \cos x \, dx$

7. $\int e^x \cos x \, dx$

2. $\int \tan^{-1} x \, dx$

8. $\int x 2^x \, dx$

3. $\int x \tan^{-1} x \, dx$

9. $\int x \ln(2x) \, dx$

4. $\int \sin^{-1}(3x) \, dx$

10. $\int x^2 \ln(5x) \, dx$

5. $\int t^2 e^t \, dt$

11. $\int x^3 (x^2 + 7)^{3/2} dx$

6. $\int \frac{\ln x}{\sqrt{x}} \, dx$

12. $\int \frac{x^5}{\sqrt{x^3 + 5}} dx$

13. $\int [\ln(3x)]^2 \, dx)$

15. $\int \cos(\ln x) \, dx$

14. $\int \ln \sqrt{x} \, dx$

16. $\int e^x \sin^{-1} e^x dx$

Answers to Selected Problems

1. $\cos x + x \sin x$

2. $x \tan^{-1}(x) - \frac{1}{2} \ln\left(x^2 + 1\right)$

3. $\frac{1}{2}[(x^2 + 1) \tan^{-1} x - x]$

4. $\frac{1}{3}\sqrt{1 - 9x^2} + x \sin^{-1}(3x)$

5. $e^t \left(t^2 - 2t + 2\right)$

6. $2\sqrt{x}(\ln(x) - 2)$

7. $\frac{1}{2} e^x (\sin(x) + \cos(x))$

8. $\frac{2^x (x \ln(2) - 1)}{\ln^2(2)}$

9. $\frac{1}{2} x^2 \ln(2x) - \frac{x^2}{4}$

10. $\frac{1}{3} x^3 \ln(5x) - \frac{x^3}{9}$

11. $\frac{1}{35} \left(x^2 + 7\right)^{5/2} \left(5x^2 - 14\right)$

12. $\frac{2}{9} \left(x^3 - 10\right) \sqrt{x^3 + 5}$

13. $2x + x \ln^2(3x) - 2x \ln(3x)$

14. $1/2(-x + x \ln x)$

15. $\frac{x}{2} [\cos(\ln x) + \sin(\ln x)]$

16. $\sqrt{1 - e^{2x}} + e^x \sin^{-1}(e^x)$

Chapter 30

Improper Integrals

Chapter Summary and Goal

We defined the definite integral

$$\int_a^b f(x)dx$$

as an area under a curve over a finite interval $[a, b]$ (definition 24.1). There are many situations in which a region can become infinite in extent but the area remains finite. Such situations arise in two ways: (a) one or both of the bounds becomes infinite, i.e., $a = -\infty$ or $b = \infty$; or (b) the magnitude of the function becomes infinite. When such situations arise, the integral is said be an **improper integral**. We classify these two situations as type I and type II improper integrals. Although the two types of improper integrals are fundamentally different in how they arise and what they represent, we handle them in essentially the same manner. Determining if such integrals converge, and solving them when possible will be our main focus in this chapter.

Student Learning Objectives

The student will:

1. Learn to find the areas of regions represented by convergent improper integrals.
2. Recognize when an integrand has a singularity within the region of integration.
3. Determine whether improper integrals of either type converge or diverge.
4. Set up and solve improper integrals using the definition (with limits) when one endpoint is infinite in magnitude.
5. Set up and solve improper integrals using the definition (with limits) when both endpoints are infinite in magnitude.
6. Set up and solve improper integrals using the definition (with limits) when there is one or more singularity within the region of integration.

Integrals with an Infinite Upper and/or Lower Bounds

If the upper or lower bounds is infinite, we replace it with a limit of a definite integral with a finite limit. If the limit exists, we say the **integral converges to the limit**, and if the limit does not exists, we say the **integral diverges**.

Definition 30.1. Improper Integral with Infinite Upper Bound

We define

$$I = \int_a^\infty f(x)dx = \lim_{t \to \infty} \int_a^t f(x)dx \tag{30.1}$$

and say the **integral I converges**, if the limit on the right exists, and **diverges** if the limit does not exist.

Definition 30.2. Improper Integral with Infinite Lower Bound

We define

$$J = \int_{-\infty}^b f(x)dx = \lim_{t \to -\infty} \int_t^b f(x)dx \tag{30.2}$$

and say the **integral J converges**, if the limit on the right exists, and **diverges** if the limit does not exist.

Furthermore, we can define an integral with both infinite upper and infinite lower bounds, if both of the above limits exist.

Definition 30.3. Improper Integral with Infinite Bounds

If a is any real number, we define

$$\int_{-\infty}^\infty f(x)dx = \lim_{t \to -\infty} \int_t^a f(x)dx + \lim_{t \to \infty} \int_a^t f(x)dx \tag{30.3}$$

and say the **integral converges**, if both limits on the right exists, and **diverges** if the either of the limits do not exist.

Example 30.1. Determine if $\int_1^\infty e^{-5x}dx$ converges and if so, find its value.

Solution. The integral has an infinite upper bound so we replace the upper bound with a limit.

$$\int_1^\infty e^{-5x}dx = \lim_{t \to \infty} \int_1^t e^{-5x}dx = \lim_{t \to \infty}\left[-\frac{1}{5}e^{-5x}\right]_1^t \tag{30.4a}$$

$$= -\frac{1}{5}\lim_{t \to \infty}\left[e^{-t} - e^{-1}\right] = \frac{1}{5e} \tag{30.4b}$$

The integral converges to $1/(5e)$.   □

Example 30.2. Determine if $\int_{-\infty}^\infty \frac{1}{1+x^2}dx$ converges, and if so, find its value.

Solution. This integral has infinite upper and lower lower bounds, so we divide the integral into the sum of two integrals and take the limit of each integral separately.

$$\int_{-\infty}^\infty \frac{1}{1+x^2}dx = \lim_{t \to -\infty} \int_t^0 \frac{1}{1+x^2}dx + \lim_{t \to \infty} \int_0^t \frac{1}{1+x^2}dx \tag{30.5a}$$

$$- \lim_{t \to -\infty} \tan^{-1} x \Big|_t^0 + \lim_{t \to -\infty} \tan^{-1} x \Big|_0^t \tag{30.5b}$$

$$= \left(\tan^{-1} 0 - \lim_{t \to -\infty} \tan^{-1} t \right) + \left(\lim_{t \to \infty} \tan^{-1} t - \tan^{-1} 0 \right) \qquad (30.5c)$$

$$= \left(0 - -\frac{\pi}{2} \right) + \left(\frac{\pi}{2} - 0 \right) = \pi. \qquad (30.5d)$$

Therefore the integral converges to π. □

Example 30.3. Determine if $\displaystyle\int_1^\infty \frac{dx}{x}$ converges, and if so, find its value.

Solution. The integral has an infinite upper bound/

$$\int_1^\infty \frac{dx}{x} = \lim_{t \to \infty} \int_1^t \frac{dx}{x} = \lim_{t \to \infty} \ln|x| \Big|_1^t = \lim_{t \to \infty} (\ln t - \ln 1) = \lim_{t \to \infty} \ln t = \infty \qquad (30.6a)$$

Therefore the integral diverges. □

Example 30.4. Determine if $\displaystyle\int_1^\infty \frac{dx}{x^2}$ converges, and if so, find its value.

Solution. The integral has an infinite upper bound that we replace with a limit.

$$\int_1^\infty \frac{dx}{x^2} = \lim_{t \to \infty} \int_1^t x^{-2} dx = \lim_{t \to \infty} \left[-\frac{1}{x} \right]_1^t = \lim_{t \to \infty} \left[-\frac{1}{t} + 1 \right] = 1 \qquad (30.7a)$$

Thus this integral converges, where the integral in example 30.3 diverged. □

Power Integrals. The difference in convergence between examples 30.3 and 30.4 leads us to ask the following question: for which values of n does $\displaystyle\int_1^\infty \frac{dx}{x^n}$ converge? To answer this question we first proceed as before and get a formula in terms of limits. Suppose $n \neq -1$.

$$\int_1^\infty \frac{dx}{x^n} = \lim_{t \to \infty} \int_1^t \frac{dx}{x^n} = \lim_{t \to \infty} \int_1^t x^{-n} dx = \lim_{t \to \infty} \frac{x^{1-n}}{1-n} \Big|_1^t \qquad (30.8)$$

$$= \frac{1}{1-n} \lim_{t \to \infty} \left[\frac{1}{t^{n-1}} - 1 \right] = \frac{1}{1-n} \left[\lim_{t \to \infty} \frac{1}{t^{n-1}} - \lim_{t \to \infty} 1 \right] \qquad (30.9)$$

$$= \frac{1}{1-n} \left[\lim_{t \to \infty} \frac{1}{t^{n-1}} - 1 \right] \qquad (30.10)$$

If $n - 1 > 0$, or equivalently, if $n > 1$, then $1/t^{n-1} \to 0$ as $t \to \infty$. In this case the integral converges to $1/(n-1)$.

If $n - 1 < 0$, or equivalently, if $n < 1$, then $1/t^{n-1} \to \infty$ as $t \to \infty$. In this case the integral diverges.

The case when $n = 1$ was already covered by by example 30.3.

Theorem 30.1. Power-integral

$$\int_1^\infty \frac{dx}{x^n} = \begin{cases} \dfrac{1}{n-1}, & \text{if } n > 1 \\ \text{diverges}, & \text{if } n \le 1 \end{cases} \qquad (30.11)$$

Example 30.5. Determine if the integral $\displaystyle\int_1^\infty \frac{1}{\sqrt{x}} dx$ converges or diverges.

Solution. This is a power integral with $n = -1/2$. Since $-1/2 < 1$, by theorem 30.1 the integral diverges. □

Integrals With Singularities

If $\lim_{x \to a} f(x) = \pm \infty$ we say that $f(x)$ has a **singularity** at $x = a$. A singularity may occur anywhere in the interval of integration, or at either one of the its endpoints.

The fix for a singularity is to replace it with a limit. If the singularity occurs at an endpoint, as with

$$\int_1^3 \frac{dx}{(x-3)^2} \tag{30.13}$$

we would only replace it with one limit. As with infinite limits, if the limit exists, the integral is said to **converge to the limit**; and if the limit does not exist, the integral is said to **diverge**.

Definition 30.4. Singularity at Upper Bound

If $f(x)$ has a singularity at $x = b$ then we say

$$\int_a^b f(x)dx = \lim_{t \to b^-} \int_a^t f(x)dx \tag{30.14}$$

Example 30.6. Determine if $\displaystyle\int_1^3 \frac{dx}{(x-3)^2}$ converges or diverges, and if so, find its value.

Solution. Since $1/(x-3)^2 \to \infty$ as $x \to 3$, there is a singularity at $x = 3$. The singularity happens to also fall at the upper bound of the integral.

$$\int_1^3 \frac{dx}{(x-3)^2} = \lim_{t \to 3^-} \int_1^t \frac{dx}{(x-3)^2} = \lim_{t \to 3^-} \left[\frac{-1}{x-3} \right]_1^t \tag{30.15a}$$

$$= \lim_{t \to 3^-} \left[\frac{-1}{1-3} - \frac{-1}{t-3} \right] \tag{30.15b}$$

$$= -\frac{1}{2} + \lim_{t \to 3^-} \frac{1}{t-3} \tag{30.15c}$$

Since the last term diverges, the integral diverges. □

Definition 30.5. Singularity at Lower Bound

If $f(x)$ has a singularity at $x = a$ then we say

$$\int_a^b f(x)dx = \lim_{t \to a^+} \int_t^b f(x)dx \tag{30.16}$$

Example 30.7. Determine if $\displaystyle\int_4^{13} \frac{1}{\sqrt{x-4}}\,dx$ converges, and if so, find its value.

Solution. The integrand is discontinuous at $x = 4$, which is the lower bound of the integral.

$$\int_4^{13} \frac{1}{\sqrt{x-4}} = \lim_{t \to 4^+} \int_t^{13} \frac{dx}{\sqrt{x-4}} = \lim_{t \to 4^+} 2\sqrt{x-4} \Big|_t^{13} \tag{30.17a}$$

$$= \lim_{t \to 4^+} 2[\sqrt{13-4} - \sqrt{t-4}] \tag{30.17b}$$

$$= 2\sqrt{9} - 2 \lim_{t \to 4^+} \sqrt{t-4} = 6 \tag{30.17c}$$

Therefore the integral converges to 6.   □

Singularities that do not occur at the endpoints, but occur inside the interval, are particularly deceptive because they may be easy to miss. For example, we may correctly compute the following integral, which does not have any singularities either within the interval or at the end points.

$$\int_5^7 \frac{dx}{(x-3)^2} = \left[\frac{-1}{x-3}\right]_5^7 = -\frac{1}{4} + \frac{1}{2} = \frac{1}{4} \tag{30.18}$$

This may embolden us to go blindly ahead and compute the following.

$$\int_2^7 \frac{dx}{(x-3)^2} = \left[\frac{-1}{x-3}\right]_2^7 = -\frac{1}{4} - 1 = -\frac{5}{4} \tag{30.19}$$

Something very strange has happened: the area under a curve that is entirely positive is a negative number! Because the singularity does not occur at an end point, we don't get an obvious error error like ∞ when we compute the answer. Students often fail to recognize that the negative answer should be suspicious and indicative of a problem.

Definition 30.6. Singularity not at a Boundary

If $f(x)$ has a singularity at $x = c$, where $a < c < b$ then we define

$$\int_a^b f(x)dx = \lim_{t \to c^-} \int_a^t f(x)dx + \lim_{t \to c^+} \int_t^b f(x)dx+ \tag{30.20}$$

and say the **integral converges to the limit** if both limits exist, and we say the integral **diverges** if either limit does not exist.

Example 30.8. Determine if $\displaystyle\int_2^7 \frac{dx}{(x-3)^2}$ converges, and if so, find its value.

Solution. This is similar to example 30.6, except that the singularity is not at an endpoint.

$$\int_2^7 \frac{dx}{(x-3)^2} = \lim_{t \to 3^-} \int_2^t \frac{dx}{(x-3)^2} + \lim_{t \to 3^+} \int_t^7 \frac{dx}{(x-3)^2} \tag{30.21a}$$

$$= \lim_{t \to 3^-} \frac{-1}{x-3}\bigg|_2^t + \lim_{t \to 3^+} \frac{-1}{x-3}\bigg|_t^7 \tag{30.21b}$$

$$= \lim_{t \to 3^-} \frac{-1}{t-3} - \frac{-1}{2-3} + \frac{-1}{7-3} - \lim_{t \to 3^+} \frac{-1}{t-3} \tag{30.21c}$$

Since the two limits on the right come from different directions, they cannot be combined together. This *seems* unfortunate, since then they would nicely cancel out. However, it is not a bad thing, because if they did, we would end up with -5/4 again, as we did in example 30.6, and we know that is wrong because the area under a curve that is entirely positive cannot be negative. The difference of the two infinities diverges, as does the integral.   □

Example 30.9. Determine if $\displaystyle\int_0^1 \ln x\, dx$ converges, and if so, find its value.

Solution. From example 29.3, $\int \ln x = x \ln x - x$. However, the integrand, $\ln x$, diverges at $x = 0$.

$$\int_0^1 \ln x\, dx = \lim_{t \to 0^+} \int_t^1 \ln x\, dx = \lim_{t \to 0^+} [x \ln x - x]\bigg|_t^1 \tag{30.22a}$$

$$= \lim_{t \to 0^+} [(1 \cdot \ln 1 - 1) - (t \ln t - t)] \tag{30.22b}$$

$$= -1 - \lim_{t \to 0^+} t \ln t + \lim_{t \to 0^+} t \tag{30.22c}$$

$$= -1 - \lim_{t \to 0^+} t \ln t \tag{30.22d}$$

In example 16.8 we found that $\lim_{t \to 0^+} t \ln t = 0$. Thus $\int_0^1 \ln x = -1$ $\qquad$ □

Comparison Test for Improper Integrals.

Since there are some improper integrals that cannot be solved analytically a comparison test similar to theorem 24.8 for definite integrals is helpful.

> **Theorem 30.2. Improper Integral Comparison Test**
>
> If $f(x)$ and $g(x)$ are continuous and $f(x) \geq g(x)$ for all $x \geq a$ for some number a, then
>
> (a) If $\int_a^\infty f(x)dx$ converges, then $\int_a^\infty g(x)$ converges.
>
> (b) If $\int_a^\infty g(x)dx$ diverges, then $\int_a^\infty f(x)$ diverges.

Example 30.10. Show that $\int_1^\infty \dfrac{3 + \sin x}{x} \, dx$ diverges using the comparison test.

Solution. Since $\sin x > -1$ then $3 + \sin x > 2 > 1$ for all x, and thus

$$\frac{3 + \sin x}{x} > \frac{1}{x} \tag{30.23a}$$

for all x. In example 30.3 we showed that $\int_1^\infty \dfrac{dx}{x}$ diverged. Therefore by theorem 30.2, $\int_1^\infty \dfrac{3 + \sin x}{x} dx$ also diverges. $\qquad$ □

Example 30.11. Show that $\int_0^\infty e^{-x^2} dx$ converges.

Solution. We can split the integral up into two parts:

$$\int_0^\infty e^{-x^2} dx = \int_0^1 e^{-x^2} dx + \int_1^\infty e^{-x^2} dx \tag{30.24a}$$

In the first integral, $e^{-x^2} < 1$ and thus by theorem 24.8,

$$\int_0^1 e^{-x^2} dx < \int_0^1 1 \cdot dx = 1 \tag{30.24b}$$

Thus

$$\int_0^\infty e^{-x^2} dx < 1 + \int_1^\infty e^{-x^2} dx \tag{30.24c}$$

For $x > 1$, $e^{-x^2} < e^{-x}$. To verify this observe that for $x > 1$, $x^2 > x$ and thus

$$x^2 > x \implies -x^2 < -x \tag{30.24d}$$

$$\implies -x^2 \cdot \ln e < -x \cdot \ln e \tag{30.24e}$$

$$\implies \ln e^{-x^2} < \ln e^{-x} \tag{30.24f}$$

$$\implies e^{-x^2} < e^{-x} \tag{30.24g}$$

Since $e^{-x^2} < e^{-x}$ on $(1, \infty)$, by theorem 30.2 the integral $\int_1^\infty e^{-x^2} dx$ converges if and only if the integral $\int_1^\infty e^{-x} dx$ converges. But (compare with example 30.1),

$$\int_1^\infty e^{-x} dx = \frac{1}{e} \tag{30.24h}$$

Therefore $\int_0^\infty e^{-x^2} dx$ converges. $\square$

Exercises

In exercises 1 through 5, determine if the given integrals converge or diverge. If the integral converges, find its value. If the integral diverges, explain why.

Use a comparison test to determine if each of the integrals in exercises 6 through 10 converges or diverges.

1. $\displaystyle\int_5^\infty \frac{dx}{x^{3/2}}$ Ans: $\frac{2}{\sqrt{5}}$

6. $\displaystyle\int_1^\infty \frac{x^4 dx}{x^7 - 1}$ Ans: Diverges

2. $\displaystyle\int_6^{11} \frac{11\,dx}{\sqrt[3]{x-6}}$ Ans: $\frac{33\,5^{2/3}}{2}$

7. $\displaystyle\int_1^\infty \frac{x^3 dx}{x^3 - 1}$ Ans: Diverges

3. $\displaystyle\int_{-\infty}^\infty x^5 e^{-x^6} dx$ Ans: 0

8. $\displaystyle\int_1^\infty \frac{x^4 dx}{x^6 + 5}$ Ans: Converges

4. $\displaystyle\int_0^9 \frac{dx}{(x-5)^2}$ Ans: Diverges

9. $\displaystyle\int_1^\infty \frac{dx}{3\sqrt{x} + 2e^{5x}}$

5. $\displaystyle\int_0^5 \frac{dx}{x^{1.25}}$ Ans: Diverges

10. $\displaystyle\int_1^\infty \frac{x^4 dx}{7x^6 + 5}$

"Gang Sines"

The effects of Dr. Shapiro's course.

Chapter 31

Trigonometric Integrals

Chapter Summary and Goal

In this chapter our main goal will be the evaluation of integrals such as

$$\int \sin^m x \, \cos^n x \, dx \quad \text{and} \quad \int \sec^p x \, \tan^q x \, dx \tag{31.1}$$

where m, n, p, and q are integers, as well as

$$\int \sin mx \, \sin nx \, dx \,, \int \sin px \, \cos qx \, dx \,, \text{ and } \int \cos rx \, \cos sx \, dx, \tag{31.2}$$

where m, n, p, q, r, and s are any integers.

Student Learning Objectives

The student will:

1. Learn to reduce integrals involving powers of $\sin x$ and $\cos x$ through substitution, and to evaluate these integrals.
2. Learn to reduce integrals involving powers of $\tan x$ and $\sec x$ through substitution, and to evaluate these integrals.
3. Learn to use trigonometric identities to solve integrals involving products of $\sin mx$ and/or $\cos nx$ when $m \neq n$.

Useful Trigonometric Identities

Our primary tools will be several basic trigonometric identities. According to the Pythagorean theorem, a right triangle with hypotenuse c and sides a and b satisfies

$$a^2 + b^2 = c^2 \tag{31.3}$$

If x is the angle between a and c then then $\sin x = b/c$ and $\cos x = a/c$ (see figure 31.1). Then

$$c^2 \cos^2 x + c^2 \sin^2 = c^2 \tag{31.4}$$

285

Figure 31.1: A triangle with unit hypotenuse and central angle x has side lengths $\cos x$ (adjacent to x) and $\sin x$ (opposite to x).

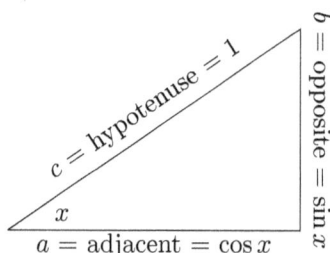

Factoring the c^2 on the left hand side of the equation and then dividing both sides of the equation by c^2, we arrive the **Pythagorean Identity** that you probably remember from your trigonometry class.

$$\cos^2 x + \sin^2 x = 1 \tag{31.5}$$

Another formula that we will have to use frequently is the **double angle formula for cosines**. We obtain this formula from the angle addition formula,

$$\cos(x + y) = \cos x \cos y - \sin x, \sin y \tag{31.6}$$

If $x = y$, this becomes

$$\cos 2x = \cos^2 x - \sin^2 x \tag{31.7}$$

Adding equations (31.7) and (31.5) and dividing by 2 gives

$$\cos^2 x = \frac{1}{2}(1 + \cos 2x) \tag{31.8}$$

Similarly, subtracting equation (31.7) from (31.5) and dividing by 2 gives

$$\sin^2 x = \frac{1}{2}(1 - \cos 2x) \tag{31.9}$$

Equations 31.8 and 31.9 are useful because they allow us to express squares of sines and cosines (e.g., $\sin^2 x$ or $\cos^2 x$) in terms of a double-angle formula ($\cos(2x)$). This, in turn, is useful, because we can easily solve $\displaystyle\int \cos 2x\, dx$ (let $u = 2x$).

Theorem 31.1. Useful Trigonometric Identites (see also Appendix A)

$\cos^2 x + \sin^2 x = 1$	(31.10)	$\cos 2x = \cos^2 x - \sin^2 x$	(31.11)
$\sec^2 x - \tan^2 x = 1$	(31.12)	$\sin 2x = 2\sin x \cos x$	(31.13)
$\csc^2 x - \cot^2 x = 1$	(31.14)	$\sin x \cos y = \dfrac{\sin(x-y) + \sin(x+y)}{2}$	(31.15)
$\cos^2 x = \frac{1}{2}(1 + \cos 2x)$	(31.16)	$\sin x \sin y = \dfrac{\cos(x-y) - \cos(x+y)}{2}$	(31.17)
$\sin^2 x = \frac{1}{2}(1 - \cos 2x)$	(31.18)	$\cos x \cos y = \dfrac{\cos(x-y) + \cos(x+y)}{2}$	(31.19)

There are similar formulas relating the tangent to the secant (and the cotangent to the cosecant). If we divide equation (31.5) by $\cos^2 x$ we

$$\frac{\cos^2 x}{\cos^2 x} + \frac{\sin^2 x}{\cos^2 x} = \frac{1}{\cos^2 x} \tag{31.20}$$

Making the substitutions $\tan x = \dfrac{\sin x}{\cos x}$ and $\sec x = \dfrac{1}{\cos x}$ and rearranging gives

$$\sec^2 x - \tan^2 x = 1 \tag{31.21}$$

Similarly (start by dividing by $\sin^2 x$),

$$\csc^2 x - \cot^2 x = 1 \tag{31.22}$$

We have summarized these identities, along with some other useful relations, in theorem 31.1.

Integrals of Even Powers of the sine or cosine

We can solve integrals of the form $\int \sin^{2m} x \, dx$ and $\int \cos^{2n} x \, dx$ by repeated applications of the identities (31.8) and (31.9).[1]

Heuristic 31.1. $\displaystyle\int \sin^{2n} x \, dx$ and $\displaystyle\int \cos^{2n} x \, dx$ **(Even Powers)**

Use the identities

$$\cos^2 x = \frac{1}{2}(1 + \cos 2x) \tag{31.23}$$

$$\sin^2 x = \frac{1}{2}(1 - \cos 2x) \tag{31.24}$$

as necessary to reduce exponents.

Example 31.1. Find $\displaystyle\int \sin^2 x \, dx$.

Solution. Here we can use the identity (31.9) once to obtain

$$\int \sin^2 x \, dx = \frac{1}{2} \int (1 - \cos 2x) \, dx \tag{31.25a}$$

$$= \frac{1}{2} \int dx - \frac{1}{2} \int \cos(2x) \, dx \tag{31.25b}$$

$$= \frac{1}{2}x - \frac{1}{4} \sin 2x \tag{31.25c}$$

$$\square$$

For powers higher than 2 we will have to make repeated substitutions.

---

[1] Starting here we may omit the constant of integration to save space. *However, just because we don't always write down the constant of integration, that doesn't mean it is not there.* We just are omitting it to go easy on the space-time continuum by reducing the expense of typesetting.

Example 31.2. Find $\int \sin^4 x \, dx$.

Solution. Since $\sin^4 x = \left(\sin^2 x\right)^2$, and since $\sin^2 x = \dfrac{1}{2}(1 - \cos 2x)$ (eq. 31.9),

$$\int \sin^4 x \, dx = \int (\sin^2 x)^2 dx = \int \left[\frac{1}{2}(1 - \cos 2x)\right]^2 dx \tag{31.26a}$$

$$= \frac{1}{4} \int (1 - 2\cos(2x) + \cos^2(2x)) \, dx \tag{31.26b}$$

$$= \frac{1}{4} \int dx - \frac{1}{2} \int \cos(2x) \, dx + \frac{1}{4} \int \cos^2(2x) dx \tag{31.26c}$$

$$= \frac{x}{4} - \frac{1}{4} \sin 2x + \frac{1}{4} \int \cos^2(2x) dx \tag{31.26d}$$

This is all fine and dandy except that in the last term we are left with another trig integral that has a $\cos^2(2x)$. We have to make a second substitution, this time using equation 31.8, but remembering to double the argument because of the $2x$.

$$\int \sin^4 x \, dx = \frac{x}{4} - \frac{1}{4} \sin 2x + \frac{1}{8} \int [1 + \cos(4x)] dx \tag{31.26e}$$

$$= \frac{x}{4} - \frac{1}{4} \sin 2x + \frac{1}{8} \int dx + \frac{1}{8} \int \cos(4x) \, dx \tag{31.26f}$$

$$= \frac{x}{4} - \frac{1}{4} \sin 2x + \frac{x}{8} + \frac{1}{32} \sin(4x) \tag{31.26g}$$

$$= \frac{3x}{8} - \frac{1}{4} \sin 2x + \frac{1}{32} \sin(4x). \tag{31.26h}$$

□

Integrals of Odd Powers of the sine or cosine

If n is an odd integer then $n = 2m + 1$, where m is some other integer. Thus integrals of odd powers of the sine or cosine will take the following forms:

$$\int \sin^{2m+1} x \, dx, \text{ or } \int \cos^{2m+1} x \, dx \tag{31.27}$$

The trick then is to factor out the "odd" factor – the one with the unitary power – and make an appropriate substitution with the even factor. The even factor – the one with the $2m$ exponent – is replaced using the pythagorean identity (eq. 31.5). To solve the cosine integral we write

$$\int \cos^{2m+1} x \, dx = \int \cos^{2m} x \cos x \, dx = \int (\cos^2 x)^m \cos x \, dx = \int (1 - \sin^2 x)^m \cos x \, dx \tag{31.28}$$

This works because $d(\sin x) = -\cos x \, dx$, and so we can follow with a substitution $u = \sin x$. This turns the integral into

$$\int (1 - u^2)^m \, du \tag{31.29}$$

which we can evaluate by multiplying out the polynomial and integrating term by term.

Heuristic 31.2. $\int \cos^{2n+1} x \ dx$ **(odd powers of $\cos x$)**

1. Factor $\cos^{2n+1} x = (\cos x) \cdot (\cos^2 x)^n$.
2. Substitute $\cos^2 x = 1 - \sin^2 x$ in the second factor.
3. Substitute $u = \sin x$ in the integral to obtain $\int (1 - u^2)^{2n} du$.
4. Multiply out the expression in u and integrate.
5. Substitute back $u = \sin x$ to get an answer in terms of x.

Example 31.3. Find $\int \cos^3 x \ dx$.

Solution. This is an odd power of $\cos x$, so we write $\cos^3 x = (\cos^2) \times (\cos x)$

$$\int \cos^3 x \ dx = \int \cos x \ \cdot \ \cos^2 x \ dx \tag{31.30a}$$

Substituting $\cos^2 x = 1 - \sin^2 x$,

$$\int \cos^3 x \ dx = \int \cos x \ (1 - \sin^2 x) \ dx \tag{31.30b}$$

$$= \int \cos x \ \ dx - \int \sin^2 x \cos x \ dx \tag{31.30c}$$

$$= \sin x - \int \sin^2 x \cos x \ dx \tag{31.30d}$$

In the last term we let $u = \sin x$. Then $du = \cos x \ dx$ and

$$\int \sin^2 x \ \cos x \ dx = \int u^2 du = \frac{1}{3} u^3 = \frac{1}{3} \sin^3 x. \tag{31.30e}$$

Thus

$$\int \cos^3 x \ dx = \sin x - \frac{1}{3} \sin^3 x. \tag{31.30f}$$

$\square$

The sine integrals are similar. We can factor out the unitary $\sin x$ and substitute:

$$\int \sin^n x \ \ dx = \int \sin 2m + 1 \ dx = \int (\sin^2 x)^m \sin x \ dx = \int (1 - \cos^2 x)^m \sin x \ dx \tag{31.31}$$

and follow with the substitution $u = \cos x$.

Heuristic 31.3. $\int \sin^{2n+1} x \ dx$ **(odd powers of $\sin x$)**

1. Factor $\sin^{2n+1} x = (\sin x) \cdot (\sin^2 x)^n$.
2. Substitute $\sin^2 x = 1 - \cos^2 x$ in the second factor.
3. Substitute $u = \cos x$ in the integral to obtain $- \int (1 - u^2)^{2n} du$.
4. Multiply out the expression in u and integrate.
5. Substitute back $u = \cos x$ to get an answer in terms of x.

Example 31.4. Solve $\int \sin^5 x \, dx$.

Solution. Here we write

$$\sin^5 x = \sin x \cdot \sin^4 x = \sin x \cdot (\sin^2 x)^2 \tag{31.32a}$$

so that we can make the simplification

$$\int \sin^5 x \, dx = \int \sin x (\sin^2 x)^2 \, dx = \int \sin x (1 - \cos^2 x)^2 \, dx \tag{31.32b}$$

As before we substitute $u = \cos x$, so that $du = -\sin x \, dx$. Thus

$$\int \sin^5 x \, dx = -\int (1 - u^2)^2 \, du = -\int (1 - 2u^2 + u^4) du \tag{31.32c}$$

$$= -u + \frac{2}{3}u^3 - \frac{1}{5}u^5 \tag{31.32d}$$

$$= -\cos x + \frac{2}{3}\cos^3 x - \frac{1}{5}\cos^5 x. \tag{31.32e}$$

$\square$

Integrating Products of Odd and Even Powers of sine and cosine

Integrals of odd and even powers of the sine and cosine have the form

$$\int \sin^{2m+1} x \cos^{2n} x \, dx, \text{ or } \int \cos^{2m+1} x \sin^{2n} x \, dx \tag{31.33}$$

for some integers m and n.   We handle these integrals by reducing the odd power as far as possible with the Pythagorean identity. This always works because we can follow the reduction with a u substitution. For example, if we have a $\sin x$ raised to an odd power multiplied by $\cos x$ raised to an even power, then for some integers m and n, it will look like

$$\int \overbrace{(\sin^{2m+1} x)}^{\text{odd power}} \cdot \overbrace{(\cos^{2n} x)}^{\text{even}} \, dx = \int (\sin x) \cdot (\sin^{2m} x) \cdot (\cos^{2n} x) \, dx \tag{31.34}$$

$$= \int (\sin x) \cdot (\sin^2 x)^m \cdot (\cos^{2n} x) \, dx \tag{31.35}$$

$$= \int \sin x (1 - \cos^2 x)^m \cos^{2n} x \, dx \tag{31.36}$$

Letting $u = \cos x$, this becomes

$$\int (\sin^{2m+1} x)(\cos^{2n} x) \, dx = -\int (1 - u^2)^m u^{2n} \, du \tag{31.37}$$

For any given values of m, the polynomials can be multiplied out, and the integral solved term by term. It is also possible to derive a general formula for the solution of this integral using the Binomial theorem (see exercise 22). However, in practice, it is easier just to apply the heuristic when needed.

Heuristic 31.4. $\int \overbrace{(\sin^{2m+1} x)}^{\text{Odd Power}} \cdot \overbrace{(\cos^{2n} x)}^{\text{Even Power}} \; dx$

1. Factor out $\sin x$ to get $\int \sin x (\sin^{2m} x)(\cos^{2n} x) \; dx$;

2. Use the $\sin^2 x = 1 - \cos^2 x$ to replace the remaining $\sin x$ factors; the resulting integral is $\int \sin x (1 - \cos^2 x)^{2m} \cos^{2n} \; dx$;

3. Substitute $u = \cos x$ to get $-\int (1-u^2)^{2m} u^{2n} du$;

4. Multiply out the polynomial in u and integrate;

5. Substitute back $u = \cos x$ to get the final answer in terms of x.

Example 31.5. Find $\int \sin^5 x \, \cos^2 x \, dx$.

Solution.   This is the product of an odd power of $\sin x$ with an even power of $\cos x$ so heuristic 31.4 applies. Factoring out $\sin x$ and then substituting $\sin^2 x = 1 - \cos^2 x$ for each remaining factor of $\sin^2 x$,

$$\int \sin^5 x \, \cos^2 x \, dx = \int \sin x \cdot (\sin^2 x)^2 \cdot \cos^2 x dx \tag{31.38a}$$

$$= \int \sin x \cdot (1 - \cos^2 x)^2 \cdot \cos^2 x \, dx \tag{31.38b}$$

Next we let $u = \cos x$. Then $du = -\sin x \, dx$.

$$\int \sin^5 x \, \cos^2 x \, dx = -\int (1-u^2)^2 u^2 du \tag{31.38c}$$

$$= -\int (1 - 2u^2 + u^4)u^2 du = -\int (u^2 - 2u^4 + u^6) \, du \tag{31.38d}$$

$$= -\frac{1}{3}u^3 + \frac{2}{5}u^5 - \frac{1}{7}u^7 = -\frac{1}{3}\cos^3 x + \frac{2}{5}\cos^5 x - \frac{1}{7}\cos^7 x. \tag{31.38e}$$

$\square$

Heuristic 31.5. $\int \overbrace{(\cos^{2m+1} x)}^{\text{Odd Power}} \cdot \overbrace{(\sin^{2n} x)}^{\text{Even Power}} \; dx$

1. Factor $\cos x$ to get $\int \cos x (\cos^{2m} x)(\sin^{2n} x) \; dx$.

2. Substitute $\cos^2 x = 1 - \sin^2 x$ to replace the remaining $\cos x$ factors; the resulting integral is $\int \cos x (1 - \sin^2 x)^{2m} \sin^{2n} \; dx$.

3. Substitute $u = \sin x$ to get $\int (1-u^2)^{2m} u^{2n} du$.

4. Multiply out the polynomial in u and integrate;.

5. Substitute back $u = \sin x$ to get the final answer in terms of x.

Example 31.6. Find $\displaystyle\int \cos^3 x\, \sin^2 x\, dx$.

Solution. This has an odd power of the cosine times an even power of the sine, so we factor out $\cos x$. In eq. $31.39c$ we let $u = \sin x$ (and $du = \cos x\, dx$).

$$\int \cos^3 x\, \sin^2 x\, dx = \int (\cos x)\,(\cos^2 x)\,(\sin^2 x)\, dx \tag{31.39a}$$

$$= \int (\cos x)\,(1 - \sin^2 x)\,(\sin^2 x)\, dx \tag{31.39b}$$

$$= \int (1 - u^2)u^2\, du = -\frac{1}{3}u^3 + \frac{1}{5}u^5 \tag{31.39c}$$

$$= -\frac{1}{3}\sin^3 x + \frac{1}{5}\sin^5 x \tag{31.39d}$$

$\square$

Integrating Products of Even Powers of the sine and cosine

A product of even powers can be reduced to a product of cosines of higher frequency arguments (i.e., $\cos(2x), \cos(4x), \cos(8x), \dots$) by making use of the substitutions $\cos^2 x = \frac{1}{2}(1 + \cos 2x)$ and $\sin^2 x = \frac{1}{2}(1 - \cos(2x))$. These reduce the integral to a polynomial in $\cos(2x)$. This polynomial can be integrated term by term as discussed in the previous sections.

Heuristic 31.6. $\displaystyle\int \overbrace{(\sin^{2n} x)}^{\text{Even Power}} \cdot \overbrace{(\cos^{2m} x)}^{\text{Even Power}}\, dx$

1. Substitute the identities

$$\cos^2 x = \frac{1}{2}(1 + \cos(2x)), \quad \sin^2 x = \frac{1}{2}(1 - \cos(2x)) \tag{31.40}$$

2. Multiply out to obtain a polynomial in $\cos(2x)$.

3. Apply the heuristic for $\displaystyle\int \cos^k u\, du$ to each term.

Example 31.7. Find $\displaystyle\int \sin^4 x\, \cos^2 x\, dx$.

Solution. This is a product of an even power power of $\sin x$ and and even power of $\cos x$. Let

$$\sin^4 x = (\sin^2 x)^2 = \left[\frac{1}{2}\left(1 - \cos(2x)\right)\right]^2 \tag{31.41a}$$

$$\cos^2 x = \frac{1 + \cos(2x)}{2} \tag{31.41b}$$

Then

$$\int \sin^4 x\, \cos^2 x\, dx = \int \frac{(1 - \cos(2x))^2}{4} \cdot \frac{1 + \cos(2x)}{2} \cdot dx \tag{31.41c}$$

$$= \frac{1}{8}\int (1 - 2\cos(2x) + \cos^2(2x)) \cdot (1 + \cos(2x)) \cdot dx \tag{31.41d}$$

$$= \frac{1}{8} \int (1 - \cos(2x) - \cos^2(2x) + \cos^3(2x)) dx \tag{31.41e}$$

$$= \frac{1}{8} \int dx - \frac{1}{8} \int \cos(2x) dx - \frac{1}{8} \int \cos^2(2x) dx + \frac{1}{8} \int \cos^3(2x) dx \tag{31.41f}$$

$$= \frac{x}{8} - \frac{\sin(2x)}{16} - \frac{1}{8} \int \cos^2(2x) dx + \frac{1}{8} \int \cos^3(2x) dx \tag{31.41g}$$

In exercise 31.1 we show that $\int \cos^2 u \; du = \frac{u}{2} + \frac{\sin(2u)}{4}$.

Letting $u = 2x$,

$$2 \int \cos^2(2x) \; dx = x + \frac{\sin(4x)}{4}. \tag{31.41h}$$

In equation 31.30f we showed that $\int \cos^3 u \, du = \sin u - \frac{\sin^3 u}{3}$.

Letting $u = 2x$,

$$2 \int \cos^3(2x) dx = \sin(2x) - \frac{\sin^3(2u)}{3}. \tag{31.41i}$$

Making these substitutions

$$\int \sin^4 x \cos^2 x \; dx = \frac{x}{8} - \frac{\sin(2x)}{16} - \frac{1}{8 \cdot 2} \left[x + \frac{\sin(4x)}{4} \right] + \frac{1}{8 \cdot 2} \left[\sin(2x) - \frac{\sin^3(2x)}{3} \right] \tag{31.41j}$$

$$= \frac{x}{16} - \frac{\sin(4x)}{64} - \frac{\sin^3(2x)}{48}. \tag{31.41k}$$

$\square$

Integrals with Tangents and Secants

The process for integrating products of powers of secants and tangents is similar to the process for products of powers of sines and cosines. Instead of the usual pythagorean identity we have

$$\sec^2 x - \tan^2 x = 1, \text{ and } \csc^2 x - \cot^2 x = 1 \tag{31.42}$$

(see eqs. 31.21 and 31.22). However, solving integrals with tangents and/or secants is complicated by two factors: (1) there is no simplification of squares formula analogous to equations 31.8 and 31.9; and (2) the derivatives of the tangent and secant are messier:

$$\frac{d}{dx} \tan x = \sec^2 x \tag{31.43}$$

$$\frac{d}{dx} \sec x = \sec x \tan x \tag{31.44}$$

Nevertheless, we are still often able to reduce things to the point where we can make an appropriate u-substitution and solve the resulting equation. Frequently the appropriate u-substitution will be either $u = \sec x$ or $u = \tan x$. It will be helpful to recall the following formulas:

$$\int \tan x \; dx = \ln|\sec x| \tag{31.45}$$

$$\int \sec x \; dx = \ln|\sec x + \tan x| \tag{31.46}$$

$$\int \sec^2 x \; dx = \tan x \tag{31.47}$$

$$\int \sec x \; \tan x \; dx = \sec x \tag{31.48}$$

All of the heuristics that we give for secants and tangents will carry over to cosecants and cotangents with very little modification.

Integrals of Powers of Tangents

Integrals of powers of the tangent can be reduced using the following general procedure. Suppose $n \geq 2$. Then

$$\int \tan^n x \; dx = \int \tan^{n-2+2} x \; dx \tag{31.49}$$

$$= \int (\tan^{n-2} x) \cdot \tan^2 x \; dx \tag{31.50}$$

$$= \int (\tan^{n-2} x) \cdot (\sec^2 x - 1) \; dx \tag{31.51}$$

$$= \int (\tan^{n-2} x) \cdot \sec^2 x \; dx - \int (\tan^{n-2} x) dx \tag{31.52}$$

The first integral on the right can be solved with the substitution $u = \tan x$. The second integral is similar to the integral we started with, *but the exponent is reduced by 2*. This works regardless of whether n is even or odd. We can express this heuristic in two different ways.

Heuristic 31.7. $\int \tan^n x \; dx$ **(Powers of the Tangent)**

1. Factor out $\tan^2 x$;
2. Substitute $\tan^2 = \sec^2 x - 1$ in that factor only.
3. Let $u = \tan x$ and integrate.

Example 31.8. Find $\int \tan^2 x \; dx$.

Solution. There is only a $\tan^2 x$ here. Thus

$$\int \tan^2 x \; dx = \int (\sec^2 x - 1) \; dx = \int \sec^2 x \; dx - \int dx = \tan x - x \tag{31.53a}$$

$\square$

Example 31.9. Find $\int \tan^3 x \; dx$.

Solution. Factoring out a single $\tan^2 x = \sec^2 x - 1$,

$$\int \tan^3 x \; dx = \int \tan x \cdot \tan^2 x \; dx = \int \tan x (\sec^2 x - 1) \; dx \tag{31.54a}$$

$$= \int \tan x \sec^2 x \, dx - \int \tan x \, dx \tag{31.54b}$$

$$= \int \tan x \sec^2 x \, dx - \ln|\sec x| \tag{31.54c}$$

In the first term we can substitute $u = \tan x$. Then

$$\int \tan^3 x \, dx = \int u \, du - \ln|\sec x| = \frac{1}{2}u^2 - \ln|\sec x| \tag{31.54d}$$

$$= \frac{1}{2}\tan^2 x + \ln|\cos x|. \tag{31.54e}$$

$\square$

Example 31.10. Find $\int \tan^4 x \, dx$.

Solution. To apply heuristic 31.7 we factor out $\tan^2 x = \sec^2 x - 1$.

$$\int \tan^4 x \, dx = \int \tan^2 x \cdot \tan^2 x \, dx \tag{31.55a}$$

$$= \int (\sec^2 x - 1) \tan^2 x \, dx \tag{31.55b}$$

$$= \int \sec^2 x \tan^2 x \, dx - \int \tan^2 x \tag{31.55c}$$

$$= \int \sec^2 x \tan^2 x \, dx - \tan x + x. \tag{31.55d}$$

The result of example 31.8 has been used in the last step. In the remaining integral we substitute $u = \tan x$ (step 3 of heuristic 31.7). $du = \sec^2 x \, dx$ and

$$\int \tan^4 x \, dx = \int u^2 \, du - \tan x + x = \frac{1}{3}u^3 - \tan x + x \tag{31.55e}$$

$$= \frac{1}{3}\tan^3 x - \tan x + x \tag{31.55f}$$

$\square$

As an alternative to Heuristic 31.7, we can iterate on eq. 31.52 as needed.

Heuristic 31.8. $\int \tan^n x \, dx$ **(Powers of the Tangent - Alternative Heuristic)**

Iterate

$$\int \tan^n x \, dx = \int \tan^{n-2+2} x \, dx = \int (\tan^{n-2} x) \cdot \sec^2 x \, dx - \int (\tan^{n-2} x) \, dx \tag{31.56}$$

as necessary to solve the integral, using the substitution $u = \tan x$.

Example 31.11. Demonstrate Heuristic 31.8 by finding $\int \tan^n \, dx$ for $n = 2, 3, 4, 5, 6$.

Solution. For reference, we recall equation 25.5:

$$\int \tan x \, dx = \ln|\sec x| \tag{31.57a}$$

Then from Heuristic 31.8,

$$\int \tan^2 x\, dx = \int \sec^2 x\, dx - \int dx = \tan x - x \tag{31.57b}$$

$$\int \tan^3 x\, dx = \int \tan x \sec^2 x\, dx - \int \tan x\, dx \tag{31.57c}$$

$$= \int u\, du - \ln|\sec x| = \frac{1}{2}\tan^2 x + \ln|\cos x| \tag{31.57d}$$

$$\int \tan^4 x\, dx = \int \tan^2 x \sec^2 x - \int \tan^2 x\, dx \tag{31.57e}$$

$$= \int u^2\, du - (\tan x - x) = \tag{31.57f}$$

$$\int \tan^5 x\, dx = \int \tan^3 x \sec^2 x\, dx - \int \tan^3 x\, dx \tag{31.57g}$$

$$= \int u^3\, du - \left(\frac{1}{2}\tan^2 x + \ln|\cos x|\right) \tag{31.57h}$$

$$= \frac{1}{4}\tan^4 x - \frac{1}{2}\tan^2 x + \ln|\sec x| \tag{31.57i}$$

$$\int \tan^6 x\, dx = \int \tan^4 x \sec^2 x\, dx - \int \tan^4 x\, dx \tag{31.57j}$$

$$= \int u^4\, du - \left(\frac{1}{3}\tan^3 x - \tan x + x\right) \tag{31.57k}$$

$$= \frac{1}{5}\tan^5 x - \frac{1}{3}\tan^3 x + \tan x - x \tag{31.57l}$$

This process can be continued indefinitely until the required power is reached. It is effectively the same procedure as implemented in Heuristic 31.7 except that we are working our way upwards from 1 towards n rather than downwards from n towards 1. □

Integrals of Products of Secants and Tangents

Integrals of the form

$$\int \tan^m x \sec^n x\, dx \tag{31.58}$$

cannot always be solved. We will discuss two specific situations in which a solution is known to exists: (a) when **the exponent on the secant is even**; and (b) when **both exponents are odd**.

Heuristic 31.9. Power of Secant is Even $\int (\tan^m x) \cdot (\sec^{2n} x)\, dx$

1. Factor out $\sec^2 x$.
2. In all remaining secants substitute $\sec^2 x = \tan^2 x + 1$.
3. Substitute $u = \tan x$, so that $du = \sec^2 x\, dx$, and integrate in u.
4. Back-substitute $u = \tan x$ to get an answer in x.

Example 31.12. Find $\int \tan^6 x \sec^4 x \, dx$.

Solution. The power of the secant is even, so we pull out a single factor of $\sec^2 x$ and substitute $\sec^2 x = 1 + \tan^2 x$ in the remaining secants.

$$\int \tan^6 x \cdot \sec^4 x \, dx = \int \tan^6 x \cdot \sec^2 x \cdot \sec^2 x \, dx \tag{31.59a}$$

$$= \int \tan^6 x \cdot (1 + \tan^2 x) \cdot \sec^2 x \, dx \tag{31.59b}$$

$$= \int \tan^6 x \cdot \sec^2 x \, dx + \int \tan^8 x \cdot \sec^2 x \, dx \tag{31.59c}$$

Next, we make the usual substitution of $u = \tan x$ and $du = \sec^2 x \, dx$.

$$\int \tan^6 x \cdot \sec^4 x \, dx = \int u^6 \, du + \int u^8 \, du \tag{31.59d}$$

$$= \frac{1}{7} u^7 + \frac{1}{9} u^9 = \frac{1}{7} \tan^7 x + \frac{1}{9} \tan^9 x. \tag{31.59e}$$

$\square$

Heuristic 31.10. Odd/Odd Powers in $\int (\tan^{2m+1} x) \cdot (\sec^{2n+1} x) \, dx$

1. Factor out one pair: $\sec x \tan x$ giving $\int (\tan^{2m} x) \cdot (\sec^{2n} x) \cdot (\sec x \tan x) \, dx$.
2. Substitute $\tan^{2m} x = (\tan^2 x)^m = (\sec^2 x - 1)^m$.
3. Let $u = \sec x$. Then $du = \sec x \tan x \, dx$, giving $\int (u^2 - 1)^m u^n du$;
4. Multiply out the polynomial in u and integrate.
5. Back-substitute $u = \sec x$ to get an answer in terms of x.

Example 31.13. Find $\int \tan^5 x \sec^7 x \, dx$.

Solution. This is a product of odd powers, so we factor out $\sec x \tan x$ to use as the derivative of $\sec x$. This gives

$$\int \tan^5 x \sec^7 x \, dx = \int \tan^4 x \sec^6 x \overbrace{(\sec x \tan x \, dx)}^{d(\sec x)} \tag{31.60a}$$

$$= \int \tan^4 x \sec^6 x \, d(\sec x) \tag{31.60b}$$

$$= \int (\sec^2 x - 1)^2 \sec^6 x \, d(\sec x) \tag{31.60c}$$

$$= \int (\sec^4 x - 2 \sec^2 x + 1) \sec^6 x \, d(\sec x) \tag{31.60d}$$

$$= \int (\sec^{10} x - 2 \sec^8 x + \sec^6 x) \left(\sec x \right) \tag{31.60e}$$

$$= \frac{1}{11} \sec^{11} x - \frac{2}{9} \sec^9 x + \frac{1}{7} \sec^7 x + C \quad \square \tag{31.60f}$$

$\square$

Other Integrals Involving the secant and tangent

In general there are no rules. We use integration by parts, substitution, or pulling a rabbit out of a hat.

Example 31.14. Find $\int \sec x \, dx$.

Solution. There is only one way to do this: pulling a rabbit out of the hat. We multiply the integrand by 1, where

$$1 = \frac{\sec x + \tan x}{\sec x + \tan x} \tag{31.61a}$$

Then we let $u = \sec x + \tan x$, so that

$$du = (\sec x \tan x + \sec^2 x)dx = \sec x(\tan x + \sec x)dx \tag{31.61b}$$

to give us the following result:

$$\int \sec x \, dx = \int \sec x \cdot \frac{\sec x + \tan x}{\sec x + \tan x} \cdot dx \tag{31.61c}$$

$$= \int \frac{du}{u} = \ln|u| = \ln|\sec x + \tan x| \tag{31.61d}$$

$\square$

Example 31.15. Find $\int \sec^3 x \, dx/$

Solution.   Use integration by parts with $u = \sec x$ and $dv = \sec^2 x \, dx$.   Then $du = \sec x \tan x \, dx$ and $v = \tan x$. With these substitutions,

$$\int \sec^3 x \, dx = uv - \int v \, du \tag{31.62a}$$

$$= \sec x \tan x - \int (\tan x) \cdot (\sec x \tan x \, dx) \tag{31.62b}$$

$$= \sec x \tan x - \int \tan^2 x \cdot \sec x \, dx \tag{31.62c}$$

Making a Pythagorean substitution $\tan^2 x = \sec^2 x - 1$,

$$\int \sec^3 x \, dx = \sec x \tan x - \int (\sec^2 x - 1) \cdot \sec x \, dx \tag{31.62d}$$

$$= \sec x \tan x - \int \sec^3 x \, dx + \int \sec x \, dx \tag{31.62e}$$

The middle term on the right is identical to the term on the left, but opposite in sign. We add it to both sides of the equation.

$$2 \int \sec^3 x \, dx = \sec x \tan x + \int \sec x \, dx \tag{31.62f}$$

$$= \sec x \tan x + \ln|\sec x + \tan x| \tag{31.62g}$$

Dividing by 2,

$$\int \sec^3 x \, dx = \frac{1}{2} \left(\sec x \tan x + \ln|\sec x + \tan x| \right) \tag{31.62h}$$

$\square$

Exercises

Find each of the trigonometric integrals in exercises 1 through 18.

1. $\int \cos^2 x \, dx$ Ans: $\frac{x}{2} + \frac{1}{4}\sin 2x$

2. $\int \cos^4(3x) \, dx$

3. $\int \sin^3(5x) \, dx$ Ans: $\frac{1}{60}\cos 15x - \frac{3}{20}\cos 5x$

4. $\int \cos^5(x+7) \, dx$

   Ans: $\frac{5}{8}\sin(x+7) + \frac{5}{48}\sin(3(x+7)) + \frac{1}{80}\sin(5(x+7))$

5. $\int_0^{\pi/2} \sin^2(3x) \, dx$ Ans: $\pi/4$

6. $\int_0^{\pi} \cos^4 x \, dx$ Ans: $3\pi/8$

7. $\int \sin^2 x \cos^2 x \, dx$ Ans: $\frac{x}{8} - \frac{1}{32}\sin 4x$

8. $\int \sin x \cos^3 x \, dx$

9. $\int \sin^4 x \cos^5 x \, dx$

   Ans: $\frac{3}{128}\sin(x) - \frac{1}{192}\sin(3x) - \frac{1}{320}\sin(5x) + \frac{1}{1792}\sin(7x) + \frac{1}{2304}\sin(9x)$

10. $\int \tan^2(\pi x) dx$ Ans: $\frac{1}{\pi}\tan \pi x - x$

11. $\int_{-1}^{1} \tan^4(\pi x/4) \, dx$ Ans: $2 - 16/(3\pi)$

12. $\int \sec x \tan^2 x \, dx$

13. $\int \sec^3 x \tan x \, dx$

14. $\int_0^{\pi/4} \sec^3 x \tan^3 x \, dx$ Ans: $(2/15)(1 + \sqrt{2})$

15. $\int \sec x \tan^3 x \, dx$ Ans: $\frac{1}{3}\sec^3 x - \sec x$

16. $\int \sec^7 x \tan^9 x \, dx$

   Ans: $\frac{1}{15}\sec^{15} x - \frac{4}{13}\sec^{13} x + \frac{6}{11}\sec^{11}(x) - \frac{4}{9}\sec^9 x + \frac{1}{7}\sec^7 x$

17. $\int \sec^4 x \tan^7 x \, dx$

   Ans: $\frac{1}{10}\sec^{10} x - \frac{3}{8}\sec^8 x + \frac{1}{2}\sec^6 x - \frac{1}{4}\sec^4 x$

18. $\int \sec^4 x \tan^4 x \, dx$

   Ans: $\frac{2}{35}\tan x + \frac{1}{7}\tan x \sec^6 x - \frac{8}{35}\tan x \sec^4 x + \frac{1}{35}\tan x \sec^2 x$

Use equations (31.15) through (31.19) to simplify and solve the following trigonometric integrals.

19. $\int \cos(2x) \sin(12x) \, dx$

   Ans: $-\frac{1}{20}\cos 10x - \frac{1}{28}\cos 14x$

20. $\int_0^{\pi/2} \sin(3x) \sin(6x) \, dx$ Ans: $-2/9$

21. $\int \sin(100x) \cos(101x) \, dx$

   Ans: $\frac{1}{2}\cos x - \frac{1}{402}\cos 201x$

---

22. Use the binomial theorem (equation A.17) to expand the $(1 - u^2)^m$ factor in equation 31.37 and integrate to find a general formula for $\int \sin^{2m+1} x \cos^{2n} x \, dx$ as a function of m and n.

23. Make an appropriate u-substitution in the first term of equation 31.52 and solve the integral to find a general formula for $\int \tan^n x \, dx$ as a function of $\int \tan^{n-2} x \, dx$.

Chapter 32

Trigonometric Substitutions

Chapter Summary and Goal

In chapter 31 we solved integrals that had trigonometric functions in the integrand. Trigonometric functions can also be used as substitutions in algebraic integrals such as

$$\int \frac{dx}{x^2 + 9} \tag{32.1}$$

For example, if we let $x = 3\tan\theta$ in (32.1) then

$$\int \frac{dx}{x^2 + 9} = \int \frac{3\sec^2\theta}{9(\tan^2\theta + 1)} = \frac{1}{3}\int d\theta = \frac{1}{3}\theta = \frac{1}{3}\tan^{-1}\left(\frac{x}{3}\right), \tag{32.2}$$

where the relationship $\sec^2\theta - \tan^2\theta = 1$ has been used in the third step. We will focus on this type of substitution in this chapter.

Student Learning Objectives

The student will:

1. Recognize integrals that can be solved with trigonometric or hyperbolic substitutions.
2. Make the appropriate substitution and simplify.
3. Solve the integrals after the substitutions have been made.
4. Back substitute and solve the for the solution in terms of the original variable.

Summary of Heuristics

Heuristic 32.1. Heuristics for Trigonometric Substitutions

If the integral contains	Use this substitution:	
$a^2 - x^2$ or $\sqrt{a^2 - x^2}$	$x = a\sin u$	$dx = a\cos u \, du$
$x^2 - a^2$ or $\sqrt{x^2 - a^2}$	$x = a\sec u$	$dx = a\sec u \tan u \, du$
$a^2 + x^2$ or $\sqrt{x^2 + a^2}$	$x = a\tan u$	$dx = a\sec^2 u \, du$

301

Integrals with $a^2 - x^2$ and $\sqrt{a^2 - x^2}$

Integrals involving expressions of the form $a^2 - x^2$ can often be solved with the substitutions

$$x = a \sin u \qquad (32.3)$$

This works because

$$dx = a \cos u \; du \qquad (32.4)$$

and as a result

$$a^2 - x^2 = a^2 - a^2 \sin^2 u = a^2 (1 - \sin^2 u) = a^2 \cos^2 u \qquad (32.5)$$

This means that many integrals can be reduced to integrals of $\cos^n u$, which we learned how to solve in chapter 31. As usual, rather than memorizing the formulas it is easier to derive the appropriate equation each time.

Figure 32.1: A triangular visualization of the substitution $x = a \sin u$ on a right triangle with unit hypotenuse. The side opposite to angle u has length x/a.

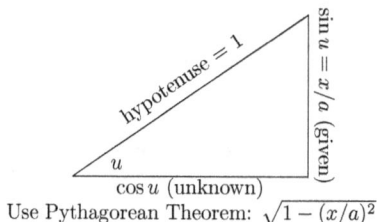

Use Pythagorean Theorem: $\sqrt{1 - (x/a)^2}$

The substitution (32.3) and calculation (32.5) can be visualized geometrically, as we did in figure 12.2. Imagine a right triangle with unit hypotenuse and one side of length x/a. Let the angle opposite this side of length x/a be called u. Then $\sin u = x/a$, as shown in figure 32.1.

Example 32.1. Solve $\displaystyle \int \sqrt{49 - x^2} \; dx$.

Solution. Let $x = 7 \sin u$. Then $dx = 7 \cos u \; du$, and

$$\int \sqrt{49 - x^2} \; dx = \int \sqrt{49 - (7 \sin u)^2} \cdot 7 \cdot \cos u \; du \qquad (32.6a)$$

$$= 7 \int \sqrt{49 - 49 \sin^2 u} \cdot \cos u \; du \qquad (32.6b)$$

$$= 7 \int \sqrt{49(1 - \sin^2 u)} \cdot \cos u \; du \qquad (32.6c)$$

$$= 7 \int \sqrt{49 \cos^2 u} \cdot \cos u \; du \qquad (32.6d)$$

$$= 7 \int 7 \cdot \cos u \cdot \cos u \; du \qquad (32.6e)$$

$$= 49 \int \cos^2 u \; du \qquad (32.6f)$$

Making the substitution $\cos^2 u = (1 + \cos(2u))/2$ (heuristic 31.1),

$$\int \sqrt{49 - x^2} \; dx = \frac{49}{2} \int [1 + \cos(2u)] \; du = \frac{49}{2} u + \frac{49}{4} \sin 2u \qquad (32.6g)$$

To finish the problem we must return to the original variable, x. Since $x = 7\sin u$, the u in the first term in the last step gives

$$u = \sin^{-1}\left(\frac{x}{7}\right) \tag{32.6h}$$

To simplify the second term in (32.6g), we use sine double angle formula:

$$\sin 2u = 2\sin u \cos u \tag{32.6i}$$

From figure 32.1

$$\cos u = \sqrt{1 - (x/7)^2} \tag{32.6j}$$

Putting this all together,

$$\int \sqrt{49 - x^2}\ dx = \frac{49}{2}\sin^{-1}\left(\frac{x}{7}\right) + \left(\frac{49}{4}\right)\left(\frac{2x}{7}\right)\sqrt{1 - (x/7)^2} \tag{32.6k}$$

$$= \frac{49}{2}\sin^{-1}\left(\frac{x}{7}\right) + \left(\frac{x}{2}\right)\sqrt{49 - x^2} \tag{32.6l}$$

□

Example 32.2. Solve $\displaystyle\int \frac{\sqrt{36 - x^2}}{x^2}\ dx$.

Solution. Since the integral involves $36 - x^2$, we substitute $x = 6\sin u$. Thus $dx = 6\cos u\ du$. The integral becomes

$$\int \frac{\sqrt{36 - x^2}}{x^2}\ dx = \int \frac{\sqrt{36 - 36\sin^2 u}}{36\sin^2 u} \cdot 6\cos u\ du \tag{32.7a}$$

$$= \int \frac{\sqrt{36(1 - \sin^2 u)}}{6\sin^2 u} \cdot \cos u\ du \tag{32.7b}$$

$$= \int \frac{6\sqrt{\cos^2 u}}{6\sin^2 u} \cdot \cos u\ du \tag{32.7c}$$

$$= \int \frac{\cos^2 u}{\sin^2 u} du \tag{32.7d}$$

$$= \int \cot^2 u\ du \tag{32.7e}$$

$$= \int (\csc^2 u - 1)\ du \tag{32.7f}$$

$$= \int \csc^2 u\ du - \int du \tag{32.7g}$$

$$= -\cot u - u = -\frac{\cos u}{\sin u} - \sin^{-1}\left(\frac{x}{6}\right). \tag{32.7h}$$

Calculation of $\cos u$ is illustrated geometrically in figure 32.2, which shows that

$$\cos u = \sqrt{1 - \sin^2 u} = \sqrt{1 - \frac{x^2}{36}} = \frac{\sqrt{36 - x^2}}{6} \tag{32.7i}$$

Since $\sin u = x/6$,

$$\int \frac{\sqrt{36 - x^2}}{x^2}\ dx = -\frac{\sqrt{36 - x^2}}{x} - \sin^{-1}\frac{x}{6} \tag{32.7j}$$

□

Figure 32.2: The calculation of the $\cos u$ in example 32.2.

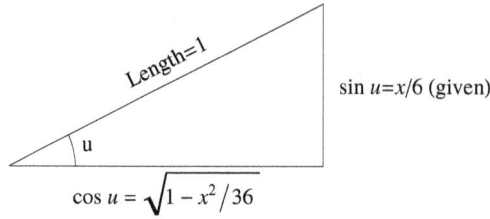

Integrals involving $x^2 - a^2$ and $\sqrt{x^2 - a^2}$

Integrals involving $x^2 - a^2$ are subtly different than integrals involving $a^2 - x^2$. *They are rarely the same thing multiplied by a -1*, especially when a square root is involved. Consider, for example, $\sqrt{9 - x^2}$ and $\sqrt{x^2 - 9}$. While $x = 3 \sin u$ will work in the first case, it will not work in the second integral because we end up with $\sqrt{-1}$ (try it out!).

Instead we will use the substitution $x = a \sec u$. Then $dx = a \sec u \tan u \, du$. Since

$$x^2 - a^2 = a^2 \sec^2 u - a^2 = a^2(\sec^2 u - 1) = a^2 \tan^2 u \tag{32.8}$$

the integral becomes something we can solve with heuristics we have previously discussed.

Figure 32.3: A triangular visualization of the substitution $x = a \sec u$ on a right triangle with unit hypotenuse. The side adjacent to angle u has length $\cos u = 1/\sec u = a/x$. By the Pythagorean theorem, the remaining side has length $\sin u = \sqrt{1 - (a/x)^2} = \sqrt{(x^2 - a^2)/x^2} = \sqrt{[(x^2 - a^2)/a^2]/(x^2/a^2)} = \sqrt{[(x^2 - a^2)/a^2]/\sec^2 u}$. Therefore $\tan^2 u = \sin^2 u / \cos^2 u = \sin^2 u \sec^2 u = (x^2 - a^2)/a^2$ (eq. 32.8).

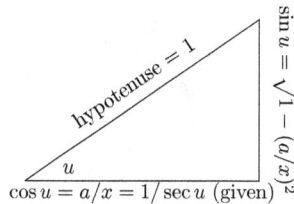

Example 32.3. Solve $\displaystyle\int \frac{dx}{\sqrt{x^2 - 16}}$.

Solution. Let $x = 4 \sec z$. Then $dx = 4 \sec z \tan z \, dz$, and therefore

$$\int \frac{dx}{\sqrt{x^2 - 16}} = \int \frac{4 \sec z \tan z \, dz}{\sqrt{16 \sec^2 z - 16}} \tag{32.9a}$$

$$= \int \frac{4 \sec z \tan z \, dz}{4\sqrt{\sec^2 z - 1}} \tag{32.9b}$$

$$= \int \frac{\sec z \tan z \, dz}{\tan z} \tag{32.9c}$$

$$= \int \sec z \, dz \tag{32.9d}$$

$$= \ln|\sec z + \tan z| \qquad (32.9\mathrm{e})$$

From the original substitution, $\sec z = x/4$; from the Pythagorean identity for the tangent ($\sec^2 z - \tan^2 z = 1$),

$$\tan z = \sqrt{\sec^2 z - 1} = \sqrt{\frac{x^2}{16} - 1} = \frac{1}{4}\sqrt{x^2 - 16} \qquad (32.9\mathrm{f})$$

Substituting into (32.9g) gives

$$\int \frac{dx}{\sqrt{x^2 - 16}} = \ln\left|\frac{x}{4} + \frac{1}{4}\sqrt{x^2 - 16}\right| = \ln\left|x + \sqrt{x^2 - 16}\right| - \ln 4 \qquad (32.9\mathrm{g})$$

Because the integral has an arbitrary constant of integration added to it, the $\ln 4$ is normally omitted (as are similar terms), and thus we would arrive at

$$\int \frac{dx}{\sqrt{x^2 - 16}} = \ln\left|x + \sqrt{x^2 - 16}\right| \qquad (32.9\mathrm{h})$$

as a simplified solution in terms of the original variable x. $\qquad\qquad\square$

Integrals involving $x^2 + a^2$ and $\sqrt{x^2 + a^2}$

For these integrals we can often get by with the substitution $x = a\tan u$. This works by combining the Pythagorean identity $\sec^2 u - \tan^2 u = 1$ with the differential $dx = a\sec^2 u\, du$. For example,

$$\sqrt{a^2 + x^2} = \sqrt{a^2 + a^2\tan^2 u} = \sqrt{a^2(1 + \tan^2 u)} = a\sec u \qquad (32.10)$$

and so we end up with integrals involving powers of tangents and secants.

Example 32.4. Find $\displaystyle\int \frac{dx}{x^2\sqrt{9 + x^2}}$.

Solution. Let $x = 3\tan u$. Then $dx = 3\sec^2 u\, du$. The integral becomes

$$\int \frac{dx}{x^2\sqrt{9 + x^2}} = \int \frac{3\sec^2 u\, du}{9\tan^2 u\sqrt{9 + 9\tan^2 u}} \qquad (32.11\mathrm{a})$$

$$= \frac{1}{3}\int \frac{\sec^2 u\, du}{\tan^2 u\sqrt{9(1 + \tan^2 u)}} \qquad (32.11\mathrm{b})$$

$$= \frac{1}{9}\int \frac{\sec^2 u\, du}{\tan^2 u \cdot \sec u} \qquad (32.11\mathrm{c})$$

$$= \frac{1}{9}\int \frac{\sec u\, du}{\tan^2 u} \qquad (32.11\mathrm{d})$$

$$= \frac{1}{9}\int \frac{1}{\cos u} \cdot \frac{\cos^2 u}{\sin^2 u}\, du \qquad (32.11\mathrm{e})$$

$$= \frac{1}{9}\int \frac{\cos u\, du}{\sin^2 u} \qquad (32.11\mathrm{f})$$

This integral can be solved with the substitution $z = \sin u$, so that $dz = \cos u\, du$.

$$\int \frac{dx}{x^2\sqrt{9 + x^2}} = \frac{1}{9}\int \frac{dz}{z^2} = -\frac{1}{9z} = -\frac{1}{9\sin u} \qquad (32.11\mathrm{g})$$

To return to the original variable x, we use the definition $x = 3 \tan u$. Thus

$$\sec^2 u = 1 + \tan^2 u = 1 + \frac{x^2}{9} = \frac{9 + x^2}{9} \tag{32.11h}$$

But

$$\cos^2 u = \frac{1}{\sec^2 u} = \frac{9}{9 + x^2} \tag{32.11i}$$

so that

$$\sin^2 u = 1 - \cos^2 u = 1 - \frac{9}{9 + x^2} = \frac{x^2}{9 + x^2} \tag{32.11j}$$

and therefore

$$\frac{1}{\sin u} = \sqrt{\frac{1}{\sin^2 u}} = \frac{\sqrt{9 + x^2}}{x} \tag{32.11k}$$

The integral is then

$$\int \frac{dx}{x^2 \sqrt{9 + x^2}} = -\frac{\sqrt{9 + x^2}}{9x} \tag{32.11l}$$

which is obtained by substituting (32.11k) into (32.11g).   □

Some integrals can be found either using a trigonometric substitution or using a different u-substitution. This is illustrated in examples 32.5 and 32.6

Example 32.5. Find $\displaystyle\int \frac{x \, dx}{\sqrt{x^2 + 16}}$ with a non-trigonometric u-substitution.

Solution. Looking for the simplest substitution, we try $u = x^2 + 16$. Then $du = 2x \, dx$ and

$$\int \frac{x dx}{\sqrt{x^2 + 16}} = \frac{1}{2} \int \frac{du}{\sqrt{u}} = \sqrt{u} = \sqrt{x^2 + 16} \tag{32.12a}$$

This illustrates how the heuristic $x = 4 \tan u$, which applies in this case, was not required to solve the integral.   □

Example 32.6. Repeat example 32.5, this time using a trigonometric substitution.

Solution. The usual heuristic for $x^2 + a^2$ is $x = 4 \tan u$. Then $dx = 4 \sec^2 u \, du$ and the integral becomes

$$\int \frac{x dx}{\sqrt{x^2 + 16}} = \int \frac{4 \tan u \cdot 4 \sec^2 u \, du}{\sqrt{16 \tan^2 u + 16}} \tag{32.13a}$$

$$= 16 \int \frac{\tan u \sec^2 u \, du}{\sqrt{16(\tan^2 u + 1)}} \tag{32.13b}$$

$$= 4 \int \frac{\tan u \sec^2 u \, du}{\sec u} \tag{32.13c}$$

$$= 4 \int \sec u \tan u \, du \tag{32.13d}$$

$$= 4 \sec u \tag{32.13e}$$

To recover the original variable,

$$16 \sec^2 u = 16(\tan^2 u + 1) = 16 \tan^2 u + 16 = x^2 + 16 \tag{32.13f}$$

Taking the square root gives $4 \sec u = \sqrt{x^2 + 16}$. Substituting this result into (32.13e), and the integral becomes

$$\int \frac{x dx}{\sqrt{x^2 + 16}} = \sqrt{x^2 + 16} \tag{32.13g}$$

which is the same answer we found in equation 32.12a. Thus it is not *wrong* to use the trigonometric substitution, just more difficult.   □

Hyperbolic Substitutions

It is also possible to use hyperbolic functions instead of trigonometric functions to solve integrals. We can make use of the hyperbolic identity

$$\cosh^2 x - \sinh^2 x = 1 \tag{32.14}$$

The substitution $x = a \sinh u$ can sometimes help simplify the expressions of the form

$$\sqrt{x^2 + a^2} = \sqrt{a^2 \sinh^2 u + a^2} = a \cosh u \tag{32.15}$$

Similarly, the substitution $x = a \cosh u$ can sometimes help simplify expressions of the form

$$\sqrt{x^2 - a^2} = \sqrt{a^2 \cosh^2 u - a^2} = a \sinh u \tag{32.16}$$

Example 32.7. Solve $\displaystyle\int x^3 \sqrt{x^2 - 2} \, dx$.

Solution. Let $x = \sqrt{2} \cdot \cosh u$; then $dx = \sqrt{2} \cdot \sinh u \, du$, and

$$\int x^3 \sqrt{x^2 - 2} \, dx = \int \overbrace{\left(\sqrt{2} \cdot \cosh u\right)^3}^{x^3} \cdot \overbrace{\left(\sqrt{2} \sinh u\right)}^{\sqrt{x^2 - 2}} \cdot \overbrace{\left(\sqrt{2} \sinh u \, du\right)}^{dx} \tag{32.17a}$$

$$= 2^{5/2} \int \cosh^3 u \sinh^2 u \, du \tag{32.17b}$$

$$= 2^{5/2} \int \cosh u \cdot \cosh^2 u \cdot \sinh^2 u \, du \tag{32.17c}$$

$$= 2^{5/2} \int \cosh u \cdot (1 + \sinh^2 u) \cdot \sinh^2 u \, du \tag{32.17d}$$

$$= 2^{5/2} \int \cosh u \cdot \sinh^2 u \, du + 2^{5/2} \int \cosh u \cdot \sinh^4 u \, du \tag{32.17e}$$

In each term substitute $z = \sinh u$; then $dz = \cosh u \, du$, and

$$\int x^3 \sqrt{x^2 - 2} \, dx = 2^{5/2} \int z^2 dz + 2^{5/2} \int z^4 dz \tag{32.17f}$$

$$= 2^{5/2} \left(\frac{1}{3} z^3 + \frac{1}{5} z^5\right) \tag{32.17g}$$

$$= 2^{5/2} \left(\frac{1}{3} \sinh^3 u + \frac{1}{5} \sinh^5 u\right) \tag{32.17h}$$

Since $\cosh u = (x/\sqrt{2})$,

$$\sinh^2 u = 1 + \cosh^2 u = 1 + \frac{x^2}{2} \tag{32.17i}$$

Substituting back into equation 32.17h,

$$\int x^3 \sqrt{x^2 - 2} \, dx = 2^{5/2} \left[\frac{1}{3} \left(1 + \frac{x^2}{2}\right)^{3/2} + \frac{1}{5} \left(1 + \frac{x^2}{2}\right)^{5/2}\right] \tag{32.17j}$$

$$= \frac{2}{3} \left(2 + x^2\right)^{3/2} + \frac{1}{5} \left(2 + x^2\right)^{5/2} \tag{32.17k}$$

$\square$

Completing the Squares Before a Substitution

Sometimes it is not possible to make a substitution into one of the standard forms right away, but a substitution can be made after completing the squares. The process usually requires multiple substitutions: (1) complete the squares; (2) make a u-substitution; (3) make a trig-substitution; (4) integrate; (4) invert the trig-substitution to get the u variable; (5) invert the u value to get the original variable.

Example 32.8. Find $\displaystyle\int \frac{x\,dx}{\sqrt{5+4x-x^2}}$.

Solution. First, complete the squares in the argument of the square root:

$$5 + 4x - x^2 = 5 - (x^2 - 4x) \tag{32.18a}$$

$$= 5 - (x^2 - 4x + 4) + 4 \tag{32.18b}$$

$$= 9 - (x-2)^2 \tag{32.18c}$$

Then

$$\int \frac{x\,dx}{\sqrt{5+4x-x^2}} = \int \frac{x\,dx}{\sqrt{9-(x-2)^2}} \tag{32.18d}$$

Let $u = x - 2$, so that $du = dx$.

$$\int \frac{x\,dx}{\sqrt{5+4x-x^2}} = \int \frac{(u+2)du}{\sqrt{9-u^2}} = \int \frac{u\,du}{\sqrt{9-u^2}} + 2\int \frac{du}{\sqrt{9-u^2}} \tag{32.18e}$$

Each of these integrals can be solved with the trigonometric substitution $u = 3\sin\theta$ ($du = 3\cos\theta\,d\theta$). Furthermore, since

$$\sqrt{9-u^2} = \sqrt{9-9\sin^2\theta} = 3\sqrt{1-\sin^2\theta} = 3\cos\theta \tag{32.18f}$$

the integral becomes

$$\int \frac{x\,dx}{\sqrt{5+4x-x^2}} = \int \frac{3\sin\theta \cdot 3\cos\theta\,d\theta}{3\cos\theta} + 2\int \frac{3\cos\theta\,d\theta}{3\cos\theta} \tag{32.18g}$$

$$= 3\int \sin\theta\,d\theta + 2\int d\theta = -3\cos\theta + 2\theta \tag{32.18h}$$

Since $\sin\theta = u/3$, and from (32.18f), $\cos\theta = \left(\sqrt{9-u^2}\right)/3$,

$$\int \frac{x\,dx}{\sqrt{5+4x-x^2}} = -\sqrt{9-u^2} + 2\sin^{-1}\frac{u}{3} \tag{32.18i}$$

To get the original x variable we back-substitute $u = x - 2$,

$$\int \frac{x\,dx}{\sqrt{5+4x-x^2}} = -\sqrt{9-(x-2)^2} + 2\sin^{-1}\left(\frac{x-2}{3}\right) \tag{32.18j}$$

$$= -\sqrt{5+4x-x^2} + 2\sin^{-1}\left(\frac{x-2}{3}\right) \tag{32.18k}$$

$\square$

Exercises

1. $\displaystyle\int \frac{\sqrt{16x^2 - 9}}{x}\, dx$

2. $\displaystyle\int \frac{dx}{x^4\sqrt{25 - x^2}}$

3. $\displaystyle\int \frac{dx}{x^2\sqrt{x^2 - 4}}$

4. $\displaystyle\int e^{2x}\sqrt{1 + e^x}\, dx$

5. $\displaystyle\int \frac{dx}{e^x\sqrt{e^{2x} - 16}}$

6. $\displaystyle\int \frac{x^5}{(x^2 + 49)^{3/2}}\, dx$

7. $\displaystyle\int \frac{x}{\sqrt{12 - 9x^2 - 4x}}\, dx$

8. $\displaystyle\int \frac{dx}{\sqrt{x^2 - 4x}}$

9. $\displaystyle\int_0^1 \frac{dx}{(1 + x^2)^2}$

10. $\displaystyle\int_0^1 \sqrt{36 - 25x^2}\, dx$

11. $\displaystyle\int \frac{2^x}{4^x + 4}\, dx$

Answers to Selected Exercises

1. $\sqrt{16x^2 - 9} + 3\tan^{-1}\left(\dfrac{3}{\sqrt{16x^2 - 9}}\right)$

2. $-\dfrac{\sqrt{25 - x^2}\left(2x^2 + 25\right)}{1875x^3}$

3. $\dfrac{\sqrt{x^2 - 4}}{4x}$

4. $\dfrac{2}{15}\sqrt{e^x + 1}\left(e^x + 3e^{2x} - 2\right)$

5. $\dfrac{1}{16}e^{-x}\sqrt{e^{2x} - 16}$

6. $\dfrac{x^4 - 196x^2 - 19208}{3\sqrt{x^2 + 49}}$

7. $\dfrac{2}{27}\sin^{-1}\left(\dfrac{-9x - 2}{4\sqrt{7}}\right) - \dfrac{1}{9}\sqrt{-9x^2 - 4x + 12}$

8. $\dfrac{2\sqrt{x - 4}\sqrt{x}\log\left(\sqrt{x - 4} + \sqrt{x}\right)}{\sqrt{(x - 4)x}}$

9. $\dfrac{2 + \pi}{8}$

10. $\dfrac{\sqrt{11}}{2} + \dfrac{18}{5}\sin^{-1}\left(\dfrac{5}{6}\right)$

11. $-\dfrac{\tan^{-1}\left(2^{1-x}\right)}{2\log(2)}$

Chapter 33

Integrating Rational Functions

Chapter Summary and Goal

Rational functions – quotients of polynomials – can usually be decomposed into sums of simpler rational functions that can be easily integrated. If the degree of the numerator is larger than the degree of the denominator, we will use **polynomial division** to replace the integrand with the sum of a polynomial and a remainder. If the degree of the denominator is larger than the degree of the numerator, we can use the **method of partial fractions** to simplify the integrand. We will discuss each of these methods in this chapter.

Student Learning Objectives

The student will:

1. Learn to recognize when rational division and partial fractions can be applied.
2. Learn to integrate with rational division.
3. Learn to perform a partial fraction decomposition of a rational function.
4. Learn to integrate with partial fractions.

Rational Functions

Definition 33.1. Rational Function

A **rational function** $r(x)$ is the quotient of two polynomials $p(x)$ and $q(x)$

$$r(x) = \frac{p(x)}{q(x)} = \frac{a_0 + a_1 x + a_2 x^2 + \cdots + a_n x^n}{b_0 + b_1 x + b_2 x^2 + \cdots + b_m x^m} \tag{33.1}$$

where $a_0, a_1, \ldots, a_n$ and $b_0, b_1, \ldots, b_m$ are known constants.
The **degree of the numerator**, $\deg(p(x))$ is the power of the largest exponent in the numerator, e.g., the number n in equation 33.1.
The **degree of the denominator**, $\deg(q(x))$, is the power of the largest exponent in the denominator, e.g, the number m in equation 33.1.

An integral of a rational function will have the form

$$\int \frac{a_0 + a_1 x + a_2 x^2 + \cdots + a_n x^n}{b_0 + b_1 x + b_2 x^2 + \cdots + b_m x^m} \, dx \tag{33.2}$$

where $a_0, a_1, \ldots, a_n$ and $b_0, b_1, \ldots, b_m$ are given constants. The approach used to solve equation 33.2 will depend on the values of m and n. If $n \geq m$ then we will use **polynomial division.** If $n < m$, we will factor the denominator and use the **method of partial fractions.**

Polynomial Division

When the degree of the numerator is greater than or equal to the degree of the denominator we can do polynomial division. The technique is similar to long division that you learned in grammar school, with the place-holders (e.g., the ones, tens, hundreds, thousands, etc.) replaced by terms in the polynomials (e.g., the constant, x terms, x^2 term, x^3 term, etc).

Example 33.1. Divide $\dfrac{x^2 + 2x + 2}{2x + 1}$ using polynomial long division.

Solution. Our goal is find the following quotient:

$$2x + 1 \overline{)\ x^2 + 2x + 2} \tag{33.3a}$$

Looking at the highest power terms, we find an $2x$ in the divisor and an x^2 in the dividend. Since x goes into x^2 a total of $x/2$ times, we write $x/2$ above the vinculum[1]

$$2x + 1 \overline{)\ \overset{\frac{1}{2}x}{x^2 + 2x + 2}} \tag{33.3b}$$

We then multiply the $x/2$ times the negative of the divisor, $-2x - 1$, and write the result beneath the dividend.

$$2x + 1 \overline{)\begin{array}{l} \overset{\frac{1}{2}x}{x^2 + 2x + 2} \\ -x^2 - \frac{1}{2}x \end{array}} \tag{33.3c}$$

Subtracting the two bottom lines term by term,

$$2x + 1 \overline{)\begin{array}{l} \overset{\frac{1}{2}x}{x^2 + 2x + 2} \\ \underline{-x^2 - \frac{1}{2}x} \\ \frac{3}{2}x + 2 \end{array}} \tag{33.3d}$$

Now the process is repeated. The highest term in the divisors is $2x$, and the highest term in the dividend is $3x/2$. Dividing, $(\frac{3}{2}x)/(2x) = 3/4$. This is the next term in the quotient. We write $3/4$ above the viniculum, copy $-(3/4) \times (2x + 1)$ to the bottom, and subtract

$$2x + 1 \overline{)\begin{array}{l} \overset{\frac{1}{2}x + \frac{3}{4}}{x^2 + 2x + 2} \\ \underline{-x^2 - \frac{1}{2}x} \\ \frac{3}{2}x + 2 \\ \underline{-\frac{3}{2}x - \frac{3}{4}} \\ \frac{5}{4} \end{array}} \tag{33.3e}$$

---

[1]The *long division symbol* is, literally, a right parenthesis ")" followed by a *vinculum*, which is a horizontal line above the dividend, as in divisor)dividend. There is no word for "long division symbol" in English (or any other language) that I know of.

No further division can be done at this point because the number we found in the last step, 5/4, has degree 0, which is smaller than the degree of the divisor. This number is the **remainder**. Thus the quotient is

$$\frac{x^2 + 2x + 2}{2x + 1} = \frac{x}{2} + \frac{3}{4} + \text{ with remainder } \frac{5}{4} \tag{33.3f}$$

The remainder itself is a fraction of the divisor, so we can rewrite the result like this:

$$\frac{x^2 + 2x + 2}{2x + 1} = \frac{x}{2} + \frac{3}{4} + \frac{5}{4(2x + 1)} \tag{33.3g}$$

You should verify that this decomposition is correct by collecting the fractions on the right hand side of the equation over a common denominator. □

Heuristic 33.1. Integrating Rational Functions: Long Division

To solve

$$\int \frac{a_0 + a_1 x + a_2 x^2 + \cdots + a_n x^n}{b_0 + b_1 x + b_2 x^2 + \cdots + b_m x^m} \, dx \tag{33.4}$$

where $n \geq m$, simplify the integrand using long division and then integrate term by term.

Example 33.2. Find $\dfrac{x^2 + 2x + 2}{2x + 1} \, dx$.

Solution. From equation 33.3g,

$$\int \frac{x^2 + 2x + 2}{2x + 1} \, dx = \int \frac{x}{2} \, dx + \int \frac{3}{4} \, dx + \int \frac{5}{4(2x + 1)} \, dx \tag{33.5a}$$

$$= \frac{1}{4}x^2 + \frac{3}{4}x + \frac{5}{8} \ln|8x + 4|. \tag{33.5b}$$

□

Example 33.3. Solve $\displaystyle\int \frac{x^3 + x^2}{x - 1} \, dx$.

Solution. The numerator has degree 3 and the denominator has degree, so long division is called for.

$$
\begin{array}{r}
x^2 + 2x + 2 \\
x - 1{\overline{\smash{\big)}\,}} \; x^3 + x^2 \\
\underline{-\,x^3 + x^2} \\
2x^2 \\
\underline{-\,2x^2 + 2x} \\
2x \\
\underline{-\,2x + 2} \\
2
\end{array}
\tag{33.6a}
$$

The integral then becomes

$$\int \frac{x^3 + x^2}{x - 1} \, dx = \int (x^3 + 2x + 2) \, dx + \int \frac{2}{x - 1} \, dx \tag{33.6b}$$

$$= \frac{1}{4}x^4 + x^2 + 2 + 2\ln|x - 1| \tag{33.6c}$$

□

Example 33.4. Integrate $\displaystyle\int \frac{10x^2 + 2x + 1}{5x^2 - x + 5}\, dx$.

Solution. The numerator and denominator have the same degree. Using polynomial division,

$$
\begin{array}{r}
2 \\
5x^2 - x + 5 \overline{)\ 10x^2 + 2x\ + 1} \\
-\ 10x^2 + 2x - 10 \\
\hline
4x\ - 9
\end{array}
\tag{33.7a}
$$

Therefore

$$
\int \frac{10x^2 + 2x + 1}{5x^2 - x + 5}\, dx = \int 2\, dx + \int \frac{4x - 9}{5x^2 - x + 5}\, dx
\tag{33.7b}
$$

The denominator in the second term has no real roots, but we can complete the square:

$$
5x^2 - x + 5 = 5\left(x^2 - \frac{x}{5}\right) + 5 = 5\left(x^2 - \frac{x}{5} + \frac{1}{100}\right) + 5 - \frac{5}{100}
\tag{33.7c}
$$

$$
= 5\left[\left(x - \frac{1}{10}\right)^2 + \frac{99}{100}\right]
\tag{33.7d}
$$

Substituting back into the integral,

$$
\int \frac{10x^2 + 2x + 1}{5x^2 - x + 5}\, dx = 2x + \frac{4}{5}\int \frac{x\,dx}{\left(x - \frac{1}{10}\right)^2 + \frac{99}{100}} - \frac{9}{5}\int \frac{dx}{\left(x - \frac{1}{10}\right)^2 + \frac{99}{100}}
\tag{33.7e}
$$

If we let $u = x - 1/10$ and $a = \dfrac{\sqrt{99}}{10}$,

$$
\int \frac{10x^2 + 2x + 1}{5x^2 - x + 5}\, dx = 2x + \frac{4}{5}\int \frac{(u + \frac{1}{10})du}{u^2 + a^2} - \frac{9}{5}\int \frac{du}{u^2 + a^2}
\tag{33.7f}
$$

$$
= 2x + \frac{4}{5}\int \frac{u\,du}{u^2 + a^2} - \frac{84}{50}\int \frac{du}{u^2 + a^2}
\tag{33.7g}
$$

$$
= 2x + \frac{4}{5}\cdot\frac{1}{2}\ln|u^2 + a^2| - \frac{43}{25}\cdot\frac{10}{\sqrt{99}}\tan^{-1} u
\tag{33.7h}
$$

$$
= 2x + \frac{2}{5}\ln\left|\left(x - \frac{1}{10}\right)^2 + \frac{99}{100}\right| - \frac{86}{15\sqrt{11}}\tan^{-1}\left(\frac{x - 1/10}{\sqrt{99}/10}\right)
\tag{33.7i}
$$

$$
= 2x + \frac{2}{5}\ln\left|5x^2 - x + 5\right| - \frac{86}{15\sqrt{11}}\tan^{-1}\left(\frac{10x - 1}{3\sqrt{11}}\right)
\tag{33.7j}
$$

$\square$

Why does long division work? The unique factorization theorem for polynomials is an application of the factorization theorem for integers.

Theorem 33.1. Factorization Theorem for Integers

If n is any integer and n is any positive integer, then there exist unique integers q (called the **quotient**) and r (called the **remainder**) such that

$$
\frac{n}{d} = q + \frac{r}{d}, \text{ or equivalently } n = d \times q + r
\tag{33.8}
$$

where $0 \le r < d$.

Thus when you divide 11 by 4 you get 2 with a remainder of 3:

$$\frac{11}{4} = 2 + \frac{3}{4} \text{ or equivalently } 11 = 4 \times 2 + 3 \tag{33.9}$$

Theorem 33.2. Unique Factorization Theorem

Let $N(x)$ and $D(x)$ be polynomials with $\deg(D(x)) \leq \deg(N(x))$. Then there exist unique polynomials $Q(x)$ (the **quotient**) and $R(x)$ (the **remainder**) such that

$$\frac{N(x)}{D(x)} = Q(x) + \frac{R(x)}{Q(x)}, \text{ or equivalently } N(x) = D(x) \times Q(x) + R(x) \tag{33.10}$$

where $\deg(R(x)) < \deg(D(x))$

The algorithm for finding the quotient, and the remainder, is an extension of Euclid's algorithm for find the greatest common divisor (GCD) of two integers.[2]

Heuristic 33.2. Euclidean Algorithm for Polynomial Division

Let $\ell(p(x))$ be a function that returns the coefficient of the highest degree term in the polynomial $p(x)$.

> **Input**: Polynomials $N(x)$, $D(x)$, where $D(x) \neq 0$
> **Initialize**:[a] $Q(x) \leftarrow 0$, $R(x) \leftarrow N(x)$, $d \leftarrow \deg(D(x))$, $c \leftarrow \ell(D(x))$
> **while** $\deg(R(x)) \geq d$ **do**:
>> Calculate the next term and write it above the viniculum:
>> $$S(x) \leftarrow \frac{\ell(R(x))}{c} x^{\deg(R(x))-d}$$
>>
>> Update the current value $Q(x)$ by adding term just calculated:
>> $$Q(x) \leftarrow Q(x) + S(x)$$
>>
>> Update the remainder calculation at the bottom:
>> $$R(x) \leftarrow R(x) - S(x)D(x)$$
>
> **Output**: Quotient $Q(x)$, Remainder $R(x)$.

---
[a]In algorithms we use the notation $x \leftarrow y$ to mean "the symbol x now represents value equal to the quantity y. The arrow is used because one may think of x as a computer memory location that we are copying the value of y into after it is computed.

---

[2]GCD(m, n) is the largest integer that divides into both m and n with zero remainder.

The Fundamental Theorem of Algebra and Partial Fractions

When the degree of the denominator is larger than the degree of the numerator, polynomial division will not help. Instead, we must factor the denominator. Factorization is possible because of the **fundamental theorem of algebra**,[3] which says that every polynomial with real coefficents can be factored into a product of linear factors (factors like $x - r$, where r is a root) and quadratic factors that do not have roots (factors like $Ax^2 + Bx + C$ where $B^2 - 4AC < 0$).

Theorem 33.3. Fundamental Theorem of Algebra

Let

$$p(x) = a_0 + a_1 x + a_2 x^2 + \cdots + a_n x^n \qquad (33.11)$$

be a polynomial of degree n with real coefficients $a_0, a_2, \ldots, a_n$. Then $p(x)$ may be factored into linear and quadratic terms as

$$p(x) = (x - r_1)^{j_1}(x - r_2)^{j_2} \cdots (x - r_M)^{j_M} \times$$
$$(A_1 x^2 + B_1 x + C_1)^{k_1}(A_1 x^2 + B_2 x + C_2)^{k_2} \cdots (A_n x^2 + B_n x + C_n)^{k_N} \qquad (33.12)$$

for some integers j_M (possibly zero) and k_N (possibly zero)[a], where $r_1, r_2, \ldots, r_M$ are real roots with multiplicities $j_1, j_2, \ldots, j_M$; $A_1, B_1, C_1, \ldots, A_n, B_n, C_n$ are real constants; $k_1, k_2, \ldots, k_N$ are the multiplicities of the quadratic terms; and

$$\underbrace{j_1 + j_2 + \cdots + j_M}_{\text{number of real roots}} + 2 \underbrace{(k_1 + k_2 + \cdots + k_N)}_{\text{number of quadratic factors}} = n \qquad (33.13)$$

such that each quadratic $A_k x^2 + B_k x + C_k$ has no real root.

---

[a]At least one of j_M and k_M will be non-zero.

This helps us because it means we can factor complicated denominators:

$$\frac{1}{q(x)} = \frac{1}{(x - r_1)^{j_1} \cdots (x - r_M)^{j_M}(A_1 x^2 + B_1 x + C_1)^{k_1} \cdots (A_n x^2 + B_n x + C_n)^{k_N}} \qquad (33.14)$$

In a **partial fractions decomposition**, we re-write this as a sum of inverse quotients of every single term in the denominator, *taking every single power into account*. Thus we posit that:

$$\begin{aligned}
\frac{p(x)}{q(x)} = {} & \frac{a_{11}}{x - r_1} + \frac{a_{12}}{(x - r_1)^2} + \cdots + \frac{a_{1,j_1}}{(x - r_1)^{j_1}} \\
& + \cdots \\
& + \frac{a_{M,1}}{x - r_M} + \frac{a_{M,2}}{(x - r_M)^2} + \cdots + \frac{a_{M,j_M}}{(x - r_M)^{j_M}} \\
& + \frac{b_{1,1}x + c_{1,1}}{(A_1 x^2 + B_1 x + C_1)} + \cdots + \frac{b_{1,k_1}x + c_{1,k_1}}{(A_1 x^2 + B_1 x + C_1)^{k_1}} \\
& + \cdots \\
& + \frac{b_{n,1}x + c_{n,1}}{(A_n x^2 + B_n x + C_n)} + \cdots + \frac{b_{n,k_n}x + c_{n,k_n}}{(A_n x^2 + B_n x + C_n)^{k_n}}
\end{aligned} \qquad (33.15)$$

---

[3]The Fundamental Theorem of Algebra was originally proposed (but not proven) by Albert Girard in 1629 and later by Rene Descartes in 1639. An erroneous proof was published by Jean le Rond d'Alembert in 1746. (This is the same d'Alembert who is well known for his contributions to mechanics and wave theory.) Carl Friedrich Gauss proved it for polynomials up through degree 6 in 1799. Gauss generalized his proof to all polynomials with real coefficients in 1816, and all polynomials with complex coefficients in 1849. It was Gauss who actually discovered the errors in d'Alembert's earlier proof. The version stated here is weaker than the most general theorem because we are not interested in complex coefficients.

where the coefficients in the numerator are unknowns that we must determine. The proof that the decomposition given by equation 33.15 depends on some results from linear algebra whose proof[4] are beyond the scope of this text:

1. The set S of functions formed by the terms on the right-hand side of equation 33.15 is **linearly independent**.[5]
2. The set S has **dimension**[6] n.
3. The set S is a **basis**[7] for the set of polynomials of degree n.

Because of equation 33.13, there are exactly n unknown constants $a_{11}, a_{12}, \ldots, c_{n,k_n}$. The algorithm for determining these coefficients is given in heuristic 33.3.

Heuristic 33.3. Partial Fractions Decomposition

To solve for the coefficients in 33.15,

1. Cross multiply to put all the terms over a common denominator.
2. Since the common denominator must equal $q(x)$, the numerators on both sides of the equation must be equal.
3. Set the two numerators equal and forget about the denominators. This give a very large equation that depends on the n unknown coefficients and x.
4. Do one of the following.
   (a) Collect all the terms in the equation into a single polynomial $K_1 + K_2 x + K_3 x^2 + \cdots + K_n x^n = 0$. Set the coefficient of each power of x equal to zero. This will give n equations in the n unknown constants; or
   (b) Pick n different values of x, and evaluate the equation at each of these n values. This will also give n equations in n unknowns.
5. Solve the system of n equations for the n unknowns.

The question of how to set up equation 33.15 is sometimes quite daunting. Due to theorem 33.3, however, there are only four possibilities: every factor is either linear, quadratic, linear raised to a power, or quadratic raised to a power. We summarized these cases, and how to handle them, in heuristic 33.4.

---

[4] Advanced students may find an accessible proof by William T. Bradley and William T. Cook (2012) *Two Proofs of the Existence and Uniqueness of the Partial Fraction Decomposition.* **International Mathematical Forum, 7**(31): 1517-1535. For an introduction to Linear Algebra, see Gilbert Strang (2006) **Linear Algebra and its Applications**, Brooks Cole.

[5] A set of functions $\{f_1, f_2, \ldots, f_n\}$ is called linearly independent if there is no set of constants $a_1, a_2, \ldots, a_n$ (not all zero) such that $a_1 f_1 + a_2 f_2 + \cdots + a_n f_n = 0$.

[6] The dimension of a vector space is the number of elements in a basis.

[7] A basis of a vector space $\mathcal{V}$ is a subset $S \subset \mathcal{V}$ such that every element v in $\mathcal{V}$ can be written as linear combinations of basis elements, i.e., $v = c_1 s_1 + c_2 s_2 + \cdots + c_n s_n$, where $c_1, c_2, \ldots$ are constants and $s_1, s_2, \cdots \in S$.

Heuristic 33.4. Partial Fractions Setup

Factor in Denominator	Terms to Include in Partial Fraction Expansion
$(x - r)$	$\dfrac{A}{x - r}$
$(x - r)^n$, $n > 1$, integer	$\dfrac{A_1}{x - r} + \dfrac{A_2}{(x - r)^2} + \cdots + \dfrac{A_n}{(x - r)^n}$
$(ax^2 + bx + c)$, with $b^2 < 4ac$	$\dfrac{Bx + C}{ax^2 + bx + c}$
$(ax^2 + bx + c)^n$, $n > 1$, $b^2 < 4ac$	$\dfrac{B_1x + C_1}{ax^2 + bx + c} + \dfrac{B_2x + C_2}{(ax^2 + bx + c)^2} + \cdots + \dfrac{B_nx + C_n}{(ax^2 + bx + c)^n}$

Denominator is a Product of Simple Linear Factors

In the simplest situations, the denominator can be factored into a product of non-repeated linear terms. If $\deg(p) < \deg(q)$ and $r_1, r_2, \ldots$ are real constants,[8] then the partial fractions decomposition is

$$\frac{p(x)}{q(x)} = \frac{p(x)}{(x - r_1)(x - r_2)(x - r_3) \cdots} = \frac{A}{x - r_1} + \frac{B}{x - r_2} + \frac{C}{x - r_3} + \cdots \tag{33.16}$$

for some numbers $A, B, C, \ldots$.

Example 33.5. Find $\displaystyle\int \frac{5x + 1}{(x + 1)(x + 2)} \, dx$.

Solution.   The integrand is a rational function. The degree of the numerator is 1 and the degree of the denominator is 2, so since $1 < 2$, we can use partial fractions. The denominator is a product of simple linear factors, so we can write the integrand as

$$\frac{5x + 1}{(x + 1)(x + 2)} = \frac{A}{x + 1} + \frac{B}{x + 2} \tag{33.17a}$$

Cross-multiplying the fraction on the right-hand side,

$$\frac{5x + 1}{(x + 1)(x + 2)} = \frac{A(x + 2) + B(x + 1)}{(x + 1)(x + 2)} \tag{33.17b}$$

Examining this equation we see that the denominators are identical. For the equation to be true, the numerators must also be equal to one another.

$$5x + 1 = A(x + 2) + B(x + 1) \tag{33.17c}$$

Equation 33.17c has two unknown constants, A and B, that we want to find, and depends on a variable x. We can solve for the two unknowns in either of two ways.

Method 1 to find A, B: Since (33.17c) must hold for all values of x, it will be true for any particular value we plug in for x . We use this fact to "pick" arbitrary values of x that will give us two equations in the two two unknowns A and B.

The general heuristic is to look for values of x that make either A or B drop out of

---

[8] $r_1, r_2, \ldots$ are the roots of the denominator.

the equation. For example, when $x = -1$, the term multiplying B is zero, and the equation will only depend on A:

$$5(-1) + 1 = A(-1 + 2) + B(-1 + 1) \tag{33.17d}$$

$$-4 = A \tag{33.17e}$$

Similarly, when $x = -2$, the term multiplying A is zero, and the equation will only depend on B.

$$5(-2) + 1 = A(-2 + 2) + B(-2 + 1) \tag{33.17f}$$

$$-9 = -B \tag{33.17g}$$

Therefore $A = 4$ and $B = 9$.

Method 2 to find A, B: An alternative method is to equate the coefficients of like powers of x in equation 33.17c.[9] Collecting terms in powers of x in (33.17c)

$$0 = (A + B - 5)x + (2A + B - 1) \tag{33.17h}$$

Because this must hold for all x, the coefficient of every power of x must be zero. Thus Equating the coefficients of x on both sides of (33.17c) gives

$$A + B = 5 \tag{33.17i}$$

$$2A + B = 1 \tag{33.17j}$$

Subtracting equation 33.17i from 33.17j gives $A = -4$. Substituting $A = -4$ into 33.17j gives $B = 9$.

Solving the Integral: Using $A = -4$ and $B = 9$ in our partial fraction decomposition eq. 33.17a converts the integral as to

$$\int \frac{5x + 1}{(x + 1)(x + 2)} \, dx = -4 \int \frac{dx}{x + 1} + 9 \int \frac{dx}{x + 2} \tag{33.17k}$$

$$= -4 \ln |x + 1| + 9 \ln |x + 2| \tag{33.17l}$$

$\square$

---

[9]This works because the set of polynomials $1, x, x^2, \ldots$ are linearly independent.

Denominators with Repeated Linear Factors

A factor like $(x - r)^k$ is called **repeated linear factors**, because we can think of is as really being $(x - r)$ repeated k times:

$$(x - r)^k = \underbrace{(x - r)(x - r) \cdots (x - r)}_{\text{same simple factor repeated k times}} \tag{33.18}$$

For each repeated factor in the denominator, we have to consider the possibility that any possible power of $(x - r)$, $(x - r)^2$, $(x - r)^3$, ..., $(x - r)^k$ may contribute to the partial fraction expansion, because each of these factors are really in the repeated factor.

Thus for **each factor** of $(x - r)^k$, we include the following terms in the expansion

$$\frac{A_1}{(x - r)} + \frac{A_2}{(x - r)^2} + \frac{A_3}{(x - r)^3} + \cdots + \frac{A_k}{(x - r)^k} \tag{33.19}$$

These terms are then added to whatever other terms we have in the partial fraction expansion.

Example 33.6. Find $\displaystyle\int \frac{5x + 1}{(x + 1)(x + 2)^2}\, dx$.

Solution. The integrand is a rational function with numerator of degree 1 and denominator degree 3. Since $1 < 3$, the method of partial fractions may be used. There is one simple linear factor $(x + 1)$, and one linear factor that is repeated $(x + 2)^2$. Therefore the decomposition is

$$\frac{5x + 1}{(x + 1)(x + 2)^2} = \frac{A}{x + 1} + \overbrace{\frac{B}{x + 2} + \frac{C}{(x + 2)^2}}^{\text{repeated linear factor}} \tag{33.20a}$$

Cross-multiplying and equating numerators,

$$5x + 1 = A(x + 2)^2 + B(x + 1)(x + 2) + C(x + 1) \tag{33.20b}$$

When $x = -1$, the second and third terms go to zero:

$$5(-1) + 1 = A(-1 + 2)^2 \implies A = -4 \tag{33.20c}$$

When $x = -2$, the first and second terms become zero:

$$5(-2) + 1 = C(-2 + 1) \implies C = 9 \tag{33.20d}$$

To find B, we substitute the values $A = -4$ (from eq. 33.20c) and $C = 9$ (eq. 33.20d) into equation 33.20b.

$$5x + 1 = -4(x + 2)^2 + B(x + 1)(x + 2) + 9(x + 1) \tag{33.20e}$$

Since this equation must hold for all values of x we are free to choose any value of x we have not already chosen in order to find B; the simplest choice is $x = 0/$

$$1 = -4(2^2) + B(1)(2) + 9(1) = -7 + 2B \implies B = 4 \tag{33.20f}$$

Substituting the values for A, B and C into (33.20a) and integrating,

$$\int \frac{5x + 1}{(x + 1)(x + 2)^2}\, dx = -4 \int \frac{dx}{x + 1} + 4 \int \frac{dx}{x + 2} + 9 \int \frac{dx}{(x + 2)^2} \tag{33.20g}$$

$$= -4 \ln |x + 1| + 4 \ln |x + 2| - \frac{9}{x + 2} \tag{33.20h}$$

$$= 4 \ln \left| \frac{x + 2}{x + 1} \right| - \frac{9}{x + 2} \tag{33.20i}$$

$\square$

Denominators with Quadratic Factors

Factors of the form $ax^2 + bx + c$, where $a \neq 0$ and $b^2 < 4ac$ are called **simple quadratic factors**. The condition $b^2 - 4ac$ means that these expressions do not have real roots. The expressions are **simple** because they are not repeated (raised to a power). The restriction $a \neq 0$ ensures that these expressions are not simple linear factors. For every simple quadratic factor in the denominator, we will need to include a term of the form

$$\frac{Ax + B}{ax^2 + bx + c} \tag{33.21}$$

in the partial fraction expansion. If there are multiple simple quadratic factors multiplied times one another, we must include one term for each one. If there are also linear or repeated linear factors in the denominator, we must account for those using the methods we have already discussed.

Example 33.7. Find $\displaystyle\int \frac{5x + 1}{(x + 1)(x^2 + 2)} \, dx$.

Solution. The numerator has degree 1 and the denominator has degree $3 > 1$, so partial fractions can be used. The denominator has one simple linear factor and one simple quadratic factor.

$$\frac{5x + 1}{(x + 1)(x^2 + 2)} = \overbrace{\frac{A}{x + 1}}^{\text{for simple linear factor}} + \underbrace{\frac{Bx + C}{x^2 + 2}}_{\text{for simple quadratic factor}} \tag{33.22a}$$

Cross multiplying and equating the numerators gives

$$5x + 1 = A(x^2 + 2) + (Bx + C)(x + 1) \tag{33.22b}$$

Setting $x = -1$ gives us

$$-4 = 3A \implies A = -\frac{4}{3} \tag{33.22c}$$

Setting $x = 0$ and substituting $A = -4/3$ gives

$$1 = -\frac{4}{3} \cdot (0 + 2) + (0 + C)(0 + 1) = -\frac{8}{3} + C \implies C = \frac{11}{3} \tag{33.22d}$$

Substituting this back into the equation (33.22b) gives

$$5x + 1 = -\frac{4}{3}(x^2 + 2) + Bx(x + 1) + \frac{11}{3}(x + 1) \tag{33.22e}$$

To find B we may select any x we have not previously used, such as $x = 1$.

$$6 = -\frac{4}{3}(3) + B(1)(1 + 1) + \frac{11}{3}(1 + 1) = -4 + 2B + \frac{22}{3} \tag{33.22f}$$

$$B = \frac{4}{3} \tag{33.22g}$$

Hence

$$\int \frac{5x + 1}{(x + 1)(x^2 + 2)} \, dx = -\frac{4}{3} \int \frac{dx}{x + 1} + \frac{4}{3} \int \frac{x \, dx}{x^2 + 2} + \frac{11}{3} \int \frac{dx}{x^2 + 2} \tag{33.22h}$$

$$= -\frac{4}{3} \ln|x + 1| + \frac{2}{3} \ln|x^2 + 2| + \frac{11}{3\sqrt{2}} \tan^{-1} \frac{x}{\sqrt{2}} \tag{33.22i}$$

$\square$

Denominators with Repeated Quadratic Factors

A factor such as $(ax^2 + bx + c)^k$, where $a \neq 0$ and $b^2 < 4ac$ is called a **repeated quadratic factor**, it is as really being $(ax^2 + bx + c)$ repeated k times:

$$(ax^2 + bx + c)^k = \underbrace{(ax^2 + bx + c)(ax^2 + bx + c) \cdots (ax^2 + bx + c)}_{\text{same simple factor repeated k times}} \qquad (33.23)$$

For each repeated factor in the denominator, we have to consider the possibility that any possible power of $(ax^2 + bx + c)$, $(ax^2 + bx + c)^2$, $(ax^2 + bx + c)^3$, ..., $(ax^2 + bx + c)^k$ may contribute to the partial fraction expansion, because each of these factors are really in the repeated factor.

Thus for **each factor** of $(ax^2 + bx + c)^k$, we include the following terms in the expansion

$$\frac{B_1 x + C_1}{(ax^2 + bx + c)} + \frac{B_2 x + C_2}{(ax^2 + bx + c)^2} + \frac{B_3 x + C_3}{(ax^2 + bx + c)^3} + \cdots + \frac{B_k x + C_k}{(ax^2 + bx + c)^k} \qquad (33.24)$$

If there are simple or repeated linear factors, or other simple quadratic factors in the denominator, terms for them must also be added to this expansion as previously discussed.

Example 33.8. Find $\displaystyle\int \frac{dx}{x(x^2 + 1)^2}$.

Solution. The integrand is a rational function with degree 0 in the numerator and degree 5 in the denominator. The denominator is the product of a simple linear factor and a quadratic factor squared. We can decompose it as

$$\frac{1}{x(x^2 + 1)^2} = \frac{A}{x} + \overbrace{\frac{Bx + C}{x^2 + 1} + \frac{Dx + E}{(x^2 + 1)^2}}^{\text{for repeated quadratic factor}} \qquad (33.25a)$$

Combining terms over a common denominator,

$$\frac{1}{x(x^2 + 1)^2} = \frac{A \cdot (x^2 + 1)^2 + x \cdot (Bx + C) \cdot (x^2 + 1) + x \cdot (Dx + E)}{x(x^2 + 1)^2} \qquad (33.25b)$$

Equating numerators,

$$1 = A(x^2 + 1)^2 + x(x^2 + 1)(Bx + C) + x(Dx + E) \qquad (33.25c)$$

The only easy substitution (that kills off a lot of terms) is $x = 0$, which gives us $A = 1$. This reduces equation 33.25d to

$$1 = (x^2 + 1)^2 + x(x^2 + 1)(Bx + C) + x(Dx + E) \qquad (33.25d)$$

To find the other coefficients we substitute x at four other values and get four equations in the four unknowns B, C, D, E. Using $x = 1, -1, 2$, and 2, in order, gives the equations

$$-3 = 2B + 2C + D + E \qquad (33.25e)$$
$$-3 = 2B - 2C + D - E \qquad (33.25f)$$
$$-24 = 20B + 10C + 4D + 2E \qquad (33.25g)$$
$$-24 = 20B - 10C + 4D - 2E \qquad (33.25h)$$

Adding equations 33.25e and 33.25f gives

$$-6 = 4B + 2D \qquad (33.25i)$$

Dividing by 2,

$$-3 = 2B + D \tag{33.25j}$$

Adding equations 33.25g and 33.25h gives

$$-48 = 40B + 8D \tag{33.25k}$$

Dividing by 8 gives

$$-6 = 5B + D \tag{33.25l}$$

Subtraction equation 33.25j from 33.25l gives

$$3B = -3 \implies B = -1 \tag{33.25m}$$

Substitution back into equation 33.25l gives $D = -1$.

Substituting these results back into either set of the original four equations leads to $C = E = 0$. Thus

$$\frac{1}{x(x^2+1)^2} = \frac{1}{x} - \frac{x}{x^2+1} - \frac{x}{(x^2+1)^2} \tag{33.25n}$$

The integral becomes

$$\int \frac{1}{x(x^2+1)^2} = \int \frac{dx}{x} - \int \frac{dx}{x^2+1} - \int \frac{x\,dx}{(x^2+1)^2} \tag{33.25o}$$

$$= \ln|x| - \tan^{-1} x - \frac{1}{2}\int \frac{du}{u^2} \tag{33.25p}$$

where $u = 1 + x^2$ in the last term. Since

$$\int \frac{du}{u^2} = -\frac{1}{u} = -\frac{1}{1+x^2} \tag{33.25q}$$

we end up with

$$\int \frac{1}{x(x^2+1)^2} = \ln|x| - \tan^{-1} x + \frac{(1/2)}{1+x^2} \tag{33.25r}$$

$\square$

Example 33.9. Find $\displaystyle\int \frac{dx}{x^3(x+1)}$.

Solution. The degree of the denominator is 4, which is higher than the degree of the numerator, so partial fractions will work. The factor of x^3 in the denominator is not a cubic but is a simple linear factor repeated three times as $(x+0) \cdot (x+0) \cdot (x+0)$. Thus we only have a product of linear factors, and not a quadratic factor.

represents repeated linear factor x

$$\frac{1}{x^3(x+1)} = \overbrace{\frac{A}{x} + \frac{B}{x^2} + \frac{C}{x^3}} + \frac{D}{x+1} \tag{33.26a}$$

$$= \frac{Ax^2(x+1) + Bx(x+1) + C(x+1) + Dx^3}{x^3(x+1)} \tag{33.26b}$$

Equating numerators,

$$1 = Ax^2(x+1) + Bx(x+1) + C(x+1) + Dx^3 \tag{33.26c}$$

Setting $x = 0$ gives $C = 1$.

Setting $x = -1$ gives $D = -1$.

Setting $x = 1$ gives $2A + 2B = 0$ or $A = -B$

Setting $x = 2$ gives $12A + 6B - 6 = 0$. Substituting $A = -B$ gives $6A = 6$ or $A = 1$. Back-substituting with A gives $B = -1$.

The partial fractions decomposition is thus

$$\frac{1}{x^3(x+1)} = \frac{1}{x} - \frac{1}{x^2} + \frac{1}{x^3} - \frac{1}{x+1} \qquad \text{(33.26d)}$$

Integrating,

$$\int \frac{dx}{x^3(x+1)} = \int \frac{dx}{x} - \int \frac{dx}{x^2} + \int \frac{dx}{x^3} - \int \frac{dx}{x+1} \qquad \text{(33.26e)}$$

$$= \ln|x| + \frac{1}{x} - \frac{1}{2x^2} - \ln|x+1| \qquad \text{(33.26f)}$$

$\square$

Exercises

Use the method of partial fractions to solve each of the of the following integrals.

1. $\displaystyle\int \frac{6+x}{(x+1)(x-3)}\, dx$

2. $\displaystyle\int \frac{2}{x(3x+4)(2x+7)}\, dx$

3. $\displaystyle\int \frac{x+1}{x(25x+36)}\, dx$

4. $\displaystyle\int \frac{2x+1}{x^2(x+8)}\, dx$

5. $\displaystyle\int \frac{dx}{x^2+16x+64}$

6. $\displaystyle\int \frac{x+1}{(x+3)^2(x+8)}\, dx$

7. $\displaystyle\int \frac{x+1}{(x+3)^2(x^2+2)}\, dx$

8. $\displaystyle\int \frac{x+1}{(x+2)^2(x+3)^2}\, dx$

9. $\displaystyle\int \frac{x+1}{(x^2+4x+5)(x+3)}\, dx$

10. $\displaystyle\int \frac{x^2}{(2x^2+3)(4x^2+5)}\, dx$

11. $\displaystyle\int \frac{x^3}{(x^2+16)(x^2+25)^2}\, dx$

12. $\displaystyle\int \frac{x^3-2}{x^2+3x+2}\, dx$

13. $\displaystyle\int \frac{3x-2}{x^2-4x+20}\, dx$

14. $\displaystyle\int_{-3}^{3} \frac{dx}{(x+9)(x^2+16)}$

Answers to Selected Exercises

1. $\dfrac{9}{4}\ln(x-3) - \dfrac{5}{4}\ln(x+1)$

2. $\dfrac{\ln(x)}{14} + \dfrac{4}{91}\ln(2x+7) - \dfrac{6}{52}\ln(3x+4)$

3. $\dfrac{\ln(x)}{36} + \dfrac{11}{900}\ln(25x+36)$

4. $-\dfrac{1}{8x} + \dfrac{15}{64}\ln\dfrac{x}{x+8}$

5. $-\dfrac{1}{x+8}$

6. $\dfrac{2}{5(x+3)} + \dfrac{7}{25}\ln\dfrac{x+3}{x+8}$

7. $\dfrac{2}{11(x+3)} - \dfrac{1}{121}\ln(x+3) +$
$\dfrac{19\tan^{-1}\left(\frac{x}{\sqrt{2}}\right)}{121\sqrt{2}} + \dfrac{1}{242}\ln(2+x^2)$

8. $\dfrac{1}{x+2} + \dfrac{2}{x+3} + 3\ln\dfrac{x+2}{x+3}$

9. $\dfrac{1}{2}(\ln(x(x+4)+5) - 2\ln(x+3))$

10. $\dfrac{\sqrt{6}}{4}\tan^{-1}\left(\sqrt{\dfrac{2}{3}}x\right) - \dfrac{\sqrt{5}}{4}\tan^{-1}\left(\dfrac{2x}{\sqrt{5}}\right)$

11. $-\dfrac{25}{18\left(x^2+25\right)} + \dfrac{8}{81}\ln\dfrac{x^2+25}{x^2+16}$

12. $\dfrac{x^2}{2} - 3x - 3\ln(x+1) + 10\ln(x+2)$

13. $\dfrac{3}{2}\ln\left(x^2-4x+20\right) + \tan^{-1}\left(\dfrac{x-2}{4}\right)$

14. $\dfrac{1}{388}\left[\ln(16) + 18\tan^{-1}\dfrac{3}{4}\right]$

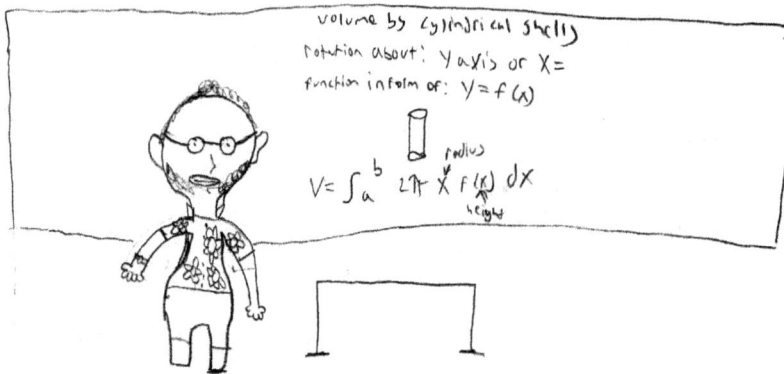

volume by cylindrical shells
rotation about: y axis or x =
function in form of: $y = f(x)$

$$V = \int_a^b 2\pi \, x \, f(x) \, dx$$

radius
height

Chapter 34

Numerical Integration

Chapter Summary and Goal

There are many problems in the sciences and engineering where we need to find a numerical solution but are faced with an intractable integral. Examples include finding areas and volumes; solving for stresses and strains; and calculating probability density. The statistical error function, for example, is given by

$$\text{erf}(x) = \frac{2}{\sqrt{\pi}} \int_0^x e^{-t^2} dt \tag{34.1}$$

but the integral

$$\int e^{-t^2} dt \tag{34.2}$$

can not be solved analytically by any known technique. Thus a "brute force" approach of solving the integral and then plugging in the value will not work in these situations. The alternative method of **numerical integration** will be the focus of this chapter.

Student Learning Objectives

The student will:

1. Understand the difference between analytic and numerical representations of functions.
2. Understand the concept of numerical integration.
3. Understand the relationship between step size and computation error.
4. Understand that different techniques have different inherent computational errors.
5. Be able to compute integrals numerically using various techniques such as the midpoint rule, trapezoidal rule, and Simpson's method.

Numerical Representation of Functions

The simplest methods for numerical integration are based on adding up the areas of rectangles that fill the region under the curve. These are similar to the Riemann sum, but subtly different. When we found the Riemann sum, we had an analytic representation of the function, that is, a formula.

Definition 34.1. Analytic Representation of a Function

An analytic representation of a function is a formula $y = f(x)$ that gives the value of $f(x)$ at a point x in the domain.

When we represent a function numerically (e.g., in a computer), we do not use an analytic representation.[1] Instead, we represent the function numerically, by evaluating it at a sequence of points

$$a = x_0 < x_1 < x_2 < \cdots < x_n = b \tag{34.3}$$

In general, the points can be irregularly spaced. In a more advanced discussion, we might want to have the points closer together in regions where the function has high curvature (a lot of squiggles) and have very few points in regions of low curvature (where the function is very nearly a straight line). For our purposes, it will be sufficient to make the points all equally spaced. Then we will define

$$h = \frac{b - a}{n} \tag{34.4}$$

and therefore

$$x_0 = a \tag{34.5}$$
$$x_1 = a + h \tag{34.6}$$
$$x_2 = a + 2h \tag{34.7}$$
$$\vdots \tag{34.8}$$
$$x_n = a + nh = b \tag{34.9}$$

in general

$$x_i = a + ih, \quad i = 0, \ldots, n \tag{34.10}$$

The **numerical representation** of a function on an interval $f(x)$ is then the collection of points $(x_i, f(x_i))$, for $i = 0$ to n.

Definition 34.2. Numerical Representation of a Function

Let $f(x)$ be defined on an interval $[a, b]$. Then its **numerical representation** on $[a, b]$ with **step size** $h = (b - a)/n$ is given by the set of points

$$\{(x_0, f(x_0)), (x_1, f(x_1)), (x_2, f(x_2)), \ldots, (x_n, f(x_n))\} \tag{34.11}$$

where $x_j = a + jh$ (so that $a = x_0$ and $b = x_n$). Notationally, we denote $f_k = f(x_k)$ so that the representation becomes

$$\{(x_0, f_0), (x_1, f_1), (x_2, f_2), \ldots, (x, f_n)\} \tag{34.12}$$

In order to actually calculate an integral numerically, we will try to derive some inspiration from the analysis we performed in deriving the Riemann Sum. There are some slight differences, but the concepts remain the same. When we calculated the Riemann Sum, we *defined* the defined

---

[1]In the separate discipline of **symbolic computation** there are symbolic representations, but these are slightly different. We will assume that there is no symbolic representation available in this chapter. Computer programs like *Mathematica* use a blending of symbolic and numerical representation. Integration functions like scipy.integrate in python, integral in *Matlab*, and libraries like quadpack and the Gnu Scientific Library GSL all incorporate numerical integration.

the definite integral as (see eq. 24.6)

$$\int_a^b f(x)dx = \lim_{n \to \infty} \sum_{i=1}^n hf(x_i^*) \tag{34.13}$$

where the x_i and h are defined as are is in definition 34.2, and and the points x_i^* all satisfy

$$x_i \le x_i^* \le x_{i+1} \tag{34.14}$$

This worked because each term gave the area of a small rectangle

$$dA_i = (\text{width}) \times (\text{height}) = hf(x_i^*) = f_i^* \tag{34.15}$$

where in the last term we are following the notation use in definition 34.2 of writing $f(x_i^*) = f_i^*$ just like we write $f(x_i) = f_i$.

In a numerical *approximation*, we must terminate the sum at some finite value of n. So one possible numerical approximation of the integral is

$$\int_a^b f(x)dx \approx \sum_{i=i}^n dA_i = \sum_{i=1}^n hf_i^* \tag{34.16}$$

If we always place x_i^* at the left-endpoint of the interval, so that $x_i^* = x_i, i = 0, \ldots, n-1$, we get the **upper-left-hand-corner approximation**. In this approximation, the height of each rectangle is calculated at its top left hand corner, and the integration formula is

$$\int_a^b f(x)dx \approx h(f_0 + f_1 + \cdots + f_{n-1}) \tag{34.17}$$

If we place x_i^* at the right-endpoint of the interval, so that $x_i^* = x_{i+1}, i = 0, \ldots, n-1$, we get the **upper-right-hand-corner approximation**. In this approximation, the height of each rectangle is calculated at its top right hand corner, and the integration formula is

$$\int_a^b f(x)dx \approx h(f_1 + f_2 + \cdots + f_n) \tag{34.18}$$

If we try to place x_i^* in the middle of the each interval (call this point $\overline{x_i}$), then

$$\overline{x}_i = \frac{x_{i-1} + x_i}{2} \tag{34.19}$$

But if we blindly plug this into equation 34.16 we obtain

$$\int_a^b f(x)dx \approx \sum_{i=0}^n hf(\overline{x}_i) \tag{34.20}$$

which is a problem, because we don't know $f(\overline{x_i})$. This is one place where the analogy with Riemann Sums does not carry over without some modification. We need to take into account the fact that we only know f at the following points:

$$(x_0, f_0), (x_1, f_1), (x_2, f_2), \ldots, (x_n, f_n) \tag{34.21}$$

To get something analogous to the midpoint rule (which is what happens with Riemann sums), we double the step size! This will work only if n is even. Then

The midpoint of $[0, 2h]$ is h

The midpoint of $[2h, 4h]$ is $3h$

The midpoint of $[4h, 6h]$ is $5h$

$\vdots$

The midpoint of $[(n-2)h, nh]$ is $(n-1)h$

and the width of each one of these intervals is $2h$. Thus the integral can be approximated by

$$\int_a^b f(x)\, dx \approx 2h(f_1 + f_3 + f_5 + \cdots f_{n-1}) \tag{34.22}$$

This gives the **midpoint rule**, where the height of each rectangle is measured at the middle of an interval, but the interval is twice is wide as the step size at which the function is known.

Figure 34.1: Approximation of $\int x^2 e^{-x}$ with the midpoint rule for $n = 8$ (left, 4 boxes) and $n = 20$ (right, 10 boxes). See example 34.1.

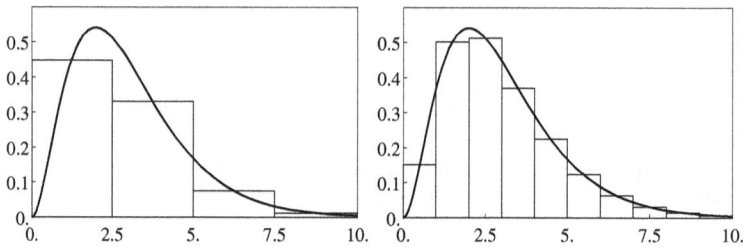

Example 34.1. Find $\displaystyle\int_0^{10} x^2 e^{-x}\, dx$ numerically with the midpoint rule, first using $n = 8$ and then using $n = 20$. Compare your result with the exact integral.

Solution. The two approximations are illustrated in figure 34.1. The exact solution is

$$\int_0^{10} x^2 e^{-x}\, dx = e^{-x}\left[-2 - 2x - x^2\right]_0^{10} = 2 - 122e^{-10} \approx 1.99446 \tag{34.23a}$$

For $\underline{n = 8}$ we have $h = (10 - 0)/8 = 1.25$. From the midpoint formula,

$$\int_0^{10} x^2 e^{-x}\, dx = 2h(f_1 + f_3 + f_5 + f_7) \tag{34.23b}$$

Values are calculated in the following table.

x_0	x_1	x_2	x_3	x_4	x_5	x_6	x_7	x_8
0.	1.25	2.5	3.75	5.	6.25	7.5	8.75	10.
	f_1		f_3		f_5		f_7	
	0.44766		0.33072		0.07541		0.01232	

Therefore

$$\int_0^{10} x^2 e^{-x}\, dx \approx 2(0.125)(0.44766 + 0.33072 + 0.07541 + 0.012132) \approx 2.165 \tag{34.23c}$$

Doing Calculus

For $n = 20$ we have $h = (20 - 0)/20 = 0.5$, so that

$$x_0 = 0, \ x_1 = 0.5, \ x_2 = 1, \ \ldots, \ x_{20} = 10 \qquad (34.23\text{d})$$

The midpoints are at $x_1, x_3, x_5, \ldots$. The integral approximation is

$$\int_0^{10} x^2 e^{-x} \, dx \approx 2h(f_1 + f_3 + f_5 + \cdots + f_{19}) \qquad (34.23\text{e})$$

Since $2h = 2(0.5) = 1$,

$$\int_0^{10} x^2 e^{-x} \, dx \approx f(0.5) + f(1.5) + \cdots + f(9.5) \qquad (34.23\text{f})$$

$$\approx 0.15163 + 0.50204 + 0.51303 + 0.36992$$
$$+ \, 0.22496 + 0.12363 + 0.06352$$
$$+ \, 0.03111 + 0.01470 + 0.00676 \qquad (34.23\text{g})$$
$$\approx 2.001 \qquad (34.23\text{h})$$

In summary, for $n = 8$, the integral is ≈ 2.165; for $n = 20$ the integral is ≈ 2.001; and the exact answer is ≈ 1.995. $\qquad\square$

Example 34.1 illustrates the typical situation with numerical integration: by decreasing the step size (h) we can get closer to the correct answer at the expense of more calculation. Two common measures of error are the **relative error** and the **percentage error**.

Definition 34.3. Relative Error

$$\text{Relative Error} = \left| \frac{\text{Actual Value - Approximate Value}}{\text{Actual Value}} \right| \qquad (34.24)$$

Definition 34.4. Percentage Error

$$\text{Percentage Error} = 100 \times \left| \frac{\text{Actual Value - Approximate Value}}{\text{Actual Value}} \right| \qquad (34.25)$$

The relative error measures the error as a fraction of the expected value, while the percentage error calculates the the same number as a percentage of the expected value

Example 34.2. Calculate the relative and percentage errors in example 34.1.

Solution. The exact value of the integral was ≈ 1.995.
For $n = 8$ we calculated a value of 2.165, giving an error of

$$\text{relative error} \approx \frac{2.165 - 1.995}{1.995} \approx 0.086 \approx 8.6\% \qquad (34.26\text{a})$$

For $n = 20$ we calculated a value of 2.001 giving an error of

$$\text{relative error} \approx \frac{2.001 - 1.995}{1.995} \approx 0.036 \approx 3.6\% \qquad (34.26\text{b})$$

$$\square$$

Figure 34.2: Percentage error versus number of intervals for the integral in example 34.1.

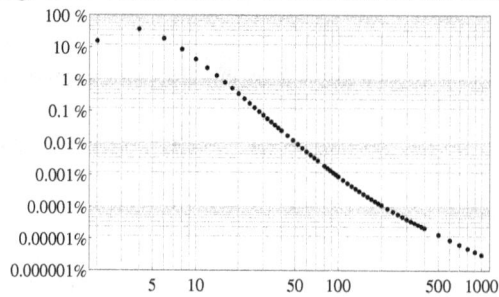

In general the error will decrease with step step (h). Since this is equivalent to increasing the number n of intervals, it means that to get better accuracy we need to do a lot more work. Figure 34.2 illustrates this in a numerical experiment in which example 34.1 was repeated a large number of times on a computer, increasing the number of intervals up to a 1000.

Method 34.1. Rectangular Approximations for Integration

Let $x_0 = a$, $h = (b-a)/n$, $x_j = a + ih$, $f_i = f(x_i)$. The fixed step size approximations for the integral using rectangles are

$$\int_a^b f(x)dx \approx \begin{cases} h(f_0 + f_1 + \cdots + f_{n-1}), & \text{left endpoint rule} \\ h(f_1 + f_2 + \cdots + f_n), & \text{right endpoint rule} \\ 2h(f_1 + f_3 + \cdots + f_{n-1}) & \text{midpoint rule, } n \text{ must be even} \end{cases} \qquad (34.27)$$

Trapezoidal Rule

Examining figures 34.2 and 34.3a it would seem that we can do better than rectangles. For example, we could just connect the points (x_i, f_i) on the curve with straight lines and find the areas of the individual trapezoids (fig. 34.3b). The area of a trapezoid is

$$\text{area} = \frac{(\text{base})_1 + (\text{base})_2}{2} \times (\text{height}) \tag{34.28}$$

Let the points at the base of the trapezoid dA_i be x_i and x_{i+1}. The two "bases" of the

Figure 34.3: Concept of the Trapezoidal rule. The error we make on the left seems to disappear on the right when we approximate the top of each vertical bar by connecting the two top corners.

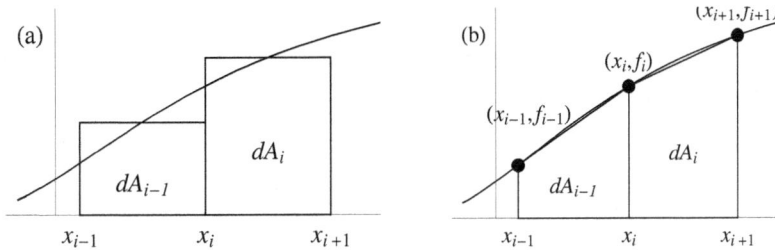

trapezoids are $f(x_i)$ and $f(x_{i+1})$ and the "height" of the trapezoid is the step size $h = x_{i+1} - x_i$. So the area of the ith trapezoid is

$$dA_i = \frac{1}{2}h(f_i + f_{i+1}) \tag{34.29}$$

If we number the left-most trapezoid as dA_0 (between x_0 and x_1) and the rightmost as dA_{n-1} (between x_{n-1} and x_n), then

$$\int_a^b f(x)dx \approx \sum_{i=0}^{n-1} dA_i = \sum_{i=0}^{n-1} \frac{h}{2}(f_i + f_{i+1}) \tag{34.30}$$

$$= \frac{h}{2}\big[\ \overbrace{(f_0}^{\text{once}} + \overbrace{f_1) + (f_1}^{\text{twice}} + \overbrace{f_2) + (f_2}^{\text{twice}} + \cdots + \underbrace{f_{n-1}) + (f_{n-1}}_{\text{twice}} + \underbrace{f_n)}_{\text{once}}\ \big] \tag{34.31}$$

This result is known as the **trapezoidal rule**.

Method 34.2. Trapezoidal Rule

Let $x_0 = a$, $h = (b-a)/n$, $x_j = a + ih$, and $f_i = f(x_i)$

$$\int_a^b f(x)dx \approx \frac{h}{2}[f_0 + 2f_1 + 2f_2 + \cdots + 2f_{n-1} + f_n] \tag{34.32}$$

For computational simplicity, the trapezoidal rule is often re-written in the following manner:

$$\int_a^b f(x)dx \approx \frac{h}{2}[f_0 + 2f_1 + 2f_2 + \cdots + 2f_{n-1} + \overbrace{f_n + f_0 - f_0 + f_n - f_n}^{\text{add zero}}] \tag{34.33}$$

$$\approx \frac{h}{2}[2f_0 + 2f_1 + \cdots + 2f_n - f_0 - f_n] \tag{34.34}$$

$$\approx \frac{h}{2}[2f_0 + 2f_1 + \cdots + 2f_n] - \frac{h}{2}(f_0 + f_n) \tag{34.35}$$

$$= h\sum_{i=0}^{n} f_i - \frac{h(f_0 + f_n)}{2} \tag{34.36}$$

Method 34.3. Trapezoidal Rule (Alternate Form)

Let $x_0 = a$, $h = (b-a)/n$, $x_j = a + ih$, and $f_i = f(x_i)$

$$\int_a^b f(x)dx \approx \left(h\sum_{i=0}^{n} f_i \right) - \frac{h(f_0 + f_n)}{2} \tag{34.37}$$

Example 34.3. Repeat example 34.1 using the Trapezoidal rule with $h = 1.25$.

Solution. Since $h = 2.5$, we have $n = (b-a)/h = 8$. This gives us the following data:

x_0	x_1	x_2	x_3	x_4	x_5	x_6	x_7	x_8
0.	1.25	2.5	3.75	5.	6.25	7.5	8.75	10.

$f_0,$	f_1	f_2	f_3	f_4	f_5	f_6	f_7	f_8
0	0.44766	0.51303	0.33072	0.16845	0.07541	0.03111	0.01232	0.00454

From equation 34.37,

$$\int_0^{10} f(x)dx = h\sum_{i=0}^{8} f_i - \frac{h(f_0 + f_8)}{2} \tag{34.38a}$$

$$\approx 1.25(0.4477 + 0.5130 + 0.3307 + 0.1684 + 0.0754 + 0.0311 +$$
$$0.0121 + 0.0045) - (0.5)(1.25)(0 + 0.0045) \tag{34.38b}$$

$$\approx 1.9758 \tag{34.38c}$$

The relative error is

$$\text{error} = \frac{1.976 - 1.995}{1.995} = 0.95\% \tag{34.38d}$$

which is significantly better than the midpoint method error of 8.6% for the same step size.   □

Simpson's Rule

With the trapezoidal rule, we approximated the curve by a sequence of line segments. Another idea is to approximate it by a sequence of parabolic arcs. This helps because parabolas have some more curvature than lines, so in principle that should be better at approximating a curve. Furthermore, since we know how to integrate parabolas we should be able to easily derive an approximation formula for the integral.

Derivation of Simpson's Rule.[2] Suppose that

$$p(x) = A(x - x_i)^2 + B(x - x_{i+1}) + C \tag{34.39}$$

Then

$$\int_{x_i}^{x_{i+2}} p(x)dx = \int_{x_i}^{x_{i+2}} [A(x - x_i)^2 + B(x - x_{i+1}) + C]dx \tag{34.40}$$

$$= A\int_{x_i}^{x_{i+2}} (x - x_i)^2 dx + B\int_{x_i}^{x_{i+2}} (x - x_i)dx + C\int_{x_i}^{x_{i+2}} dx \tag{34.41}$$

$$= \frac{A}{3}(x - x_i)^3 \Big|_{x_i}^{x_{i+2}} + B(x - x_{i+1})^2 \Big|_{x_i}^{x_{i+2}} + C(x - x_i) \Big|_{x_i}^{x_{i+2}} \tag{34.42}$$

$$= \frac{A}{3}\left[(x_{i+2} - x_i)^3 - (x_i - x_i)^3\right] + B\left[(x_{i+2} - x_{i+1})^2 - (x_i - x_{i+1})^2\right]$$
$$+ C\left[(x_{i+2} - x_i) - (x_i - x_i)\right] \tag{34.43}$$

$$= \frac{A}{3}\left[(2h)^3 - 0^3\right] + B\left[h^2 - (-h)^2\right] + C\left[2h - 0\right] \tag{34.44}$$

Therefore the factor multiplying B disappears and the integral becomes

$$\int_{x_i}^{x_{i+2}} p(x)dx = \frac{8Ah^3}{3} + 2Ch \tag{34.45}$$

So if we can find a parabola given by (34.39), then the area under the first two sub-intervals is given by (34.45). The integral over a larger interval can be found by subdividing it into an even number of intervals and summing the individual sub-integrals. N

The next step is to derive formulas for A and C. We want to find the equation of a parabola (34.39) that passes through the three points (x_i, f_i), (x_{i+1}, f_{i+1}), and (x_{i+2}, f_{i+2}). Since

$$x_{i+1} = x_i + h \quad \text{and} \quad x_{i+2} = x_{i+1} + h \tag{34.46}$$

we must have

$$f_i = p(x_i) = B(x_i - x_{i+1}) + C = -Bh + C \tag{34.47}$$

$$f_{i+1} = A(x_{i+1} - x_i)^2 + C = Ah^2 + C \tag{34.48}$$

$$f_{i+2} = A(x_{i+2} - x_i)^2 + B(x_2 - x_1) + C = 4Ah^2 + Bh + C \tag{34.49}$$

Equations 34.47 through 34.49 give three equations in three unknowns, A, B, and C. We only care about the values of two of them, A, and C (because B does not appear in equation 34.45), so we want to eliminate B from the system. To do this we can solve for Bh in the (34.47)

$$Bh = C - f_i \tag{34.50}$$

---

[2]This paragraph can be skipped on a first reading without loss of continuity. Go directly to method 34.4 if you must.

and then substitute it into (34.49). The remaining two equations are:

$$f_{i+1} = Ah^2 + C \tag{34.51}$$

$$f_{i+2} = 4Ah^2 + 2C - f_i \tag{34.52}$$

Multiplying (34.51) by 2 and rearranging,

$$2f_{i+1} = 2Ah^2 + 2C \tag{34.53}$$

$$f_{i+2} - f_i = 4Ah^2 + 2C \tag{34.54}$$

Subtracting (34.53) from (34.54),

$$f_{i+2} - 2f_{i+1} + f_i = 2Ah^2 \tag{34.55}$$

Substituting this result back into (34.53)

$$2C = 2f_{i+1} - 2Ah^2 \tag{34.56}$$

$$= 2f_{i+1} - (f_{i+2} - 2f_{i+1} + f_i) \tag{34.57}$$

$$= -f_{i+2} + 4f_{i+1} - f_i \tag{34.58}$$

Using (34.55) for $2Ah^2$ and (34.58) for $2C$ in (34.45) gives

$$\int_{x_i}^{x_{i+2}} p(x)dx = \frac{h}{3}\left(4Ah^2 + 3 \cdot 2C\right) \tag{34.59}$$

$$= \frac{h}{3}\left(4(f_{i+2} - 2f_{i+1} + f_i) + 3(-f_{i+2} + 4f_{i+1} - f_i)\right) \tag{34.60}$$

$$= \frac{h}{3}\left(f_{i+2} + 4f_{i+1} + f_i\right) \tag{34.61}$$

If there are an even number of steps in the interval $[a, b]$ such that $a = x_0 < x_1 < \cdots < x_n = b$, where n is even, then we can approximate the function on each interval in the same manner:

$$\int_a^b f(x)dx \approx \int_{x_0}^{x_2} p_0(x)dx + \int_{x_2}^{x_4} p_2(x)dx + \cdots + \int_{x_{n-2}}^{x_n} p_{n-2}(x)dx \tag{34.62}$$

$$= \frac{h}{3}[(f_0 + 4f_1 + \overset{2}{\overbrace{f_2) + (f_2}} + \overset{4}{\overbrace{4f_3}} + \overset{2}{\overbrace{f_4) + (f_4}} + \overset{4}{\overbrace{4f_5}} + \overset{2}{\overbrace{f_6) + \cdots}}$$

$$\cdots + \underbrace{(f_{n-4} + \underbrace{4f_{n-3}}_{4} + \underbrace{f_{n-2}) + (f_{n-2}}_{2} + \underbrace{4f_{n-1}}_{4} + f_n)]}_{} \tag{34.63}$$

Noticing the $\{1, 4, 2, 4, ..., 2, 4, 1\}$ coefficient pattern we obtain Simpson's Rule.

Method 34.4. Simpson's Rule

Let $x_0 = a$, $h = (b-a)/n$, where $x_j = a + ih$, and $f_i = f(x_i)$

$$\int_a^b f(x)dx \approx \frac{h}{3}[f_0 + 4f_1 + 2f_2 + 4f_3 + \cdots + 2f_{n-2} + 4f_{n-1} + f_n] \tag{34.64}$$

Example 34.4. Repeat example 34.1 using Simpson's rule with $n = 8$.

Solution. Since $n = 4$ then $h = (b - a)/n = 1.25$. The function values we need to use are the same as those already tabulated in example 34.3. The approximation formula is

$$\int_0^{10} e^{-x} x^2 dx \approx \frac{h}{3}[f_0 + 4f_1 + 2f_2 + 4f_3 + 2f_4 + 4f_5 + 2f_6 + 4f_7 + f_8] \tag{34.65a}$$

$$\approx \frac{0.125}{3} \times [0 + 4 \cdot 0.4476 + 2 \cdot 0.5130 + 4 \cdot 0.3307 + 2 \cdot 0.1684 +$$

$$4 \cdot 0.0754 + 2 \cdot 0.0311 + 4 \cdot 0.0121 + 0.0045] \tag{34.65b}$$

$$\approx 2.039 \tag{34.65c}$$

The relative error for Simpson's method is

$$\text{error} = \frac{2.039 - 1.995}{1.995} = 2.21\% \tag{34.65d}$$

which is better than the midpoint rule with the same step size but worse than the trapezoidal rule. In most cases Simpson's rule is much better. It seems that this is not one of those cases. Can you figure why this might be? □

Math people are crazy people.

Exercises

In exercises 1 and 2, (a) draw a picture illustrating the curve fit; (b) write down the formula for approximating the area under the curve; (c) calculate the area numerically; and (d) compare with the exact value of the integral.

1. $\displaystyle\int_0^\pi \sin x \, dx$ using $n = 2$ for (i) Midpoint; (ii) Trapezoidal; (iii) Simpson's Rule.

2. $\displaystyle\int_0^\pi /2 \cos x \, dx$ usin $n = 2$ for (i) Midpoint; (ii) Trapezoidal; (iii) Simpson's Rule.

In exercise 3 through 11, estimate the integral using (a) the midpoint rule; (b) the trapezoidal rule; and (c) Simpson's rule for n=8 and n=20. Calculate the relative and percentage error in each case by comparing the exact integral given.

3. $\displaystyle\int_3^6 x^2 \, dx = 63$

4. $\displaystyle\int_{15}^{20.} (x-3) \, dx = 72.5$

5. $\displaystyle\int_{3.}^6 x e^{-x} \, dx \approx 0.181797$

6. $\displaystyle\int_1^2 \frac{e^{-x}}{x} \, dx \approx 0.170483$

7. $\displaystyle\int_0^{1.} e^{-x^2} \, dx \approx 0.746824$

8. $\displaystyle\int_0^{1.} e^{x^2} \, dx \approx 16.4526$

9. $\displaystyle\int_{-10}^{10} \frac{1}{e^{-x}+1} \, dx \approx 10$

10. $\displaystyle\int_{-1}^1 \frac{1}{x^2+1.} \, dx \approx 1.5708$

11. $\displaystyle\int_0^{4.} \frac{x^2}{x^4+1} \, dx \approx 0.860916$

Chapter 35

Arc Length

Chapter Summary and Goal

In this chapter we will calculate the length of curves. The length of a curve is called its **arc length**. The process is similar to what we have done previously. We will approximate the curve by plotting points on the curve and then connecting them by line segments. We measure the length of each line segment using the distance formula, sum these lengths to give us an expression for the arc length that depends on the number of segments, and then take the limit as $n \to \infty$. The result, as before, is a Riemann Sum, and thus we obtain a formula for the arc length as a definite integral.

Student Learning Objectives

The student will:

1. Understand the relationship between arc length and the definite integral.
2. Understand the derivation of the arc length formula.
3. Learn how to calculate the arc length of a function $f(x)$ on an interval $[a, b]$.

A Formula for Arc Length

Let $f(x)$ be defined and integrable on an interval $[a, b]$, and define the numbers

$$a = x_0 \leq x_1 \leq x_2 \leq \cdots \leq x_n = b \tag{35.1}$$

as we did in chapter 34. let

$$y_0 = f(x_0), \ y_1 = f(x_1), \ \ldots, y_n = f(x_n) \tag{35.2}$$

Then we have defined $n + 1$ points on the curve of $f(x)$,

$$P_0 = (x_0, y_0) \tag{35.3}$$
$$P_1 = (x_1, y_1) \tag{35.4}$$
$$\cdots \tag{35.5}$$
$$P_n = (x_n, y_n) \tag{35.6}$$

as illustrated in figure 35.1.

Figure 35.1: To measure the arc length we approximate the curve by finite line segments and then take the limit as the number of line segments $\to \infty$.

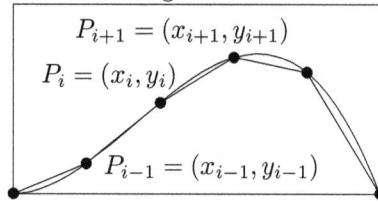

The sequence of n line segments $\overline{P_0 P_1}$, $\overline{P_1 P_2}$, ..., $\overline{P_{n-1} P_n}$ roughly approximates the shape of the curve of $f(x)$. Let us denote the length of the segment $\overline{P_{i-1} P_i}$ by $|P_{i-1} P_i|$. Then by the distance formula (definition 20.1),

$$|P_{i-1} P_i| = \text{distance from } P_{i-1} \text{ to } P_i \tag{35.7}$$

$$= \text{distance from } (x_{i-1}, y_{i-1}) \text{ to } (x_i, y_i) \tag{35.8}$$

$$= \sqrt{(x_i - x_{i-1})^2 + (y_i - y_{i-1})^2} \tag{35.9}$$

$$= \sqrt{(\Delta x_i)^2 + (\Delta y_i)^2} \tag{35.10}$$

where

$$\Delta x_i = x_i - x_{i-1} \tag{35.11}$$

$$\Delta y_i = y_i - y_{i-1} \tag{35.12}$$

Define $\overline{L}$ to be the total length of all the line segments. This is

$$\overline{L} = \sum_{i=i}^{n} |P_{i-1} P_i| = \sum_{i=i}^{n} \sqrt{(\Delta x_i)^2 + (\Delta y_i)^2} \tag{35.13}$$

$$= \sum_{i=i}^{n} \sqrt{(\Delta x_i)^2} \sqrt{\frac{(\Delta x_i)^2 + (\Delta y_i)^2}{(\Delta x_i)^2}} \tag{35.14}$$

$$= \sum_{i=i}^{n} \Delta x_i \sqrt{1 + \frac{(\Delta y_i)^2}{(\Delta x_i)^2}} \tag{35.15}$$

$$= \sum_{i=i}^{n} \Delta x_i \sqrt{1 + \left(\frac{\Delta y_i}{\Delta x_i}\right)^2} \tag{35.16}$$

If we now let $\Delta x_i \to 0$ (and hence $n \to \infty$), we get a Riemann Sum. We **define** the **arc length** L by the limit

$$L = \lim_{n \to \infty} \overline{L} = \lim_{n \to \infty} \Delta x_i \sqrt{1 + \left(\frac{\Delta y_i}{\Delta x_i}\right)^2} = \int_a^b \sqrt{1 + \left(\frac{dy}{dx}\right)^2} \, dx \tag{35.17}$$

This gives us theorem 35.1.

Theorem 35.1. Arc Length Formula

Let $f(x)$ be integrable on (a, b). Then the **arc length** L of f on (a, b), or the length of the curve of $f(x)$ on the interval (a, b), is given by

$$L = \int_a^b \sqrt{1 + f'(x)^2} \, dx \tag{35.18}$$

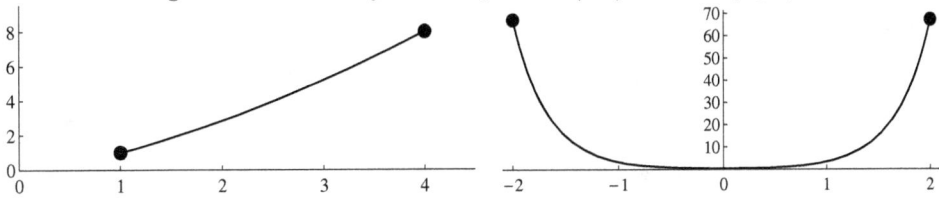

Example 35.1. Find the length of the curve of $y^2 = x^3$ from the point $(1,1)$ to the point $(4,8)$.

Solution. Solving for $y = f(x)$ we find $y = x^{3/2}$. Hence $y' = \frac{3}{2}\sqrt{x}$. Therefore

$$L = \int_1^4 \sqrt{1 + \left(\frac{3}{2}\sqrt{x}\right)^2}\,dx = \int_1^4 \sqrt{1 + \frac{9x}{4}}\,dx \tag{35.19a}$$

To solve the integral, let $u = 1 + \dfrac{9x}{4}$. Then $du = \dfrac{9dx}{4}$. When $x = 1$, $u = 13/4$, and when $x = 4$, then $u = 10$. Thus

$$L = \frac{9}{4}\int_{13/4}^{10} \sqrt{u}\,du = \frac{8}{27}\, u^{3/2}\Big|_{13/4}^{10} \tag{35.19b}$$

$$= \frac{8}{27}\left[10^{3/2} - \left(\frac{13}{4}\right)^{3/2}\right] \tag{35.19c}$$

$$= \frac{1}{27}[10\sqrt{10} - 13\sqrt{13}] \approx 7.63371 \tag{35.19d}$$

$\square$

Example 35.2. Find the length of the portion of the circle $x^2 + y^2 = a^2$ that is in the first quadrant.

Solution.   From elementary geometry, we know that this is the equation of a circle, and that a circle has circumference $2\pi a$. Since this particular circle is centered at the origin, only one quarter of it is in the first quadrant. Hence the total arc length should be $\pi a/2$.

Solving the equation for y gives

$$y = \sqrt{a^2 - x^2} \tag{35.20a}$$

Differentiating and using the chain rule,

$$y' = \frac{-x}{\sqrt{a^2 - x^2}} \tag{35.20b}$$

The arc length is

$$L = \int_0^a \sqrt{1 + (y')^2}\,dx = \int_0^a \sqrt{1 + \frac{x^2}{a^2 - x^2}}\,dx \tag{35.20c}$$

$$= \int_0^a \sqrt{\frac{a^2}{a^2 - x^2}}\,dx = a\int_0^a \frac{dx}{\sqrt{a^2 - x^2}} \tag{35.20d}$$

Let $x = a\sin u$; then $dx = a\cos u\,du$. When $x = 0$, $u = 0$; when $x = a$, $u = \pi/2$.

$$L = a\int_0^{\pi/2} \frac{a\cos u\,du}{\sqrt{a^2 - a^2\sin^2 u}} = a\int_0^{\pi/2} du = \frac{\pi a}{2} \tag{35.20e}$$

as expected. $\square$

Example 35.3. Find the length of the catenary $y = \dfrac{1}{3}\cosh(3x)$ from $x = -2$ to $x = 2$.

Solution. Differentiating, $y' = \sinh 3x$ and therefore

$$L = \int_{-1}^{1} \sqrt{1 + (y')^2}\, dx = \int_{-1}^{1} \sqrt{1 + \sinh^2 3x}\, dx = \int_{-1}^{1} \sqrt{\cosh^2 3x}\, dx \qquad (35.21a)$$

$$= \int_{-1}^{1} \cosh 3x\, dx = 2 \int_{0}^{1} \cosh 3x\, dx \qquad (35.21b)$$

where the last step follows because the integrand is even. Integrating,

$$L = \frac{2}{3} \sinh 3x \Big|_{0}^{1} = \frac{2}{3}(\sinh 3 - \sinh 0) = \frac{2}{3}\sinh 3 \approx 6.6786 \qquad (35.21c)$$

$\square$

Example 35.4. Find the length of the curve $x = \dfrac{2}{3}(y-1)^{3/2}$ on the interval $1 < y < 2$.

Solution. Since the equation is written as $x = f(y)$, the derivative is $x'(y)$:

$$\frac{dx}{dy} = \sqrt{y - 1} \qquad (35.22a)$$

and the length formula is $\displaystyle\int_{a}^{b} \sqrt{1 + x'(y)^2}\, dy$, so that the arc length is

$$L = \int_{1}^{2} \sqrt{1 + \left(\sqrt{y-1}\right)^2}\, dy = \int_{1}^{2} \sqrt{y}\, dy = \frac{2}{3} y^{\frac{3}{2}} \Big|_{1}^{2} \qquad (35.22b)$$

$$= \frac{2}{3}(2^{3/2} - 1) = \frac{2}{3}(2\sqrt{2} - 1). \qquad (35.22c)$$

$\square$

Example 35.5. Find the length of the curve $y = e^{-x}$ from $x = 0$ to $x = 10$.

Solution. Since $y' = -e^{-x}$,

$$L = \int_{0}^{10} \sqrt{1 + e^{-2x}}\, dx \qquad (35.23a)$$

To solve this integral let $e^{-x} = \sinh u$. Then

$$\cosh u\, du = -e^{-x} dx \qquad (35.23b)$$

Solving for dx,

$$dx = -\frac{\cosh u\, du}{e^{-x}} = -\frac{\cosh u\, du}{\sinh u} \qquad (35.23c)$$

Furthermore,

$$\sqrt{1 + e^{-2x}} = \sqrt{1 + \sinh^2 u} = \sqrt{\cosh^2 u} = \cosh u \qquad (35.23d)$$

When $x = 0$, $\sinh u = e^{-0} = 1$, or $u = \operatorname{arcsinh}(1)$. When $x = 10$, $\sinh u = e^{-10}$, or $u = \operatorname{arcsinh}(e^{-10})$. Therefore, substituting (35.23c) and (35.23d) into the integral (35.23a) gives

$$L = -\int_{\operatorname{arcsinh}(1)}^{\operatorname{arcsinh}(e^{-10})} \frac{\cosh^2 u}{\sinh u}\, du \qquad (35.23e)$$

$$= -\int_{\text{arcsinh}(1)}^{\text{arcsinh}(e^{-10})} \frac{1 + \sinh^2 u}{\sinh u}\, du \tag{35.23f}$$

$$= -\int_{\text{arcsinh}(1)}^{\text{arcsinh}(e^{-10})} (\operatorname{csch} u + \sinh u)\, du \tag{35.23g}$$

$$= \left[-\ln \tanh \frac{u}{2} - \cosh u \right]_{\text{arcsinh}(1)}^{\text{arcsinh}(e^{-10})} \tag{35.23h}$$

$$= -\ln\left(\tanh \frac{\text{arcsinh}(e^{-10})}{2} \right) - \cosh\left(\text{arcsinh}(e^{-10})\right)$$

$$+ \ln \tanh\left(\frac{\text{arcsinh}(1)}{2} \right) + \cosh\left(\text{arcsinh}(1)\right) \tag{35.23i}$$

To simplify this use the following. To find $\cosh\left(\text{arcsinh}(1)\right)$, let $y = \text{arcsinh}(1)$. Then $\sinh y = \sinh(\text{arcsinh})(1) = 1$. Hence[1] $\cosh y = \sqrt{1 + \sinh^2 y} = \sqrt{2}$, i.e., $\cosh\left(\text{arcsinh}(1)\right) = \sqrt{2}$.

By a similar argument, $\cosh\left(\text{arcsinh}(e^{-10})\right) = \sqrt{1 + e^{-20}} \approx 1$ (to 8 digits).

To find $\ln \tanh\left(\dfrac{\text{arcsinh}(1)}{2} \right)$ we start with the half angle formula

$$\tanh \frac{\theta}{2} = \frac{\cosh \theta - 1}{\sinh \theta} \tag{35.23j}$$

Letting $\theta = \text{arcsinh}(1)$,

$$\ln \tanh\left(\frac{\text{arcsinh}(1)}{2} \right) = \ln \frac{\cosh \text{arcsinh}(1) - 1}{\sinh \text{arcsinh}(1)} = \ln \frac{\sqrt{2} - 1}{1} = \ln(\sqrt{2} - 1) \tag{35.23k}$$

$$\approx -0.881 \tag{35.23l}$$

Letting $\theta = \text{arcsinh}(e^{-10})$,

$$\ln \tanh\left(\frac{\text{arcsinh}(e^{-10})}{2} \right) = \ln \frac{\cosh \text{arcsinh}(e^{-10}) - 1}{\sinh \text{arcsinh}(e^{-10})} = \ln \frac{\sqrt{1 + e^{-20}} - 1}{e^{-10}} \tag{35.23m}$$

$$= -10.693 \tag{35.23n}$$

Thus $L \approx 10.693 - 1 - 0.881 + 1.414 \approx 10.226$ $\qquad\square$

---

[1]This follows from the identity $\cosh^2 y - \sinh^2 y = 1$.

Exercises

Set up the integral to find the length of each curve in exercises 1 through 3, but do not solve the integral.

1. $y = x^3$ on $(5, 8)$

2. $y = \cos x$ on $(0, 2\pi)$

3. $y = \ln(x + x^2)$ on $(1, 2)$

Find the length of each curve between the specified points in exercises 3 through 6.

3. $y = x^{3/2} - 2$, on $(7, 10)$
   Ans: $\frac{1}{27}\left(94\sqrt{94} - 67\sqrt{67}\right)$

4. $y = \ln(x^2 - 1)$, on $(3, 4)$
   Ans: $1 + 2\tanh^{-1}(3) - 2\tanh^{-1}(4)$

5. $y = 3x^{4/3} - \frac{3}{32}x^{2/3}$, on $(1,2)$
   Ans: $\frac{3}{32}\left(-33 + 64\sqrt[3]{2} + 2^{2/3}\right)$

6. $\ln \cos x$ on $(3\pi/4, \pi)$
   Ans: $(1/2)\ln(3 + 2\sqrt{2})$

Chapter 36

Parametric Equations

Chapter Summary and Goal

A **parametric representation** of a curve in the (x, y) plane describes it in terms of a third variable, called the parameter[1] (e.g, t). In this way it is possible to describe more complicated figures than those that can be expressed as simple functions of a single variable $y = f(x)$. This is because each coordinate is allowed to changed independently of the other as a function of the parameter t. Time dependent **paths** (or **trajectories**) provide a good example. Think of tracing your route to school on a map; draw a dot for every minute of your commute. Label the dots with the time. Each dot has a map coordinate (x, y) as well as a time that you wrote next to it. You can convert this data set to a list of (t, x) values and a list of (t, y) values. These are your parametrizations, which are functions: $x(t)$ and $y(t)$.

Student Learning Objectives

The student will:

1. Learn to represent simple curves parametrically.
2. Learn how to convert between the Cartesian and parametric representation of a curve.
3. Calculate the 1st and 2nd derivatives of a parametric curve.

Parametric Representations

A relationship between two variables x and y may be defined **parametrically** in terms of a third variable t. where both $x(t)$ and $y(t)$ are functions of t, as illustrated in fig. 36.1.

Definition 36.1. Parametric Representation

A **parametric representation** of a curve represents the (x, y) coordinates of the points of curve as functions of a third variable t.

$$(x, y) = (x(t),\ y(t)) \tag{36.1}$$

The variable t is called a **parameter**.

---

[1]Not to be confused with parameters in most other fields of mathematics, where the term parameter is generally reserved for a fixed constant!

Figure 36.1: Concept of a parameteric representation of a curve. Think of the curve as a path that we travel in time. The black dots mark where we are at various times. The arrows connect the timeline at the top to the coordinates on our path at various times.

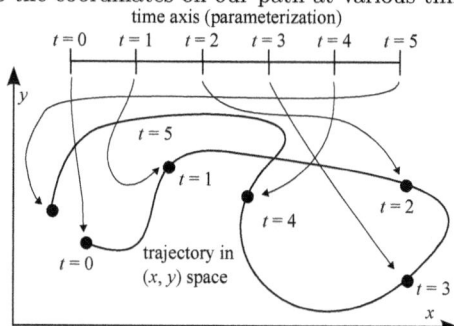

Example 36.1. Show that the unit circle with center at the origin is described by the

$$(x, y) = (\cos t, \sin t), \ 0 \le t < 2\pi \tag{36.2}$$

Solution. To demonstrate that (36.2) is the unit circle we must show two things: (1) that (36.2) satisfies the equation of the unit circle; and (2) that every point on the unit circle satisfies (36.2).

(36.2) satisfies the unit circle: By substitution, every point on (36.2) satisfies

$$x^2 + y^2 = \cos^2 t + \sin^2 t = 1 \tag{36.3a}$$

Since $x^2 + y^2 = 1$ is the equation of the unit circle, this means every point described by the parametrization lies on the unit circle.

The unit circle satisfies (36.2): This is more difficult. It requires showing that for every point (x, y) on the unit circle, there is some number t such that $x = \cos t$ and $y = \sin t$. To prove that this is the case, we must find t. If $P = (x, y)$ is a point on the unit circle, and we draw a line connecting P to the origin, its central angle will be $t = \arctan y/x$. We are tempted to use this value for t. The problem with this is that there are two possible values to pick in the range $[0, 2\pi]$, depending upon whether y is in the upper half plane or lower half plane. Otherwise we will not include the entire circle. We resolve this by the following rule:

$$t = \begin{cases} \arctan(y/x), & \text{pick } t \text{ in } [0, \pi] \text{ if } y \ge 0 \\ \arctan(y/x), & \text{pick } t \text{ in } [\pi, 2\pi] \text{ if } y < 0 \end{cases} \tag{36.3b}$$

This will ensure that we get the entire circle.   □

The parametrization in example 36.1 is not unique, because the sine and cosine are periodic. Other, equally good, parametrizations include

$$(x, y) = (\cos t, \sin t), \ \frac{\pi}{4} \le t < \frac{9\pi}{4} \tag{36.4}$$

and

$$(x, y) = (\cos(4t), \sin(4t)), \ 0 \le t < \frac{\pi}{2} \tag{36.5}$$

The point is that a parametrization is not unique.

Example 36.2. Sketch the the curve defined parametrically by

$$(x, y) = (4t - t^2, t + 1) \tag{36.6}$$

by plotting points.

Solution. In the table on the left we have calculated the coordinates for several values of t.

t	x	y
-1	-5	0
0	0	1
1	3	2
2	4	3
3	3	4
4	0	5
5	-5	6

Figure 36.2. Sketch of the curve defined parametrically by equation 36.6.

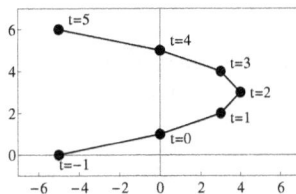

The curve is sketched in figure 36.2. While it looks somewhat like a parabola with central axis $y = 3$, we will not be able to verify that until we complete example 36.3.

$\square$

Example 36.3. Find an algebraic formula for the curve defined by equation 36.6.

Solution. We are given

$$(x, y) = (4t - t^2, t + 1) \tag{36.8a}$$

Solving the y equation for t gives

$$t = y - 1 \tag{36.8b}$$

Substituting this result into the x equation gives

$$x = 4t - t^2 = 4(y - 1) - (y - 1)^2 \tag{36.8c}$$
$$= 4y - 4 - y^2 + 2y - 1 \tag{36.8d}$$
$$= y^2 + 6y - 5 \tag{36.8e}$$
$$= (y - 1)(y - 5) \tag{36.8f}$$

The curve is a parabola that intersects the y axis at 1 and 5; the axis of symmetry is $y = 3$, as previously suspected. $\square$

Example 36.4. Eliminate the parameter to find the Cartesian equation of the curve described by

$$(x, y) = (e^t - 1, e^{2t}) \tag{36.9}$$

Describe the curve does represented by this parametrization.

Solution. Solving the x equation for e^t gives

$$e^t = x + 1 \tag{36.10a}$$

Substituting this into the y equation,

$$y = e^{2t} = (e^t)^2 = (x + 1)^2 \tag{36.10b}$$

This is the equation of a parabola with vertex at $(-1, 0)$, axis of symmetry $x = -1$, and opening upwards. $\square$

Example 36.5. Eliminate the parameter to find the Cartesian equation parametrized by

$$(x, y) = (\tan^2 \theta, \ \sec \theta) \tag{36.11}$$

Describe the curve does represented by this parametrization.

Solution. Using the identity $\sec^2 \theta - \tan^2 \theta$ we obtain

$$1 = \sec^2 \theta - \tan^2 \theta = y^2 - x \tag{36.12a}$$

Therefore

$$x = y^2 - 1 \tag{36.12b}$$

This is a parabola that opens to the right, with vertex at $(-1, 0)$ and axis of symmetry along the x axis. It crosses the y axis at $(0, 1)$ and $(0, -1)$. $\qquad\square$

Example 36.6. Express the equation of a circle with radius a and center at coordinates (h, k) parametrically.

Solution. We have already determined that one way of representing the unit circle is

$$(x, y) = (\cos t, \ \sin t), \ 0 \le t < 2\pi \tag{36.13a}$$

If we multiply the values of x and y by a we get a circle of radius a:

$$(x, y) = (a \cos t, \ a \sin t), \ 0 \le t < 2\pi \tag{36.13b}$$

To verify that this gives a circle of radius a centered at the origin, calculate

$$x^2 + y^2 = (a \cos t)^2 + (a \sin t)^2 = a^2 (\sin^2 t + \cos^2 t) = a^2 \tag{36.13c}$$

To shift the center from (0,0) to (h, k), we add h to x and k to y

$$(x, y) = (h + a \cos t, \ k + a \sin t), \ 0 \le t < 2\pi \tag{36.13d}$$

Then

$$(x - h)^2 + (y - k)^2 = (a \cos t)^2 + (a \sin t)^2 \tag{36.13e}$$
$$= a^2 (\sin^2 t + \cos^2 t) \tag{36.13f}$$
$$= a^2 \tag{36.13g}$$

which is the equation of a circle of radius a and center (h, k). $\qquad\square$

Example 36.7. Eliminate the parameter to find an equation for

$$(x, y) = \cos t, \ \cos t \sin t), \ 0 \le t \le 2\pi \tag{36.14}$$

in Cartesian coordinates (figure **??**).

Solution. Squaring the equation for y, using the Pythagorean identity, and then substituting the equation for x gives

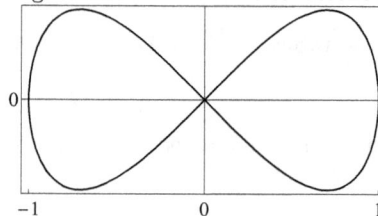

Figure 36.3: Lemniscate of Gerono.

$$y^2 = \cos^2 t \sin^2 t \tag{36.15a}$$
$$= (\cos^2 t)(1 - \cos^2 t) \tag{36.15b}$$
$$= x^2(1 - x^2) \tag{36.15c}$$
$$= x^2 - x^4 \tag{36.15d}$$

This curve is sometimes called the lemniscate of Gerono. $\square$

Example 36.8. The **lemniscate of Bernoulli** is described by the parametrization

$$(x, y) = \left(\frac{a\sqrt{2}\cos t}{1 + \sin^2 t}, \frac{a\sqrt{2}\cos t \sin t}{1 + \sin^2 t} \right), \ 0 \le t \le 2\pi \tag{36.16}$$

where a is a fixed constant. Find an equivalent representation in Cartesian coordinates.

Solution. We observe that the $\sin t$ factors out of the y equation, and what remains is x:

$$y = \frac{a\sqrt{2}\cos t}{1 + \sin^2 t} \cdot \sin t = x \sin t \tag{36.17a}$$

Therefore,

$$y = x \sin t \tag{36.17b}$$

Using equation 36.17b and a trigonometric identity $(\sin^2 t + \cos^2 t = 1)$,

$$x^2 - y^2 = x^2 - x^2 \sin^2 t = x^2(1 - \sin^2 t) = x^2 \cos^2 t \tag{36.17c}$$

Similarly,

$$x^2 + y^2 = x^2 + x^2 \sin^2 t = x^2(1 + \sin^2 t) \tag{36.17d}$$

But from the x equation in the parametrization (36.17a)

$$x(1 + \sin^2 t) = a\sqrt{2}\cos t \tag{36.17e}$$

Using 36.17e in 36.17d

$$x^2 + y^2 = xa\sqrt{2}\cos t \tag{36.17f}$$

Squaring both sides of the equation and substituting (36.17c)

$$(x^2 + y^2)^2 = 2a^2 x^2 \cos^2 t = 2a^2(x^2 - y^2) \tag{36.17g}$$

This is the standard form for lemniscate of Bernoulli in Cartesian coordinates,

$$(x^2 + y^2)^2 = 2a^2(x^2 - y^2) \tag{36.17h}$$

This curve is shown in figure 36.4 for several values of the parameter a. □

Figure 36.4: The lemniscate of Bernoulli $(x^2 + y^2)^2 = 2a^2(x^2 - y^2)$ for $a = 1, 2, 3, 4, 5$ ($a = 5$ is the largest lemniscate). The shape is rounder and more like a figure-8 than the Lemniscate of Gerono.

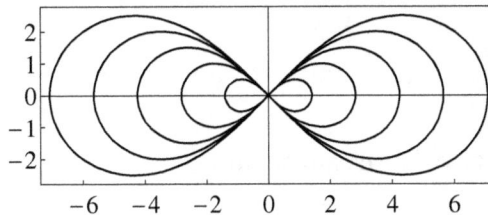

Figure 36.5: Geometry of the cycloid.

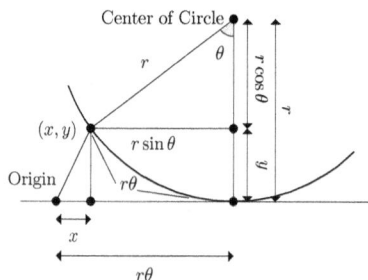

Example 36.9. A **cycloid** describes the curve followed by a point on the rim of a circle of radius r that is rolled along the positive x-axis, starting from the origin. Find a parametric representation.

Solution. The geometry is described in figure 36.5. Let the angle θ be measured at the center of the circle, and let the origin be at the point shown. The distance from the origin to the point directly beneath the center of the circle is $r\theta$ radians (the arc length of the circle between the indicated dots). Thus

$$x = r\theta - r\sin\theta \qquad\qquad (36.18\text{a})$$

The height of the center of circle above the origin as shown is r, and thus

$$y = r - r\cos\theta \qquad\qquad (36.18\text{b})$$

The point (x, y) makes one revolution as $0 \le t \le 2\pi$ (fig 36.6). $\qquad\qquad\square$

Figure 36.6: The cycloid is formed a a point on the rim of a circle (hollow white circle dots) rolls.

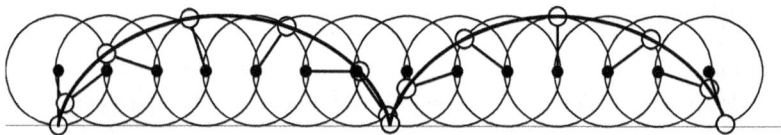

Finding a Parametric Representation

Unfortunately there is no easy, general method that always works – finding a representation usually requires a bit of trial and error. Often we can find a representation by looking at the form of the equation and knowing something about algebra.

If the curve we want to represent is a function of a single variable $y = f(x)$ then this works:

$$(x, y) = (t,\ f(t)) \qquad\qquad (36.19)$$

Example 36.10. Find a parametric representation for $y = (x-1)^2$ on the interval $(0,3)$.

Solution. This curve is described by a function of a single variable so we can use

$$(x,y) = (t,\, (t-1)^2),\, 0 \le t \le 3 \tag{36.20a}$$

$\square$

If there is hyperbolic symmetry and the curve is symmetric about the axes, we can use one of the identities $\sec^2 t - \tan^2 t = 1$ or $\cosh^2 t - \sinh^2 t = 1$.

Example 36.11. Find a parametric representation for the right hyperbola

$$\frac{(x-a)^2}{h^2} - \frac{(y-b)^2}{k^2} = 1. \tag{36.21}$$

Solution. <u>Method 1</u>. Use the hyperbolic identity $\cosh^2 t - \sinh^2 t = 1$ and make the associations

$$\frac{x-a}{h} = \cosh t \tag{36.22a}$$

$$\frac{y-b}{k} = \sinh t \tag{36.22b}$$

Solving for x and y,

$$(x,y) = (a + h\cosh t,\, b + k\sinh t) \tag{36.22c}$$

<u>Method 2</u>. Use the trigonometric identity $\sec^2 t - \tan^2 t = 1$ and make the associations

$$\frac{x-a}{h} = \sec t \tag{36.22d}$$

$$\frac{y-b}{k} = \tan t \tag{36.22e}$$

Solving for x and y,

$$(x,y) = (h + a\sec t,\, k + b\tan t). \tag{36.22f}$$

$\square$

For circles or ellipses we can repeat example 36.11 using the identity $\sin^2 t + \cos^2 t = 1$. If the object is rotated about the origin by an angle θ, its new coordinates are given by

$$(x,y) \to (x\cos\theta + y\sin\theta, x\sin\theta - y\cos\theta) \tag{36.23}$$

Thus the equation of an ellipse centered at the origin but rotated by an angle θ is

$$\frac{(x\cos\theta + y\sin\theta)^2}{a^2} + \frac{(x\sin\theta - y\cos\theta)^2}{b^2} = 1 \tag{36.24}$$

where θ is a fixed constant.

Example 36.12. Find a parametric form for the equation of an ellipse with semi-major axis $a = 2$ and semi-minor axis $b = \sqrt{3}$, rotated by an $\pi/6$ about the origin.

Solution. From equation 36.24, we start by making the association

$$x\cos\theta + y\sin\theta = a\cos t \tag{36.25a}$$

$$x\sin\theta - y\cos\theta = b\sin t \tag{36.25b}$$

To get equations for $x(t)$ and $y(t)$ we multiply $(36.25a)$ by $\cos\theta$ and $(36.25b)$ by $\sin\theta$,

$$x\cos^2\theta + y\sin\theta\cos\theta = a\cos t\cos\theta \tag{36.25c}$$

$$x\sin^2\theta - y\sin\theta\cos\theta = b\sin t\sin\theta \tag{36.25d}$$

Adding equations (36.25c) and (36.25d),

$$x = a \cos t \cos \theta + b \sin t \sin \theta \tag{36.25e}$$

Next, we multiply (36.25a) by $\sin \theta$ and (36.25b) by $-\cos \theta$,

$$x \cos \theta \sin \theta + y \sin^2 \theta = a \cos t \sin \theta \tag{36.25f}$$

$$-x \cos \theta \sin \theta + y \cos^2 \theta = -b \sin t \cos \theta \tag{36.25g}$$

Adding (36.25f) and (36.25g) gives

$$y = a \cos t \sin \theta - b \sin t \cos \theta \tag{36.25h}$$

The parametric representation is

$$x = 2 \cos t \cos \left(\frac{\pi}{6} \right) + \sqrt{3} \sin t \sin \left(\frac{\pi}{6} \right) = \sqrt{3} \cos t + \frac{\sqrt{3}}{2} \sin t \tag{36.25i}$$

$$y = 2 \cos t \sin \left(\frac{\pi}{6} \right) - \sqrt{3} \sin t \cos \left(\frac{\pi}{6} \right) = \cos t - \frac{3}{2} \sin t \tag{36.25j}$$

$\square$

Example 36.13. Lissajous plots arise in electronics, where they are used to demonstrate the relationship between waveforms on an oscilloscope screen (fig. 36.7). The phase shift between audio channels, for example, can be represented visually this way. A Lissajous Curve is described by the parametrization[2]

$$(x, y) = (\sin(jt + \delta),\ \sin(kt)) \tag{36.26}$$

Plot the Lissajous curve for various combinations of j and k.

Solution. When $\delta = \pi/2$ and $j = k$, they are unit circles (main diagonal of left grid of figure 36.7), and as δ decreases, the circles shrink to ellipses symmetric about the line $y = x$ (main diagonal of right grid of figure 36.7). In the limit ($\delta = 0$, not shown), they become line segments. For $k/j = 2$, the are parabolas. By slowly varying the value of δ in an animation the the Lissajous figure appears as a rotating 3D object. $\square$

Calculus on Parametric Curves

The derivative dy/dx can be calculated using the chain rule,

$$\frac{dy}{dt} = \frac{dy}{dx} \frac{dx}{dt} \implies \frac{dy}{dx} = \frac{dy/dt}{dx/dt} = \frac{y'(t)}{x'(t)} \tag{36.28}$$

where **the prime always refers to differentiation with respect to t.**

Similarly, we calculate the second derivative as

$$\frac{d^2 y}{dx^2} = \frac{d}{dx} \left(\frac{dy}{dx} \right) = \frac{d}{dt} \left(\frac{dy}{dx} \right) \frac{dt}{dx} = \frac{\frac{d}{dt} \frac{dy}{dx}}{dx/dt} \tag{36.29}$$

Here dy/dx in the numerator of the result is the derivative calculated in (36.28),

---

[2] **Spirographs** (exercise 20) look similar to Lissajous plots, but are constrained to a circle, and are constructed by rolling one circle inside of another.

Figure 36.7: Lissajous figures, as described by equation 36.26. Left: $\delta = \pi/2$; right, $\delta = \pi/6$.

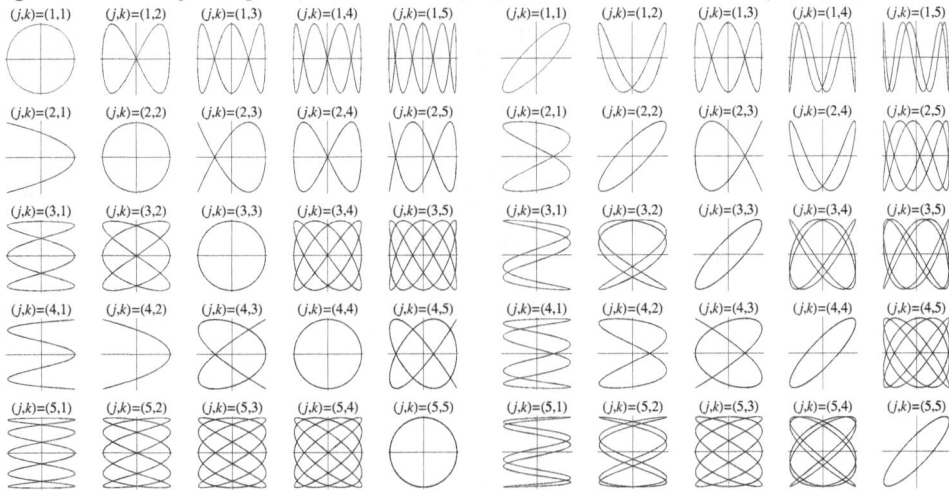

Theorem 36.1. Parametric Derivatives

If $(x(t), y(t))$ is the parametric representation of a curve then

$$\frac{dy}{dx} = \frac{dy/dt}{dx/dt} = \frac{y'(t)}{x'(t)} \tag{36.30}$$

$$\frac{d^2y}{dx^2} = \frac{\dfrac{d}{dt}\left(\dfrac{dy/dt}{dx/dt}\right)}{dx/dt} = \frac{(y'(t)/x'(t))'}{x'(t)} \tag{36.31}$$

where the prime in $x'(t)$ and $y'(t)$ means differentiation with respect to t.

Example 36.14. Compare the slopes of the lemniscates of Gerono and Bernoulli (with $a = 1/\sqrt{2}$) at the origin.

Solution. For the lemniscate of Gerono,

$$(x, y) = (\cos t, \, \cos t \sin t), \, 0 \le t \le 2\pi \tag{36.32a}$$

The curve passes through the origin $(0, 0)$ when $\cos t = 0$, i.e., $t = \pi/2$ or $t = 3\pi/2$. The derivative is

$$\frac{dy}{dx} = \frac{y'}{x'} = \frac{\cos^2 t - \sin^2 t}{-\sin t} = -\frac{\cos(2t)}{\sin t} \tag{36.32b}$$

At $t = \pi/2$ the slope is $m = 1$, and at $t = 3\pi/2$, the slope is $m = -1$.

For the lemniscate of Bernoulli, the origin also occurs at $t = \pi/2$ or $t = 3\pi/2$. The parametrization at $a = 1/\sqrt{2}$ is

$$(x, y) = \left(\frac{\cos t}{1 + \sin^2 t}, \, \frac{\cos t \sin t}{1 + \sin^2 t} = \frac{\sin(2t)}{2(1 + \sin^2 t)} \right) \tag{36.32c}$$

Differentiating,

$$x' = \frac{(1 + \sin^2 t)(\cos t)' - (\cos t)(1 + \sin^2 t)'}{(1 + \sin^2 t)^2} \tag{36.32d}$$

Doing Calculus

$$= \frac{(1 + \sin^2 t)(-\sin t) - (\cos t)(-2 \sin t \cos t)}{(1 + \sin^2 t)^2} \tag{36.32e}$$

$$= -\sin t \cdot \frac{1 + \sin^2 t - 2\cos^2 t}{(1 + \sin^2 t)^2} \tag{36.32f}$$

$$x'(\pi/2) = -(1) \cdot \frac{1 + 1^2 - 0}{(1 + 1^2)^2} = -\frac{1}{2} \tag{36.32g}$$

$$x'(3\pi/2) = -(-1) \cdot \frac{1 + (-1)^2 - 0}{(1 + (-1)^2)^2} = \frac{1}{2} \tag{36.32h}$$

$$y' = \frac{2(1 + \sin^2 t)(\sin(2t))' - (\sin 2t)(2(1 + \sin^2 t))'}{4(1 + \sin^2 t)^2} \tag{36.32i}$$

$$= \frac{2(1 + \sin^2 t)(2 \cos(2t)) - (\sin 2t)(2)(2 \sin t \cos t)}{4(1 + \sin^2 t)^2} \tag{36.32j}$$

$$y'(\pi/2) = \frac{2(1 + 1^2)(2 \cdot -1) - 0 \cdot 2 \cdot 2 \cdot 1 \cdot 0}{4(1 + 1^2)^2} = -\frac{1}{2} \tag{36.32k}$$

$$y'(3\pi/2) = \frac{2(1 + (-1)^2)(2 \cdot -1) - 0 \cdot 2 \cdot 2 \cdot (-1) \cdot 0}{4(1 + (-1)^2)^2} = -\frac{1}{2} \tag{36.32l}$$

Thus $dy/dx = 1$ or -1 for both lemniscates at the origin (figure 36.8).   □

Figure 36.8: Comparison of the lemniscate of Bernoulli with $a = 1/\sqrt{2}$ (dashed line) with the lemniscate of Gernono (solid line). Both lemniscates have the same slope at the origin (example 36.14).

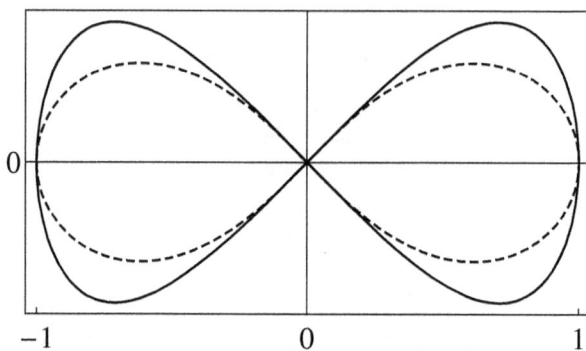

Example 36.15. Find the equation of the tangent line to cycloid

$$(x, y) = (3(t - \sin t), \; y = 3(1 - \cos t) \tag{36.33}$$

at $t = \pi/2$.

Solution. The slope is given by the derivative at at $t = \pi/2$.

$$\frac{dy}{dx} = \frac{dy/dt}{dx/dt} = \frac{(3(1 - \cos t))'}{(3(t - \sin t))'} = \frac{3 \sin t}{3 - 3 \cos t} = \frac{\sin t}{1 - \cos t} \tag{36.34a}$$

The slope at $t = \pi/2$ is then

$$m = \frac{\sin(\pi/2)}{1 - \cos(\pi/2)} = 1 \tag{36.34b}$$

To get a point, we substitute $t = \pi/2$ into the parametric equations

$$x_1 = 3\left(\frac{\pi}{2} - \sin\frac{\pi}{2}\right) = 3\left(\frac{\pi}{2} - 1\right) \tag{36.34c}$$

$$y_1 = 3\left(1 - \cos\frac{\pi}{2}\right) = 3 \tag{36.34d}$$

Using the point-slope equation for the tangent line,

$$y = y_1 + m(x - x_1) = 3 + 1 \cdot \left(x - \frac{3\pi}{2} + 3\right). = 6 - \frac{3\pi}{2} + x \tag{36.34e}$$

$\square$

Areas in Parametric Coordinates

In chapter 24 we derived a formula for the area under the curve of a function $y = \phi(x)$ by adding up rectangles of height $\phi(x_i^*)$ and width Δx

$$A = \int_a^b \phi(x)dx = \lim_{n\to\infty} \sum_{i=1}^n \Delta x\, \phi(x_i^*) \tag{36.35}$$

and called this the definite integral (equation 24.5).

When $\phi(x)$ is parametrized by $(x, y) = (f(t), g(t))$ on some interval $a \le t \le b$, then the height $\phi(x_i^*)$ of the rectangle is given by $y = g(t_i^*)$, and its width is given by (equation 8.37)

$$\Delta x = \frac{df(t_i^*)}{dt}\Delta t = f'(t_i^*)\Delta t, \tag{36.36}$$

where t_i^* is the value of t such that $x_i^* = f(t_i^*)$, and $y_i^* = g(t_i^*)$. The Riemann sum (and hence the area formula) becomes

$$A = \lim_{n\to\infty} \sum_{i=1}^n (f'(t_i^*)\Delta t)\, g(t_i^*) = \int_a^b g(t)f'(t)dt \tag{36.37}$$

Theorem 36.2. Area in Parametric Coordinates

The area under a curve $(x, y) = (f(t), g(t))$, $a \le t \le b$, and above the x axis is

$$A = \int_a^b g(t)f'(t)dt = \int_a^b y(t)x'(t)dt \tag{36.38}$$

Depending on the direction of integration, the limits might have to be reversed.

An easy way to remember this is to think of the area as

$$\text{Area} = \int_{x(t=a)}^{x(t=b)} y\, dx = \int_{x(t=a)}^{x(t=b)} y \cdot \left(\frac{dx}{dt}\right) dt = \int_a^b y(t)x'(t)dt \tag{36.39}$$

Example 36.16. Find the area of the lemniscate of Gerono (equation 36.32a).

Solution. By symmetry, we can find the total area by finding the area in the first quadrant and multiplying by 4.

The parametrization is given by

$$(x, y) = \cos t, \cos t \sin t) \tag{36.40a}$$

When $t = 0$, the curve is at $(1, 0)$; and when $t = \pi/2$, the curve is at the origin and has traced out one quarter of the figure-eight. Because the path from $(1,0)$ to $(0,0)$ moves from the right to the left, to get a positive area, we need to reverse the order of the limits. Thus

$$A = 4 \int_{\pi/2}^0 y(t) x'(t) dt = 4 \int_{\pi/2}^0 (\cos t \, \sin t)(\cos t)' \, dt \tag{36.40b}$$

$$= -4 \int_{\pi/2}^0 \sin^2 t \, \cos t \, dt = 4 \int_0^{\pi/2} \sin^2 t \, \cos t \, dt \tag{36.40c}$$

$$= \frac{4}{3} \left[\sin^3 \left(\frac{\pi}{2} \right) - \sin^3 0 \right] = \frac{4}{3} \tag{36.40d}$$

$\square$

Example 36.17. Find the area under one arch of the cycloid

$$(x, y) = (a(t - \sin t), \ a(1 - \cos t)). \tag{36.41}$$

Solution. One arch of the cycloid corresponds to $0 \le t \le 2\pi$. Since

$$dx = \frac{dx}{dt} \cdot dt = a(1 - \cos t) dt \tag{36.42a}$$

the area is

$$A = \int_{t=0}^{t=2\pi} y(t) \, dx(t) = \int_{t=0}^{t=2\pi} a(1 - \cos t) \, a(1 - \cos t) \, dt \tag{36.42b}$$

$$= a^2 \int_0^{2\pi} (1 - \cos t)(1 - \cos t) \, dt \tag{36.42c}$$

$$= a^2 \int_0^{2\pi} (1 - 2\cos t + \cos^2 t) \, dt \tag{36.42d}$$

$$= a^2 \left(\overbrace{\int_0^{2\pi} dt}^{=2\pi} - 2 \overbrace{\int_0^{2\pi} \cos t \, dt}^{=0} + \int_0^{2\pi} \cos^2 t \, dt \right) \tag{36.42e}$$

$$= a^2 \left(2\pi + 0 + \frac{1}{2} \int_0^{2\pi} (1 + \cos(2t)) \, dt \right) \tag{36.42f}$$

$$= a^2 \left(2\pi + \pi + \overbrace{\int_0^{2\pi} \cos(2t) \, dt}^{=0} \right) = 3\pi a^2 \tag{36.42g}$$

$\square$

Arc Length in Parametric Equations

In deriving theorem 35.1 we found that sum of lengths of the line segments approximating any curve was

$$\overline{L} = \sum_{i=i}^{n} |P_{i-1}P_i| = \sum_{i=i}^{n} \sqrt{(\Delta x_i)^2 + (\Delta y_i)^2} \tag{36.43}$$

We did not impose the requirement that y be a function of x until later in the derivation. If, instead, we assume that x and y depend parametrically on a third variable t, then we can write

$$\overline{L} = \sum_{i=i}^{n} \sqrt{(\Delta x_i)^2 + (\Delta y_i)^2} \cdot \frac{\Delta t_i}{\Delta t_i} \tag{36.44}$$

where t_i is the value of the parameter at each point (x_i, y_i). Since $\Delta t_i = \sqrt{(\Delta t_i)^2}$,

$$\overline{L} = \sum_{i=i}^{n} \sqrt{\left(\frac{\Delta x_i}{\Delta t_i}\right)^2 + \left(\frac{\Delta y_i}{\Delta t_i}\right)^2} \cdot \Delta t_i \tag{36.45}$$

Taking the limit as $n \to \infty$ and $\Delta t_i \to 0$, we get a Riemann Sum in t, where $a \le t \le b$,

$$L = \lim_{\substack{n \to \infty \\ \Delta t_i \to 0}} \sum_{i=i}^{n} \sqrt{\left(\frac{\Delta x_i}{\Delta t_i}\right)^2 + \left(\frac{\Delta y_i}{\Delta t_i}\right)^2} \cdot \Delta t_i = \int_a^b \sqrt{\left(\frac{dx}{dt}\right)^2 + \left(\frac{dy}{dt}\right)^2}\, dt \tag{36.46}$$

Theorem 36.3. Arc Length Formula, Parametric Form

The length of a curve described parametrically by the functions $x(t)$ and $y(t)$ on the interval $a \le t \le b$ is

$$L = \int_a^b \sqrt{(x'(t))^2 + (y'(t))^2}\, dt \tag{36.47}$$

where differentiation is with respect to t.

Example 36.18. Find the circumference of a circle of radius r.

Solution. Use the parametrization

$$(x, y) = (r\cos t,\, r\sin t),\, 0 \le t \le \pi \tag{36.48a}$$

Then since r is a fixed constant,

$$(x', y') = (-r\sin t, r\cos t) \tag{36.48b}$$

$$L = \int_0^{2\pi} \sqrt{x'(t)^2 + y'(t)^2}\, dt \tag{36.48c}$$

$$= \int_0^{2\pi} \sqrt{r^2 \sin^2 t + r^2 \cos^2 t}\, dt \tag{36.48d}$$

$$= \int_0^{2\pi} \sqrt{r^2(\sin^2 t + \cos^2 t)}\, dt = r\int_0^{2\pi} dt = 2\pi r \tag{36.48e}$$

$\square$

Example 36.19. Find the arc length of one arch of the cycloid,

$$(x, y) = (a(t - \sin t), \, y = a(1 - \cos t)). \tag{36.49}$$

Solution. One arch of the cycloid is $0 \le t \le 2\pi$. Since

$$(x'(t), y'(t)) = (a(1 - \cos t), \, a \sin t) \tag{36.50a}$$

the arc length is

$$L = \int_0^{2\pi} \sqrt{x'(t)^2 + y'(t)^2} dt \tag{36.50b}$$

$$= \int_0^{2\pi} \sqrt{a^2(1 - \cos t)^2 + a^2 \sin^2 t} \; dt \tag{36.50c}$$

$$= a \int_0^{2\pi} \sqrt{1 - 2\cos t + \cos^2 t + \sin^2 t} \; dt \tag{36.50d}$$

$$= a \int_0^{2\pi} \sqrt{2 - 2\cos t} \; dt \tag{36.50e}$$

$$= a\sqrt{2} \int_0^{2\pi} \sqrt{1 - \cos t} \; dt \tag{36.50f}$$

$$= a\sqrt{2} \int_0^{2\pi} \sqrt{2 \cdot \frac{1}{2}(1 - \cos t)} \; dt \tag{36.50g}$$

$$= 2 \int_0^{2\pi} \sqrt{\sin^2 \frac{t}{2}} \; dt \tag{36.50h}$$

$$= 2a \int_0^{2\pi} \sin \frac{t}{2} \; dt \tag{36.50i}$$

$$= -4a \cos \frac{t}{2} \Big|_0^{2\pi} = 8a \tag{36.50j}$$

$\square$

Exercises

In exercises 1 through 10 eliminate the parameter to find representations for each of the given curves as a single equation in Cartesian coordinates.

1. $x = 1 + t$, $y = 1 - t$
   Ans: $y = 1 - x$

2. $x = 6 + t$, $y = t^2$
   Ans: $y = x^2 - 12x + 36$

3. $x = 1 + t^2$, $y = t - 2$
   Ans: $x = y^2 + 4y + 5$

4. $x = \cos t$, $y = 2 \sin t$

5. $x = \dfrac{3at}{1 + t^3}$, $y = \dfrac{3at^2}{1 + t^3}$
   Ans: $x^3 + y^3 = 3axy$
   (Folium of Descartes)

6. $x = e^{2t} + 1$, $y = e^t$

7. $x = e^t + 1$, $y = e^{2t} - 1$

8. $x = a \cos^3 t$, $y = a \sin^3 t$
   Ans: $x^{2/3} + y^{2/3} = a^{2/3}$
   (Hypocycloid)

9. $x = \cos(2t) \cos t$, $y = \cos(2t) \sin t$
   Ans: $(x^2 + y^2)^{3/2} = x^2 - y^2$
   (4-petalled rose)

10. $x = 2a \cot t$, $a(1 - \cos(2t)$
    Ans: $y = 8a^3/(x^2 + 4a^2)$
    (Witch of Agnesi)

Find the first and second derivatives dy/dx and d^2y/dx^2 as a function of t in exercises 11 through 12.

11. $x = 3t + 5$, $y = \sin^2(5t)$

12. $x = \cos^3(t)$, $y = 3 \sin^2(t)$

13. Find the equation of the line tangent to the parametric curve $x = t^3 - 25t$, $y = 25t^2 - t^4$ at the points where $t = 5$ and $t = -5$.

14. Find the equation of the line tangent to the curve parametrized by $x = e^t$, $y = (t - 9)^2$ at the point $(1, 81)$.

15. Find the area of the region in the first quadrant that is below the parametric curve $x = t^3 + 6t$, $y = 2t - t^2$ and above the x axis.

16. Find a formula for the area of a circle using the area formula (theorem 36.2).

17. Find a parametrization for an ellipse, centered at the origin, with semi-major axis a that is parallel to the x axis and semi-major axis b that is parallel to the y axis.

18. Use the results of exercise 17 to find a formula for the area of an ellipse.

19. Find an equation for the perimeter of an ellipse using the result of exercise 17.

Figure 36.9: The spirograph for $h = .2$, $k = .74$.

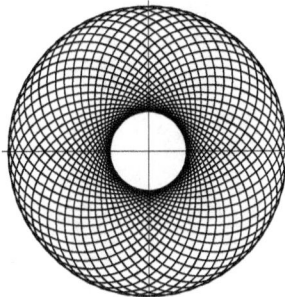

20. (Exploration*) The **spirograph** is be parametrized by

$$x(t) = (1 - k) \cos t + hk \cos(1 - k)t/k)$$
$$y(t) = (1 - k) \sin t - hk \sin((1 - k)t/k)$$

where k, h are fixed parameters in the range $(0, 1)$. Use a numerical plotting program to explore for $h, k = .1, .2, \ldots, .9$. Fix $h = 0.2$ and vary $k = 0.70, 0.71, \ldots, 0.80$. What do you observe about the sensitivity of the parameters?

Figure 36.10: Harmonograph plot with $A = 1.1$, $B = 1.25$, $C = .5$, $D = 1.25$, $c_1 = c_2 = c_3 = c_4 = 0.001$, $\alpha = \beta = 0.999$, $\gamma = 1.99$, $\delta = 0.5$.

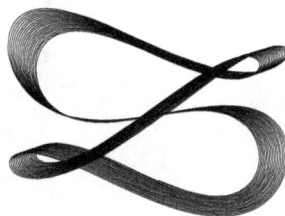

21. (Exploration*) The **harmonograph** was invented in the 19th to draw complicated figures; it consists of two pendulums that move back and forth along perpendicular axes, whose dynamics are controlled by the equations

$$x = Ae^{-c_1 t} \sin(\alpha t) + Be^{-c_2 t} \sin(\beta t)$$
$$y = Ce^{-c_3 t} \sin(\gamma t) + De^{-c_4 t} \sin(\delta t)$$

If you have access to a numerical plotting program explore what happens when you vary the different parameters.

Doing Calculus

Chapter 37

Polar Coordinates

Chapter Summary and Goal

In **polar coordinates**, the location of a point P in the plane is described by a pair of numbers (r, θ). Here r is the distance from the origin O, and θ is the angle between the ray $\overrightarrow{OP}$ and the positive x axis (see figure 37.1). Describing points and curves by polar coordinates is just as valid as our usual representation using (x, y) pairs, which is known as either **Cartesian coordinates** or **rectangular coordinates**. In this chapter we will learn to convert between the two coordinate representations. We will then learn how to do calculus in polar coordinates, finding values for derivatives, arc lengths, and areas.

Student Learning Objectives

The student will:

1. Convert between polar and rectangular coordinates.
2. Find the polar representation of a point or function given in rectangular coordinates.
3. Find the Cartesian representation of a point or function given in polar coordinates.
4. Calculate the slope (dy/dx) and second derivative (d^2y/dx^2) in polar coordinates.
5. Set up integrals for arc length and area in polar coordinates.
6. Calculate simple areas and arc lengths in polar coordinates.

Converting Between Polar and Rectangular Coordinates

Normally we describe the locations of points in the plane by their **rectangular coordinates**.

Definition 37.1. Rectangular (Cartesian) Coordinates

Each point in the plane is represented by a pair of numbers (x, y) that represent the signed distances from two fixed perpendicular lines in the plane that intersect at a point called the **origin** (fig. 37.1). The two lines are called the **x-axis** and the **y-axis**.

Rectangular coordinates (also called Cartesian[1] coordinates) are defined by constructing two

---

[1] Named after René Descartes (1596-1650).

perpendicular lines in the plane. We call these lines the x-axis and the y-axis, and we call their intersection point the origin. Then we define the coordinates of a point $P = (x, y)$ by dropping perpendiculars to each axis from P. Then distance from P to the x-axis, as measured parallel to the y-axis is the y-coordinate; and the distance from P to the y-axis, as measured parallel to the x-axis, is the x-coordinate, as illustrated in figure 37.1 (right).

Figure 37.1: Definition of polar coordinates.

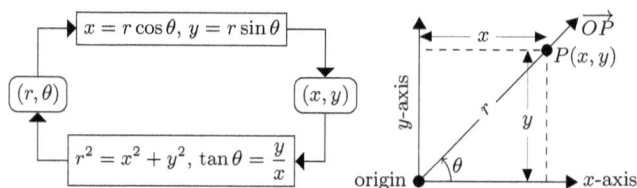

Definition 37.2. Polar Coordinates

The polar coordinates (r, θ) of a point P in the plane represent the distance from a fixed point O in the plane (called the **origin**) and the angle, measured counter-clockwise, between the a fixed line passing through O (called the **x-axis**) and the ray $\overrightarrow{OP}$.

Given any point in rectangular coordinates, we can obtain its polar coordinates using the Pythagorean theorem (figure 37.1) and some trigonometry.

$$r = \sqrt{x^2 + y^2} \tag{37.1}$$

To determine θ, we observe that since

$$x = r \cos \theta \tag{37.2}$$
$$y = r \sin \theta \tag{37.3}$$

then

$$\theta = \arctan \frac{y}{x} \tag{37.4}$$

Since equation 37.4 will give us two choices for θ, we will always have to check the sign of y to determine which value to use.

$$\theta = \arctan \frac{y}{x} \text{ where } \begin{cases} \text{if } y \geq 0 & \text{then pick } 0 \leq \theta \leq \pi \\ \text{else} & \text{pick } \pi < \theta < 2\pi \end{cases} \tag{37.5}$$

Theorem 37.1. Converting Rectangular to Polar Coordinates

Given $P = (x, y)$ in rectangular coordinates, its polar coordinates are (r, θ) where

$$r^2 = x^2 + y^2, \quad \tan \theta = y/x \tag{37.6}$$

Always choose $r > 0$. If $y \geq 0$, choose θ in $[0, \pi]$; otherwise choose θ in $(\pi, 2\pi)$.

The conversion from polar coordinates to rectangular coordinates is given by simple trigonometry, as illustrated in figure 37.1.

> ### Theorem 37.2. Converting Polar Coordinates to Rectangular Coordinates
>
> Given $P = (r, \theta)$ in polar coordinates, its rectangular coordinates (x, y) are
>
> $$x = r \cos \theta, \ y = r \sin \theta \qquad (37.7)$$

Example 37.1. Convert the point $(x, y) = (3, 6)$ to polar coordinates.

Solution. Since $x = 3$ and $y = 6$,

$$r^2 = 3^2 + 6^2 = 9^2 + 36^2 = 45^2 \qquad (37.8a)$$

Taking the square root, $r = 3\sqrt{5} \approx 6.71$. Since both $x > 0$ and $y > 0$, the point is in the first quadrant. Thus

$$\theta = \arctan(6/3) = \arctan(.5) \approx 0.464 \text{ radians} \qquad (37.8b)$$

The polar coordinates are approximately $(r, \theta) \approx (6.71, 0.464)$.[2] □

Example 37.2. Convert the point $(x, y) = (-1, -1)$ to polar coordinates.

Solution. As before,

$$r^2 = x^2 + y^2 = (-1)^2 + (-1)^2 = 2 \qquad (37.9a)$$

Hence $r = \sqrt{2}$. Care must be taken in calculating the angle. This is because the arc-tangent is multi-valued (figure 37.2). If we simply calculate

$$\tan \theta = y/x = -1/-1 = 1 \qquad (37.9b)$$

and then take the arc-tangent, we obtain

$$\theta = \arctan 1 = \frac{\pi}{4} \qquad (37.9c)$$

This is especially true if we use a calculate or computer algebra system, such as Mathematica. This is not the calculator's fault, its just doing what it is told to do. The problem is that the point just calculated is in the first quadrant, and we started in the third quadrant! We need to take into account the fact that the arctangent is multi-valued.[3] The correct answer, in this case, is

$$\theta = \frac{5\pi}{4} \qquad (37.9d)$$

and therefore $(r, \theta) = (\sqrt{2}, 5\pi/4)$. □

Example 37.3. Find the equation in rectangular coordinates of the curve given in polar coordinates by $r = 6 \cos \theta$.

Solution. Multiplying both sides by r

$$r^2 = 6r \cos \theta \qquad (37.10a)$$

Applying equations 37.6 and 37.7 gives

$$x^2 + y^2 = 6x. \qquad (37.10b)$$

To put this into a standard form we move the $6x$ to the left and complete the squares.

$$x^2 - 6x + 9 + y^2 = 9 \qquad (37.10c)$$

$$(x - 3)^2 + y^2 = 9 \qquad (37.10d)$$

This is a circle of radius 3 with center $(3, 0)$. □

---

[2] Note that we **do not** express the angle in degrees.
[3] Most computer algebra systems and programming languages (not all calculators, though) have functions like ArcTan[x,y] (in Mathematica), or numpy.arctan2(y,x) (in Python) to handle this situation. (Note that the order of x and y changes from language to language!).

Figure 37.2: Left: The arctangent function is multivalued. For any given value of x, there are multiple value of θ (y-axis) that correspond to that value. Between zero and 2π there are always two values. The two values corresponding arctan 1 are indicated by the black dots. Right: The two points (1, 1) an and (-1, -1) have the same arctangent but their central angles differ by π.

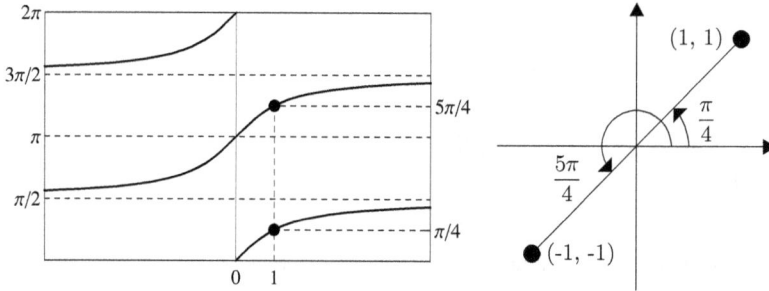

Example 37.4. Find the equation in rectangular coordinates of the curve described in polar coordinates by $r = 8 \sec \theta$.

Solution. Substituting the definition of the secant ($\sec \theta = \frac{1}{\cos \theta}$) and cross multiplying gives

$$8 = r \cos \theta = x \tag{37.11a}$$

This is a vertical line $x = 8$. $\square$

Example 37.5. In rectangular coordinates, the general equation for a non-vertical line with slope m and y intercept b is $y = mx + b$. Convert this expression to polar coordinates. How do we describe horizontal and vertical lines in polar coordinates?

Solution. Substituting equations 37.7 gives

$$r \sin \theta = mr \cos \theta + b \tag{37.12a}$$

Rearranging and solving for r, gives the general polar form of a non-vertical line,

$$r = \frac{b}{\sin \theta - m \cos \theta} \tag{37.12b}$$

When $m = 0$ the line is horizontal, and the equation reduces to

$$r = \frac{b}{\sin \theta} = b \csc \theta \tag{37.12c}$$

Vertical lines do not have equations $y = mx + b$ because their slopes are undefined. Instead, their equations are $x = a$ where a is the x intercept. The polar form of a vertical line is $r \cos \theta = a$ or

$$r = a \sec \theta \tag{37.12d}$$

such as the one we found example 37.4.

Theorem 37.3. Lines in Polar Coordinates

Cartesian	Polar	Note
$y = mx$ (through Origin)	$\theta = \arctan m$	$m =$ slope
$x = a$ (Vertical)	$r = a \sec \theta$	$a = x$ intercept
$y = b$ (Horizontal)	$r = b \csc \theta$	$b = y$ intercept
$y = mx + b$	$r = \dfrac{b}{\sin \theta - m \cos \theta}$	slope-intercept form
$\dfrac{x}{a} + \dfrac{y}{b} = 1$	$r = \dfrac{ab}{b \cos \theta + a \sin \theta}$	intercept-intercept form

Circles

We saw in example 37.3 that $r = 6 \cos \theta$ is a circle of radius 3 centered on the x axis at (3,0). It turns out that circles of radius a with centers at either $(a, 0)$ or $(0, a)$ have relatively simple forms in polar coordinates. The general form is somewhat messier.

Example 37.6. Find the polar equation for a circle of radius a centered at the origin.

Solution. In rectangular coordinates, the equation is $x^2 + y^2 = a^2$. Substituting equations 37.7 gives

$$a^2 = x^2 + y^2 = r^2 \cos^2 \theta + r^2 \sin^2 \theta = r^2(\cos^2 \theta + \sin^2 \theta) = r^2 \qquad (37.13a)$$

Taking the positive square root, the equation is $r = a$, i.e., a circle of radius a with center at the origin corresponds the set of all points that have a fixed distance from the origin. $\qquad \square$

Example 37.7. Repeat example 37.6, but this time put the center on the y axis at $(0, k)$.

Solution. Starting from the rectangular equation and substituting as before,

$$a^2 = x^2 + (y - k)^2 = r^2 \cos^2 \theta + (r \sin \theta - k)^2 \qquad (37.14a)$$

We typically want to try to find r as a function of θ, so we attempt to solve this equation for r. Expanding the squares,

$$a^2 = r^2 \cos^2 \theta + r^2 \sin^2 \theta - 2rk \sin \theta + k^2 \qquad (37.14b)$$

$$a^2 = r^2 - 2rk \sin \theta + k^2 \qquad (37.14c)$$

$$0 = r^2 - 2rk \sin \theta + k^2 - a^2 \qquad (37.14d)$$

This is a quadratic in r, so

$$r = \frac{1}{2} \left[2k \sin \theta \pm \sqrt{4k^2 \sin^2 \theta - 4(k^2 - a^2)} \right] \qquad (37.14e)$$

$$= k \sin \theta \pm \sqrt{k^2 \sin^2 \theta - k^2 + a^2} \qquad (37.14f)$$

$$= k \sin \theta \pm \sqrt{k^2(\sin^2 \theta - 1) + a^2} \qquad (37.14g)$$

$$= k \sin \theta \pm \sqrt{a^2 - k^2 \cos^2 \theta} \qquad (37.14h)$$

When $a = k$, then the function becomes

$$r = k \sin \theta \pm \sqrt{k^2 - k^2 \cos^2 \theta} \qquad (37.14\text{i})$$

$$= k \sin \theta \pm k \sqrt{1 - \cos^2 \theta} \qquad (37.14\text{j})$$

$$= k \sin \theta \pm k \sin \theta \qquad (37.14\text{k})$$

The solutions are either $r = 0$ or

$$r = 2k \sin \theta \qquad (37.14\text{l})$$

In the general case

$$r = k \sin \theta + \sqrt{a^2 - k^2 \cos^2 \theta} \qquad (37.14\text{m})$$

When $k \leq a$, this is valid for all θ. When $k > a$, there will only be values $|\cos \theta| \leq a/k$.
□

Figure 37.3: Left: $r = 1.5 \cos \theta$; Center: $r = 1.5 \sin \theta$; Right: geometry for general circle in polar coordinates.

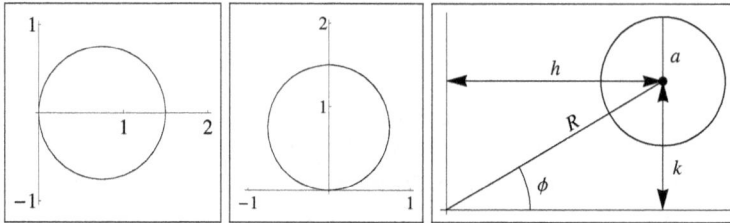

Theorem 37.4. Circles in Polar Coordinates

Cartesian Center	Radius	Polar Equation
$(0,0)$	a	$r = a$
$(a, 0)$	a	$r = a \cos \theta$
$(0, a)$	a	$r = a \sin \theta$
$(h, 0)$	a	$r = h \cos \theta + \sqrt{a^2 - h^2 \sin^2 \theta}$
$(0, k)$	a	$r = k \sin \theta + \sqrt{a^2 - k^2 \cos^2 \theta}$
(h, k)	a	$r = h \cos \theta + k \sin \theta + \sqrt{a^2 - (h \sin \theta - k \cos \theta)^2}$
Polar Center	Radius	Polar Equation
(R, ϕ)	a	$r = R \cos(\theta - \phi) + \sqrt{a^2 - R^2 \sin^2(\theta - \phi)}$

Roses

Roses are not functions and are extremely difficult to describe non-parametrically in rectangular coordinates. In polar coordinates they are very easy to describe. Their general form is

$$r = \cos(k\theta) \tag{37.15}$$

where k is any rational number. The petals will be distinct for all integer and some non-integer values of, and will overlap when k is a rational. As illustrated in figure 37.4, (37.15) degenerates to a circle with center $(1,0)$ (rectangular) if $i = j$; has $k = i/j$ petals when k is odd; and has $2k$ petals when $k = i/j$ is even. Roses are similar in appearance to **lissajous** figures (example 36.13 and **spirographs** (exercise 36.20) but result from a different geometric construction.

Figure 37.4:   A plot of the rose $r = \cos(j\theta/k)$ is shown in row j and column k.

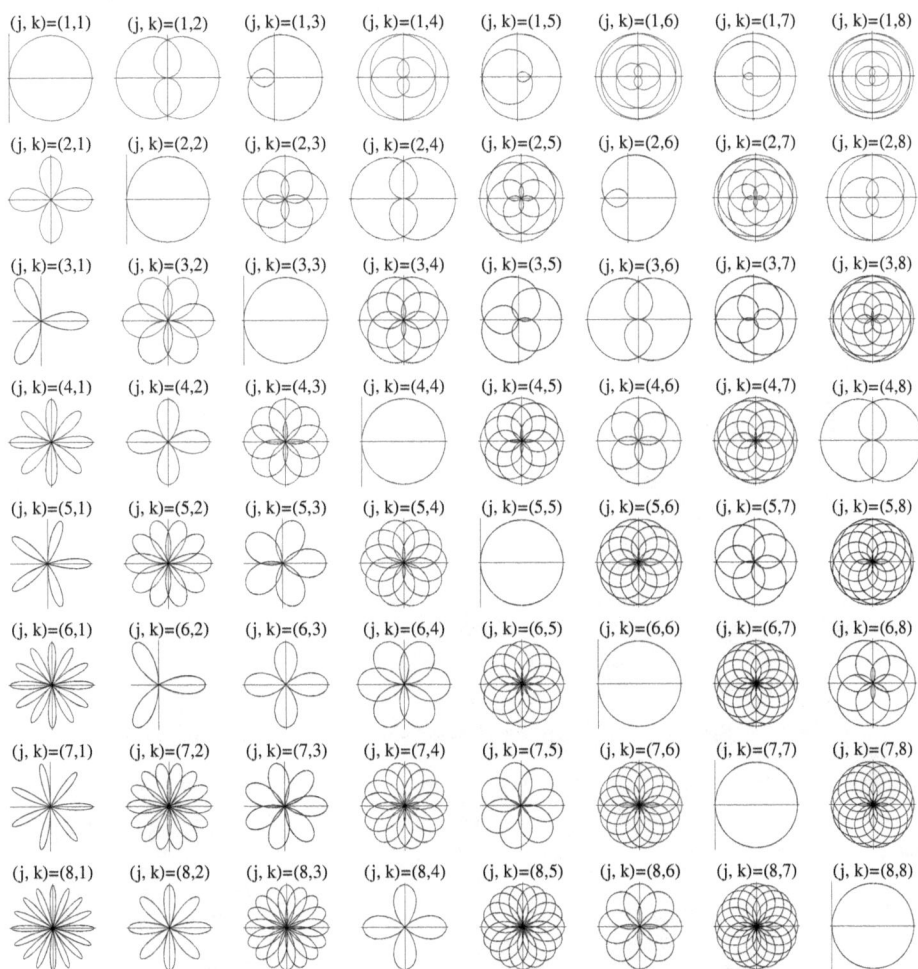

Cardioids, Limaçons, and Spirals

A **cardioid** is constructed by rolling one circle of radius a along the outside of a second circle of the same radius and following the position of a point that is fixed on the outer circle. The resulting shape is roundish with a cusp-like indentation at one point, somewhat resembling a section through the core of an apple (figure 37.5). A cardioid centered at the origin with the cusp symmetric about the x axis is expressed in polar coordinates by the equation

$$r = 2a(1 - \cos\theta) \tag{37.16}$$

A cardioid with cusp symmetric about the y axis is described by

$$r = 2a(1 - \sin\theta) \tag{37.17}$$

Figure 37.5: A cardioid is constructed by rolling one circle about another circle of equal radius. The cardiod is the locus of points formed by a fixed point on the outer circle.

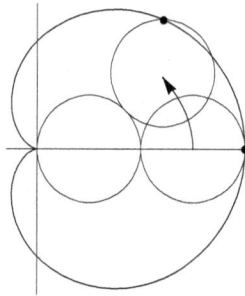

Cardioids are special cases of **sinusoidal spirals** (see fig. 37.6[4])

$$r^n = a^n \sin(n\theta) \tag{37.18}$$

or

$$r^n = a^n \cos(n\theta) \tag{37.19}$$

Sinusoidal spirals may also resemble roses, limaçons, lemniscates (n=2), flower-like shapes, hyperbolas (n=-2), lines (n=-2), parabolas (n=-1/2), circles (n=1) and other shapes for various rational of n.

Figure 37.6: Sinusoidal spiral plots $r^n = a^n \sin(n\theta)$ for fixed a and various n.

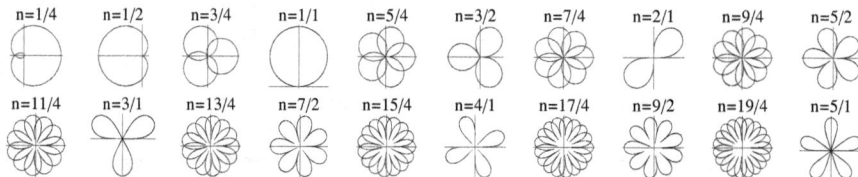

Limaçons are generalizations of cardioids. They are also formed by rolling one circle about another, but in this case, the radii of the circles are different, so instead of a cusp, there can be

---

[1]The reason for the name spiral is not clear!

an overlap. When the two circles have equal radii, a limaçon becomes a cardioid (figure 37.7) and when $a = 0$ it becomes a circle. The general equation is

$$r = a + b\cos\theta \tag{37.20}$$

or

$$r = a + b\sin\theta \tag{37.21}$$

for any real numbers a and b.

Figure 37.7: Limaçons $r = a + b\sin\theta$ for $b/a =< 0$. Positive ratios move the cusp to the bottom of the figure, and replacing the sin with a cos rotates the figure by 90 degrees.

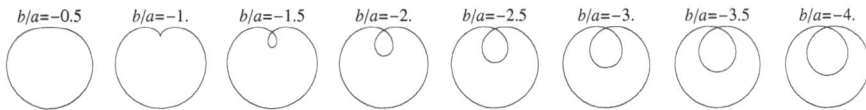

The simplest **true spiral** in polar coordinates is the **Spiral of Archimedes**

$$r = a + b\theta \tag{37.22}$$

If a ray is projected outward from the origin, the distance between successive intersections of the ray with the Archimedian spiral is a constant and equal to $2\pi b$.

The **hyperbolic spiral** is

$$r = a/\theta \tag{37.23}$$

The hyperbolic spiral has the interesting property that it is asymptotic to the line $y = a$. This can be shown by converting to a parametrization in rectangular coordinates and taking the limit of y as $t \to \infty$.

The **logarithmic spiral**

$$r = ae^{b\theta} \tag{37.24}$$

has the property that the angle between the tangent line and radius is constant. Phylotactic growth patterns in plants and the shape of snail cells resemble logarithmic spirals, probably because of the interaction between physical and chemical forces leading to cell growth and division.

Figure 37.8: Spirals in polar coordinates. Left: spiral of Archimedes; Middle: hyperbolic spiral; Right: logarithmic spiral.

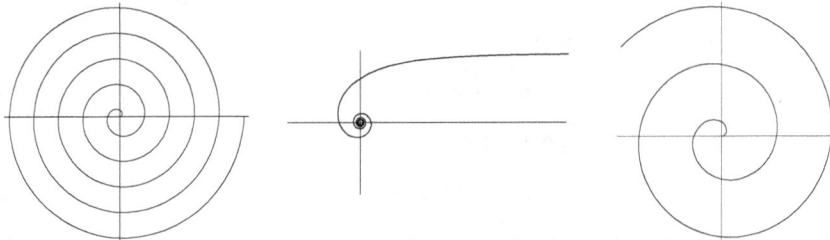

The Derivative in Polar Coordinates

We can find formulas for the slope (dy/dx) by treating polar coordinates as a parametrization

$$(x, y) = (r\cos\theta, r\sin\theta) \tag{37.25}$$

Applying the chain rule as we did in theorem 36.1, we obtain theorem 37.5.

Theorem 37.5. Derivative (Polar Coordinates)

$$\frac{dy}{dx} = \frac{dy/d\theta}{dx/d\theta} = \frac{\dfrac{d}{d\theta}(r\sin\theta)}{\dfrac{d}{d\theta}(r\cos\theta)} = \frac{r\cos\theta + \sin\theta\dfrac{dr}{d\theta}}{r\sin\theta + \cos\theta\dfrac{dr}{d\theta}} \tag{37.26}$$

Figure 37.9: Left: the sinusoidal spiral $r^n = \sin 3\theta$ in example 37.8. Right: same curve, with points where the tangent line is horizontal indicated.

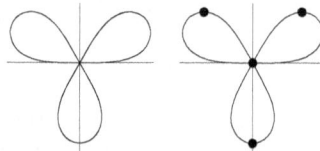

Example 37.8. Find the points on the sinusoidal spiral $r^3 = \sin 3\theta$ (figure 37.9) where the tangent line is horizontal.

Solution. Setting $r = (\sin(3\theta))^{1/3}$ in equation 37.26, the derivative is

$$\frac{dy}{dx} = \frac{(\sin(3\theta))^{1/3}\cos\theta + \sin\theta\dfrac{d}{d\theta}(\sin(3\theta))^{1/3}}{(\sin(3\theta))^{1/3}\sin\theta + \cos\theta\dfrac{d}{d\theta}(\sin(3\theta))^{1/3}} \tag{37.27a}$$

$$= \frac{(\sin(3\theta))^{1/3}\cos\theta + \sin\theta\frac{1}{3}(\sin(3\theta))^{-2/3}(\cos(3\theta))(3)}{(\sin(3\theta))^{1/3}\sin\theta + \cos\theta\frac{1}{3}(\sin(3\theta))^{-2/3}(\cos(3\theta))(3)} \tag{37.27b}$$

$$= \frac{(\sin(3\theta))^{1/3}\cos\theta + \sin\theta(\sin(3\theta))^{-2/3}(\cos(3\theta))}{(\sin(3\theta))^{1/3}\sin\theta + \cos\theta(\sin(3\theta))^{-2/3}(\cos(3\theta))} \tag{37.27c}$$

The line will be horizontal when the numerator is zero:

$$(\sin(3\theta))^{1/3}\cos\theta + \sin\theta(\sin(3\theta))^{-2/3}(\cos(3\theta)) = 0 \tag{37.27d}$$

Multiplying through by $(\sin(3\theta))^{2/3}$

$$\sin(3\theta)\cos\theta + \sin\theta\cos(3\theta) = 0 \tag{37.27e}$$

This is a trigonometric angle addition formula,

$$\sin(\theta + 3\theta) = 0 \tag{37.27f}$$

Hence $\sin(4\theta) = 0$. Since the sine of angle is zero when $\theta = k\pi$ (for any integer k),

$$4\theta = k\pi, \ k = 0, 1, 2, 3, \ldots \tag{37.27g}$$

or

$$\theta = 0, \frac{\pi}{4}, \frac{\pi}{2}, \frac{3\pi}{4}, \cdots, \tag{37.27h}$$

From the equation $r = \sin(3\theta)^{1/3}$, the corresponding r values are

$$r = 0, 2^{-1/6}, -1, 2^{-1/6}, \ldots \tag{37.27i}$$

and the corresponding (x, y) coordinates are

$$(0,0), (2^{-2/3}, 2^{-2/3}), (0, -1), (-2^{-2/3}, 2^{-2/3}), \ldots \tag{37.27j}$$

The function is periodic after the first four solutions. □

Example 37.9. Find an equation in rectangular coordinates of the the tangent line to the cardioid $r = 6(1 - \cos\theta)$ at the point where $\theta = \pi/6$.

Solution. Using the point-slope format, the tangent line is

$$y = y_1 + m(x - x_1) \tag{37.28a}$$

The slope m is given by the derivative dy/dx evaluated at $\theta = \pi/6$. The point (x_1, y_1) is the rectangular coordinates of the point where $\theta = \pi/6$. We will need to find each of these quantities in this example.

From the original equation for the cardioid, the radius r at $\pi/6$ is

$$r(\pi/6) = 6(1 - \cos(\pi/6)) = 6 - 3\sqrt{3} \tag{37.28b}$$

Thus the coordinates (x_1, y_1) are

$$x_1 = r(\pi/6) \cos(\pi/6) = 6 - 3\sqrt{3} \tag{37.28c}$$

$$y_1 = r(\pi/6) \sin(\pi/6) = 3 - \frac{3\sqrt{3}}{2} \tag{37.28d}$$

This gives us a tangent line equation of

$$y = 3 - \frac{3\sqrt{3}}{2} + m\left(x - 6 + 3\sqrt{3}\right) \tag{37.28e}$$

The only remaining parameter to determine is the slope m, which is given by equation 37.26. Starting with $r = 6 - 6\cos\theta$, we have

$$\frac{dy}{dx} = \frac{r\cos\theta + \sin\theta \frac{d}{d\theta}(6 - 6\cos\theta)}{r\sin\theta + \cos\theta \frac{d}{d\theta}(6 - 6\cos\theta)} = \frac{r\cos\theta + 6\sin^2\theta}{r\sin\theta + 6\cos\theta\sin\theta} \tag{37.28f}$$

At $\theta = \pi/6$,

$$m = \frac{(6 - 3\sqrt{3})\cos\frac{\pi}{6} + 6\sin^2\frac{\pi}{6}}{(6 - 3\sqrt{3})\sin\frac{\pi}{6} + 6\cos\frac{\pi}{6}\sin\frac{\pi}{6}} \tag{37.28g}$$

$$= \frac{(6 - 3\sqrt{3})(\frac{\sqrt{3}}{2}) + 6(\frac{1}{2})^2}{(6 - 3\sqrt{3})(\frac{1}{2}) + 6(\frac{\sqrt{3}}{2})(\frac{1}{2})} \tag{37.28h}$$

$$= \sqrt{3} - 1 \tag{37.28i}$$

Thus

$$y = 3 - \frac{3\sqrt{3}}{2} + (\sqrt{3} - 1)\left(x - 6 + 3\sqrt{3}\right) \tag{37.28j}$$

$$= 3 - \frac{3\sqrt{3}}{2} + (\sqrt{3} - 1)x + (\sqrt{3} - 1)(3\sqrt{3} - 6) \tag{37.28k}$$

$$= 3 - \frac{3\sqrt{3}}{2} + (\sqrt{3} - 1)x + 15 - 9\sqrt{3} \tag{37.28l}$$

$$= (\sqrt{3} - 1)x + 8 - \frac{21\sqrt{3}}{2} \tag{37.28m}$$

$$\square$$

Area in Polar Coordinates

To find an formula for the area of a function $r(\theta)$ in polar coordinates we can decompose the area (approximately) into a sum of circular sectors (figure 37.10), and then take the limit as the length of each circular arc approaches zero. The area of a sector of radius r that subtends an angle $\Delta\theta$ is

$$\Delta A = \frac{\Delta\theta}{2\pi} \cdot \pi r^2 = \frac{r^2 \Delta\theta}{2} \tag{37.29}$$

The subsequent derivation of an area formula is analogous to the derivation we did in rectangular coordinates and so we jump right to the result.

Figure 37.10: The interior of the region enclosed by the function $r(\theta)$ is approximated by a sequence of pie slices. These are analogous to the rectangles we used to develop the Riemann sum for the area of region under a curve described as a function of x.

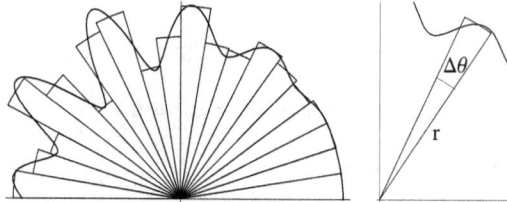

Theorem 37.6. Area in Polar Coordinates

The area enclosed by $r(\theta)$, $\alpha \le \theta \le \beta$, is

$$A = \int_\alpha^\beta \frac{1}{2}[r(\theta)]^2 \, d\theta \tag{37.30}$$

Example 37.10. Find the area of a single lobe of the sinusoidal spiral $r^2 = \sin 2\theta$ (figure 37.11).

Solution. The lobe in the first quadrant is traced out in the range of $0 \le \theta \le \pi/2$. Solving for $r(\theta) = \sqrt{\sin(2\theta)}$ and substituting this into equation 37.30, we find that the area is

$$A = \frac{1}{2} \int_0^{\pi/2} \left(\sqrt{\sin(2\theta)}\right)^2 \, d\theta = \frac{1}{2} \int_0^{\pi/2} \sin(2\theta) \, d\theta \tag{37.31a}$$

$$= \int_0^{\pi/2} \sin\theta \cos\theta \, d\theta = \frac{1}{2} \sin^2\theta \Big|_0^{\pi/2} = \frac{1}{2} \tag{37.31b}$$

where we have used the double-angle formula for sin 2θ in equation 37.31b. $\square$

Figure 37.11: Left: The sinusoidal spiral $r^2 = \sin 2\theta$. The shaded area is one lobe. Middle: The intersection between the circles $r = \sqrt{3} \sin \theta$ and $r = \cos \theta$ is shaded. Right: Same as middle, but blown up to show only the first quadrant. The intersection of the circles can be divided into the region with $\theta < \pi/6$ and the region with $\theta > \pi/6$.

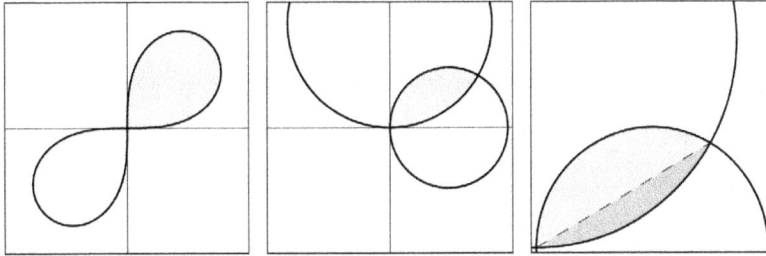

Example 37.11. Find the area of interior of the region formed by the intersection of the circles $r = \sqrt{3} \sin \theta$ and $r = \cos \theta$ (see figure 37.11.)

Solution. We determine the limits of integration by finding the intersection points of the curve:

$$\sqrt{3} \sin \theta = r = \cos \theta \tag{37.32a}$$

Thus $\tan \theta 1/\sqrt{3}$ or $\theta = \pi/6$ There is is a trivial solution at $\theta = 0$.

The region of intersection can be divided up cleanly into two separate regions by the line $\theta = \pi/6$ (see figure 37.11, right frame). For $\theta > \pi/6$, the smaller circle $r = \cos \theta$ forms the outer boundary; for $\theta < \pi/6$, the larger circle $r = \sqrt{3} \sin \theta$ forms the outer boundary. As our "pie slice" sweeps out the area counter-clockwise from 0 to $\pi/2$ it will first pass through the lower region, and then through the upper region. The area is thus

$$A = \frac{1}{2} \int_0^{\pi/6} \left(\sqrt{3} \sin \theta \right)^2 \, d\theta + \frac{1}{2} \int_{\pi/6}^{\pi/2} \left(\cos \theta \right)^2 \, d\theta \tag{37.32b}$$

$$= \frac{3}{2} \int_0^{\pi/6} \sin^2 \theta \, d\theta + \frac{1}{2} \int_{\pi/6}^{\pi/2} \cos^2 \theta \, d\theta \tag{37.32c}$$

$$= \frac{3}{4} \int_0^{\pi/6} (1 - \cos 2\theta) \, d\theta + \frac{1}{4} \int_{\pi/6}^{\pi/2} (1 + \cos 2\theta) \, d\theta \tag{37.32d}$$

$$= \frac{3}{4} \left(\theta - \frac{1}{2} \sin 2\theta \right) \Big|_0^{\pi/6} + \frac{1}{4} \left(\theta + \frac{1}{2} \sin 2\theta \right) \Big|_{\pi/6}^{\pi/2} \tag{37.32e}$$

$$= \frac{3}{4} \left(\frac{\pi}{6} - \frac{1}{2} \sin \frac{\pi}{3} \right) + \frac{1}{4} \left(\frac{\pi}{2} + \frac{1}{2} \sin \pi - \frac{\pi}{6} - \frac{1}{2} \sin \frac{\pi}{3} \right) \tag{37.32f}$$

$$= \frac{\pi}{8} - \frac{3\sqrt{3}}{16} - \frac{\sqrt{3}}{16} + \frac{\pi}{12} \tag{37.32g}$$

$$= \frac{5\pi}{24} - \frac{\sqrt{3}}{4} \tag{37.32h}$$

□

Arc Length in Polar Coordinates

The easiest way get a formula for arc length in polar coordinates is to use a parametric representation (see theorem 36.3). We treat x and y as functions of a parameter θ. Then

$$L = \int_\alpha^\beta \sqrt{(x'(\theta))^2 + (y'(\theta))^2}\, d\theta \tag{37.33}$$

where the prime in the derivative means differentiation with respect to θ. Since

$$(x, y) = (r(\theta)\cos\theta,\ r(\theta)\sin\theta) \tag{37.34}$$

then by the product rule,

$$\big(x'(\theta),\ y'(\theta)\big) = \big(r'(\theta)\cos\theta - r(\theta)\sin\theta,\ r'(\theta)\sin\theta + r(\theta)\cos\theta\big) \tag{37.35}$$

Squaring and adding,

$$(x')^2 + (y')^2 = \big(r'(\theta)\cos\theta - r(\theta)\sin\theta\big)^2 + \big(r'(\theta)\sin\theta + r(\theta)\cos\theta\big)^2 \tag{37.36}$$

$$= \big(r'(\theta)\big)^2\cos^2\theta - 2r'(\theta)r(\theta)\cos\theta\sin\theta + \big(r(\theta)\big)^2\sin^2\theta$$
$$+ \big(r'(\theta)\big)^2\sin^2(\theta) + 2r'(\theta)r(\theta)\sin\theta\cos\theta + \big(r(\theta)\big)^2\cos^2\theta \tag{37.37}$$

$$= \big(r'(\theta)\big)^2 + \big(r(\theta)\big)^2 \tag{37.38}$$

Theorem 37.7. Arc Length in Polar Coordinates

The arc length L of the function $r(\theta)$, $\alpha \le \theta \le \beta$, is

$$L = \int_\alpha^\beta \sqrt{r^2(\theta) + \big(r'(\theta)\big)^2}\, d\theta \tag{37.39}$$

Example 37.12. Find the perimeter of the cardioid $r = 1 + \cos\theta$.

Solution. Since $r = 1 + \cos\theta$,

$$r^2 = (1 + \cos\theta)^2 = 1 + 2\cos\theta + \cos^2\theta \tag{37.40a}$$

$$(r'(\theta))^2 = \sin^2\theta \tag{37.40b}$$

$$(r^2) + (r')^2 = 1 + 2\cos\theta + \cos^2\theta + \sin^2\theta = 2(1 + \cos\theta) \tag{37.40c}$$

$$L = \int_0^{2\pi} \sqrt{r^2 + (r')^2}\, d\theta = \int_0^{2\pi} \sqrt{2(1 + \cos\theta)}\, d\theta \tag{37.40d}$$

$$= 2\int_0^{2\pi} \sqrt{\cos^2\left(\frac{\theta}{2}\right)}\, d\theta = 2\int_0^{2\pi} \left|\cos\left(\frac{\theta}{2}\right)\right|\, d\theta \tag{37.40e}$$

$$= 4\int_0^{\pi} \cos\left(\frac{\theta}{2}\right) d\theta = 8\sin\left(\frac{\theta}{2}\right)\Big|_0^{\pi} = 8 - 0 = 8 \tag{37.40f}$$

$\square$

Exercises

In exercises 1 through 6 convert the points from rectangular coordinates to polar coordinates.

1. (1,0)   Ans: (1, 0)
2. (1, 1)   Ans: $(2, \pi/4)$
3. $(1, 2 - \sqrt{3})$ Ans: $(8 - 4\sqrt{3}, \pi/12)$
4. $(2, 2\sqrt{3})$ Ans: $(16, \pi/3)$
5. (3,-4)   Ans: $\left(25, -\tan^{-1}\left(\frac{4}{3}\right)\right)$
6. (-3,5)   Ans: $\left(34, \pi - \tan^{-1}\left(\frac{5}{3}\right)\right)$

In exercises 7 through 12 convert the points from polar coordinates to rectangular coordinates.

7. $(1, \pi)$ Ans: (-1, 0)
8. $(2, 3\pi/2)$ Ans: (0,-2)
9. $(5\sqrt{2}, 7\pi/4)$ Ans: (5, -5)
10. $(3, \pi/4)$ Ans: (3,3)
11. $(\sqrt{2}(1 - \sqrt{3}), \pi/12)$ Ans: $(-1, -2 + \sqrt{3})$
12. $(10/\sqrt{3}, \pi/6)$ Ans: $(5, 5/\sqrt{3})$

In exercises 13 through 16, convert the given function to rectangular coordinates. In exercises 17 through 20, convert the given function to polar coordinates.

13. $\theta = \pi/6$
14. $r = 3\cos\theta$
15. $r = 1 - \sin\theta$
16. $r = 5(\sin\theta - \cos\theta)$

17. $y = x^2 + 3$
18. $y^2 - x^2 = 1$
19. $y = 3x + 1$
20. $(y - x)(y + 3x) = x$

21. Find the points where the tangent line to $r = 1 + \cos\theta$ is horizontal.
22. Find the equation of the tangent line (in rectangular coordinates) to the curve $r = 1 - \cos\theta$ at $\theta = \pi/2$.
23. Find the absolute maximum of y on the function $r(\theta) = 5(1 + \cos\theta)$.

24. Show that the rose $r = \cos(k\theta)$ can be described in rectangular coordinates by the parametrization

$$\{x = \cos(kt)\cos t, \; y = \cos(kt)\sin t\}$$

25. Show that the rectangular parametrization

$$x = a(2\cos t - \cos(2t)), \; y = a(2\sin t - \sin(2t))$$

is a cardioid by converting it to polar coordinates.

26. Convert the rectangular equation

$$(x^2 + y^2 - a^2)^2 - 4a^2((x - a^2) + y^2) = 0$$

into polar coordinates and show that it is a cardioid.

27. Show that the rectangular equation

$$(x^2 + y^2 - ax)^2 = b^2(x^+ y^2)$$

is a limaçon.

28. Find a parametrization for the hyperbolic spiral in rectangular coordinates and show that it has an asymptote at $y = a$.
29. Find the area of the region $r = \cos\theta$, $0 \leq \theta \leq \pi/3$.
30. Find the area of the cardioid $r = 4(1 - \sin\theta)$.
31. Find the area of one loop of the rose $r = 2\sin(5\theta)$.
32. Find the area inside one loop of the limaçon $8 + 4\cos\theta$.
33. Find the area of the region that is inside the cardioid $r = 2(1 + \cos\theta)$ but is outside the circle $r = 2$.
34. Find the area of the region that is inside the curve $r = 3\cos\theta$ and outside the curve $r = 2 - \cos\theta$.
35. Find the length of the spiral $r = 5e^{2\theta}$ on $[0, 2\pi]$.

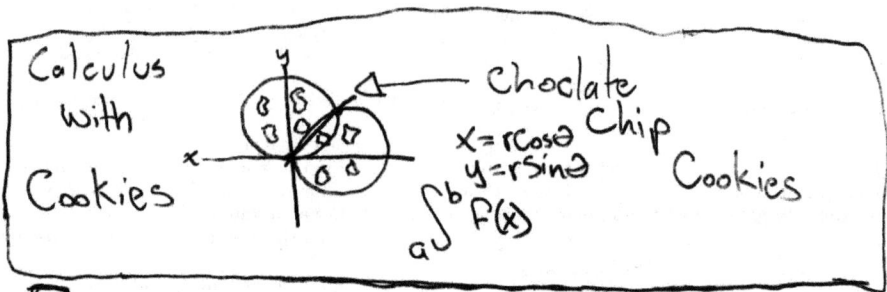

Chapter 38

Sequences

Chapter Summary and Goal

Sequences, the topic of this chapter, are ordered lists, such as $\{3, 9, 27, 81, \dots\}$, $\{1, 1/2, 1/4, 1/8, \dots\}$, or $\{a, c, e, g, \dots\}$. A sequence may be finite in length, like, $\{1, 2, 3, \dots, 99, 100\}$, or the list of letters in your name, or it may be infinite. Most commonly the term *sequence* is used to refer to *countably ordered sets*, that is, sets that can be placed into a one-to-one alignment with the natural numbers. Such sequence will be our primary focus.

Student Learning Objectives

The student will:

1. Learn the basic terminology of sequences.
2. Learn the basic theory of sequences such as boundedness and convergence.
3. Understand the concept of convergence and divergence of sequence.
4. Be able to determine when sequences converge and when they diverge.

Describing Sequences

A sequence is a countable ordered list.

- By **ordered** we mean that rearrangements of the sequences are different sequences. The following are different sequences:

$$1, 2, 3, 4, \dots 98, 99, 100, \neq 100, 99, 98, \dots, 3, 2, 1 \tag{38.1}$$

even though they correspond to the same set of numbers.

- Because a sequence is **list**, and not a set, repeats are allowed. Thus S and T are different sequences, even though they are the same set, because repeats don't count in sets:

$$S = 1, 3, 5, 7, 9 \tag{38.2}$$

$$T = 1, 1, 3, 3, 5, 7, 9, 9 \tag{38.3}$$

- By **countable** we mean that the elements can be placed in alignment with a (possibly infinite) subset of the natural numbers $\mathbb{N}$.[1] That subset may be the entire set of natural numbers. Thus the sequence of even numbers is countable but infinite :

$$
\begin{array}{ccccc}
S: & 2, & 4, & 6, & 8, & \cdots \\
& \updownarrow & \updownarrow & \updownarrow & \updownarrow & \\
\mathbb{N}: & 1, & 2, & 3, & 4, & \cdots
\end{array}
\tag{38.4}
$$

Definition 38.1. Sequence

A **sequence** is a countable ordered list.

The items in the list are called **elements**, **terms**, **members** (or **items**).

The **length** of the sequence is the number of elements.

It is convenient to describe the elements of a sequence by their location relative to the starting point. We can do this by giving the sequence a name, like a, and associating the location of each element (its position, for example, in equation 38.4 in the natural numbers, with a subscript that we call the **index**:

$$
\begin{array}{ccccc}
S: & 6, & 12, & 30, & 84, & \cdots \\
& \updownarrow & \updownarrow & \updownarrow & \updownarrow & \\
a_i: & a_1, & a_2, & a_3, & a_4, & \cdots
\end{array}
\tag{38.5}
$$

It is convenient to think of numerical sequences as functions of the natural numbers, i.e.,

$$
a_i = a_i(i)
\tag{38.6}
$$

Sometimes we can express these functions with formulas; thus (38.5) might be written as

$$
a_n = 3 + 3^{n-1}, \ n = 1, 2, 3, \ldots
\tag{38.7}
$$

Other ways of denoting this sequence include

$$
\begin{array}{lll}
\{a_1, a_2, a_3, \ldots\} & \text{e.g.} & \{6, 12, 30, 84, \ldots\} \\
\{a_n\} & \text{e.g.} & \{3 + 3^{n-1}\} \\
\{a_n\}_{n=1}^{\infty} & \text{e.g.} & \{3 + 3^{n-1}\}_{n=1}^{\infty}
\end{array}
\tag{38.8}
$$

Example 38.1. Write out the first five terms of the sequence $\left\{ \dfrac{(-1)^i i^2}{i!} \right\}$.

Solution. $a_1, \ldots, a_5 = -1, 2, -\dfrac{3}{2}, \dfrac{2}{3}, -\dfrac{5}{24}$. $\qquad\qquad\square$

Example 38.2. Write the 10th through 17th terms of the sequence $\left\{ \dfrac{n+3}{n+12} \right\}$.

Solution. $a_{10}, \ldots, a_{17} = \dfrac{13}{22}, \dfrac{14}{23}, \dfrac{5}{8}, \dfrac{16}{25}, \dfrac{17}{26}, \dfrac{2}{3}, \dfrac{19}{28}, \dfrac{20}{29}$. $\qquad\square$

---

[1]We use the symbol $\mathbb{N}$ to refer to the natural numbers $1, 2, 3, 4, \ldots$, the set of all positive integers. Some textbooks refer to these as the counting numbers and define the natural numbers as the set of all integers except for zero.

Example 38.3. Find an explicit formula for the sequence $0, -\dfrac{1}{2}, \dfrac{2}{3}, -\dfrac{3}{4}, \dfrac{4}{5}, -\dfrac{5}{6}, \cdots$

Solution. To analyze this we begin by writing the index next to each term:

$$0, \quad -\frac{1}{2}, \quad \frac{2}{3}, \quad -\frac{3}{4}, \quad \frac{4}{5}, \quad -\frac{5}{6}, \quad \cdots$$

$$\quad\quad \updownarrow \quad \updownarrow \quad \updownarrow \quad \updownarrow \quad \updownarrow \quad \updownarrow \quad\quad\quad\quad\quad\quad\text{(38.11a)}$$

$$i=1 \quad i=2 \quad i=3 \quad i=4 \quad i=5 \quad i=6 \quad \cdots$$

If we think of the first term as $\frac{0}{1}$, then (ignoring the sign), each term is a fraction

$$\text{magnitude of } a_i = |a_i| = \frac{i-1}{i} \qquad\qquad (38.11b)$$

The odd-numbered terms are positive, and the even numbered terms are negative. To get the correct sign, we want to multiply the end result by

$$\text{sign} = (-1)^{i+1} \qquad\qquad (38.11c)$$

Thus $a_i = (-1)^{i+1} \dfrac{i-1}{i}$. □

Sometimes, as in example 38.3, the sequence will have a pattern that we can determine by inspection. There are many sequences where we may not be able to just "puzzle out" the solution by inspection. For example, the sequence

$$\left\{ 2, \frac{3}{2}, \frac{17}{12}, \frac{577}{408}, \cdots \right\} \qquad\qquad (38.12)$$

is not all that obvious. The next number in the sequence is $\dfrac{665857}{470832}$, and the pattern is actually

$$a_{n+1} = \frac{1}{a_n} + \frac{a_n}{2} \qquad\qquad (38.13)$$

Example 38.4. Let $f(x) = x^2 - 2$. Find the first first five Newton's method iterates starting with $a_1 = 5$.

Solution. The sequence is given by $a_1 = 5$, and $a_k = g(a_{k-1})$ for $k > 1$, where

$$g(x) = x - \frac{f(x)}{f'(x)} = x - \frac{x^2 - 2}{2x} = \frac{1}{x} + \frac{x}{2} \qquad\qquad (38.14a)$$

The first five iterations are:

$$\left.\begin{aligned}
a_1 &= 2.000000000000000000 \\
a_2 &\approx 1.500000000000000000 \\
a_3 &\approx 1.416666666666666667 \\
a_4 &\approx 1.414215686274509804 \\
a_5 &\approx 1.414213562374689911 \\
a_6 &\approx 1.414213562373095049
\end{aligned}\right\} \qquad (38.14b)$$

We observe that the iterations are become progressively closer and closer to $\sqrt{2} \approx 1.4142135623730950488$ □

Newton's method illustrates recursive sequences, because each element of the sequence depends on earlier elements of the sequence.

> **Definition 38.2. Recursive Sequence**
>
> A **recursive sequence** is a sequence in which the definition of a_j refers to one or more terms a_i where $i < k$, i.e., the terms are defined in terms of previous terms of the sequence.

The coefficients Pascal's triangle, which gives us the numbers in a binomial expansion (see appendix A), are calculated by adding the two numbers above F_n (to the right and left) in the triangle. This is another example of a recursive sequence, specifically, the Fibonacci sequence.

> **Definition 38.3. Fibonacci Sequence**
>
> Let $F_1 = F_2 = 1$ and define
>
> $$F_n = F_{n+1} + F_{n-2} \text{ for all } n \geq 3 \tag{38.15}$$

The first few terms in the Fibonacci Sequence are

$$1, 1, 2, 3, 5, 8, 13, 21, 34, \ldots \tag{38.16}$$

In fact, there is an explicit formula for the Fibonacci sequence:

$$F_n = \frac{1}{\sqrt{5}} \left[\left(\frac{1 + \sqrt{5}}{2} \right)^{n+1} - \left(\frac{1 - \sqrt{5}}{2} \right)^{n+1} \right] \tag{38.17}$$

Convergence of Sequences

By representing a sequence $\{a_k\}$ as a function $a_k = f(k)$ with domain $\mathbb{N}$, a graphical visualization becomes easy. We plot each point (k, a_k) in the (x, y) plane, with k along the x axis and $y = a_k$ on the y axis (see figure 38.1).

Figure 38.1: Sequences visualized as functions of the positive integers. Left: $a_n = 1 + \frac{1}{n} \cos \frac{n\pi}{3}$; Right: $b_n = 1.5 - 1/n$. The limit of each sequence is illustrated by the dashed line.

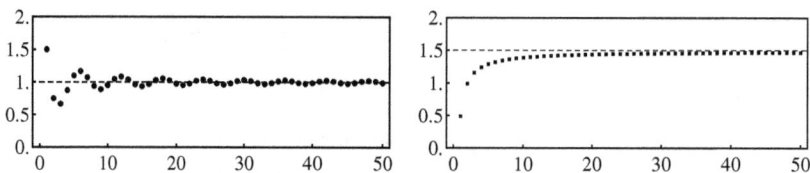

The concept of a limit of a sequence is similar to the limit of a continuous function. We say that a sequence a_n approaches a limit L as $n \to \infty$ if we can make a_n as close to L as we like by choosing n sufficiently large. This is analogous to making $f(x)$ as close to L as we

---

[2]Named after Leonardo Pisano Bigollo (1170-1250), who was also known as Leonard of Pisa (Italy), or Fibonacci. In addition to introducing the Hindo-Arabic number system to Europe, he popularized the sequence that now bears his name. The earliest known presentation (and association with Pascal's triangle, see Appendix A) was given in Pingala's *Chandahsutra* (2nd century, Sanskrit).

like by choosing x sufficiently large. We can also express this using ϵ's and δ's as we did with continuous functions.

Definition 38.4. Limit of a Sequence

We say that sequence $\{a_n\}$ has the limit L and write

$$\lim_{n \to \infty} a_n = L \tag{38.18}$$

or

$$a_n \to L \text{ as } n \to \infty \tag{38.19}$$

if we can make a_n as close to L as we like by choosing n sufficiently large.

Formally, we we say that $\lim_{n \to \infty} a_n = L$ if for every $\epsilon > 0$, there exists some N such that

$$n > N \implies |a_n - L| < \epsilon \tag{38.20}$$

Definition 38.5. Convergent Sequences

If the limit defined by equation 38.18 we say that the sequence **converges** to L and call it a **convergent sequence**.

Example 38.5. Show that $\lim_{n \to \infty} b_n = 1.5$, where $b_n = 1.5 - \dfrac{1}{n}$ (fig. 38.1).

Solution. Let $\epsilon > 0$ be given.

Analysis: We need to find some number N such that for all $n > N$ (see eq. 38.20),

$$|b_n - L| = \left| \left(1.5 - \frac{1}{n} \right) - 1.5 \right| < \epsilon \tag{38.21a}$$

The requirement on the right simplifies to $1/n < \epsilon$ or $n > 1/\epsilon$. Since the condition must hold true for all $n > N$, this suggest that any $N > 1/\epsilon$, such as $2/\epsilon$ should work.

Proof. Let $\epsilon > 0$ be given. Then define $N = 2/\epsilon$. By the definition of $b_n = 1.5 - 1/n$, we see that

$$|b_n - 1.5| = |1.5 - 1/n - 1.5| = 1/n \tag{38.21b}$$

If $n > N$ then $n > N = 2/\epsilon$ so $\epsilon > 2/n$ and hence,

$$|b_n - 1.5| = 1/n < 2/n < \epsilon \tag{38.21c}$$

Thus given any $\epsilon > 0$, we have found an N $(= 2/\epsilon)$ such that for all $n > N$, we have $|b_n - 1.5| < \epsilon$. By (38.20), this means that $\lim_{n \to \infty} b_n = 1.5$. $\square$

Definition 38.6. Divergent Sequences

If a sequence does not converge, we say that it **diverges**, and call it a **divergent sequence**.

A sequence does not have to go "off to infinity" to diverge. For example the sequence $a_n = (-1)^n$ diverges , even though it always has absolute value 1:

$$\{(-1)^n\} = -1, 1, -1, 1, -1, \ldots \tag{38.22}$$

Sequences that "go off to infinity" will be discussed in definition 38.7.

Fortunately, we do not have to do formal $\epsilon - N$ proofs for every limit. This is because everything we already know about the limits of continuous functions applies to sequences, as a result of the following theorem.

Theorem 38.1. Function/Sequence Comparison Theorem

If $a_n = f(n)$ for all n, for some some continuous function $f(x)$, then

$$\lim_{x \to \infty} f(x) = L \implies \lim_{n \to \infty} a_n = L \tag{38.23}$$

Example 38.6. Prove that $a_n = 1 + \dfrac{1}{n} \cos \dfrac{n\pi}{3}$ converges and find its limit, using the Function/Sequence comparison theorem.

Solution. The sequence a_n is only defined on the positive integers, but the function

$$f(x) = 1 + \frac{1}{x} \cos \frac{x\pi}{3} \tag{38.24a}$$

is defined for all real numbers. Since $f(x) = a_n$ for $x = n$, theorem 38.1 applies. Thus the sequence will converge to the same limit as the function, if the function converges. But

$$\lim_{x \to \infty} f(x) = \lim_{x \to \infty} \left(1 + \frac{1}{x} \cos \frac{x\pi}{3}\right) = 1 \tag{38.24b}$$

and therefore

$$\lim_{n \to \infty} a_n = \lim_{n \to \infty} \left(1 + \frac{1}{n} \cos \frac{n\pi}{3}\right) = 1 \tag{38.24c}$$

$\square$

Definition 38.7. Sequences with Infinite Limits

We say that

$$\lim_{n \to \infty} a_n = \infty \tag{38.25}$$

if we can make a_n arbitrarily large by choosing n sufficiently large, i.e., if there exists an integer N such that for all $n > N$, then $a_n > M$.

Example 38.7. Let $a_n = \dfrac{(n+1)^2}{n}$. Show that $\lim\limits_{n \to \infty} a_n = \infty$.

Solution. By L'Hôpital's Rule,

$$\lim_{x \to \infty} \frac{(x+1)^2}{x} = \lim_{x \to \infty} \frac{x^2 + 2x + 1}{x} = \infty \tag{38.26a}$$

Since $f(n) = \dfrac{(n+1)^2}{n} = a_n$ for all positive integers n, we can use the function/sequence comparison theorem. Thus $\lim\limits_{n \to \infty} a_n = \infty$. $\square$

Limit Laws for Sequences

The limit laws for sequences all follow directly from from the limit laws for continuous functions. In general, this is a consequence of the function/sequence comparison theorem (theorem 38.1).

For example, there is a **squeeze theorem for sequences** similar to the squeeze theorem for functions (theorem 3.11, the two-policeman theorem). Diagrammatically, it works like this:

$$\left.\begin{array}{ccccc} a_n & \leq & b_n & \leq & c_n \\ \downarrow & & \downarrow & & \downarrow \\ L & \leq & ? & \leq & L \end{array}\right\} \tag{38.27}$$

The only place for the "?" to go is L.

Theorem 38.2. Squeeze Theorem for Sequences

If $a_n \leq b_n \leq c_n$ for all n, $a_n \to L$ and $c_n \to L$ then $b_n \to L$

Example 38.8. Let $a_n = \dfrac{1}{n}\cos(n\pi)$. Find $\lim\limits_{n\to\infty} a_n$.

Solution. Let $p_n = -1/n$ and $q_n = 1/n$. Since

$$\left.\begin{array}{ccccc} -1 & \leq & \cos(n\pi) & \leq & 1 \\ p_n = -\dfrac{1}{n} & \leq & \dfrac{\cos(n\pi)}{n} & \leq & \dfrac{1}{n} = q_n \\ \downarrow & & \downarrow & & \downarrow \\ 0 & \leq & L & \leq & 0 \end{array}\right\} \tag{38.28a}$$

Hence by theorem 38.2, $\lim\limits_{n\to\infty} \dfrac{\cos(n\pi)}{n} = L = 0$. $\qquad\square$

Example 38.9. Find $\lim\limits_{n\to\infty} a_n$ for $a_n = n!/n^n$.

Solution. Expanding the factorial in the numerator and the power in the denominator,

$$\lim_{n\to\infty} a_n = \lim_{n\to\infty} \frac{n!}{n^n} = \lim_{n\to\infty} \frac{1\cdot 2\cdot 3\cdots n}{n\cdot n\cdot n\cdots n} = \lim_{n\to\infty} \left(\frac{1}{n}\right)\left(\frac{2}{n}\right)\left(\frac{3}{n}\right)\cdots\left(\frac{n}{n}\right) \tag{38.29a}$$

But

$$\underbrace{\left(\frac{1}{n}\right)}_{<1}\cdot\underbrace{\left(\frac{2}{n}\right)}_{<1}\cdot\underbrace{\left(\frac{3}{n}\right)}_{<1}\cdots\left(\frac{n}{n}\right) \leq \frac{1}{n}\cdot 1\cdot 1\cdots 1 = \frac{1}{n} \tag{38.29b}$$

Thus if we define $\{q_n\} = \{1/n\}$, we have that $a_n \leq q_n$ for all n.

Furthermore, we can define the sequence $\{p_n\} = \{0, 0, \ldots\}$. Then since $q_n \to 0$ and $p_n \to 0$ as $n \to \infty$, we have:

$$\begin{array}{ccccc} p_n = 0 & \leq & a_n = \dfrac{n!}{n^n} & \leq & q_n = \dfrac{1}{n} \\ \downarrow & & \downarrow & & \downarrow \\ 0 & \leq & L & \leq & 0 \end{array} \tag{38.29c}$$

Thus by the squeeze theorem, $n!/n^n \to L = 0$. $\qquad\square$

> **Theorem 38.3. Absolute Converge of Sequences**
>
> $$|a_n| \to 0 \implies a_n \to 0 \tag{38.30}$$

Example 38.10. Show that $\lim\limits_{n\to\infty} \dfrac{(-1)^n}{2^n} = 0$

Solution. This follows from theorem 38.3 because

$$\lim_{n\to\infty} \frac{1}{2^n} = 0 \tag{38.31a}$$

Since $1/2^n = |(-1)^n/2^n|$, then $(-1)^n/2^n \to 0$. $\square$

The Sum and Difference Rule for Sequences is analogous to theorem 3.3.

> **Theorem 38.4. Sum and Difference of Sequences**
>
> If both $\lim\limits_{n\to\infty} a_n$ and $\lim\limits_{n\to\infty} b_n$ exist, then
>
> $$\lim_{n\to\infty} (a_n + b_n) = \lim_{n\to\infty} a_n + \lim_{n\to\infty} b_n \tag{38.32}$$
>
> $$\lim_{n\to\infty} (a_n - b_n) = \lim_{n\to\infty} a_n - \lim_{n\to\infty} b_n \tag{38.33}$$

Example 38.11. Find $\lim\limits_{n\to\infty} \left((n+1)^{1/n} + \left(\dfrac{1}{n} + 1 \right)^n \right)$.

Solution. If we define $a_n = (n+1)^{1/n}$ and $b_n = (1/n + 1)^n$ then

$$\lim_{n\to\infty} \left((n+1)^{1/n} + \left(\frac{1}{n} + 1 \right)^n \right) = \lim_{n\to\infty} (a_n + b_n) \tag{38.34a}$$

$$= \lim_{n\to\infty} a_n + \lim_{n\to\infty} b_n \tag{38.34b}$$

$$= \lim_{x\to\infty} (x+1)^{1/x} + \lim_{x\to\infty} \left(\frac{1}{x} + 1 \right)^x \tag{38.34c}$$

We have used theorems 38.4 and 38.1. The first limit is 1 (exercise 16.24) and the second limit is e (see equation 15.81). Hence $\lim\limits_{n\to\infty} \left((n+1)^{1/n} + \left(\dfrac{1}{n} + 1 \right)^n \right) = 1 + e$. $\square$

The Constant Multiple Rule for Sequences is analogous to theorem 3.4 for functions.

> **Theorem 38.5. Constant Multiple Rule for Sequences**
>
> $$\lim_{n\to\infty} ca_n = c \lim_{n\to\infty} a_n \tag{38.35}$$

Example 38.12. Find $\lim\limits_{n\to\infty} \dfrac{3n}{n+1}$.

Solution. By the constant multiple rule, the 3 comes out of the limit.

$$\lim_{n\to\infty} \frac{3n}{n+1} = 3 \lim_{n\to\infty} \frac{n}{n+1} = 3 \cdot 1 = 3 \tag{38.36a}$$

$n/(n+1) \to 1$ by theorem 38.1, since $x/(x+1) \to 1$ as $x \to \infty$. $\square$

There is a Product of Sequences Rule, analogous to theorem 3.5.

Theorem 38.6. Limit of a Product for Sequences

If $\{a_n\}$ and $\{b_n\}$ both converge then

$$\lim_{n\to\infty} (a_n b_n) = \left(\lim_{n\to\infty} a_n\right)\left(\lim_{n\to\infty} b_n\right) \tag{38.37}$$

Example 38.13. Find $\displaystyle\lim_{n\to\infty} \frac{n^2(1+n)^{1/n}}{5n^2 + 2}$.

Solution. By theorem 38.1,

$$\lim_{n\to\infty} \frac{n^2}{5n^2 + 2} = \frac{1}{5} \tag{38.38a}$$

because $x^2/(5x^2 + 2) \to 1/5$; and (from exercise 16.24),

$$\lim_{n\to\infty} (1+n)^{1/n} = 1 \tag{38.38b}$$

Using theorem 38.6,

$$\lim_{n\to\infty} \frac{n^2(1+n)^{1/n}}{5n^2 + 2} = \lim_{n\to\infty} \frac{n^2}{5n^2 + 2} \cdot \lim_{n\to\infty} (1+n)^{1/n} = \frac{1}{5} \cdot 1 = \frac{1}{5}. \tag{38.38c}$$

$\square$

As with theorem 3.6, when the limit of the denominator is non-zero, the limit of a quotient is the quotient of the limits.

Theorem 38.7. Limit of a Quotient for Sequences

If $\{a_n\}$ and $\{b_n\}$ both converge and $\displaystyle\lim_{n\to\infty} b_n \neq 0$ then

$$\lim_{n\to\infty} \frac{a_n}{b_n} = \frac{\lim_{n\to\infty} a_n}{\lim_{n\to\infty} b_n} \tag{38.39}$$

Example 38.14. Find $\displaystyle\lim_{n\to\infty} \left(\frac{3}{2}\right)^n \left(\frac{1 + 5 \cdot 2^n}{1 + 7 \cdot 3^n}\right)$.

Solution. Manipulating the argument of the limit,

$$\left(\frac{3}{2}\right)^n \left(\frac{1 + 5 \cdot 2^n}{1 + 7 \cdot 3^n}\right) = \frac{3^n \cdot (1 + 5 \cdot 2^n)}{2^n \cdot (1 + 7 \cdot 3^n)} = \frac{1 + 5 \cdot 2^n}{2^n} \cdot \frac{3^n}{1 + 7 \cdot 3^n} \tag{38.40a}$$

$$= \frac{\dfrac{1}{2^n} + \dfrac{5 \cdot 2^n}{2^n}}{\dfrac{1}{3^n} + \dfrac{7 \cdot 3^n}{3^n}} = \frac{\left(\dfrac{1}{2}\right)^n + 5}{\left(\dfrac{1}{3}\right)^n + 7} \tag{38.40b}$$

Since $\left(\dfrac{1}{2}\right)^n + 5 \to 5$, and $\left(\dfrac{1}{3}\right)^n + 7 \to 7$, theorem 38.7 applies. Consequently

$$\lim_{n\to\infty} \left(\frac{3}{2}\right)^n \left(\frac{1 + 5 \cdot 2^n}{1 + 7 \cdot 3^n}\right) = \lim_{n\to\infty} \frac{\left(\dfrac{1}{2}\right)^n + 5}{\left(\dfrac{1}{3}\right)^n + 7} = \frac{\displaystyle\lim_{n\to\infty} \left(\dfrac{1}{2}\right)^n + 5}{\displaystyle\lim_{n\to\infty} \left(\dfrac{1}{3}\right)^n + 7} = \frac{5}{7}. \tag{38.40c}$$

$\square$

The limit of a power is the power of the limit.

> ### Theorem 38.8. Limit of a Power for Sequences
>
> If $\{a_n\}$ converges then
> $$\lim_{n\to\infty} (a_n)^p = \left(\lim_{n\to\infty} a_n\right)^p \qquad (38.41)$$

Example 38.15. Find $\lim_{n\to\infty} \left(1 + \dfrac{a}{n}\right)^n$.

Solution. Recall from equation 15.81 that $\lim_{n\to\infty} (1 + 1/n)^n = e$. We make the substitution $m = n/a$; then $m \to \infty$ as $n \to \infty$. By the laws of exponents,

$$\lim_{n\to\infty} \left(1 + \frac{a}{n}\right)^n = \lim_{n\to\infty} \left(\left(1 + \frac{a}{n}\right)^{n/a}\right)^a \qquad (38.42\text{a})$$

$$= \lim_{m\to\infty} \left[\left(1 + \frac{1}{m}\right)^m\right]^a \qquad (38.42\text{b})$$

$$= \left[\lim_{m\to\infty} \left(1 + \frac{1}{m}\right)^m\right]^a = e^a \qquad (38.42\text{c})$$

$\square$

The composition rule tells us that if we apply a function $f(x)$ to every element of a convergent sequence,

$$a_1, a_2, a_3, \cdots \to a \qquad (38.43)$$

the resulting sequence will converge to the function of the limit

$$f(a_1), f(a_2), f(a_3), \cdots \to f(a) \qquad (38.44)$$

> ### Theorem 38.9. Limit of Composition for Sequences
>
> If $\lim_{n\to\infty} a_n = a$ for some number L, then
> $$\lim_{n\to\infty} f(a_n) = f\left(\lim_{n\to\infty} a_n\right) = f(a) \qquad (38.45)$$

Example 38.16. Find $\lim_{n\to\infty} a_n$ for $a_n = e^{\cos(n\pi/6)/n^2}$.

Solution. Since $\dfrac{\cos(n\pi/6)}{n^2} \to 0$ converges to a number in the domain of e^x,

$$\lim_{n\to\infty} a_n = \lim_{n\to\infty} e^{\cos(n\pi/6)/n^2} = \exp\left\{\lim_{n\to\infty} \frac{\cos(n\pi/6)}{n^2}\right\} = e^0 = 1 \qquad (38.46\text{a})$$

$\square$

Other Properties of Sequences

Definition 38.8. Increasing and Decreasing

We call a sequence is **increasing** if $a_n < a_{n+1}$ for all n.

We call a sequence is **decreasing** if $a_n > a_{n+1}$ for all n.

A sequence is called **monotonic** if it is either increasing or decreasing.

Example 38.17. $a_n = \dfrac{7}{n^2 + 4}$ is decreasing because

$$a_n = \frac{7}{n^2 + 4} > \frac{7}{(n+1)^2 + 4} = a_{n+1} \quad \square \tag{38.47}$$

Example 38.18. Determine whether $a_n = \dfrac{n+3}{3n-2}$ is monotonic increasing, monotonic decreasing, or neither.

Solution. Computing $a_1 = 4$ and $a_2 = 5/4$, we hypothesize that the sequence is decreasing.

$$\text{Hypothesis: } a_{n+1} < a_n \text{ for all n} \tag{38.48a}$$

If this hypothesis is correct, then

$$\frac{(n+1)+3}{3(n+1)-2} < \frac{n+3}{3n-2} \tag{38.48b}$$

$$\iff \frac{n+4}{3n+1} < \frac{n+3}{3n-2} \tag{38.48c}$$

$$\iff (n+4)(3n-2) < (3n+1)(n+3) \tag{38.48d}$$

$$\iff 3n^2 + 10n - 8 < 3n^2 + 10n + 3 \tag{38.48e}$$

$$\iff -8 < 3 \tag{38.48f}$$

Since is $-8 < 3$ and all of the steps are reversible, then the first step is true, hence our hypothesis that the sequence is decreasing was correct. Hence the sequence is monotonically decreasing. $\square$

Definition 38.9. Boundedness of Sequences

We call a sequence **Bounded Above** if there is some number M such that $a_n \le M$ for all $n \ge 1$.

We call a sequence **Bounded Below** if there is some number m such that $a_n \ge m$ for all $n \ge 1$.

Theorem 38.10. Convergence of Bounded Monotonic Sequences

Every Bounded Monotonic Sequence Converges, in the following sense:

1. If the sequence is bounded above and increasing, then it converges; or

2. If the sequence is bounded below and decreasing then it converges.

Example 38.19. Show that $a_n = \dfrac{2^n}{n!}$ converges.

Solution. We will do this by demonstrating that (a) a_n is decreasing and (b) bounded below.
The sequence is decreasing:

$$\frac{a_{n+1}}{a_a} = \frac{2^{n+1}}{(n+1)!} \cdot \frac{n!}{2^n} = \frac{2}{n+1} < 1 \qquad (38.49a)$$

where the last statement holds for all $n > 2$. Hence

$$a_{n+1} < a_n \qquad (38.49b)$$

Therefore the sequence is decreasing.
The sequence is bounded below: Since $a_n > 0$ for all n, the sequence is bounded below.
Thus the sequence is monotonically decreasing and bounded below. By Theorem 38.10,
the sequence converges. $\qquad\qquad\square$

Exercises

In exercises 1 through 4, find a formula for the specified sequence.

1. $\left\{ 240, 160, \dfrac{320}{3}, \dfrac{640}{9}, \dfrac{1280}{27}, \ldots \right\}$
   ans: $\left\{ 5 \cdot 2^{n+3} 3^{2-n} \right\}$

2. $\left\{ -\dfrac{1}{5}, \dfrac{3}{25}, -\dfrac{9}{125}, \dfrac{27}{625}, -\dfrac{81}{3125}, \ldots \right\}$
   ans: $\left\{ \left(-\dfrac{1}{5} \right)^n 3^{n-1} \right\}$

3. $\{1, -1, 0, 1, -1, 0, \ldots\}$ ans: $\left\{ \dfrac{2}{\sqrt{3}} \sin \dfrac{2n\pi}{3} \right\}$

4. $\left\{ 0, \dfrac{1}{5}, \dfrac{1}{5}, \dfrac{3}{17}, \dfrac{2}{13}, \ldots \right\}$ ans: $\dfrac{n-1}{n^2+1}$

Write the first 5 terms of each of sequences specified in exercises 5 through 12.

5. $\left\{ (2)^{-3n} (3)^{5n/2} \right\}$

6. $\left\{ \dfrac{(-1)^{n+3} 2^{n-2}}{(n+4)^2} \right\}$

7. $\left\{ \dfrac{\ln n}{n} \right\}$

8. $\left\{ \dfrac{\sqrt{n}}{n^2+2} \right\}$

9. $\left\{ n \cos \dfrac{n\pi}{4} \right\}$

10. $\left\{ \dfrac{\sin^2 n}{n} \right\}$

11. $\left\{ 3 - \dfrac{1}{n^n} \right\}$

12. $\left\{ \dfrac{(-1)^{n-1} \ln n}{n!} \right\}$

13. For each of the sequences in exercises 5 through 12, determine if the sequence is monotonic. If it is, determine if it is increasing, or decreasing; bounded; and if it converges.

14. Show that the explicit formula for the Fibonacci sequence (38.17) agrees with the recursive definition (38.15) for the first 10 members of the sequence.

15. Find $\displaystyle\lim_{n\to\infty} \dfrac{F_{n+1}}{F_n}$ for the Fibonacci sequence. What does the result suggest to you, if anything?

Determine whether each of the sequences in exercises 16 through 24 converges or diverges. If it converges, find the limit.

16. $\left\{ \dfrac{\ln n}{n} \right\}$ ans: 0

17. $\left\{ n^2 e^{-n} \right\}$ ans: 0

18. $\left\{ \dfrac{5n}{6+7n} \right\}$ ans: 5/7

19. $\left\{ 1 - \dfrac{1}{2^n} \right\}$ ans: 1

20. $\left\{ \ln \dfrac{n+1}{n} \right\}$ ans: 0

21. $\left\{ n^{2/3} \right\}$ ans: ∞

22. $\left\{ \dfrac{n^n}{n!} \right\}$ ans: ∞

23. $\left\{ \left[\dfrac{3n+7}{4n+6} \right]^{1/2} \right\}$ ans: $\dfrac{\sqrt{3}}{2}$

24. $\left\{ \dfrac{n}{\sqrt{n+1} - n} \right\}$ ans: -1

Chapter 39

Series

Chapter Summary and Goal

A **series** is the sum of the terms in a sequence. In this chapter we will study the properties of sequences, such as how to determine whether or not they converge.

Student Learning Objectives

The student will:

1. Understand the relationships between sequences and series.
2. Understand the concept of partial sums and convergence of series.
3. Understand the the basic properties of series.
4. Be able to identify and find the sum of a geometric series.
5. Be able to use the integral test to determine convergence and estimate remainders.
6. Be able to apply the comparison tests for convergence and divergence.
7. Understand the distinction between absolute and conditional convergence.
8. Be able apply to the alternating series test for convergence and estimate remainders.
9. Be able to use the ratio and root tests to determine convergence.

Convergence of Series

A **series** is a sum of terms, such as the sum of terms in a sequence. A **partial sum** of any series is obtained by adding up leading terms of that series. We label the partial sums by the number of terms that are included. For example, if we start with the following **harmonic sequence**,

$$1, \frac{1}{2}, \frac{1}{3}, \frac{1}{4}, \frac{1}{5}, \cdots \tag{39.1}$$

Then the first few partial sums of the sequence (39.1) are

$$1^{\text{st}} \text{ partial sum } = 1$$

$$2^{\text{nd}} \text{ partial sum } = 1 + \frac{1}{2} = \frac{3}{2}$$

$$3^{\text{rd}} \text{ partial sum } = 1 + \frac{1}{2} + \frac{1}{3} = \frac{11}{6} \tag{39.2}$$

$$4^{\text{th}} \text{ partial sum } = 1 + \frac{1}{2} + \frac{1}{3} + \frac{1}{4} = \frac{25}{12}$$

Definition 39.1. Partial Sum

Let $\{a_n\}$ be any infinite sequence. Then

$$S_n = \sum_{n=1}^{n} a_n \tag{39.3}$$

is called a **partial sum** of the sequence.

For any series we can define the **sequence of partial sums**

$$S_1, S_2, S_3, S_4, \ldots \tag{39.4}$$

For example, the sequence of partial sums for the sequence in equation 39.1 begins with

$$S_1, \ S_2, \ S_3, \ S_4, \ S_5, \ \cdots = 1, \ \frac{3}{2}, \ \frac{11}{6}, \ \frac{25}{12}, \ \frac{137}{60}, \ \cdots \tag{39.5}$$

There is no guarantee that the sequence of numbers $\{S_i\}$ will converge to a finite number. It is just like any other sequence: sometimes it will converge to a finite number; sometimes it will converge to $\pm\infty$; and sometimes it will bounce all over the place. Hopefully the following three examples will convince you that any of these are possible.

Example 39.1. Show that the sequence of partial sums of $\{(-1)^n\}$ does not converge.

Solution. We will calculate the first several terms. The original sequences $\{a_n\}$ bounces back and forth between -1 and 1:

$$\{a_n\} = -1, \ 1, \ -1, \ 1, \ \ldots \tag{39.6a}$$

Here are the first few partial sums:

$$S_1 = a_1 = -1 \tag{39.6b}$$
$$S_2 = a_1 + a_2 = -1 + 1 = 0 \tag{39.6c}$$
$$S_3 = a_1 + a_2 + a_3 = -1 + 1 - 1 = -1 \tag{39.6d}$$
$$S_4 = a_1 + a_2 + a_3 + a_4 = -1 + 1 - 1 + 1 = 0 \tag{39.6e}$$

$$\vdots$$

Thus the sequence of partial sums is

$$\{S_n\} = -1, \ 0, \ -1, \ 0, \ \ldots \tag{39.6f}$$

Clearly $\lim_{n\to\infty} S_n$ does not exist. $\blacksquare$

Example 39.2. Show that the sequence of partial sums for $a_n = 1$ converges to infinity.

Solution. We have

$$\{a_n\} = 1, 1, 1, 1, 1, \ldots \tag{39.7a}$$

Thus

$$S_1 = 1 \tag{39.7b}$$
$$S_2 = 1 + 1 = 2 \tag{39.7c}$$
$$S_3 = 1 + 1 + 1 = 3 \tag{39.7d}$$

$$\vdots$$

$$S_n = \underbrace{1 + \cdots + 1}_{n \text{ times}} = n \tag{39.7e}$$

Since $\lim\limits_{n \to \infty} n = \infty$, we conclude that $S_n \to \infty$. $\qquad\square$

Example 39.2 illustrates the Principle of Archimedes very nicely (Axiom 8.1[1]). As a corollary, any sequence that does not converge to zero, but instead converges to some limit L (in this case, $L = 1$), will have an nth partial sum of at least nL, and $nL \to \infty$ as $n \to \infty$.

The converse is not necessarily true. Convergence of a sequence to zero will not ensure that its partial sums converge. An example is given by the harmonic sequence in equation 39.1:

$$\lim_{n \to \infty} \frac{1}{n} = 0 \tag{39.8}$$

but

$$\sum_{n=1}^{\infty} \frac{1}{n} = \infty \tag{39.9}$$

We will demonstrate this later in example 39.10.

Example 39.3. Show that the sequence of partial sums for $a_n = 2^{-n}$ converges to 1.

Solution. We have

$$\{a_n\} = \frac{1}{2}, \frac{1}{4}, \frac{1}{8}, \frac{1}{16}, \cdots \tag{39.10a}$$

Thus

$$S_n = \frac{1}{2} + \frac{1}{2^2} + \frac{1}{2^3} + \cdots + \frac{1}{2^n} \tag{39.10b}$$

$$S_{n+1} = \frac{1}{2} + \frac{1}{2^2} + \frac{1}{2^3} + \cdots + \frac{1}{2^n} + \frac{1}{2^{n+1}} \tag{39.10c}$$

Multiplying equation 39.10c by 2,

$$2S_{n+1} = 1 + \frac{1}{2} + \frac{1}{2^2} + \cdots + \frac{1}{2^n} = 1 + S_n \tag{39.10d}$$

Dividing by 2,

$$S_{n+1} = \frac{1}{2} + \frac{S_n}{2} \tag{39.10e}$$

But

$$S_{n+1} = S_n + \frac{1}{2^{n+1}} \tag{39.10f}$$

Setting the two expressions for S_{n+1} equal,

$$\frac{1}{2} + \frac{S_n}{2} = S_n + \frac{1}{2^{n+1}} \tag{39.10g}$$

---

[1]The Principle of Archimedes is true for the real numbers.

Solving for S_n,

$$S_n = 1 - \frac{1}{2^n} \tag{39.10h}$$

Thus

$$\lim_{n\to\infty} S_n = \lim_{n\to\infty} \left(1 - \frac{1}{2^n}\right) = 1 \tag{39.10i}$$

$\square$

In the last three examples we have illustrated three different situations

1. The sequence of partial sums bounces back and forth between two different numbers. When this happens, we say the **series diverges**.

2. The sequence of partial sums becomes infinite. When this happens, we also say that the **series diverges**.

3. The sequence of partial sums converges to a finite number. When this happens, we say the **series converges**.

Definition 39.2. Convergence of a Series

Let $\{a_n\}$ be any sequence, and let $S_1, S_2, \ldots$ be its sequence of partial sums. Then if the limit

$$\lim_{n\to\infty} S_n = \lim_{n\to\infty} \sum_{n=1}^{n} a_n = S \tag{39.11}$$

exists and is finite then we say that **the series converges** and write $S = \sum_{n=1}^{\infty} a_n$. If the sequence does not converge, we say that **the series diverges**.

Theorem 39.1. First Divergence Test

Let a_n be a sequence. If $\lim_{n\to\infty} a_n$ does not exist, then **the series** $\sum_{n=0}^{\infty} a_n$ **diverges**.

Theorem 39.2. Second Divergence Test

Let a_n be a sequence. If $\lim_{n\to\infty} a_n \neq 0$, then **the series** $\sum_{n=0}^{\infty} a_n$ **diverges**, even though the sequence converges.

As we saw in example 39.3, we can sometimes calculate the sum of a series by algebraic manipulation. Sometimes we only have to manipulate the terms of the series themselves, not look at the series globally, as we did in example 39.3. In example 39.4 we illustrate a **telescoping series**, in which successive terms in the series cancel out, allowing us to calculate the sum.

Example 39.4. Show that $\displaystyle\sum_{n=1}^{\infty} \frac{1}{n(n+1)}$ converges, and calculate the sum.

Solution. By the method of partial fractions we can find that $\dfrac{1}{n(n+1)} = \dfrac{1}{n} - \dfrac{1}{n+1}$.

Thus the Nth partial sum is

$$\sum_{n=1}^{N} \frac{1}{n(n+1)} = \sum_{n=1}^{N} \left(\frac{1}{n} - \frac{1}{n+1} \right) \tag{39.12a}$$

$$= \left(1 - \frac{1}{2} \right) + \left(\frac{1}{2} - \frac{1}{3} \right) + \left(\frac{1}{3} - \frac{1}{4} \right) + \cdots \left(\frac{1}{n} - \frac{1}{n+1} \right) \tag{39.12b}$$

$$= \left(1 - \frac{1}{n+1} \right) \to 1 \text{ as } N \to \infty \tag{39.12c}$$

Thus the series converges and $\sum_{n=1}^{\infty} a_n = 1$. $\qquad\square$

The constant multiple rule tells us that when a series converges the distributive property still holds over every term, and we can pull out any constant factor. Thus, for example,

$$\sum_{k=1}^{\infty} \frac{3}{2^k} = 3 \sum_{k=1}^{\infty} \frac{1}{2^k} = 3 \tag{39.13}$$

Theorem 39.3. Constant Multiple Rule for Series

If $\displaystyle\sum_{n=1}^{\infty} a_n$ converges, and C is any constant, then $\displaystyle\sum_{n=1}^{\infty} C a_n = C \sum_{n=1}^{\infty} a_n$

Similarly, if two series converge, we are allowed to add them term by term. Thus

$$\sum_{n=1}^{\infty} \left(\frac{2}{3^n} + \frac{3}{4^n} \right) = \sum_{n=1}^{\infty} \frac{2}{3^n} + \sum_{n=1}^{\infty} \frac{3}{4^n} \tag{39.14}$$

Theorem 39.4. Sum/Difference Rule for Series

If $\displaystyle\sum_{n=1}^{\infty} a_n$ and $\displaystyle\sum_{n=1}^{\infty} b_n$ both converge, then they can be added (or subtracted) term by term:

$$\sum_{n=1}^{\infty} (a_n \pm b_n) = \sum_{n=1}^{\infty} a_n \pm \sum_{n=1}^{\infty} b_n \tag{39.15}$$

Geometric Series

A geometric series is a series of the form

$$a + ar + ar^2 + ar^3 + \cdots = a \sum_{n=1}^{\infty} r^{n-1} \tag{39.16}$$

where $a \neq 0$.

> **Theorem 39.5. Sum of a Geometric Series.**
>
> If $|r| < 1$ and $a \neq 0$ then
>
> $$\sum_{n=1}^{\infty} ar^{n-1} = a + ar + ar^2 + \cdots = \frac{a}{1-r} \tag{39.17}$$

Proof. The proof is a generalization of the calculation we did in example 39.3.

$$S_n = a + ar + ar^2 + ar^3 + \cdots + ar^{n-1} \tag{39.18a}$$
$$rS_n = ar + ar^2 + \cdots + ar^{n-1} + ar^n \tag{39.18b}$$

Subtracting gives

$$S_n - rS_n = a - ar^n = a(1 - r^n) \tag{39.18c}$$

Factoring the S_n on the left hand side and dividing by $1 - r$,

$$S_n = \frac{a(1 - r^n)}{1 - r} \tag{39.18d}$$

Taking the limit as $n \to \infty$,

$$\lim_{n \to \infty} S_n = \lim_{n \to \infty} \frac{a(1 - r^n)}{1 - r} = \begin{cases} \dfrac{a}{1-r} & \text{if } |r| < 1 \\ \text{diverges} & \text{otherwise} \end{cases} \tag{39.18e}$$

as stated in the theorem. $\square$

Example 39.5. Find the sum of the series $3 + \dfrac{3}{4} + \dfrac{3}{16} + \dfrac{3}{64} + \ldots$

Solution. This is a geometric series

$$S = 3 + 3 \cdot \frac{1}{4} + 3 \cdot \frac{1}{4^2} + 3 \cdot \frac{1}{4^3} + \cdots \tag{39.19a}$$

so $a = 3$ and $r = 1/4$. Since $r < 1/4$ the series converges and $S = \dfrac{a}{1-r} = \dfrac{3}{1 - 1/4} = 4$.
$\square$

Example 39.6. Determine whether the series

$$S = \sum_{n=1}^{\infty} 2^{4n-3} 5^{-n} \tag{39.20}$$

converges or diverges.

Solution. Manipulating the a_n,

$$a_n = 2^{4n-3}5^{-n} = \frac{(2^4)^n}{2^3 5^n} = \frac{1}{8}\left(\frac{16}{5}\right)^n \tag{39.21a}$$

This is a geometric series with $a = 1/8$ and $r = 16/5 > 1$ so it diverges. $\qquad\square$

Example 39.7. Determine if the series

$$S = \frac{21}{4} + \frac{63}{16} + \frac{189}{64} + \frac{567}{256} + \frac{1701}{1024} + \cdots \tag{39.22}$$

converges or diverges, and if it converges find its sum.

Solution. We begin by factoring to look for a pattern. First, we observe the powers of 2 in the denominator, and that the numerators are all divisible by 7. So we factor the 7, and write the powers of 2.

$$\frac{21}{4} + \frac{63}{16} + \frac{189}{64} + \frac{567}{256} + \frac{1701}{1024} + \cdots = 7\left[\frac{3}{2^2} + \frac{9}{2^4} + \frac{27}{2^6} + \frac{81}{2^8} + \frac{243}{2^{10}} + \cdots\right] \tag{39.23a}$$

$$= 7\left[\frac{3^1}{2^2} + \frac{3^2}{2^4} + \frac{3^3}{2^6} + \frac{3^4}{2^8} + \frac{3^5}{2^{10}} + \cdots\right] \tag{39.23b}$$

$$\tag{39.23c}$$

$$= 7\left[\frac{3^1}{(2^2)^1} + \frac{3^2}{(2^2)^2} + \frac{3^3}{(2^2)^3} + \frac{3^4}{(2^2)^4} + \frac{3^5}{(2^2)^5} + \cdots\right] \tag{39.23d}$$

$$= 7\left[\left(\frac{3}{4}\right) + \left(\frac{3}{4}\right)^2 + \left(\frac{3}{4}\right)^3 + \left(\frac{3}{4}\right)^4 + \left(\frac{3}{4}\right)^5 + \cdots\right] \tag{39.23e}$$

$$= \frac{21}{4}\left[1 + \left(\frac{3}{4}\right) + \left(\frac{3}{4}\right)^2 + \left(\frac{3}{4}\right)^3 + \left(\frac{3}{4}\right)^4 + \left(\frac{3}{4}\right)^5 + \cdots\right] \tag{39.23f}$$

This is a geometric series with $a = 21/4$ and $r = 3/4 < 1$. The sum is

$$S = \frac{a}{1-r} = \frac{21/4}{1-3/4} = 21. \tag{39.23g}$$

$$\square$$

One example of geometric series that we use everyday is decimal representations of fractions with repeating digits. Representing $1/3$ as a decimal, we write

$$\frac{1}{3} \approx 0.333\ldots \tag{39.24}$$

or

$$\frac{1}{3} \approx 0.33\overline{3} \tag{39.25}$$

where the ellipsis ($\ldots$) or the overline ($\overline{3}$) indicates that the 3 is repeated indefinitely. What we really mean, of course, is

$$\frac{1}{3} \approx 3 \cdot 10^{-1} + 3 \cdot 10^{-2} + 3 \cdot 10^{-3} + \cdots \tag{39.26}$$

This is a geometric series with $a = 3 \cdot 10^{-1}$ and $r = 10^{-1}$. The sum, of course, is

$$S = \frac{a}{1-r} = \frac{3 \cdot 10^{-1}}{1 - 10^{-1}} = \frac{.3}{1 - .1} = \frac{.3}{.9} = \frac{1}{3} \tag{39.27}$$

You were probably told in elementary school that every repeating decimal is actually an exact fraction of integers. This is true. The algorithm illustrated in example 39.8.

Example 39.8. Find the fraction represented by $3.142\,857\,\overline{142\,857}\ldots$ as a ratio of two integers.

Solution. The technique we will use is to eliminate the repeating digits by lining them up and then subtracting them out. We observe that there is a 6-digit repeat, $142\,857$. Thus if

$$x = 3.142\,857\,\overline{142\,857}\cdots \tag{39.28a}$$

and we multiply x by 10^6, it will shift all the digits over six places.

$$1\,000\,000\,x = 3\,142\,857.142\,857\,\overline{142\,857}\cdots \tag{39.28b}$$

This lines up the repeating digits precisely with the original number. So if we subtract, everything to the right of the decimal point cancels out.

$$1\,000\,000\,x - x = 3\,142\,857.142\,857\,\overline{142\,857}\cdots$$
$$-3.142\,857\,\overline{142\,857}\cdots \tag{39.28c}$$
$$999\,999\,x = 3\,142\,854.0 \tag{39.28d}$$

Thus we have an exact ratio of integers:

$$x = \frac{3\,142\,854}{999\,999} = \frac{22}{7} \tag{39.28e}$$

□

This method also works even when some of the digits to the right of the decimal point are not repeating.

Example 39.9. Write the repeating decimal $5.111\,67\,67\,\overline{67}\cdots$ as a fraction of integers.

Solution. We want to move the digits over so that the repeats exactly match up. There is a two digit repeat, so we only have to multiply by 100.

$$x = 5.111\,67\,67\,\overline{67}\cdots \tag{39.29a}$$
$$100x = 511.167\,67\,67,\,\overline{67}\cdots \tag{39.29b}$$

Subtracting,

$$99x = 511.167 - 5.111 = 506.056 \tag{39.29c}$$

and therefore the fraction of integers is

$$x = \frac{506.056}{99} = \frac{506\,056}{99\,000} = \frac{63\,257}{12\,375}. \tag{39.29d}$$

□

The Integral Test and p-Series

If we think of the terms a_n of an infinite series $\sum_{n=1}^{\infty} a_n$ as points on a continuous curve of some function $f(x)$ at the integers $x_n = n$, then the series gives the upper left-corner approximation[2] to the integral

$$\int_1^{\infty} f(x) \; dx \approx \sum_{n=1}^{\infty} f(x_n)\Delta x = \sum_{n=1}^{\infty} a_n \tag{39.30}$$

where $\Delta x = 1$. Therefore the series and the integral will converge or diverge together: if either converges, the other converges; and if either diverges, the other diverges.

Figure 39.1: The series $\sum_{n=1}^{\infty} a_n$ and the integral $\int_1^{\infty} f(x)dx$ converge or diverge together.

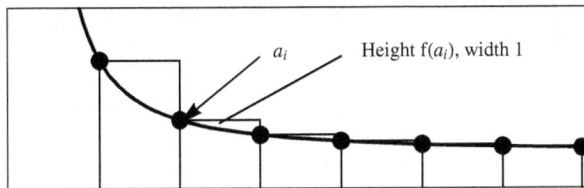

Theorem 39.6. Integral Test

Let $a(x)$ be a continuous function such that $a(x) = a_n$ when $x = n$. Then

If $\displaystyle\int_1^{\infty} a(x)dx$ converges, then $\displaystyle\sum_{n=1}^{\infty} a_n$ converges.

If $\displaystyle\int_1^{\infty} a(x)dx$ diverges, then $\displaystyle\sum_{n=1}^{\infty} a_n$ diverges.

Example 39.10. Show that the **Harmonic Series** $\displaystyle\sum_{n=1}^{\infty} \frac{1}{n}$ diverges.

Solution. To use the integral test, we calculate the following integral:

$$\int_1^{\infty} \frac{1}{x} \; dx = \ln |x| \Big|_1^{\infty} \to \infty. \tag{39.31a}$$

Since $\displaystyle\int_1^{\infty} \frac{1}{x} \; dx$ diverges, by the integral test, $\displaystyle\sum_{n=1}^{\infty} \frac{1}{n}$ diverges. $\qquad\square$

Example 39.11. Show that $\displaystyle\sum_{n=1}^{\infty} \frac{1}{n^2}$ converges.

Solution. To apply the integral test, we calculuate

$$\int_1^{\infty} \frac{1}{x^2} dx = -\frac{1}{x} \Big|_1^{\infty} = 1 \tag{39.32a}$$

---

[2]think in terms of Riemann sums; see figure 39.1.

Since $\displaystyle\int_1^\infty \frac{dx}{x^2}$ converges, by the integral test $\displaystyle\sum_{n=1}^\infty \frac{1}{n^2}$ converges.   □

Example 39.12. Examine the convergence of $\displaystyle\sum_{n=1}^\infty \frac{1}{n^2+1}$.

Solution. To apply the integral test we must calculate the integral

$$\int_1^\infty \frac{dx}{1+x^2} = \tan^{-1} x \Big|_1^\infty = \tan^{-1}(\infty) - \tan^{-1}(1) \tag{39.33a}$$

$$= \frac{\pi}{2} - \frac{\pi}{4} = \frac{\pi}{4} \tag{39.33b}$$

which is finite, which tells us that the series converges.   □

Definition 39.3. p-Series

A **p-Series** is any series of the form $\displaystyle\sum_{n=1}^\infty \frac{1}{n^p}$ where p is a constant

---

Examples of p-series are

$$1 + \frac{1}{2} + \frac{1}{3} + \frac{1}{4} + \dots \tag{39.34}$$

$$1 + \frac{1}{4} + \frac{1}{9} + \frac{1}{16} + \dots \tag{39.35}$$

$$1 + \frac{1}{8} + \frac{1}{27} + \frac{1}{64} + \dots \tag{39.36}$$

Series (39.34) is called a **harmonic series**. We saw in example 39.10 that the harmonic series diverges. We also saw in example 39.11 that series (39.35) converges. Thus p-series converge for some values of p, and diverge for other values of p. We can ask the following question: for which values of p does a p-series converge?

We can use the integral test to answer this question. The crucial integral is

$$\int_1^\infty \frac{dx}{x^p} = \lim_{a\to\infty} \frac{1}{1-p} x^{1-p} \Big|_1^a = \frac{1}{1-p} \lim_{a\to\infty} \left(a^{1-p} - 1\right) \tag{39.37}$$

If $p > 1$: then $1 - p < 0$. In this case $a^{1-p} \to 0$ as $a \to \infty$, and we get

$$\int_1^\infty \frac{dx}{x^p} = \frac{1}{p-1} \text{ for } p > 1 \tag{39.38}$$

which is finite. Therefore the series converges when $p > 1$.

If $p < 1$: then $1 - p > 0$. In this case $a^{1-p} \to \infty$. Thus the integral, and hence the series, diverges when $p < 1$.

If $p = 1$ the series diverges. We determined this in example 39.10.

Theorem 39.7. p-Series Test

The p-series $\displaystyle\sum_{n=1}^\infty \frac{1}{n^p}$ converges if $p > 1$ and diverges if $p \leq 1$.

Estimation of the Remainder

We can sub-divide any infinite series S after its nth term into a partial sum S_n (definition 39.1) and a **remainder** R_n,

$$S = S_n + R_n \tag{39.39}$$

The formula for the remainder term will always depend on where we truncate the partial sum.

Definition 39.4. Remainder Term

Let S_n be the nth partial sum of the series $S = \sum_{n=1}^{\infty} a_n$. Then the remainder is

$$R_n = \sum_{n=1}^{\infty} = a_{n+1} + a_{n+2} + \cdots \tag{39.40}$$

Theorem 39.8. Series Remainder Theorem

Let a_k be a positive, decreasing function such that $\sum_{n=1}^{\infty} a_n$ converges. Then

$$\sum_{k=1}^{\infty} a_k = \sum_{k=1}^{n} a_k + R_n \tag{39.41}$$

where

$$\int_{n+1}^{\infty} a(x)dx \le R_n \le \int_{n}^{\infty} a(x)dx \tag{39.42}$$

Proof. Let a_n be a monotonically decreasing sequence, and let $a(n)$ be its continuous extension.

Since the sequence is monotonically decreasing, then the boxes drawn using the upper left hand corner to approximate the integral will **overestimate** the area under the curve so that

$$R_n = a_{n+1} + a_{n+2} + \cdots \ge \int_{n+1}^{\infty} a(x)dx \tag{39.43a}$$

In (39.43a), the term a_{n+1} represents the area from $x = n + 1$ to $x = n + 2$, because its **upper left hand corner** is at $(n+1, a_{n+1})$, and hence the lower limit of the integral is at $n + 1$.

If we use the **upper right hand corner**, the term a_{n+1} will be at the upper right hand corner of the box from $x = n$ to $x = n + 1$. The upper upper left hand corner of this box will be at (n, a_n). Thus the box represented by a_n is larger than the box represented by a_n. Again using the upper left hand corner, but starting at n, we conclude that these boxes underestimate the area.

$$R_n = a_{n+1} + a_{n+2} + \cdots \le \int_{n}^{\infty} a(x)dx \tag{39.43b}$$

Thus

$$\int_{n+1}^{\infty} a(x)dx \le R_n \le \int_{n}^{\infty} a(x)dx \tag{39.43c}$$

$\square$

Example 39.13. How may terms in the series $\sum_{n=1}^{\infty} \dfrac{1}{n^2}$ do we need to ensure that the sum is accurate to 0.01?

Solution. To ensure an accuracy of 0.01 we need to make sure the remainder term $R_n \leq 0.01$. From equation 39.42 we have the requirement that

$$\int_n^{\infty} \frac{dx}{x^2} \leq 0.01 \tag{39.44a}$$

But

$$\int_n^{\infty} \frac{dx}{x^2} = -\frac{1}{x}\Big|_n^{\infty} = \frac{1}{n} \tag{39.44b}$$

Thus

$$\frac{1}{n} \leq 0.01 \text{ or } n \geq 100 \tag{39.44c}$$

So we will need at least 100 terms. $\qquad\square$

Example 39.14. Estimate $\sum_{n=1}^{\infty} \dfrac{1}{n^2}$ using the first five terms and use the error series remainder theorem to estimate a bounds on the actual value of the series.

Solution. The first five terms are

$$\sum_{n=1}^{5} \frac{1}{n^2} = 1 + \frac{1}{4} + \frac{1}{9} + \frac{1}{16} + \frac{1}{25} \tag{39.45a}$$

$$\approx 1.0000 + 0.2500 + 0.1111 + 0.0625 + 0.0400 \approx 1.4636 \tag{39.45b}$$

Thus $S_5 \approx 1.4636$.

The actual sum is $S = S_5 + R_5$ where by (39.42)

$$\int_{5+1}^{\infty} \frac{dx}{x^2} \leq R_5 \leq \int_5^{\infty} \frac{dx}{x^2} \tag{39.45c}$$

Integrating,

$$-\frac{1}{x}\Big|_6^{\infty} \leq R_5 \leq -\frac{1}{x}\Big|_5^{\infty} \tag{39.45d}$$

$$0.1667 \approx \frac{1}{6} \leq R_5 \leq \frac{1}{5} = 0.2000 \tag{39.45e}$$

Adding S_5 all the way through the inequality ,

$$0.1667 + S_5 \leq R_5 + S_5 \leq 0.2 + S_5 \tag{39.45f}$$

Using $S_5 = 1.4636$ on the left and right side of the inequality and $S = S_5 + R_5$ in the center,

$$1.6303 \leq S \leq 1.6636 \tag{39.45g}$$

to four digits. The actual sum is $\sum_{n=1}^{\infty}(1/n^2) = \pi^2/6 \approx 1.6449$. $\qquad\square$

Comparison Tests

The comparison tests only tell us if a series converges or diverges. Unlike a geometric series, these tests do not give us a values for the series if it converges, and unlike the integral test, they will not give us a useful bound for the remainder. If the series does converge, they will sometimes give us an upper bound on the value but this will rarely be useful.

In figure 39.2 we illustrate three monotonically decreasing sequences that satisfy

$$0 \leq c_n \leq b_n \leq a_n \tag{39.46}$$

for all n. Supposed first that the series in the middle ($\sum_{n=1}^{\infty} b_n$) converges. Then since

$$0 \leq c_n \leq b_n \tag{39.47}$$

we can add the sequence term by term

$$0 \leq \sum_{n=1}^{\infty} c_n \leq \sum_{n=1}^{\infty} b_n < \infty \tag{39.48}$$

The last term follows because $\sum_{n=1}^{\infty} b_n$ converges. Thus $0 \leq \sum_{n=1}^{\infty} c_n < \infty$, i.e., $\sum_{n=1}^{\infty} c_n$ converges because its sum is not infinite.

Theorem 39.9. Comparison Test (for Convergence)

In $0 \leq c_n \leq b_n$ for all n and $\displaystyle\sum_{n=1}^{\infty} b_n$ converges, then $\displaystyle\sum_{n=1}^{\infty} c_n$ converges

---

Figure 39.2: Illustration of the comparison test with $a_n \geq b_n \geq c_n \geq 0$ for all n. Sequence $a_n = \blacklozenge$; $b_n = \blacktriangle$; $c_n = \bullet$.

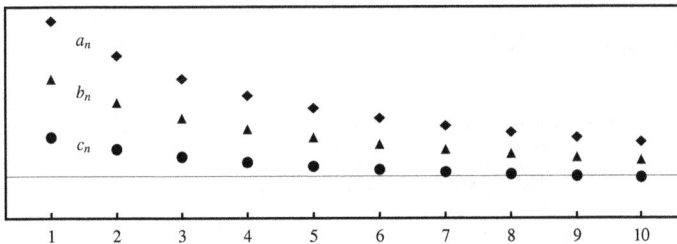

Example 39.15. Show that $\displaystyle\sum_{n=1}^{\infty} \frac{\sqrt{n}}{n^3 + 3n + 4}$ converges.

Solution. First we observe that a_n is positive and that for all positive n, $n^3 + 3n + 4 > n^3$. Thus

$$\frac{1}{n^3 + 3n + 4} < \frac{1}{n^3} \tag{39.49a}$$

Multiplying through by $\sqrt{n}$ gives a_n,

$$a_n = \frac{\sqrt{n}}{n^3 + 3n + 4} < \frac{\sqrt{n}}{n^3} = \frac{1}{n^{5/2}}. \tag{39.49b}$$

Now we define a new sequence by $b_n = 1/n^{5/2}$. By the p-test (theorem 39.7),

$$\sum_{n=1}^{\infty} b_n = \sum_{n=1}^{\infty} \frac{1}{n^{5/2}} \tag{39.49c}$$

converges because $5/2 > 1$. Hence since $0 \leq a_n \leq b_n$ for all n and $\displaystyle\sum_{n=1}^{\infty} b_n$ converges, by theorem 39.9, $\displaystyle\sum_{n=1}^{\infty} a_n$ converges. $\square$

We can also get a test for divergence by comparing sequences. Returning to figure 39.2, suppose that

$$\sum_{n=1}^{\infty} b_n = \infty \tag{39.50}$$

Then since $b_n \leq a_n$ for all n,

$$\infty = \sum_{n=1}^{\infty} b_n \leq \sum_{n=1}^{\infty} a_n \tag{39.51}$$

Thus $\displaystyle\sum_{n=1}^{\infty} a_n$ diverges.

Theorem 39.10. Comparison Test (for Divergence)

If $0 \leq a_n \leq b_n$ for all n, and $\displaystyle\sum_{n=1}^{\infty} a_n$ diverges then $\displaystyle\sum_{n=1}^{\infty} b_n$ diverges.

Example 39.16. Show that $\displaystyle\sum_{n=1}^{\infty} \frac{\sqrt{n^3+1}}{n^2}$ diverges.

Solution. First we observe that for all positive n,

$$\sqrt{n^3+1} = n^{3/2} + 1 \geq n + 1 > n \tag{39.52a}$$

hence

$$a_n = \frac{\sqrt{n^3+1}}{n^2} > \frac{n}{n^2} = \frac{1}{n} \tag{39.52b}$$

Let $b_n = 1/n$. But $\displaystyle\sum_{n=1}^{\infty} b_n = \sum_{n=1}^{\infty} \frac{1}{n}$ is the harmonic series, which diverges (example 39.10). Thus $a_n > b_n$ for all n, where $\displaystyle\sum_{n=1}^{\infty} b_n$ diverges. By theorem 39.10, $\displaystyle\sum_{n=1}^{\infty} a_n$ also diverges. $\square$

Finally, if the ratio of the corresponding terms in two sequences approaches a **finite, non-zero limit**, then their corresponding series will either diverge or converge together.

Theorem 39.11. Limit Comparison Test

If $\displaystyle\sum_{n=0}^{\infty} a_n$ and $\displaystyle\sum_{n=0}^{\infty} b_n$ are series with positive terms and

$$\lim_{n\to\infty}\frac{a_n}{b_n} = C \qquad (39.53)$$

where $C > 0$ and $C \neq \infty$, then either both series converge or both series diverge.

Example 39.17. Show that $\displaystyle\sum_{n=1}^{\infty}\frac{\sqrt{n}}{3n+4}$ diverges.

Solution. Let $b_n = 1/\sqrt{n}$. By the p-test, since $\displaystyle\sum_{n=1}^{\infty} b_n = \sum_{n=1}^{\infty} n^{-1/2}$ is a p-series with $p = -1/2 < 1$, so by the p-test, it diverges. Furthermore,

$$\lim_{n\to\infty}\frac{a_n}{b_n} = \lim_{n\to\infty}\left(\frac{\sqrt{n}}{3n+4}\Big/\frac{1}{\sqrt{n}}\right) = \lim_{n\to\infty}\left(\frac{\sqrt{n}}{3n+4}\cdot\frac{\sqrt{n}}{1}\right) = \lim_{n\to\infty}\frac{n}{3n+4} = \frac{1}{3} \quad (39.54\text{a})$$

which is finite and non-zero. By Theorem 39.11, $\displaystyle\sum_{n=1}^{\infty} a_n$ and $\displaystyle\sum_{n=1}^{\infty} b_n$ converge or diverge together. Since $\displaystyle\sum_{n=1}^{\infty} b_n$ diverges, we conclude that $\displaystyle\sum_{n=1}^{\infty} a_n$ diverges. $\qquad\square$

Example 39.18. Show that $\displaystyle\sum_{n=1}^{\infty}\frac{3^{n-1}}{4^{n+2}+2}$ converges.

Solution. Let $b_n = (3/4)^n$. We pick this sequence because it resembles the format of a_n. Further, we know that $\displaystyle\sum_{n=1}^{\infty} b_n = \sum_{n=1}^{\infty}\left(\frac{3}{4}\right)^n$ is a geometric series with $r = \frac{3}{4} < 1$, and thus it converges. But

$$\frac{a_n}{b_n} = \left(\frac{3^{n-1}}{4^{n+2}+2}\right)\Big/\left(\frac{3^n}{4^n}\right) = \frac{3^{n-1}}{4^{n+2}+2}\cdot\frac{4^n}{3^n} \qquad (39.55\text{a})$$

$$= \frac{1/3}{(4^{n+2}/4^n)+2/4^n} = \frac{1/3}{1/16+2/4^n} \qquad (39.55\text{b})$$

$$\to \frac{1/3}{1/16} = \frac{16}{3} \qquad (39.55\text{c})$$

Since the limit is finite and non-zero, then according to theorem 39.11, either both $\displaystyle\sum_{n=1}^{\infty} a_n$ and $\displaystyle\sum_{n=1}^{\infty} b_n$ must converge, or both must diverge. Since $\displaystyle\sum_{n=1}^{\infty} b_n$ converges, we conclude that $\displaystyle\sum_{n=1}^{\infty} a_n$ also converges. $\qquad\square$

Alternating Series

We have seen examples of series where the sign changes *with each term*. Such series are called **alternating series**. Sometimes these series converge,

$$\left\{\frac{(-1)^n}{2^n}\right\} = 1, \ -\frac{1}{2}, \ \frac{1}{4}, \ -\frac{1}{8}, \ \frac{1}{16}, \ \cdots \tag{39.56}$$

Some of these series do not converge:

$$\{(-1)^n\} = -1, \ 1, \ -1, \ 1, \ 1, \ 1, \ \cdots \tag{39.57}$$

or

$$\left\{(-3)^{n+1}\right\} = 3, \ -9, \ 81, \ -243, \ \cdots \tag{39.58}$$

The fundamental property of these series is that they **change sign with every term**. It is convenient to write them as

$$a_n = (-1)^{n-1}b_n \tag{39.59}$$

where $b_n = |a_n| > 0$. If the first term is actually negative, we generally redefine the series coefficients to fit this format.

Definition 39.5. Alternating Series

An **alternating series** is any series of the form $\displaystyle\sum_{n=1}^{\infty}(-1)^{n-1}b_n$ where $b_n > 0$ for all n

---

Alternating sequences have the special property that if $|a_n|$ is monotonically decreasing (remember that $|a_n| > 0$) then $\displaystyle\sum_{n=1}^{\infty}a_n = \sum_{n=1}^{\infty}(-1)^{n-1}|a_n|$ converges.

This does not guarantee that $\displaystyle\sum_{n=1}^{\infty}|a_n|$ converges; in fact, $\displaystyle\sum_{n=1}^{\infty}|a_n|$ often diverges when $\displaystyle\sum_{n=1}^{\infty}a_n$ converges.

Theorem 39.12. Alternating Series Test

If (1) $|a_{n+1}| < |a_n|$ for all n, and (2) $a_n \to 0$ and $n \to \infty$, then $\displaystyle\sum_{n=1}^{\infty}(-1)^{n-1}|a_n|$ converges.

Example 39.19. Alternating Harmonic Series. Show that

$$\sum_{n=1}^{\infty}\frac{(-1)^{n-1}}{n} = 1 - \frac{1}{2} + \frac{1}{3} - \frac{1}{4} + \cdots \tag{39.60}$$

converges.

Solution. We have

$$|a_n| = \frac{1}{n} > \frac{1}{n+1} = |a_{n+1}| \tag{39.61a}$$

for all n, and

$$\lim_{n\to\infty} a_n = \lim_{n\to\infty}\frac{(-1)^{n-1}}{n} = 0 \tag{39.61b}$$

Doing Calculus

Therefore by the alternating series test (theorem 39.12), $\displaystyle\sum_{n=1}^{\infty} \frac{(-1)^{n-1}}{n}$ converges. $\square$

Proof. (Proof of Alternating Series Test[3]). Let $S = \sum_{n=1}^{\infty} a_n$. Then since

$$|a_1| \geq |a_2| \geq |a_3| \cdots 0 \tag{39.62a}$$

then the sum is

$$S = \underbrace{\overbrace{\underbrace{|a_1| - |a_2|}_{\geq 0}}^{S_2} + \underbrace{|a_3| - |a_4|}_{\geq 0}}_{S_4}{}^{S_6} + \underbrace{|a_5| - |a_6|}_{\geq 0} + \cdots \tag{39.62b}$$

i.e.,

$$S_2 = |a_1| - |a_2| \geq 0 \tag{39.62c}$$
$$S_4 = S_2 + |a_3| - |a_4| \geq S_2 \tag{39.62d}$$
$$S_6 = S_4 + |a_5| - |a_6| \geq S_4 \tag{39.62e}$$

$$\vdots$$

hence

$$S_2 \leq S_4 \leq S_6 \leq S_8 \leq \cdots \leq S_{2n} \leq \cdots \tag{39.62f}$$

But we can regroup the terms in equation 39.62b as follows:

$$S = \overbrace{|a_1| - \underbrace{\overbrace{(|a_2| - |a_3|)}}_{\geq 0}}^{S_7} - \underbrace{(|a_4| - |a_5|)}_{\geq 0} - \underbrace{(|a_6| - |a_7|)}_{\geq 0} - \cdots \geq |a_1| \tag{39.62g}$$

The last inequality follows because we are subtracting a bunch of positive terms from $|a_1|$ so we must be getting smaller than $|a_1|$. Therefore the sequence $S_2, S_4, S_6, \ldots$ is increasing and bounded by some number S. Thus $S_{2n} \to S$. To prove that $S_n \to S$ we need to find the limit of the odd partial sums S_{2n+1} as well:

$$\lim_{n\to\infty} S_{2n+1} = \lim_{n\to\infty} (S_{2n} + |a_{2n+1}|) = \lim_{n\to\infty} S_{2n} + \lim_{n\to\infty} |a_{2n+1}| = S + 0 = S \tag{39.62h}$$

Thus since $S_{2n} \to S$ and $S_{2n+1} \to S$, all the partial sums go to S, i.e, $S_n \to S$. Thus the series converges to some number S. $\square$

To show that a series converges using the alternating series test, we need to demonstrate two facts: (1) that $|a_n|$ is decreasing; and (2) that $|a_n| \to 0$. Sometimes it is difficult to demonstrate that the sequence is decreasing using algebraic techniques, and so we can resort to calculus: if the function $f(x)$ is decreasing, where $f(n) = a_n$ at the integers $x = n$, then we know that $|a_n|$ is decreasing.

---

[3]This proof can be skipped without loss of continuity.

Example 39.20. Determine the convergence of $\displaystyle\sum_{n=1}^{\infty} \frac{(-1)^{n-1}(n+1)}{(n+2)^2}$.

Solution. First we show that a_n is decreasing. Let

$$f(x) = \frac{x+1}{(x+2)^2} \tag{39.63a}$$

Then

$$f'(x) = \frac{(x+2)^2(1) - (x+1)(2)(x+2)}{(x+2)^4} = \frac{(x+2)(x+2-2x-2)}{(x+2)^4} \tag{39.63b}$$

$$= \frac{-x}{(x+2)^3} < 0 \tag{39.63c}$$

where the inequality holds for all positive x (whether or not it holds for negative x is irrelevant because we are only interested in the positive integers). Thus f is decreasing, and therefore $|a_n|$ is decreasing: $|a_{n+1}| < |a_n|$.

Next, we calculate the limit

$$\lim_{n\to\infty} |a_n| = \lim_{n\to\infty} \frac{n+1}{(n+2)^2} = 0 \tag{39.63d}$$

Thus the sequence converges by the alternating sequence test.   □

In the proof of the alternating series test (Theorem 39.12) we showed that S is larger than all the even partial sums (eq. 39.62b), and smaller than each of the odd partial sums (eq. 39.62g):

$$S_{2n} \le S \le S_{2n+1} \tag{39.64}$$

Thus the sum will always fall between two consecutive partial sums.

$$S_k \le S \le S_{k+1} = S_k + |a_{k+1}| \tag{39.65}$$

Therefore

$$|S - S_k| \le |a_{k+1}| \tag{39.66}$$

i.e., **the error made in truncated an alternating series (in absolute value) is bounded by the (absolute value of) the first omitted term.**

Theorem 39.13. Alternating Series Estimation Theorem

If $\displaystyle\sum_{n=1}^{\infty}(-1)^{n-1}|a_n|$ converges then

$$\sum_{n=1}^{\infty}(-1)^{n-1}|a_n| = \underbrace{\overbrace{\sum_{k=1}^{n}(-1)^k|a_k|}^{n^{th}\ \text{partial sum}} + \underbrace{\overbrace{|a_{n+1}|}^{R_n}}_{\text{remainder}}}_{S_n} \tag{39.67}$$

Example 39.21. How many terms are required to ensure that the sum $\displaystyle\sum_{n=1}^{\infty} \frac{(-1)^n 2^n}{n!}$ converges to 5 decimal places?

Solution. First we need to verify that the series converges. But

$$|a_n| = \frac{2^n}{n!} \geq \frac{2^n}{n!} \cdot \frac{2}{n+1} = \frac{2^{n+1}}{(n+1)!} = |a_{n+1}| \tag{39.68a}$$

because $2/(n+1) < 1$ for $n > 1$. Hence $|a_n|$ is decreasing. To show that $|a_n| \to 0$, for $n > 3$,

$$|a_n| = \frac{2^n}{n!} = \underbrace{\frac{2}{n} \cdot \frac{2}{n-1} \cdot \frac{2}{n-2} \cdot \frac{2}{n-3} \cdots \frac{2}{3} \cdot \frac{2}{2}}_{\text{each of these is } <2/3} \cdot \frac{2}{1} \tag{39.68b}$$

$$< \left(\frac{2}{3}\right)^{n-1} \cdot 2 = \left(\frac{2}{3}\right)^n \cdot \left(\frac{2}{3}\right)^{-1} \cdot 2 = 3\left(\frac{2}{3}\right)^n \tag{39.68c}$$

$$\tag{39.68d}$$

Therefore

$$\lim_{n \to \infty} |a_n| = 3 \lim_{n \to \infty} \left(\frac{2}{3}\right)^n = 0 \tag{39.68e}$$

Thus the by the alternating series test, the series converges.

To determine how many terms we need for an accuracy of 5 decimal places, we need to find the first term that is smaller in absolute value than 10^{-5}, i.e., we must find n such that

$$\frac{2^n}{n!} \leq 0.00001 \tag{39.68f}$$

This type of equation is not directly solvable. The answer can only be found by calculating terms.

n	2^n	$n!$	$2^n/n!$	$2^n/n! < 0.00001$
1	2	1	2	False
2	4	2	2	False
3	8	6	$\frac{4}{3}$	No
4	16	24	$\frac{2}{3}$	No
5	32	120	$\frac{4}{15}$	No
6	64	720	$\frac{4}{45}$	No
7	128	5040	$\frac{8}{315}$	No
8	256	40320	$\frac{2}{315}$	No
9	512	362880	$\frac{4}{2835}$	No
10	1024	3628800	$\frac{4}{14175}$	No
11	2048	39916800	$\frac{8}{155925}$	No
12	4096	479001600	$\frac{4}{467775}$	Yes

Thus 12 terms are required, since $2^{11}/11! = \dfrac{8}{155925} \approx 0.000\,051$ while $2^{12}/12! = \dfrac{4}{467775} \approx 0.000\,00855$.

$\square$

Definition 39.6. Absolute Convergence

$\displaystyle\sum_{n=1}^{\infty} a_n$ is said to **converge absolutely** if $\displaystyle\sum_{n=1}^{\infty} |a_n|$ converges.

Definition 39.7. Conditional Convergence

$\displaystyle\sum_{n=1}^{\infty} a_n$ is **conditionally convergent** if it converges but $\displaystyle\sum_{n=1}^{\infty} |a_n|$ does not converge.

Example 39.22. We have discussed how the **alternating harmonic series** $\displaystyle\sum \frac{(-1)^n}{n}$ converges (example 39.19), but the harmonic series $\displaystyle\sum \frac{1}{n}$ diverges (example 39.10). Thus the alternating harmonic series converges bit does not converge absolutely, i.e., it is conditionally convergent. $\square$

Example 39.23. The series $S = \displaystyle\sum_{n=1}^{\infty} \frac{(-1)^n}{n^2}$ meets the conditions of the alternating series test

and therefore it converges. Furthermore, it converges absolutely, because $\displaystyle\sum_{n=1}^{\infty} \frac{1}{n^2}$ converges

(e.g., p-test, with $p = 2$). $\square$

The concept of absolute convergence applies to a more general class of series than just alternating series. Theorem 39.14 is very helpful in this regard.

Theorem 39.14. Absolute Convergence Implies Convergence

If $\displaystyle\sum_{n=1}^{\infty} |a_n|$ converges then $\displaystyle\sum_{n=1}^{\infty} a_n$ converges.

Example 39.24. Determine the nature of the convergence of $\displaystyle\sum_{n=1}^{\infty} \frac{\cos n}{n^2}$.

Solution. This series has both positive and negative terms but is not an alternating series because the sign does not change with every term. For example, the first few terms are:

$$\sum_{n=1}^{\infty} \frac{\cos n}{n^2} = \cos 1 + \frac{\cos 2}{4} + \frac{\cos 3}{9} + \cos 416 + \cos 525 + \cos 636 + \cdots \tag{39.69a}$$

$$\approx 0.5403 - 0.1040 - 0.1099 - 0.04085 + 0.01134 + 0.02667 + \cdots \tag{39.69b}$$

However, $0 \le |\cos n| < 1$. Equality never holds for $n > 0$, because $|\cos n| = 1$ requires n to be a multiple of π and that is not possible for any integer values of n. Therefore

$$\frac{|\cos n|}{n^2} < \frac{1}{n^2} \tag{39.69c}$$

for all positive n.

By the p-test (theorem 39.7) we know that $\displaystyle\sum_{n=1}^{\infty} \frac{1}{n^2}$ converges.

By the series comparison test (theorem 39.9), since

$$0 \le \left| \frac{\cos n}{n^2} \right| < \frac{1}{n^2} \tag{39.69d}$$

we know that $\displaystyle\sum_{n=1}^{\infty} \left| \frac{\cos n}{n^2} \right|$ converges.

Hence $\displaystyle\sum_{n=1}^{\infty} \frac{\cos n}{n^2}$ converges absolutely (definition 39.6).

Hence $\displaystyle\sum_{n=1}^{\infty} \frac{\cos n}{n^2}$ converges (theorem 39.14). $\qquad\qquad\square$

Ratio Tests

The ratio test determines whether or not a series converges by examining the ratio of two consecutive elements of a series. It is not always conclusive. When it is inconclusive you need to use a different method to determine if the series converges.

Theorem 39.15. Ratio Test

Let $L = \lim\limits_{n\to\infty} |a_{n+1}/a_n|$ Then

1. If $L < 1$ then $\sum\limits_{n=1}^{\infty} a_n$ converges absolutely.

2. If $L > 1$ then $\sum\limits_{n=1}^{\infty} a_n$ diverges.

3. If $L = 1$ then the test is inconclusive.

Proof. (Proof of the ratio test.[4]) Define the series S by

$$S = \sum_{n=1}^{\infty} (-1)^{n-1} |a_n|, \tag{39.70a}$$

and define the ratio L by

$$L = \lim_{n\to\infty} \left| \frac{a_{n+1}}{a_n} \right| \tag{39.70b}$$

If this limit exists, then $L \geq 0$, because of the absolute value. One of the following cases three cases must hold: $(1) 0 \leq L < 1$; $(2)\ L > 1$; or $(3)\ L = 1$. We will consider each of these three cases in turn.

Case 1: If $0 \leq L < 1$ then pick any number r such that $L < r < 1$. Since $|a_{n+1}/a_n| \to L$ as as $n \to \infty$, then by the formal definition of a limit (def 38.4), there is some number N, such that for all $n > N$, the ratio will be less than r,

$$\left| \frac{a_{n+1}}{a_n} \right| < r \text{ for all } n > N \tag{39.70c}$$

Therefore

$$|a_{N+1}| < r|a_N| \tag{39.70d}$$

$$|a_{N+2}| < r|a_{N+1}| < r^2|a_N| \tag{39.70e}$$

$$|a_{N+3}| < r|a_{N+2}| < r^3|a_N| \tag{39.70f}$$

$$\vdots$$

$$|a_{N+p}| < r^p|a_N| \tag{39.70g}$$

The series

$$R = \sum_{p=1}^{\infty} r^p |a_N| \tag{39.70h}$$

---

[4] The proof may be omitted on a first reading without any loss of continuity.

is a converging geometric series with ratio $r < 1$. By the comparison test the series

$$S_T = \sum_{p=N+1}^{\infty} |a_p| \tag{39.70i}$$

also converges, since every term in S_T is smaller than a term in the series R, because the corresponding terms are smaller:

larger: $R = \sum_{p=1}^{\infty} r^p |a_N| = \quad r|a_N| \quad + \quad r^2|a_{N+1}| \quad + \quad r^3|a_{N+3}| \quad + \quad \cdots$

smaller: $S_T = \sum_{p=N+1}^{\infty} |a_p| = \quad |a_{N+1}| \quad + \quad |a_{N+2}| \quad + \quad |a_{N+3}| \quad + \quad \cdots$

$$\tag{39.70j}$$

Since $S_T = \sum_{p=1}^{N} |a_p|$ converges, it is a finite number, and hence

$$\sum_{n=1}^{\infty} |a_n| = S_n + S_T = \overbrace{\sum_{p=1}^{N} |a_p|}^{S_n} + \overbrace{\sum_{p=N+1}^{\infty} |a_p|}^{S_T} \tag{39.70k}$$

converges.

Case 2: If $L > 1$ then for some integer N the ratio $|a_{n+1}/a_n|$ will satisfy

$$\left| \frac{a_{n+1}}{a_n} \right| > 1 \tag{39.70l}$$

for all $n > N$, i.e.,

$$|a_{n+1}| > |a_n| \tag{39.70m}$$

for all $n > N$. The sequence is increasing and positive and hence does not not go to zero. Hence the series diverges.

Case 3: If $L = 1$ the test limit tells us nothing.   □

Example 39.25. Apply the ratio test to $\sum_{n=1}^{\infty} \dfrac{(-2)^{3n}}{n!}$ to determine whether or not it converges.

Solution. The ratio is

$$\left| \frac{a_{n+1}}{a_n} \right| = \left(\frac{(-2)^{3(n+1)}}{(n+1)!} \right) \Big/ \left(\frac{(-2)^{3n}}{n!} \right) \tag{39.71a}$$

$$= \frac{(-1)^n \cdot 2^{3n} \cdot 2^3}{(-1)^n \cdot 2^{3n}} \cdot \frac{n!}{(n+1)n!} \tag{39.71b}$$

$$= \frac{8}{n+1} \to 0 < 1 \tag{39.71c}$$

Therefore the series converges.   □

Example 39.26. For what values of K does the series $1 + 2K + 3K^2 + 4K^3 + \cdots$ converge?

Solution. The general term is $a_n = nK^{n-1}$. Therefore the ratio test gives

$$\lim_{n \to \infty} \left| \frac{a_{n+1}}{a_n} \right| = \lim_{n \to \infty} \frac{(n+1)K^n}{nK^{n-1}} = K \lim_{n \to \infty} \frac{n+1}{n} = K \tag{39.72a}$$

The ratio test requires $|K| < 1$ for convergence, and tells us that if $|K| > 1$, the series will diverge. It tells us nothing about the series when $K = 1$. When $K = 1$, the series is the sum of all the positive integers, which is infinite. When $K = -1$, the terms are not approaching zero. So the series only converges for $|K| < 1$.   □

Root Test

The root test looks at the limit of the nth root as $n \to \infty$; mind boggling as that concept sounds, the number is often very easy to compute. If you can remember the ratio test, you can remember the root test, because the rule is pretty much the same: if the limit is less than 1, it converges.

Theorem 39.16. Root Test

Let $r = \lim\limits_{n \to \infty} (a_n)^{1/n}$ Then

1. If $r < 1$ then $\sum\limits_{n=1}^{\infty} a_n$ converges absolutely.

2. If $r > 1$ then $\sum\limits_{n=1}^{\infty} a_n$ diverges.

3. If $r = 1$ then the test is inconclusive.

Proof. (Proof of the Root Test). Let $L = \lim\limits_{n \to \infty} \sqrt[n]{|a_n|}$. Either $L < 1$, $L > 1$, or $L = 1$.

Case 1. $L < 1$. Pick some number r such that $L < r < 1$. Then by the formal definition of a limit of a series, for some (possibly very large) number N, for all $n > N$,

$$\sqrt[n]{|a_n|} < r \tag{39.73a}$$

Raising both sides of the inequality to the power of n,

$$|a_n| < r^n \tag{39.73b}$$

for all $n > N$. Since $r < 1$, the series

$$R = \sum_{n=N+1}^{\infty} r^n \tag{39.73c}$$

converges – it is a geometric series with $a = r^N$ and ratio $r < 1$. Furthermore, and since $|a_n| < r^n$ for all $n > N$,

$$S_T = \sum_{n=N+1}^{\infty} |a_n| \tag{39.73d}$$

also converge (by comparison with R). Since S_n (the partial sums of S) is always a finite number we conclude that

$$\sum_{n=1}^{\infty} a_n = S_n + S_T = S_n + \sum_{n=N+1}^{\infty} |a_n| \tag{39.73e}$$

must also converge.

Case 2: $L > 1$ then there is some N such that $\sqrt[n]{|a_n|} > 1$ for all $n > N$. Then $|a_n| > 1$. This means that $|a_n|$ does not go to zero. Thus the series must diverge.

Case 3: $L = 0$. The limit tells us nothing.   □

Example 39.27. Determine whether or not $\displaystyle\sum_{n=1}^{\infty} \frac{(-1)^n (3n+4)^n}{(5n+6)^n}$ converges.

Solution. Using the root test,

$$|a_n|^{1/n} = \left(\frac{(3n+4)^n}{(5n+6)^n} \right)^{1/n} = \frac{3n+4}{5n+6} = \frac{3}{5} < 1 \tag{39.74a}$$

Therefore the series converges. □

Example 39.28. Derive the p-test (theorem 39.7) using the root test.

Solution. Let $S = \displaystyle\sum_{n=1}^{\infty} \frac{1}{n^p}$. We ask the following question: for what values of p does S converge? To use the ratio test, we calculate

$$\lim_{n\to\infty} (a_n)^{1/n} = \lim_{n\to\infty} \left(\frac{1}{n^p} \right)^{1/n} = \lim_{n\to\infty} \left(\frac{1}{n^{1/n}} \right)^p \tag{39.75a}$$

$$= \left(\lim_{n\to\infty} \frac{1}{n^{1/n}} \right)^p \tag{39.75b}$$

$$= \left(1 \Big/ \lim_{n\to\infty} n^{1/n} \right)^p \tag{39.75c}$$

To find $L = \displaystyle\lim_{n\to\infty} n^{1/n}$ we use calculate $\ln L$:

$$\ln L = \lim_{n\to\infty} \ln n^{1/n} = \lim_{n\to\infty} \frac{\ln n}{n} \to \frac{\infty}{\infty} \tag{39.75d}$$

$$\stackrel{H}{=} \lim_{n\to\infty} \frac{1/n}{1} = 0 \tag{39.75e}$$

Therefore (since $\ln L \to 0$), $L \to e^0 = 1$. Therefore

$$\lim_{n\to\infty} (a_n)^{1/n} = 1^p = 1 \tag{39.75f}$$

The root test is inconclusive. The p-test cannot be derived from the root test! □

Example 39.29. Apply the root test to the geometric series $\displaystyle\sum_{n=1}^{\infty} (8/5)^n$.

Solution. Since $a_n = (8/5)^n$, the root test gives

$$\lim_{n\to\infty} (a_n)^{(}1/n) = \lim_{n\to\infty} ((8/5)^n)^{(}1/n) = \frac{8}{5} > 1 \tag{39.76a}$$

Therefore the series diverges. □

Exercises

Determine if the series in exercises 1 through 9 converges or diverges. If the series converges, find its sum.

1. $1 - \dfrac{1}{2} + \dfrac{1}{4} - \dfrac{1}{8} + \dfrac{1}{16} + \cdots$

2. $\dfrac{4}{3} + \dfrac{8}{9} + \dfrac{16}{27} + \dfrac{32}{81} + \cdots$

3. $\displaystyle\sum_{n=1}^{\infty} \dfrac{1+n}{n}$

4. $\displaystyle\sum_{n=1}^{\infty} (-7)^{n+2} \cdot (2)^{n-1}$

5. $\displaystyle\sum_{n=1}^{\infty} 2^n \cdot (3)^{-(n+2)}$

6. $\displaystyle\sum_{n=1}^{\infty} \sqrt[n]{n}$

7. $\displaystyle\sum_{n=1}^{\infty} 3^n \cdot (2)^{-(n+2)}$

8. $\displaystyle\sum_{n=1}^{\infty} \dfrac{(-2)^4 n}{16^n}$

9. $\displaystyle\sum_{n=1}^{\infty} \ln \dfrac{n}{n+1}$

Use the integral test to determine whether the series given in exercises 10 through 19 diverge or converge.

10. $\displaystyle\sum_{n=1}^{\infty} \dfrac{\ln n}{n}$

11. $\displaystyle\sum_{n=1}^{\infty} \dfrac{1}{1+n}$

12. $\displaystyle\sum_{n=1}^{\infty} \dfrac{1}{n\sqrt{n}}$

13. $\displaystyle\sum_{n=1}^{\infty} \sqrt{n^{-5}}$

14. $\displaystyle\sum_{n=1}^{\infty} ne^{-n}$

15. $\displaystyle\sum_{n=1}^{\infty} ne^{-n^2}$

16. $\displaystyle\sum_{n=1}^{\infty} \dfrac{1}{n \ln n}$

17. $\displaystyle\sum_{n=1}^{\infty} \dfrac{1}{n^2 + 36}$

18. $\displaystyle\sum_{n=1}^{\infty} \dfrac{n}{n^2 + 36}$

19. $\displaystyle\sum_{n=1}^{\infty} \dfrac{n}{(n^2 + 4)^2}$

Use the comparison tests in exercises 20 through 20.

20. $\displaystyle\sum_{n=1}^{\infty} \dfrac{2n + 7}{5n^4 + 6n + 2}$

21. $\displaystyle\sum_{n=1}^{\infty} \dfrac{1}{\sqrt{n(n+1)}}$

22. $\displaystyle\sum_{n=1}^{\infty} \dfrac{2^n}{n!}$

23. $\displaystyle\sum_{n=1}^{\infty} \dfrac{n^4}{n^5 - 4}$

24. $\displaystyle\sum_{n=1}^{\infty} \dfrac{\ln n}{n}$

25. $\displaystyle\sum_{n=1}^{\infty} \dfrac{\ln n}{n^2}$

26. $\displaystyle\sum_{n=1}^{\infty} \dfrac{1}{n!}$

27. $\displaystyle\sum_{n=1}^{\infty} \dfrac{3\sqrt[5]{n^4} + 12}{3n^2}$

28. $\displaystyle\sum_{n=1}^{\infty} \dfrac{1}{n\sqrt{n+2}}$

29. $\displaystyle\sum_{n=1}^{\infty} \dfrac{n^2}{3^n}$

In exercises 30 through 41 use the alternating series test on the given series to test it for convergence. If the alternating series does not apply, find another test to determine if the series converges.

30. $\displaystyle\sum_{n=1}^{\infty} \dfrac{(-1)^{n+1}}{n!}$

31. $\displaystyle\sum_{n=1}^{\infty} \dfrac{(-1)^{n+1}}{\sqrt[n]{3}}$

32. $\displaystyle\sum_{n=1}^{\infty} \dfrac{(-1)^{n+1}(n+3)}{n+7}$

33. $\displaystyle\sum_{n=1}^{\infty} \dfrac{(-1)^{n+1} n^2}{n^2 + 4n + 5}$

34. $\displaystyle\sum_{n=1}^{\infty} \dfrac{(-1)^{n+1}}{n\sqrt{n}}$

35. $\displaystyle\sum_{n=1}^{\infty} \dfrac{(-1)^{n+1}}{\sqrt{n(n+1)}}$

36. $\displaystyle\sum_{n=1}^{\infty} \dfrac{(-1)^{n+1}}{\sqrt{n}}$

37. $\displaystyle\sum_{n=1}^{\infty} \dfrac{(-1)^{n+1} \sin^2 n}{n}$

38. $\displaystyle\sum_{n=1}^{\infty} \dfrac{(-1)^{n+1} 3^n}{n!}$

39. $\displaystyle\sum_{n=1}^{\infty} \dfrac{(-1)^{n+1}}{n^3}$

40. $\displaystyle\sum_{n=1}^{\infty} \dfrac{(-1)^{n+1}}{\ln(n+1)}$

41. $\displaystyle\sum_{n=1}^{\infty} \dfrac{(-1)^{n+1} \sqrt{n}}{n + 10}$

Show that each of the following series converges. Then determine how many terms are needed for each of the following series to converge to the stated accuracy.

42. $1 - \dfrac{1}{4} + \dfrac{1}{9} - \dfrac{1}{16} + \cdots, \; 0.001$

43. $5 - \dfrac{5^3}{3!} + \dfrac{5^5}{5!} - \dfrac{5^7}{7!} + \cdots, \; 0.01$

44. $\dfrac{1}{5} - \dfrac{\sqrt{2}}{6} + \dfrac{\sqrt{3}}{7} - \dfrac{\sqrt{4}}{8} + \cdots, \; 0.01$

45. Determine if each of the series in exercises 30 through 41 converges absolutely, conditionally, or diverges.

Use the ratio or root test to determine if each of the series in exercises 46 through 51 converges or diverges. If the ratio or root test does not apply, attempt to find a different test to determine if the series converges or diverges.

46. $\displaystyle\sum_{n=1}^{\infty} \dfrac{3n^3 + 4}{n^4 + 5}$

47. $\displaystyle\sum_{n=1}^{\infty} \dfrac{\sqrt{3^n}}{2^n}$

48. $\displaystyle\sum_{n=1}^{\infty} \dfrac{n^4}{3^{k+2}} 3^{3n}$

49. $\displaystyle\sum_{n=1}^{\infty} \dfrac{n 2^n}{(n+2)!}$

50. $\displaystyle\sum_{n=1}^{\infty} \left(1 - \dfrac{1}{n}\right)^n$

51. $\displaystyle\sum_{n=1}^{\infty} \dfrac{n!}{5^n}$

Chapter 40

Power Series

Chapter Summary and Goal

In chapter 39 we looked at a series as a sum of numbers. In this chapter we consider series that depend on a variable x in a very specific way, namely, the nth term will have the form $c_n x^n$ for some constant c_n. Since these series depend on sums of powers of x they are called **Power Series**. Power series have very many useful applications and are widely used throughout the mathematical sciences.

Student Learning Objectives

The student will:

1. Understand the basic properties of power series.
2. Be able to determine the radius of convergence of a power series.
3. Learn how to calculate power series from geometric series.
4. Learn how to calculate basic Taylor and Maclaurin series from Taylor's theorem.
5. Learn how to calculate the remainder term and determine how many terms are required for a specified precision.
6. Understand how the Taylor polynomials extend the linearization with higher order approximations.

Series that Depend on a Variable x

An example of a power series is

$$f(x) = 1 - x + \frac{1}{2}x^2 - \frac{1}{8}x^3 + \frac{1}{16}x^4 + \cdots \tag{40.1}$$

For convenience, we will almost always number power series starting with the index 0 rather than 1. This will allow us to number the coefficients c_k with the same index as the exponent, and still have a constant coefficient c_0. So if we begin to label the **coefficients** (constants) in (40.1), we would start with

$$c_0 = 1,\ c_1 = -1,\ c_1 = \frac{1}{2},\ c_3 = -\frac{1}{8},\ c_4 = \frac{1}{16}x^4,\ \ldots \tag{40.2}$$

Note that minus signs are incorporated into the coefficients. Furthermore, if any terms are missing, these are incorporated into the coefficients as zeros. Thus we would dissect the series

$$f(x) = x - \frac{x^3}{3!} + \frac{x^5}{5!} - \frac{x^7}{7!} + \cdots \tag{40.3}$$

as

$$0 \cdot x^0 \quad + \quad 1 \cdot x \quad + \quad 0 \cdot x^2 \quad + \quad \frac{(-1)}{3!} x^3 \quad + \quad 0 \cdot x^4 \quad + \quad \frac{1}{5!} x^5 + \cdots$$

$$c_0 = 0 \qquad c_1 = 1 \qquad c_2 = 0 \qquad c_3 = \frac{(-1)}{3!} \qquad c_4 = 0 \qquad c_5 = \frac{1}{5!} \tag{40.4}$$

Definition 40.1. Power Series

A **power series** is any series of the form

$$f(x) = \sum_{n=0}^{\infty} c_n x^n = c_0 + c_1 x + c_2 x^2 + c_3 x^3 + c_4 x^4 + \cdots \tag{40.5}$$

where $c_1, c_2, \ldots$ are fixed constants, called **coefficients**.

All of the convergence rules that we learned in chapter 39 apply to power series.

Example 40.1. For what values of x does the power series $\sum_{n=0}^{\infty} \frac{(-1)^{n+1} x^n}{n!}$ converge?

Solution. This is same series as eq. 40.1. To study it with the ratio test we compute

$$\left| \frac{a_{n+1}}{a_n} \right| = \left| \frac{x^{n+1}}{(n+1)!} \right| \bigg/ \left| \frac{x^n}{n!} \right| = \frac{|x|}{n} \tag{40.6a}$$

To converge, the ratio test requires that

$$1 > \lim_{n \to \infty} \frac{|x|}{n} = |x| \lim_{n \to \infty} \frac{1}{n} = |x| \cdot 0 = 0 \tag{40.6b}$$

This a true statement that is independent of x. Therefore the power series converges for all values of x. □

Shifting to a general center. When we subtract a constant from the argument of a function, as in $f(x - a)$, we shift the graph of $f(x)$ a distance a to the right. Thus one might argue that the expression $f(x - a)$ is more general than the expression $f(x)$ because it incorporates $f(x)$ (when $a = 0$) as well as all possible right (when $a > 0$) and left (when $a < 0$) shifts.

Consequently, when see a formula like $y = (x - 3)^2$, we know that it has the same, identical graph as the expression $y = x^2$, but with one modification: it is shifted 3 units to the right. We see this immediately by looking at the formula.

Similarly, we can write a more general formula for a power series as

$$f(x) = \sum_{n=0}^{\infty} c_k (x - a)^k = c_0 + c_1(x - a) + c_2(x - a)^2 + c_3(x - a)^3 + \cdots \tag{40.7}$$

Equation 40.7 is the same as equation 40.5, shifted to the right by a distance of a units.

Definition 40.2. Power Series

A **power series about** a is any series of the form

$$S(x) = \sum_{n=0}^{\infty} c_n(x-a)^n \qquad (40.8)$$

where the numbers c_0, c_1, c_2, $\cdots$ are fixed constants, called the **coefficients of the power series**, and a is any fixed (possibly zero) real number called the **center of the power series**.

Example 40.2. Identify the center, a, and coefficients, c_n, of the power series $\displaystyle\sum_{n=1}^{\infty} \frac{(6x-3)^n}{n}$.

Solution. This series has the same form as equation 40.8, even though it begins at $n = 1$ and not $n = 0$. To put each term (for $n > 0$) in the form $c_n(x-a)^n$ requires a small amount of algebraic manipulation:

$$\frac{(6x-3)^n}{n} = \frac{6^n(x-3/6)^n}{n} = \frac{6^n(x-1/2)^n}{n} \qquad (40.9\mathrm{a})$$

Setting this equal to $c_n(x-a)^n$ for all n tells us that $a = \dfrac{1}{2}$ and $c_n = \dfrac{6^n}{n}$.   □

Example 40.3. For what values of x does the power series $\displaystyle\sum_{n=1}^{\infty} \frac{(6x-3)^n}{n}$ converge?

Solution.   Using the ratio test,

$$\left| \frac{a_{n+1}}{a_n} \right| = \left| \frac{(6x-3)^{n+1}}{n+1} \right| \Big/ \left| \frac{(6x-3)^n}{n} \right| = \frac{n}{n+1} \cdot |6x-3| \qquad (40.10\mathrm{a})$$

Taking the limit as $n \to \infty$,

$$\lim_{n\to\infty} \left| \frac{a_{n+1}}{a_n} \right| = \lim_{n\to\infty} \frac{n}{n+1} \cdot |6x-3| \qquad (40.10\mathrm{b})$$

$$= |6x-3| \lim_{n\to\infty} \frac{n}{n+1} \qquad (40.10\mathrm{c})$$

$$= |6x-3| \cdot 1 = |6x-3| \qquad (40.10\mathrm{d})$$

This limit mus be strictly less than 1 if the series is to converge, so we require

$$|6x-3| < 1 \qquad (40.10\mathrm{e})$$

By the definition of absolute value, this means that

$$-1 < 6x - 3 < 1 \qquad (40.10\mathrm{f})$$

Adding 3 across the inequality and dividing by 6, we obtain the result

$$\frac{1}{3} = \frac{2}{6} < x < \frac{4}{6} = \frac{2}{3} \qquad (40.10\mathrm{g})$$

This tells us that the series converges on $(1/3, 2/3)$ and diverges on $(-\infty, 1/3)$ and $(2/3, \infty)$. Because the ratio test gives a ratio of 1 at the end points, it tells us nothing about what happens when $x = 1/3$ or $x = 2/3$. To find out what happens there, we have to evaluate the series at each end point separately.

At $x = 1/3$, the series is $\displaystyle\sum_{n=1}^{\infty} \frac{(6(1/3) - 3)^n}{n} = \sum_{n=1}^{\infty} \frac{(-1)^n}{n}$. This is an alternating harmonic series, which converges.

At $x = 2/3$, the series is $\displaystyle\sum_{n=1}^{\infty} \frac{(6(2/3) - 3)^n}{n} = \sum_{n=1}^{\infty} \frac{(1}{n}$. This is a harmonic series, which diverges.

Thus the series converges only on $1/3 \leq x < 2/3$.   □

Theorem 40.1. Radius of Convergence of Power Series

Let $S(x) = \displaystyle\sum_{n=0}^{\infty} c_n (x - a)^n$ be a power series. Then one of the following is true:

1. There exists some number $R > 0$ such that the series converges for all

$$|x - a| < R \qquad (40.11)$$

   and diverges for all

$$|x - a| > R \qquad (40.12)$$

   The number R is called the **radius of convergence** of the power series; or

2. The series converges only at $x = a$; we say the radius of convergence $R = 0$; or

3. The series converges for all x; we say the radius of convergence $R = \infty$.

Definition 40.3. Interval of Convergence

The **interval of convergence** of a power series is the set of all x values for which a power series in x converges.

Example 40.4. Find the center, radius, and interval of convergence of $\displaystyle\sum_{n=1}^{\infty} \frac{(6x - 3)^n}{n}$.

Solution. In example 40.3 we found that this series converges for $1/3 \leq x < 2/3$ and diverges for all other x. The interval of convergence is therefore $[1/3, 2/3)$. The center of convergence is the center of this interval, $r = 1/2$. We can also read that directly from eq. 40.9a (the value of a in that earlier equation). The radius of convergence is the half-width of the interval, which is $1/3$.   □

Theorem 40.2. Size of Interval of Convergence

If $f(x)$ can be expanded in a power series, then the interval of convergence will fall into one of the following categories:

1. A single point.
2. The entire real line, $-\infty < x < \infty$
3. A single interval $(a - R, a + R)$ (possibly including one or both endpoints), where a is the center of convergence, and R the radius of convergence.

Representing Functions as Power Series

Almost any function can be written as a power series. One way to find a function's representation as a power series is by using geometric series, although this only works for some functions. Recall from theorem 39.5 (the sum of a geometric series) that

$$\sum_{n=1}^{\infty} ar^{n-1} = a + ar + ar^2 + \cdots = \frac{a}{1-r}, \tag{40.14}$$

so long as $|r| < 1$. This formula will still work even if we let r be a variable, such as x. If we set $a = 1$, then

$$\frac{1}{1-x} = 1 + x + x^2 + \cdots \tag{40.15}$$

Example 40.5. Find a power series representation for $\dfrac{1}{1+x^3}$.

Solution. We get this by replacing x with $-x^3$ in equation 40.15:

$$\frac{1}{1+x^3} = \frac{1}{1-(-x^3)} = 1 - x^3 + x^6 - x^9 + \cdots \tag{40.16a}$$

$\square$

Example 40.6. Find a power series representation of $\dfrac{1}{6+7x}$.

Solution.   We begin by factoring the $1/6$ so that this looks like $a/(1-x)$, where a is a constant.

$$\frac{1}{6+7x} = \frac{1}{6} \cdot \frac{1}{1-(-7/6)x} \tag{40.17a}$$

The second factor looks something like equation 40.15. We can substitute $(-7/6)x$ for x to get the following:

$$\frac{1}{6+7x} = \frac{1}{6}\left[1 + \left(-\frac{7x}{6}\right) + \left(-\frac{7x}{6}\right)^2 + \left(-\frac{7x}{6}\right)^3 + \cdots\right] \tag{40.17b}$$

$$= \frac{1}{6}\sum_{n=0}^{\infty}(-1)^n\left(\frac{7}{6}\right)^n x^n = \sum_{n=0}^{\infty}\frac{(-1)^n 7^n x^n}{6^{n+1}} \tag{40.17c}$$

$\square$

Another way we can get new power series is by differentiating or integrating old power series. This works because of some useful theory from advanced calculus that tells us that we are allowed to differentiate and integrate power series term by term.

1. The derivative of a series is the series of the derivatives.

2. The integral of a series is the series of the integrals.

Theorem 40.3. Differentiation of Power Series

If a power series converges, then it can be differentiated term by term, and

$$\frac{d}{dx}\sum_{n=0}^{\infty} c_n(x-a)^n = \sum_{n=0}^{\infty} nc_n(x-a)^{n-1} \tag{40.18}$$

Example 40.7. Find a power series for $\dfrac{1}{(1-x)^2}$.

Solution. We observe that $-1/x^2$ is the derivative of $1/x$. Generalizing this (by the chain rule), and substituting eq. 40.15

$$\frac{1}{(1-x)^2} = \frac{d}{dx}\frac{1}{1-x} = \frac{d}{dx}\left(1+x+x^2+x^3+\cdots\right) \tag{40.19a}$$

$$= 1 + 2x + 3x^2 + 4x^3 + \cdots = \sum_{n=0}^{\infty} nx^{n-1}. \tag{40.19b}$$

$\square$

Theorem 40.4. Integration of Power Series

If a power series converges, then it can be integrated term by term, and

$$\int \sum_{n=0}^{\infty} c_n(x-a)^n dx = \sum_{n=0}^{\infty} c_n \int (x-a)^n dx = C + \sum_{n=0}^{\infty} \frac{c_n}{n+1}(x-a)^{n+1} \tag{40.20}$$

where C is a general constant of integration.

Example 40.8. Find a power series for $\ln(1+x)$.

Solution. Integrating equation eq. 40.15 gives

$$\ln(1+x) = \int \frac{dx}{1+x} = \int \frac{dx}{1-(-x)} \tag{40.21a}$$

$$= \int \left(1 - x + x^2 - x^3 + x^2 - x^3\right) dx \tag{40.21b}$$

$$= C + x - \frac{x^2}{2} + \frac{x^3}{3} - \frac{x^4}{4} + \cdots \tag{40.21c}$$

$$= C + \sum_{n=1}^{\infty} \frac{(-1)^{n-1}x^n}{n} \tag{40.21d}$$

To get a value for the constant of integration we use our knowledge of natural logarithms, specifically, that that $\ln 1 = 0$. Since $\ln(1+x) = \ln 1$ when $x = 0$, we require the right hand side of the equation to equal zero when $x = 0$. Plugging zero into (40.21d) gives

$$0 = \ln 1 = \ln(1+x)|_{x=0} = \left(C + \sum_{n=1}^{\infty} \frac{(-1)^{n-1}x^n}{n}\right)\Bigg|_{x=0} = C \tag{40.21e}$$

Thus

$$\ln(1+x) = x - \frac{x^2}{2} + \frac{x^3}{3} - \frac{x^4}{4} + \cdots \tag{40.21f}$$

$\square$

Example 40.9. Find a power series for $\tan^{-1} x$.

Solution. From equation 22.30,

$$\tan^{-1} x = \int \frac{dx}{1+x^2} = \int \frac{dx}{1-(-x^2)} \tag{40.22a}$$

Expressing the integrand with equation 40.15

$$\tan^{-1} x = \int \left(1 - x^2 + x^4 - x^6 + \cdots\right) dx \tag{40.22b}$$

$$= C + x - \frac{x^3}{3} + \frac{x^5}{5} - \frac{x^7}{7} + \cdots \tag{40.22c}$$

But since $\tan^{-1} 0 = 0$,

$$0 = \tan^{-1} 0 = C + 0 = C \tag{40.22d}$$

so

$$\tan^{-1} x = x - \frac{x^3}{3} + \frac{x^5}{5} - \frac{x^7}{7} + \cdots = \sum_{n=0}^{\infty} \frac{(-1)^n x^{2n+1}}{2n+1}. \tag{40.22e}$$

$\square$

Taylor and Maclaurin Series

Recall our general definition of a power series from equation 40.5

$$f(x) = \sum_{n=0}^{\infty} c_n x^n = c_0 + c_1 x + c_2 x^2 + c_3 x^3 + c_4 x^4 + \cdots \tag{40.23}$$

If we substitute $x = 0$ into both sides of the equation, all of the terms on the right disappear except for the c_0 go away, giving us

$$c_0 = f(0) \tag{40.24}$$

If we differentiate (40.23) term by term,

$$f'(x) = c_1 + 2c_2 x + 3c_3 x^2 + 4c_4 x^3 + 5c_5 x^4 + \cdots = \sum_{n=1}^{\infty} n c_n x^{n-1} \tag{40.25}$$

Setting $x = 0$ in (40.25), all of the terms on the right except for the c_1 become zero, and we are left with

$$c_1 = f'(0) \tag{40.26}$$

Differentiating (40.25) term by term,

$$f''(x) = 2c_2 + 3 \cdot 2c_3 x + 4 \cdot 3c_4 x^2 + 5 \cdot 4c_5 x^3 + 6 \cdot 5c_6 x^4 + \cdots \tag{40.27}$$

$$= \sum_{n=2}^{\infty} n(n-1)c_n x^{n-2} \tag{40.28}$$

and therefore

$$c_2 = \frac{1}{2}f''(0) \tag{40.29}$$

Continuing the process,

$$f'''(x) = \sum_{n=3}^{\infty} n(n-1)(n-2)c_n x^{n-3} \tag{40.30}$$

$$f^{(4)}(x) = \sum_{n=4}^{\infty} n(n-1)(n-2)(n-3)c_n x^{n-3} \tag{40.31}$$

$$f^{(5)}(x) = \sum_{n=5}^{\infty} n(n-1)(n-2)(n-3)(n-4)c_n x^{n-4} \tag{40.32}$$

$$\vdots$$

$$f^{(k)}(x) = \sum_{n=k}^{\infty} n(n-1)(n-1)\cdots(n-k+1)c_n x^{n-k} \tag{40.33}$$

Substituting $x = 0$ into each of these,

$$f'''(0) = 3 \cdot 2 \cdot 1 \cdot c_3 + 0 + \cdots + 0 = 3!c_3 \tag{40.34}$$

$$f^{(4)}(0) = 4 \cdot 3 \cdot 2 \cdot 1 \cdot c_4 + 0 + \cdots + 0 = 4!c_4 \tag{40.35}$$

$$f^{(5)}(0) = \sum_{n=5}^{\infty} n(n-1)(n-2)(n-3)(n-4)c_n x^{n-4} \tag{40.36}$$

$$= 5!c_5 + 0 + \cdots + 0 = 5!c_5 \tag{40.37}$$

$$\vdots$$

$$f^{(k)}(0) = \sum_{n=k}^{\infty} n(n-1)(n-1)\cdots(n-k+1)c_n x^{n-k} \tag{40.38}$$

$$= k!c_k + 0 + \cdots + 0 = k!c_k \tag{40.39}$$

Putting all of these results together, we have

$$c_0 = \frac{f(0)}{0!} = f(0) \tag{40.40}$$

$$c_1 = \frac{f'(0)}{1!} = f'(0) \tag{40.41}$$

$$c_2 = \frac{f''(0)}{2!} \tag{40.42}$$

$$c_3 = \frac{f'''(0)}{3!} \tag{40.43}$$

$$c_4 = \frac{f^{(4)}}{4!} \tag{40.44}$$

$$\vdots$$

$$c_n = \frac{f^{(n)}}{n!} \tag{40.45}$$

Theorem 40.5. Maclaurin Series

If the function $f(x)$ is infinitely differentiable at $x = 0$ then $f(x)$ can be expanded as

$$f(x) = \sum_{n=0}^{\infty} \frac{f^{(n)}(0)}{n!} x^n = f(0) + f'(0)x + \frac{f''(0)}{2}x^2 + \frac{f'''(0)}{3!}x^3 + \cdots \tag{40.46}$$

Example 40.10. Find a Maclaurin series for $y = e^x$.

Solution. Since every derivative of e^x is e^x, this is the easiest Maclaurin series to calculate.

$$\left.\begin{aligned}
f(x) &= e^x & f(0) &= 1 \\[4pt]
f'(x) &= e^x & f'(0) &= 1 \\[4pt]
f''(x) &= e^x & f''(0) &= 1 \\[4pt]
f'''(x) &= e^x & f''(0) &= 1 \\[4pt]
\vdots & & \vdots &
\end{aligned}\right\}\tag{40.47a}$$

Therefore

$$e^x = 1 + x + \frac{x^2}{2} + \frac{x^3}{3!} + \frac{x^4}{4!} + \frac{x^5}{5!} + \cdots = \sum_{n=0}^{\infty} \frac{x^n}{n!}\tag{40.47b}$$

$\square$

Example 40.11. Find a Maclaurin series for $y = \sin x$.

Solution. This one is a little more difficult but there is a pattern after the first four derivatives.

$$\left.\begin{aligned}
f(x) &= \sin x & f(0) &= 0 \\[4pt]
f'(x) &= \cos x & f'(0) &= 1 \\[4pt]
f''(x) &= -\sin x & f''(0) &= 0 \\[4pt]
f'''(x) &= -\cos x & f'''(0) &= -1 \\[4pt]
f^{(4)}(x) &= \sin x & f^{(4)}(0) &= 0 \\[4pt]
\vdots & & \vdots &
\end{aligned}\right\}\tag{40.48a}$$

Thus all the even-powered terms are zero, and the odd powered terms alternate in sign.

$$f(x) = f(0) + f'(0)x + \frac{f''(0)}{2}x^2 + \frac{f'''(0)}{3!}x^3 + \frac{f^{(4)}(0)}{4!}x^4 + \cdots\tag{40.48b}$$

$$= x - \frac{x^3}{3!} + \frac{x^5}{5!} - \frac{x^7}{7!} + \cdots\tag{40.48c}$$

$\square$

Definition 40.4. Taylor Series

If $f(x)$ is infinitely differentiable at $x = a$ then the **Taylor Series** about $x = a$ is

$$f(x) = \sum_{n=0}^{\infty} \frac{f^{(n)}(a)}{n!}(x - a)^n\tag{40.49}$$

$$= f(a) + f'(a)(x - a) + \frac{f''(a)}{2}(x - a)^2 + \frac{f'''(a)}{3!}(x - a)^3 + \cdots\tag{40.50}$$

Remark 40.1.

A Maclaurin series is a Taylor series about about $a = 0$.

Example 40.12. Find a Taylor series for $y = e^x$ about $x = 2$.

Solution. We calculate derivatives as we did for the Maclaurin series, but now we need to evaluate them all at $x = 2$ instead of at $x = 0$.

$$\left.\begin{aligned}
f(x) &= e^x & f(2) &= e^2 \\[4pt]
f'(x) &= e^x & f'(2) &= e^2 \\[4pt]
f''(x) &= e^x & f''(2) &= e^2 \\[4pt]
f'''(x) &= e^x & f''(2) &= e^2 \\[4pt]
&\;\;\vdots & &\;\;\vdots
\end{aligned}\right\} \tag{40.51a}$$

Therefore the Taylor series for e^x about $x = 2$ is

$$e^x = f(2) + f'(2)(x-2) + \frac{f''(2)}{2}(x-2)^2 + \frac{f'''(2)}{3!}(x-2)^3 + \cdots \tag{40.51b}$$

$$= e^2 + e^2(x-2) + \frac{e^2(x-2)^2}{2!} + \frac{e^2(x-2)^3}{3!} + \cdots \tag{40.51c}$$

$\square$

Example 40.13. Find a Taylor series for $y = \sin x$ about $x = \pi/4$.

Solution. First we calculate the function and all of its derivatives at $x = \pi/4$.

$$\left.\begin{aligned}
f(x) &= \sin x & f\left(\frac{\pi}{4}\right) &= \frac{\sqrt{2}}{2} \\[6pt]
f'(x) &= \cos x & f'\left(\frac{\pi}{4}\right) &= \frac{\sqrt{2}}{2} \\[6pt]
f''(x) &= -\sin x & f''\left(\frac{\pi}{4}\right) &= -\frac{\sqrt{2}}{2} \\[6pt]
f'''(x) &= -\cos x & f'''\left(\frac{\pi}{4}\right) &= -\frac{\sqrt{2}}{2} \\[6pt]
f^{(4)}(x) &= \sin x & f^{(4)}\left(\frac{\pi}{4}\right) &= \frac{\sqrt{2}}{2} \\[6pt]
&\;\;\vdots & &\;\;\vdots
\end{aligned}\right\} \tag{40.52a}$$

The pattern is $+$, $+$, $-$, $-$, etc., and all of the derivatives have magnitude $\sqrt{2}/2$; hence all terms are present. Therefore

$$\sin(x) = f\left(\frac{\pi}{4}\right) + f'\left(\frac{\pi}{4}\right)(x-a) + f''\left(\frac{\pi}{4}\right)\frac{(x-a)^2}{2} + f'''\left(\frac{\pi}{4}\right)\frac{(x-a)^3}{3!} + \cdots \tag{40.52b}$$

$$= \frac{\sqrt{2}}{2} + \frac{\sqrt{2}}{2}\cdot\left(x-\frac{\pi}{4}\right) - \frac{\sqrt{2}}{2\cdot 2!}\cdot\left(x-\frac{\pi}{4}\right)^2 - \frac{\sqrt{2}}{2\cdot 3!}\cdot\left(x-\frac{\pi}{4}\right)^3 + \cdots \tag{40.52c}$$

$\square$

Estimating Error. If a power series is alternating, we can use the Alternating Series Estimation Theorem (theorem 39.13) to estimate the error or to determine how many terms are required for a required level of precision.

Example 40.14. If we never calculate an angle larger than 10 degrees, how many terms do we need to keep in the Maclaurin series for $\cos x$ about $x = 0$ if we want an accuracy of 10^{-6} degrees?

Solution. Since 10 degrees $= 10 \, \pi/180 \approx 0.174533$ radians, then $|x| < 10°$ is equivalent to

$$|x| \leq 0.174533 \tag{40.53a}$$

The desired precision is $10^{-6}\pi/180 \approx 1.74533 \times 10^{-8}$ radians. Thus by theorem 1.74533×10^{-8}. The series for $\cos x$ is (see eq. 40.71) is

$$\cos x = 1 - \frac{x^2}{2!} + \frac{x^4}{4!} - \frac{x^6}{6!} + \cdots \tag{40.53b}$$

By theorem 39.13 we need to find at least n terms such that the first term omitted is smaller than 10^{-6} degrees, or

$$\frac{|x^{n+1}|}{(n+1)!} \leq 1.74533 \times 10^{-8} \tag{40.53c}$$

Thus we need

$$(n+1)! \geq \frac{|x^{n+1}|}{1.74533 \times 10^{-8}} \tag{40.53d}$$

This must work for all values of x in the range $|x| \leq 0.174533$, so we use the largest value.

$$(n+1)! \geq \frac{(0.174533)^{n+1}}{1.74533 \times 10^{-8}} \tag{40.53e}$$

For $n = 2$, this gives

$$6 = 3! \geq \frac{(0.174533)^3}{1.74533 \times 10^{-8}} \approx 304\,618 \tag{40.53f}$$

which is not true. For $n = 4$, this requirements gives

$$120 = 5! \geq \frac{(0.174533)^6}{1.74533 \times 10^{-8}} \approx 9279 \tag{40.53g}$$

which is also not true. For $n = 6$, this requirement gives

$$5040 = 7! \geq \frac{(0.174533)^8}{1.74533 \times 10^{-8}} \approx 283 \tag{40.53h}$$

which is true. Thus the $n = 6$ term satisfies equation 40.53c, and may be omitted. Since the $n = 5$ term is zero, we only need to keep the series through $n = 4$. Thus

$$\cos x \approx 1 - \frac{x^2}{2!} + \frac{x^4}{4!} \tag{40.53i}$$

where x is measured in radians, with precision better that 10^{-6} degrees, so long as $|x| < 10$ degrees. $\qquad\Box$

When a series is not alternating, we can not use the alternating series test to obtain an error estimate. However, Taylor's remainder theorem gives us a way to estimate the error. The proof is typically given in more advanced classes in calculus or numerical analysis.

Theorem 40.6. Taylor's Remainder Theorem

If a Taylor series about a for $f(x)$ has a radius of convergence of R, then the error R_n made in terminating a Taylor series after n terms is bounded by

$$|R_n(x)| \leq \frac{\left|f^{(n+1)}(c)\right|}{(n+1)!} |x - a|^{n+1} \tag{40.54}$$

for some c in the in the interval $|x - a| < R$.

Example 40.15. Estimate the error ϵ in calculate using a 3 term Maclaurin series (i.e., to x^3) to estimate $y = \sqrt{x}$ on $[3, 10]$.

Solution. Since $a = 0$, the remainder theorem says that the error is bounded by

$$\epsilon \leq \frac{\left|f^{(4)}(c)\right|}{5!} |x|^4 \tag{40.55a}$$

where $f(x) = \sqrt{x}$. Differentiating 4 times,

$$f^{(4)}(x) = -\frac{15}{16x^{7/2}} \tag{40.55b}$$

The theorem does not tell us what value of c to use, just that there is some value where the formula is true. It might hold for some value where the estimate is a very small number (worst-case scenario), and it might hold where the estimate is a very large number (best-case scenario). The largest value this can possible take occurs when x takes on its smallest value. So

$$|f^{(4)}(c)| \leq \max_{3 \leq x \leq 10} \left| -\frac{15}{16x^{7/2}} \right| = \frac{15}{16 \cdot 3^{7/2}} \tag{40.55c}$$

Thus

$$\epsilon \leq \frac{15}{16 \cdot 3^{7/2}} \cdot \frac{|x|^4}{5!} \tag{40.55d}$$

Since the right hand side is at its largest when x is biggest, and this occurs at $x = 10$; and since $5! = 120$,

$$\epsilon \leq \frac{15 \cdot 10^4}{120 \cdot 16 \cdot 3^{7/2}} \approx 1.67 \tag{40.55e}$$

In fact this gives an overestimate for two reasons:

- We used a value of c that gives the maximum possible error (worst-case scenario), because the theorem did not tell us what value to use. In reality, it could be better than this.

- We also use the value of x that gave the worst error. If we are interested in $\sqrt{x}$ for $x \neq 10$, the error would be smaller.

$\square$

Taylor Polynomials. If we truncate a Taylor series after n terms we have the **Taylor Polynomials**. These give lower order approximations to a function near $x = a$. Formulas for the first few Taylor Polynomials are

$$T_1(x) = f(a) + f'(a)(x - a) \tag{40.56}$$

$$T_2(x) = f(a) + f'(a)(x - a) + \frac{1}{2}f''(a)(x - a)^2 \tag{40.57}$$

$$T_3(x) = f(a) + f'(a)(x - a) + \frac{1}{2}f''(a)(x - a)^2 + \frac{1}{3!}(a)f'''(a)(x - a)^3 \tag{40.58}$$

$$\vdots$$

Definition 40.5. Taylor Polynomial

Let $f(x)$ be n-times differentiable. Then its **Taylor Polynomial of Degree n about a** is

$$T_n(x) = \sum_{k=0}^{n} \frac{f^{(n)}(a)}{n!}(x - a)^n \tag{40.59}$$

The first Taylor polynomial T_1 is also called the **linearization** of $f(x)$ (see equation 8.43).

Example 40.16. Find the first three Taylor Polynomials for the function $f(x) = e^x$ about $x = 2$.

Solution. From example 40.12 the Taylor series for e^x about $x = 2$ is

$$e^x = e^2\left(1 + (x - 2) + \frac{(x - 2)^2}{2!} + \frac{(x - 2)^3}{3!} + \cdots\right) \tag{40.60a}$$

Thus

$$T_1(x) = e^2 + e^2(x - 2) \tag{40.60b}$$

$$T_2(x) = e^2 + e^2(x - 2) + \frac{e^2(x - 2)^2}{2!} \tag{40.60c}$$

$$T_3(x) = e^2 + e^2(x - 2) + \frac{e^2(x - 2)^2}{2!} + \frac{e^3(x - 2)^3}{3!} \tag{40.60d}$$

These are illustrated in figure 40.1. They all match the function (the solid line) well near $x = 2$ but the approximation becomes worse as $|x - 2|$ increases. In general, the approximation is better over a wider range of values if you include more terms. □

Example 40.17. Find the Taylor polynomial T_2 about $x = 27$ of $y = x^{1/3}$ and use it to estimate $\sqrt[3]{25}$.

Solution. We only need the function and first two derivatives:

$$\left.\begin{array}{ll} f(x) = x^{1/3} & f(27) = 3 \\[2mm] f'(x) = \dfrac{1}{3}x^{-2/3} & f'(27) = \dfrac{1}{27} \\[2mm] f''(x) = -\dfrac{2}{9}x^{-5/3} & f''(27) = -\dfrac{2}{2187} \end{array}\right\} \tag{40.61a}$$

Therefore the Taylor polynomial is

$$T_3(x) = 3 + \frac{1}{27}(x - 27) - \frac{2}{2187} \cdot \frac{(x - 27)^2}{2!} \tag{40.61b}$$

Figure 40.1: Taylor polynomials for $y = e^x$ about $x = 2$ (see example 40.16: linear (dotted); quadratic (short dashed); cubic (long dashed).

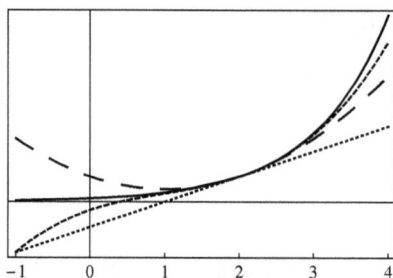

An estimate of the cube root of 25 is

$$\sqrt[3]{25} \approx 3 + \frac{1}{27}(-2) - \frac{2}{2187} \cdot \frac{(-2)^2}{2} \qquad (40.61\text{c})$$

$$\approx 3 - 0.07407 - 0.00183 \approx 2.9241 \qquad (40.61\text{d})$$

The actual cube root of 25 is ≈ 2.92402.   □

Table of Some Basic Power Series

$$\frac{1}{1 - x} = 1 + x + x^2 + x^3 + \cdots, \; |x| < 1 \qquad (40.62)$$

$$\frac{1}{1 + x} = 1 - x + x^2 - x^3 + \cdots, \; |x| < 1 \qquad (40.63)$$

$$\frac{1}{1 - x^2} = 1 + x^2 + x^4 + x^6 + \cdots, \; |x| < 1 \qquad (40.64)$$

$$\frac{1}{1 + x^2} = 1 - x^2 + x^4 - x^6 + \cdots, \; |x| < 1 \qquad (40.65)$$

$$\frac{1}{(1 - x)^2} = 1 + 2x + 3x^2 + 4x^3 + \cdots, \; |x| < 1 \qquad (40.66)$$

$$\ln(1 + x) = x - \frac{x^2}{2} + \frac{x^3}{3} - \frac{x^4}{4} + \cdots, \; |x| < 1 \qquad (40.67)$$

$$\tan^{-1} x = x - \frac{x^3}{3} + \frac{x^5}{5} - \frac{x^7}{7} + \cdots, \; |x| < 1 \qquad (40.68)$$

$$e^x = 1 + x + \frac{x^2}{2} + \frac{x^3}{3!} + \frac{x^4}{4!} + \frac{x^5}{5!} + \cdots \qquad (40.69)$$

$$\sin x = x - \frac{x^3}{3!} + \frac{x^5}{5!} - \cdots \qquad (40.70)$$

$$\cos x = 1 - \frac{x^2}{2!} + \frac{x^4}{4!} - \frac{x^6}{6!} - \cdots \qquad (40.71)$$

Exercises

Use geometric series to find a power series expansions in exercises 1 through 4.

1. $\ln(1 - x)$

2. $\dfrac{x}{(1 - x)^2}$

3. $\dfrac{2}{(1 - x)^3}$

4. $\dfrac{2x^2}{1 - x^3}$

Find Maclaurin series in exercises 5 through 12.

5. $y = \cos x$

6. $y = \tan x$

7. $y = \arcsin x$

8. $y = \sinh x$

9. $y = xe^x$

10. $y = x \sin x$

11. $\dfrac{x}{1 + x^2}$

12. $y = x \ln x$

Find a Taylor series about the given point in exercises 13 through 16.

13. $y = \cos x$; $a = \pi/3$

14. $y = \sqrt{x}$; $a = 4$

15. $y = \sqrt{x + 1}$; $a = 16$

16. $y = e^{-(x-1)^2}$; $a = 1$

17. Find the linearization of $y = \sqrt{x + 1}$ about $a = 48$ and use it to estimate $\sqrt{50}$.

18. Find the Taylor polynomial T_2 for $y = \sqrt{x + 1}$ about $a = 48$ and use it to estimate $\sqrt{50}$.

19. Estimate the number of terms needed to calculate $\sqrt{50}$ to 10 decimal places accuracy using a Taylor series for $\sqrt{x + 1}$ expanded about $x = 48$.

20. Repeat the previous problem using a Maclaurin series.

Foundations of Pure and Utter Nonsense

Chapter 41

Differential Equations

Chapter Summary and Goal

Differential equations (DE's, or ODE's[1]) are equations that contain derivatives. Examples are $y' + 3xy = x$, $xy' + x/y = y'/x$, and $y'' + 3xy' + 7y' = 0$. Usually, when we are given a differential equation, what we really want is its **solution**. If we say that $y = \phi(x)$ is a solution to a differential equation, this means that we can replace y with $\phi(x)$ everywhere in the DE, and the resulting equation is still correct.

Initial value problems (IVP's) are a special class of differential equations that are subject to constraints. These are like the examples we studied when we first learned about integration(e.g., in examples 22.7 and 22.8). When we "solve" a differential equation, we have to integrate. This gives us an arbitrary constant, and the initial condition will determine the value of the constant.

In this chapter we will learn about the basic properties of differential equations and initial value problems, and introduce elementary methods for solving simple equations.

Student Learning Objectives

The student will:

1. Understand how differential equations are classified.
2. Learn the distinction between differential equations and initial value problems.
3. Learn to solve simple first order separable, linear, and homogeneous differential equations.

First Order Differential Equations

While it is possible to define much broader classes of differential equations, we will limit our discussion to a class of equations that can be put into a **standard form**,

$$\frac{dy}{dx} = f(x, y) \tag{41.1}$$

Here $f(x, y)$ is any function that depends on both x and y. We will often use the shorthand y' to denote dy/dx, so that the general equation of a differential equation becomes

$$y' = f(x, y) \tag{41.2}$$

---

[1] Ordinary differential equations.

431

Definition 41.1. Differential Equation, Standard Form

$$y' = f(x, y) \tag{41.3}$$

Our definition (41.3) is not the most general form that a differential equation can take, but it is a useful restriction. In general differential equation can be classified by

- **Type** as **ordinary** or **partial**. Equations that contain partial derivatives (which are defined in a multivariate calculus course) are called **partial differential equations** (PDE's). The equations that we will study here will not contain any partial derivatives and are call **ordinary differential equations** (ODE's).

- **Special Types** of differential equations include **functional differential equations**, in which the derivative is a function of another function (e.g, $y' = f(x, y(x), y(z(x)))$; and **delay differential equations**, in which the derivative depends on the value of the same function at a different time (e.g., $dy/dx = y(x - a)$).

- **Order**: the order of a differential equation is the order of the highest derivative in the equation. Thus $y' = x^2 + y$ is first order, and $y'' + 2xy' + 3y = 0$ is second order.

- **Linear** or **Non-linear** depending upon the how y appears on the right hand side of the equation. To be linear, $f(x, y)$ must have the form $m(x)y + b(x)$ where $m(x)$ and $b(x)$ are *any* functions that depend only on x (or are constant). The dependence on x in $f(x, y)$ does not matter. Thus $y' = x^2 + y$ is linear but $y' = x + y^2$ is nonlinear.

We will only consider equations that can be rearranged into the form (41.3) in this chapter.

One thing that students often find confusing in the study of differential equation is that $f(x, y)$ is not the same thing as y. It is the same thing as y'.

Note that $f(x, y)$ is a function of two variables, and the relationship between x and y may be very complicated. Here are some examples of differential equations and how we would define the function $f(x, y)$.

$$y' = \underbrace{3x \sin x + \cos y}_{f(x,y)} \tag{41.4}$$

$$y' = \underbrace{x^2 y + 2x + 3}_{f(x,y)} \tag{41.5}$$

$$y' = \underbrace{(x - y)(x + y)}_{f(x,y)} \tag{41.6}$$

You will often come across the differential equation in a form in which you have to do some rearrangement to find the form of $f(x, y)$.

Example 41.1. Put $xy' + 3y = x^2 y'$ into standard form and identify $f(x, y)$.

Solution. The idea is to solve the equation analytically for y'. This is not the same thing as solving the differential equation, but it is often the first step in obtaining a solution. The process is similar to solving for y' when we did implicit differentiation in chapter 12. In this case, we can begin by moving all the terms with y' to the left side of the equation, and all the other terms to the right.

$$xy' - x^2 y' = -3y \tag{41.7a}$$

On the left, we can factor out the common y',

$$y'(x^2 - x) = 3y \tag{41.7b}$$

Dividing by $x^2 - x$, we solve for y'. The expression on the right hand side of the equation is $f(x, y)$.

$$y' = \frac{3y}{x^2 - x} \tag{41.7c}$$

Equation 41.7c is now in standard form with $f(x, y) = \dfrac{3y}{x^2 - x}$.   □

Remark 41.1. Solutions of Differential Equations

A solution of a differential equation is a **function**.

The main goal in studying differential equations is usually to obtain a solution. If finding a solution is not possible, then we will want to find out as much as possible about the behavior of the solution. To do this, we have to understand just what we mean by a solution.

A solution to a differential equation is a **function**. If $\phi(x)$ is the solution to $y' = f(x, y)$, this means that if we replace y everywhere in the equation with the formula for $\phi(x)$, then the equation will still work.

Definition 41.2. Solution of a Differential Equation

A function $\phi(x)$ is called a **solution** of $y' = f(x, y)$ on the interval $a < x < b$ if

$$\phi'(x) = f(x, \phi(x)) \tag{41.8}$$

for all x in (a, b).

Example 41.2. Show that $\phi(x) = 1 + e^{-x^2/2}$ is a solution to $y' = x(1 - y)$.

Solution. According to definition 41.2, we need to show that

$$\phi'(x) = f(x, \phi(x)) = x(1 - \phi(x)) \tag{41.9a}$$

since $f(x, y) = x(1 - y)$. The right hand side of (41.9a) gives

$$\text{RHS} = x(1 - 1 - e^{-x^2/2}) = -xe^{-x^2/2} \tag{41.9b}$$

Using the chain rule, the left hand side of (41.9a) gives

$$\text{LHS} = \frac{d}{dx}\left(1 + e^{-x^2/2}\right) = -xe^{-x^2/2} \tag{41.9c}$$

Since the LHS and the RHS are equal, we conclude that

$$\phi'(x) = f(x, \phi(x)) \tag{41.9d}$$

as required.   □

One way to "solve" a differential equation is proceed blindly forward through integration, as in the following example.

Example 41.3. Find a solution to $y' = x(1 - y)$.

Solution. The goal is to put all of the x variables (including the dx) on one side of the equation, and all of the y variables (including the dy) on the other side. First, we write the DE as

$$\frac{dy}{dx} = x(1 - y) \tag{41.10a}$$

Then we can multiply both sides of the equation by dx

$$\frac{dy}{dx} \cdot dx = x(1-y) \cdot dx \qquad (41.10\text{b})$$

Next, divide both sides by $1-y$,

$$\frac{1}{1-y} \cdot \frac{dy}{dx} \cdot dx = \frac{x(1-y)dx}{1-y} \qquad (41.10\text{c})$$

Cancelling common factors,

$$\frac{1}{1-y} \cdot dy = x \cdot dx \qquad (41.10\text{d})$$

We are allowed to integrate this equation.

$$\int \frac{1}{1-y} \cdot dy = \int x \cdot dx \qquad (41.10\text{e})$$

Integrating,

$$-\ln|1-y| + C_1 = \frac{1}{2}x^2 + C_2 \qquad (41.10\text{f})$$

Since there are two arbitrary constants, we can bring the C_1 to the right, and define $C_3 = C_2 - C_1$. Instead of calling this C_3, we just drop the subscript. This is something we do in general: when there are integrals on both sides of the equation, we can combine the constants of integration. Thus

$$-\ln|1-y| = \frac{1}{2}x^2 + C \qquad (41.10\text{g})$$

Multiplying through by -1 and observing that $-C$ is still a constant, we first call it $C_4 = -C$, and then drop the subscript (again).

$$\ln|1-y| = -\frac{1}{2}x^2 + C \qquad (41.10\text{h})$$

Exponentiating both sides of the equation,

$$1 - y = e^{\ln|1-y|} = e^{-(x^2/2)+C} = e^{-x^2/2}e^C \qquad (41.10\text{i})$$

Defining $C_5 = e^C$ and then dropping the subscript (one more time) gives

$$1 - y = Ce^{-x^2/2} \qquad (41.10\text{j})$$

or $y = 1 + Ce^{-x^2/2}$ (where $C_6 = -C$ and the subscript was dropped) demonstrating that the solution used in example 41.2 was not (completely) pulled out of the air. □

Example 41.4. Find a solution for $\dfrac{dy}{dx} = y$.

Solution. The equation is actually already written in standard form with

$$f(x,y) = y \qquad (41.11\text{a})$$

We can use the same process as in example 41.3 to find a solution. We multiply the equation through by dx, divide by y and simplify:

$$\frac{dy}{dx} \cdot dx = y \cdot dx \qquad (41.11\text{b})$$

$$\frac{1}{y}\frac{dy}{dx}\, dx = \frac{1}{y}\, y\, dx \qquad (41.11\text{c})$$

$$\frac{dy}{y} = dx \tag{41.11d}$$

$$\tag{41.11e}$$

Usually we skip the formality of actually writing the differential dx as $dx = \frac{dy}{dx} \cdot dx$ and just cross multiply; since

$$\frac{dy}{dx} = \frac{y}{1} \tag{41.11f}$$

we conclude that

$$\frac{dy}{y} = \frac{dx}{1} = dx \tag{41.11g}$$

Once we have put all of the x variables on one side of the equation, and all of the y variables on the other side of the equation, we can integrate.

$$\int \frac{dy}{y} = \int dx \tag{41.11h}$$

This gives us

$$\ln|y| = x + C \tag{41.11i}$$

where C is any constant. Exponentiating,

$$|y| = e^{x+C} = e^x e^C = C_1 e^x \tag{41.11j}$$

where $C_1 = e^C$ is any positive constant (because e^C can only take on positive values but has a range of all $x > 0$). Allowing for both positive and negative values (and dropping the subscript),

$$y = Ce^x \tag{41.11k}$$

where C is any real (nonzero) number (C is nonzero because it must have the form $\pm e^{C_2}$ for some real number C_2).

Note, however, that although the **derivation** does not allow $C = 0$, if we were to set $C = 0$, we would get a solution $y = 0$, and $y = 0$ is a solution of the differential equation (verify this by substituting y into $y' = y$). Thus a more general solution is $y = Ce^x$ where C is any real number, including, possibly, zero. □

When we "solved" the differential equation $y' = y$ in example 41.4, we obtained the collection of solutions $y = Ce^x$ where C is any nonzero constant, but missed the solution $y = 0$. This demonstrates that the straightforward process of integration does not necessarily provide all of the solutions to the problem.

Remark 41.2. Missed Solutions

Solving a differential equation by separating the variables and integrating will not necessarily give you every solution.

Example 41.5. Verify that the $y = Ce^x$, for any real number C, is a solution of $y' = y$.

Solution. In the previous example we *derived* the formula $y = Ce^x$, but to verify it according to definition 41.2 we need to plug it into the differential equation. Thus we calculate

$$y' = \frac{d}{dx}(Ce^x) = Ce^x = y \tag{41.12a}$$

Thus it satisfies the differential equation. □

In example 41.5 we showed that $y = Ce^x$ is a solution to $y' = y$ for *any* value of the constant C. Thus for example, $y = 3e^x$, $y = 0$, and $y = -27.4e^x$ are all solutions. Thus we see that unlike algebraic problems, which may have "a solution," differential equations may have multiple solutions. Thus we have been careful to use the expression "a solution" rather than "the solution."

Figure 41.1: Plots of the family of solutions of $y' = 4x - \frac{3}{2}x^2$ are shown by the dashed lines (see example 41.6). The solid line illustrates the single solution that passes through $(0, 1)$.

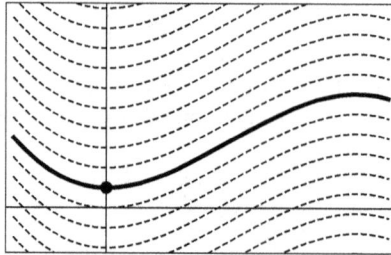

Remark 41.3. Solutions of Differential Equations

A differential equation will, in general, have an infinite number of solutions. This collection of solutions is sometimes called a **one parameter family of solutions**.

Definition 41.3. Initial Value Problem (IVP)

An **initial value problem** is a differential equation together with a constraint, usually written in the form

$$\left. \begin{array}{l} y' = f(x, y) \\ y(x_0) = y_0 \end{array} \right\} \tag{41.13}$$

where x_0 and y_0 are fixed constants.

The equation $y(x_0) = y_0$ is called the **initial condition**.

An initial value problem selects a single solution that goes through the point (x_0, y_0) from the collection of all possible solutions to the differential equation.

Example 41.6. Solve the initial value problem

$$\left. \begin{array}{l} y' = 4x - \frac{3}{2}x^2 \\ y(0) = 1 \end{array} \right\} \tag{41.14}$$

Solution. Integrating the differential equation we obtain

$$\int dy = \int \left(4x - \frac{3}{2}x^2 \right) dx \tag{41.15a}$$

$$y = 2x^2 - \frac{1}{2}x^3 + C \tag{41.15b}$$

To determine the constant, we substitute the initial condition. This says that $y = 1$ when $x = 0$,

$$1 = 2(0^2) - \frac{1}{2} \cdot (0^3) + C \tag{41.15c}$$

Hence c=1 and therefore the solution is

$$y = 2x^2 - \frac{1}{2}x^3 + 1 \tag{41.15d}$$

The single solution selected by the initial condition from the infinite family of possible solutions is illustrated in figure 41.1.   □

Separation of Variables

The types of differential equation that we have been discussing so far – ones that we can rearrange and integrate – are called **separable differential equation**. When we cross-multiply so that all the x variables are on one-side of the equation and all of the y variables are on the other side of the equation, we say that the variables have become separated. Once the variables are **separated**, we are allowed to integrate, and not before.

Definition 41.4. Separable Equation

A differential equation $y' = f(x, y)$ is called **separable** if there exist functions $p(x)$ and $q(y)$ such that

$$y' = p(x)q(y) \tag{41.16}$$

where $p(x)$ depends only on x and $q(y)$ depends only on y.

Example 41.7. Solve the initial value problem

$$\left. \begin{array}{c} \dfrac{dy}{dx} = \left(1 - \dfrac{x}{2}\right)y^2 \\ y(0) = 1 \end{array} \right\} \tag{41.17}$$

Solution. This variables can be separated as $y' = p(x)q(y)$, where $p(x) = 1 - \dfrac{x}{2}$ and $q(y) = y^2$. Cross-multiplying equation 41.17 and integrating,

$$\frac{dy}{y^2} = \left(1 - \frac{x}{2}\right)dx \tag{41.18a}$$

$$\int \frac{dy}{y^2} = \int \left(1 - \frac{x}{2}\right)dx \tag{41.18b}$$

$$-\frac{1}{y} = x - \frac{1}{4}x^2 + C \tag{41.18c}$$

The initial condition gives

$$-1 = 0 - 0 + C \tag{41.18d}$$

hence

$$\frac{1}{y} = \frac{1}{4}x^2 - x + 1 = \frac{1}{4}(x^2 - 4x + 4) = \frac{1}{4}(x - 2)^2 \tag{41.18e}$$

Solving for y, we obtain the solution of the initial value problem as $y = \dfrac{4}{(x - 2)^2}$.   □

Often with separable equations we will not be able to find an explicit expression for y as a function of x; instead, we will have to remain happy with an implicit equation that relates the two variables.

Example 41.8. Find a general solution of $\dfrac{dy}{dx} = \dfrac{x}{e^x - 2y}$.

Solution. Separating the variables and integrating,

$$(e^y - 2y)dy = xdx \tag{41.19a}$$

$$\int (e^y - 2y)dy = \int xdx \tag{41.19b}$$

$$e^y - y^2 = \frac{1}{2}x^2 + C \tag{41.19c}$$

It is not possible to solve this equation explicitly y. $\square$

Modeling Exponential Growth and Decay

One of the most commonly used mathematical models is the following: the rate of change of some variable y is proportional to itself. Since the words *rate of change* are synonymous with *derivative*, we can write this as

$$\frac{dy}{dt} = ky \tag{41.20}$$

We are using t as the independent variable because these are usually time dependent processes. The constant k may be either negative or positive, depending on the application. Examples include:

- Radioactive decay - y represents the amount of mass present after a time t.
- Bacterial population growth - y represents the number of bacteria.
- RC circuits - y represents the voltage drop across a loop.
- Charged capicitors - y represents the charge on a capictor.
- Temperature change - y represents the difference between the temperature of some object, such as a glass of iced tea, and its environment.
- Bank accounts - y represents the amount of an investment at fixed interest.

Example 41.9. Solve the initial value problem

$$\left.\begin{array}{c} y' = ky \\ y(t_0) = y_0 \end{array}\right\} \tag{41.21}$$

where t_0, y_0, and k are known constants.

Solution. Example 41.4 was a special case of this, with $k = 1$, but the same approach can be used. Separating variables,

$$\int \frac{dy}{y} = k \int dt \tag{41.22a}$$

Integrating,

$$\ln |y| = kt + C \tag{41.22b}$$

Exponentiating both sides of the equation

$$|y| = e^{kt+C} = e^{kt}e^C = c_1 e^{kt} \tag{41.22c}$$

where $C_1 = e^C > 0$ is any positive constant. Dropping the absolute value,

$$y = C_2 e^{kt} \qquad (41.22\text{d})$$

where where $C_2 = \pm C_1$. Since $y = 0$ is also a solution, even though it did not arise out of the integration, (41.22d) is valid for all real C_2. Dropping the subscript, we write the *general solution* of the differential equation as

$$y = C e^{kt} \qquad (41.22\text{e})$$

where C is any real number. Returning to the initial condition, $y_0 = C e^{kt_0}$; thus $C = y_0 e^{-kt_0}$. Substitution into eq. 41.22e gives

$$y = y_0 e^{k(t-t_0)} \qquad (41.22\text{f})$$

after some simplification. □

One thing that students often find confusing is that equation 41.22e has two constants, k, and C. These are fundamentally different:

- C is determined by initial condition (y_0), for example, the initial amount of money you deposit in the bank, the initial temperature of your iced tea, or the initial mass of uranium.

- k is determined by the problem parameters, such as the interest rate of your bank account, the heat capacity of water, or the half life or uranium.

Compound Interest. If we make an initial deposit P (called the **principal**) at a fixed interest rate r for a time t, the investment will accrue at a rate proportional to the total amount on deposit. Suppose the total amount on deposit is A; then the initial value problem is

$$\left.\begin{aligned} A' &= rA \\ A(t_0) &= P \end{aligned}\right\} \qquad (41.23)$$

The solution is given by equation 41.22f

$$A(t) = P e^{r(t-t_0)} \qquad (41.24)$$

Here interest is measured as a pure number, e.g., instead of 5% we write 0.05; time is in years; and $A(t)$ and P are in dollars. Equation 41.24 is called the **pert formula** because of the sequence of variable and parameter names on the right-hand side of the equation. If we choose the start time $t_0 = 0$ then the solution becomes $A = P e^{rt}$ (which looks like the word Pert).

Radioactive Decay. In the decay of radioactive elements, the amount (mass) of radioactive material decreases at a rate proportional to its total mass. Thus if there is N_0 grams of Iodine-125 at time $t_0 = 0$ then it will, the amount present will change at a rate

$$N' = -kN \qquad (41.25)$$

where k is a positive constant. The solution is given by equation 41.22f

$$N = N_0 e^{-kt} \qquad (41.26)$$

The **half-life** of a radioactive elmement is the amount of time it takes to decay to one half of its original mass. The half life is independent of the time when it is measured. To calculate the half life $t_{1/2}$ we set $N = N_0/2$ and solve for $t_{1/2}$. Then

$$\frac{N_0}{2} = N_0 e^{-kt_{1/2}} \qquad (41.27)$$

Dividing both sides of the equation by N_0,

$$\ln(1/2) = -kt_{1/2} \tag{41.28}$$

Solving for $t_{1/2}$,

$$t_{1/2} = -\frac{\ln 1/2}{k} = \frac{\ln 2}{k} \approx \frac{.693}{k} \tag{41.29}$$

Substitution into equation 41.26 gives

$$N = N_0 e^{-t(\ln 2)/t_{1/2}} \approx N e^{-.693t/t_{1/2}}. \tag{41.30}$$

Example 41.10. Sodium Iodide has a half-life of 59.4 days. Find the amount of time it takes to decay to less than 1/10 of 1% of its original amount.

Solution. We need to find the time it takes an amount N_0 to decay to $(1/10) \times 0.01 \times N_0 = 0.001N_0$. Thus to 3 digits accuracy

$$0.001N_0 \approx N_0 e^{-.693t/59.4} \tag{41.31a}$$

Dividing both sides of the equation by N_0,

$$0.001 \approx e^{-.693t/59.4} \tag{41.31b}$$

Taking natural logarithms of both sides of the equation and approximating $\ln .001 \approx -6.91$ gives

$$\ln .001 \approx -6.91 \approx \frac{-.693\,t}{59.4} \tag{41.31c}$$

Solving for t,

$$t \approx \frac{-6.91 \cdot 59.4}{-.693} \approx 592 \text{ days} \tag{41.31d}$$

$\square$

Exponential Relaxation

A more general statement of growth and/or relaxation problems is this: the rate of change of y is proportional to the difference between y and Z where Z is a fixed parameter. We can write this as

$$\frac{dy}{dx} = k(y - Z) \tag{41.32}$$

Example 41.11. Solve the initial value problem

$$\left.\begin{array}{l} \dfrac{dy}{dx} = k(y - Z) \\[2mm] y(t_0) = y_0 \end{array}\right\} \tag{41.33}$$

Solution. Here the approach is to make a change of variables

$$u = y - Z \tag{41.34a}$$

Then $u' = y$, since Z is a fixed constant, and $u(t_0) = y(t_0) - Z = y_0 - Z$. The problem reduces to example 41.9, and the solution is given by equation 41.22f,

$$u = u_0 e^{k(t - t_0)} = (y_0 - Z)e^{k(t - t_0)} \tag{41.34b}$$

To get the original variable y back,

$$y = Z + (y_0 - Z)e^{k(t-t_0)} \tag{41.34c}$$

where we solved explicitly for $y = u + Z$. $\square$

We are usually interested in the behavior of equation $41.34c$ over time. At $t = t_0$, the equation returns the initial condition $y = y_0$, and as $t \to \infty$, we have two possibilities, depending on the sign of k.

If $k > 0$, then $y \to \text{sign}(y_0 - Z) \times \infty$ as $t \to \infty$, i.e., the function becomes explosively large in absolute value.

If $k < 0$, then the second term decays to zero as $t \to \infty$ and $y \to Z$. It is sometimes convenient to write the solution in the following form instead when $k < 0$

$$y = Z + (y_0 - Z)e^{-|k|(t-t_0)} = Z(1 - e^{-|k|(t-t_0)}) + y_0 e^{-|k|(t-t_0)} \tag{41.35}$$

At $t = t_0$, the first term is zero and the second term dominates. As t becomes larger, the second terms becomes smaller and the first term begins to grow. The effect of the exponential is to pull the solution from the initial state $y = y_0$ to the **steady state** $y = Z$, which we denote now as y_{SS}:

$$y = y_0 e^{-|k|(t-t_0)} + y_{SS}(1 - e^{-|k|(t-t_0)}) \tag{41.36}$$

Regardless of the starting value, we say that the solution **relaxes** to the steady state, as illustrated in figure 41.2.

Definition 41.5. Steady State Solution

Let $y(t)$ be the solution of $y' = f(t, y)$. If $\lim\limits_{t \to \infty} y(t) = y_{SS}$ exists then we call $y = y_{SS}$ the **steady state** solution.

Figure 41.2: Two examples of exponential relaxation, with different initial states: one with $y(t_0) = b > y_{ss}$ and the other with $y(t_0) = a < y_{ss}$, both illustrate how the solution is pulled towards y_{ss} when $k < 0$.

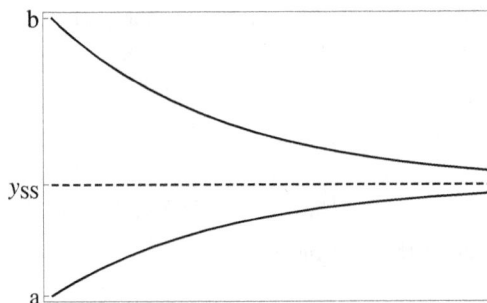

RC Circuits.   An example from engineering is given by RC-circuits (see figure 41.3). This is a circuit that contains a battery (voltage source, V); a resistor of resistance R; and a capacitor of capacity C. If there is an initial charge q on the capacitor, then the charge will decay exponentially if the voltage is set to zero. RC circuits have the following rules:

1. **Current** i is a rate of change of charge:

$$i = \frac{dq}{dt} \tag{41.37}$$

2. **Capacitors** hold a charge q that is proportional to the voltage V:

$$q = CV \tag{41.38}$$

   where C is a fixed constant that depends on the physical properties of the capacitor.

3. **Voltage Drops** (changes) **across a capacitor** occur when current is flowing. Differentiating (41.38) gives

$$i = \frac{dq}{dt} = C\frac{dV}{dt} \tag{41.39}$$

4. **Ohm's Law: The voltage drop across a resistor** R with a current i is given by

$$V = iR \tag{41.40}$$

5. **Conservation of Energy.** The total voltage drop around a loop must sum to zero.

Figure 41.3: Schematic of RC circuit.

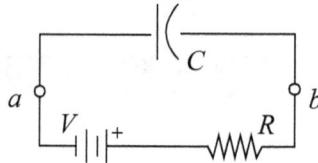

If we were to measure the voltage drop V_{ab} between points a and b in an RC circuit as illustrated in fig 41.3, these rules tell us first that the capacitor should cause the voltage along the top branch to fluctuate according to

$$i_C = C\frac{dV_{ab}}{dt} \tag{41.41}$$

where i_C is the current through the capacitor. If i_R is the current through the resistor, then the voltage difference between a and b along the bottom

$$V_{ab} = V_{batt} + i_R R \tag{41.42}$$

where V_{batt} is the battery voltage. Solving for i_R,

$$i_R = \frac{V_{ab} - V_{batt}}{R} \tag{41.43}$$

Since the total current around the loop must be zero

$$0 = i_R + i_C = \frac{V_{ab} - V_{batt}}{R} + C\frac{dV_{ab}}{dt} \tag{41.44}$$

Rearranging,

$$\frac{dV_{ab}}{dt} = -\frac{1}{RC}(V_{ab} - V_{batt}) \tag{41.45}$$

This is the same as equation 41.33 with $k = -1/RC < 0$ and $Z = V_{batt}$. The solution is given by equation 41.36

$$V_{ab} = V_0 e^{-t/(RC)} + V_{batt}(1 - e^{-t/(RC)}) \tag{41.46}$$

Doing Calculus

Regardless of the initial voltage, the voltage always tends towards the battery voltage. The current through the circuit is

$$i = C\frac{dV}{dt} = \frac{V_0 - V_{batt}}{R}e^{-t/(RC)} \tag{41.47}$$

so the current decays exponentially.

Newton's Law of Heating (or Cooling). This rule from physics says that the rate of change of the temperature T of an object (e.g., a potato) is proportional to the difference in temperatures between the object and its environment (e.g., an oven). We can write this as

$$\frac{dT}{dt} = k(T_{oven} - T) \tag{41.48}$$

Example 41.12. Suppose that an oven is set to 350 F and a potato at room temperature (70 F) is put in the oven at $t = 0$, with a baking thermometer inserted into the potato. After three minutes you observe that the temperature of the potato is 150. How long will it take the potato to reach a temperature of 350?

Solution. The initial value problem is

$$\left.\begin{array}{c} T' = k(400 - T) \\ T(0) = 70 \end{array}\right\} \tag{41.49a}$$

Dividing and integrating,

$$\int \frac{dT}{400 - T} = \int k\,dt \tag{41.49b}$$

$$-\ln(400 - T) = kt + C \tag{41.49c}$$

From the initial condition, at $t = 0$, the temperature is $T = 70$. Thus

$$C = -\ln(400 - 70) = -\ln 330 \tag{41.49d}$$

Hence

$$-\ln(400 - T) = kt - \ln 330 \tag{41.49e}$$

From the second observation, that $T = 150$ when $t = 3$,

$$-\ln(400 - 150) = 3k - \ln 330 \tag{41.49f}$$

Adding $\ln 330$ to both sides of the equation and dividing by 3,

$$k = \frac{1}{3}\left[\ln 330 - \ln(400 - 150)\right] = \frac{1}{3}\ln\frac{33}{25} \approx 0.09 \tag{41.49g}$$

From (41.49e)

$$-\ln(400 - T) = .09t - \ln 330 \tag{41.49h}$$

Multiplying by -1 and exponentiating,

$$400 - T = 330e^{-.09t} \tag{41.49i}$$

Solving for T,

$$T = 400 - 330e^{-.09t} \tag{41.49j}$$

The problem asked when the potato will reach a temperature of 350. Setting $T = 350$ and solving for t,

$$350 = 400 - 330e^{-.09t} \tag{41.49k}$$

Subtracting 400 from both sides of the equation,

$$-50 = -300e^{-.09t}$$ (41.49l)

Dividing through by -300, and taking natural logarithms of both sides of the equation,

$$-.09t = \ln \frac{-50}{-300} = \ln \frac{1}{6} \approx -1.79$$ (41.49m)

Solving for t,

$$t \approx \frac{1.79}{.09} \approx 19.9 \text{ minutes}$$ (41.49n)

$\square$

Linear Differential Equations

A differential equation is said to be linear if we can rewrite in the form

$$\frac{dy}{dx} = M(x)y + B(x)$$ (41.50)

where $M(x)$ and $B(x)$ can be any functions of x at all. They do not have to be equations of lines. The functions $M(x)$ and $B(x)$ are analogous to the constants m and b in the point-slope equation of a line

$$y = mx + b$$ (41.51)

However, we do not normally write a linear differential equation in the form (41.50). Instead, we bring the term $M(x)y$ to the left hand side, and rename both $-M(x)$ and $B(x)$ to $p(x)$ and $q(x)$ respectively:

$$y' + p(x)y = q(x)$$ (41.52)

For example, in the linear differential equation

$$y' = 3x^2 + y\cos x$$ (41.53)

$p(x) = -\cos x$ and $q(x) = 3x^2$.

Definition 41.6. Linear Differential Equation: Standard Form

A **first order linear differential equation** in standard form is given by

$$\frac{dy}{dx} + p(x)y = q(x)$$ (41.54)

where $p(x)$ and $q(x)$ are any functions that depend only on x but not on y.

To find a general method for solving a linear differential equation we look for a function $\mu(x)$ such that

$$\mu(x)\left(\frac{dy}{dx} + p(x)y\right) = (\mu(x)y)'$$ (41.55)

Why? If we multiply the left hand side of (41.54) by $\mu(x)$ then we also multiply the right hand side, and this result tells us that

$$(\mu(x)y)' = \mu(x)\left(\frac{dy}{dx} + p(x)y\right) = \mu(x)q(x)$$ (41.56)

Dropping the middle part of the equation,

$$\frac{d}{dx}(\mu(x)y) = \mu(x)q(x)$$ (41.57)

Multiplying both sides by dx and integrating,

$$\int \frac{d}{dx}(\mu(x)y)dx = \int \mu(x)q(x)dx \tag{41.58}$$

On the left we have the integral of a derivative:

$$\mu(x)y = C + \int \mu(x)q(x)dx \tag{41.59}$$

where C is the constant of integration. Dividing by $\mu(x)$, this gives us an explicit formula for the solution:

$$y = \frac{1}{\mu(x)}\left[C + \int \mu(x)q(x)dx\right] \tag{41.60}$$

If we can find a formula for $\mu(x)$, then all we have to do is plug that into (41.60) and we will have a plug-in-the-blank formula for solving any linear differential equation.

So the problem becomes one of finding $\mu(x)$. But from (41.55) we have

$$\mu(x)(y' + p(x)y) = (\mu(x)y)' \tag{41.61}$$

Multiplying out the left, and applying the product rule on the right,

$$\mu(x)y' + \mu(x)p(x)y = \mu'(x)y + \mu(x)y' \tag{41.62}$$

The term $\mu(x)y'$ appears on both sides of the equation and can be subtracted off, to give

$$\mu(x)p(x)y = \mu'(x)y \tag{41.63}$$

The factor y appears on both sides; dividing through by y gives

$$\frac{d\mu(x)}{dx} = \mu(x)p(x) \tag{41.64}$$

Dividing both sides by μ and multiplying by dx,

$$\frac{d\mu}{\mu} = p(x)dx \tag{41.65}$$

Integrating,

$$\int \frac{d\mu}{\mu} = \int p(x)dx \tag{41.66}$$

$$\ln \mu = \int p(x)dx + C \tag{41.67}$$

Since we want **any** function $\mu(x)$ that satisfies this requirement we are free to choose any value for C, such as $C = 0$. This is equivalent to choosing any initial condition we like for equation 41.64. Thus we have

$$\mu = e^{\int p(x)dx} \tag{41.68}$$

Theorem 41.1. Linear Differential Equations

The general solution of $y' + p(x)y = q(x)$ is

$$y = \frac{1}{\mu(x)}\left[C + \int \mu(x)q(x)dx\right] \tag{41.69}$$

where $\mu = e^{\int p(x)dx}$.

Example 41.13. Find a general solution of $y' + 3y = 1$.

Solution. This is linear with $p(x) = 3$ and $q(x) = 1$. Hence

$$\mu = e^{\int 3dx} = e^{\int 3dx} = e^{3x} \tag{41.70a}$$

Thus

$$y = \frac{1}{e^{3x}}\left[C + \int e^{3x} \cdot 1 \cdot dx\right] = Ce^{-3x} + \frac{1}{3} \tag{41.70b}$$

$\square$

Example 41.14. Solve the initial value problem

$$\left.\begin{array}{r} x^2 y' + xy = 1 \\ y(1) = 2 \end{array}\right\} \tag{41.71}$$

assuming $x > 0$.

Solution. To apply theorem 41.1 we need to put the differential equation in standard form. We can do this by dividing through by x^2, which is allowed by the assumption that $x > 0$. This gives

$$y' + \frac{1}{x}y = \frac{1}{x^2} \tag{41.72a}$$

Thus $p(x) = 1/x$ and $q(x) = 1/x^2$. Hence

$$\mu = e^{\int (1/x)dx} = e^{\ln x} = x \tag{41.72b}$$

Using this in theorem 41.1,

$$y = \frac{1}{x}\left[C + \int x \cdot \frac{1}{x^2}\, dx\right] = \frac{1}{x}[C + \ln x] \tag{41.72c}$$

From the initial condition $2 = \frac{1}{1}[1 + \ln 1]$ and thus $C = 2$ The solution of the IVP is then $y = \frac{1}{x}(1 + \ln x)$. $\square$

As an alternative to "plugging in" to the formula for the solution, some students prefer to re-derive the formula each time. This technique is illustrated in the second part of example 41.15.

Bernoulli Equations

A Bernoulli differential equation has the form

$$y' + p(x)y = y^n q(x) \tag{41.73}$$

where $p(x)$ and $q(x)$ are any functions of x , and n is any real number.

When $n = 0$ a Bernoulli equation is a linear equation in standard form,

$$y' + p(x)y = q(x) \tag{41.74}$$

which we have already solved in theorem 41.1.

When $n = 1$, a Bernoulli equation is both linear and separable:

$$y' + p(x)y = yq(x) \tag{41.75}$$

Bringing the $p(x)y$ term to the right hand side of the equation and factoring,

$$y' = [q(x) - p(x)]y \tag{41.76}$$

Dividing both sides by y

$$\int \frac{dy}{y} = \int [q(x) - p(x)]dx \tag{41.77}$$

Integrating and exponentiating,

$$y = C \exp \int (q(x) - p(x)) \, dx \tag{41.78}$$

In the general case, $n \neq 0, 1$, equation 41.73 is not-quite linear, because of the factor of y^n on the right-hand side. For any other value of n, Bernoulli equations can be made linear with with the substitution

$$z = y^{1-n} \tag{41.79}$$

Theorem 41.2. Solution to Bernoulli Equations

The substitution $z = y^{1-n}$ will transform the Bernoulli differential equation

$$y' + p(x)y = y^n q(x) \tag{41.80}$$

into a linear differential equation in z.

Proof. Let $z = y^{1-n}$ and differentiate with respect to x, using the chain rule. Then

$$\frac{dz}{dx} = (1-n)y^{-n}\frac{dy}{dx} \tag{41.81a}$$

Solving for $'y$,

$$\frac{dy}{dx} = \frac{1}{1-n}y^n\frac{dz}{dx}, \; n \neq 1 \tag{41.81b}$$

The restriction to $n \neq 1$ is not a problem because we have already shown how to solve the special case $n = 1$ in equation 41.78.

Substituting equation 41.81b into 41.73 gives

$$\frac{1}{1-n}y^n\frac{dz}{dx} + p(x)y = y^n q(x) \tag{41.81c}$$

Dividing by y^n,

$$\frac{1}{1-n}\frac{dz}{dx} + p(x)\underbrace{y^{1-n}}_{z} = q(x) \tag{41.81d}$$

Substituting $z = y^{1-n}$ in the second term, as indicated, and then multiplying the equation through by $1 - n$ gives

$$\frac{dz}{dx} + (1-n)p(x)z = (1-n)q(x) \tag{41.81e}$$

which is a linear ODE for z in in standard form

$$z' + P(x)z = Q(x) \tag{41.81f}$$

where $P(x) = (1-n)p(x)$ and $Q(x) = (1-n)q(x)$.   □

Heuristic 41.1. Bernoulli Equations

To solve a Bernoulli Equation, $y' + p(x)y = y^n q(x)$
1. Let $z = y^{1-n}$
2. Find z' using the chain rule.
3. Replace all occurrences of y and y' in the differential equation.
4. Simplify the result and verify that[a] $z' + (1-n)p(x)z = (1-n)q(x)$.
5. Solve the equation for z.
6. Substitute $z = y^{1-n}$ and solve for y.

---
[a]You might prefer to just skip steps 1-3 and jump immediately to this step.

Example 41.15. Solve the initial value problem

$$y' + xy = \frac{x}{y^3}, \ y(0) = 2 \tag{41.82}$$

Solution. This is a Bernoulli equation with $n = -3$. Let

$$z = y^{1-n} = y^{1-(-3)} = y^4 \tag{41.83a}$$

The initial condition on z is $z(0) = 2^4 = 16$. Differentiating equation 41.83a

$$\frac{dz}{dx} = 4y^3 \frac{dy}{dx} \tag{41.83b}$$

Hence

$$\frac{dy}{dx} = \frac{1}{4y^3}\frac{dz}{dx}. \tag{41.83c}$$

Substituting equation 41.83c into the original differential equation equation 41.82

$$\frac{1}{4y^3}\frac{dz}{dx} + xy = \frac{x}{y^3} \tag{41.83d}$$

Multiplying through by $4y^4$,

$$\frac{dz}{dx} + 4xy^4 = 4x \tag{41.83e}$$

Noting that $z = y^4$ in the second term,

$$\frac{dz}{dx} + 4xz = 4x \tag{41.83f}$$

This is a first order linear ODE in z with $p(x) = 4x$ and $q(x) = 4x$. An integrating factor is

$$\mu(x) = \exp \int 4x \, dx = e^{2x^2}. \tag{41.83g}$$

Multiplying equation $41.83f$ through by this μ gives

$$\left[\frac{dz}{dx} + 4xz\right] e^{2x^2} = 4xe^{2x^2} \tag{41.83h}$$

By construction the left hand side must be the exact derivative of $z\mu$; hence

$$\frac{d}{dx}\left[ze^{2x^2}\right] = 4xe^{2x^2} \tag{41.83i}$$

Multiplying by dx and integrating both sides of the equation,

$$ze^{2x^2} = \int \frac{d}{dx}\left[ze^{2x^2}\right] dx = \int 4xe^{2x^2} \, dx = e^{2x^2} + C \tag{41.83j}$$

From the initial condition $z(0) = 16$,

$$16e^0 = e^0 + C \tag{41.83k}$$

and therefore $C = 15$. Plugging this into the solution for z gives

$$z = 1 + 15e^{-2x^2}. \tag{41.83l}$$

Since $y = z^{1/4}$, the solution to the example is

$$y = \left[1 + 15e^{-2x^2}\right]^{1/4} \tag{41.83m}$$

□

Homogeneous Differential Equations

A differential equation is said to be **homogeneous** if it can be written in the form

$$\frac{dy}{dx} = g\left(\frac{y}{x}\right) = g(z) \tag{41.84}$$

Example 41.16. Show that $\dfrac{dy}{dx} = \dfrac{2xy}{x^2 - 3y^2}$ is homogeneous.

Solution. The equation has the form $y' = f(x, y)$ where

$$f(x, y) = \frac{2xy}{x^2 - 3y^2} = \frac{2xy}{(x^2)(1 - 3(y/x)^2)} = \frac{2(y/x)}{1 - 3(y/x)^2} = \frac{2z}{1 - 3z^2} \tag{41.85a}$$

where $z = y/x$. Hence the ODE is homogeneous.   □

Homogeneous differential equations can be solved with the substitution $z = y/x$. This works because if $z = y/z$ then $y = zx$, so by the product rule

$$y' = z + xz' \tag{41.86}$$

But if $y' = g(z)$ (from equation 41.84),

$$z + xz' = g(z) \tag{41.87}$$

Solving for z',

$$\frac{dz}{dx} = z' = \frac{g(z) - z}{x} \tag{41.88}$$

This equation is separable and can be integrated.

$$\int \frac{dz}{g(z) - z} = \int \frac{dx}{x} \tag{41.89}$$

It is generally easier to re-derive this formula each time rather than memorize it.

Heuristic 41.2. Homogeneous Differential Equations

Suppose $y' = g(y/x)$.
  1. Let $z = y/x$.
  2. Differentiate and substitute $y' = z + xz'$.
  3. The resulting differential equation is separable in z and x. Solve for z.
  4. Substitute $z = y/x$ in the solution and solve for y.

Example 41.17. Find the general solution to $y' = \dfrac{y^2 + 2xy}{x^2}$.

Solution. Let $z = y/x$. Then

$$y' = \frac{y^2 + 2xy}{x^2} = \frac{y^2}{x^2} + \frac{2xy}{x^2} = \left(\frac{y}{x}\right)^2 + 2\frac{y}{x} = z^2 + 2z \tag{41.90a}$$

Since the right hand side depends only on z and there is no x or y dependence, the differential equation is homogeneous. Thus $y = zx$ or $y' = z + xz'$. Substituting this gives

$$z + xz' = y' = z' + 2z \tag{41.90b}$$

Solving for dz/dx,

$$\frac{dz}{dx} = \frac{z^2 + 2z - z}{x} = \frac{z^2 + z}{x} \tag{41.90c}$$

Cross multiplying and integrating,

$$\int \frac{dx}{x} = \int \frac{dz}{z^2 + z} = \int \frac{dz}{z(z+1)} \tag{41.90d}$$

Using the method of partial fractions,

$$\ln|x| = \int \frac{dz}{z} - \int \frac{dz}{1+z} = \ln\left|\frac{z}{1+z}\right| + C \tag{41.90e}$$

Exponentiating both sides of the equation,

$$x = \frac{C_1 z}{1+z} \tag{41.90f}$$

where C_1 is a different constant (from C). Dropping the subscript on C_1 and cross-multiplying to solve for z,

$$(1 + z)x = Cz \tag{41.90g}$$

Distributing the $1 + z$ on the left,

$$x + xz = Cz \tag{41.90h}$$

Subtracting $Cz + x$ from both sides of the equation,

$$xz - CZ = -x \tag{41.90i}$$

Factoring the x,

$$(x - C)z = -x \tag{41.90j}$$

Solve for z:

$$z = \frac{x}{C - x} \tag{41.90k}$$

Since $z = y/x$ then $y = xz$, so that

$$y = xz = \frac{x^2}{C - x}. \tag{41.90l}$$

$\square$

Example 41.18. Solve the initial value problem

$$\frac{dy}{dx} = \frac{y}{x} - 1, \; y(1) = 2. \tag{41.91}$$

Solution. Let $z = y/x$. Then $y' = z + xz'$ (eq. 41.86) and thus

$$z + xz' = y' = z - 1 \tag{41.92a}$$

Cancelling the z terms on both sides of the equation,

$$xz' = -1 \tag{41.92b}$$

Writing $z' = dz/dx$, multiplying by dx, and integrating,

$$\int dz = - \int \frac{dx}{x} \tag{41.92c}$$

Evaluating the integral gives

$$z = -\ln|x| + C \qquad\qquad (41.92d)$$

Solving for y,

$$y = xz = x(C - \ln|x|) \qquad\qquad (41.92e)$$

Substituting the initial condition

$$2 = 1(C - \ln 1) = C \qquad\qquad (41.92f)$$

Therefore $y = x(2 - \ln|x|)$. $\square$

Exercises

Put the following differential equations into standard form and identify the function $f(x, y)$.

1. $y^2 + 3xyy' = x \cos x$
2. $\cos(x + y) = x + y' \sin x$
3. $x^2 y' + y^2 x + x^3 = y^2 y'$
4. $y'(1 - x^2) = (1 - y')e^x$

Find general solutions for each of the separable differential equations.

5. $y' = y/x$
   Ans: $y = Cx$
6. $y' = \dfrac{x^2 + 3x + 1}{y}$
7. $yy' = \dfrac{x}{y}$
   Ans: $y^3 = \dfrac{\pm 3(x^2 + 2)}{2}$
8. $\sec x \csc x\, y' = 1$
   Ans: $y = C - \frac{1}{2}\cos^2 x$

Put each of the following linear differential equation into standard form and find an integrating factor.

9. $x^2 y' + xy = 1$
10. $xy' + xy = 1$
11. $y' \cos x + y \sec x = \sin x$
12. $y' + \dfrac{y}{x} = \dfrac{1}{x}$

Solve the linear equations.

13. $x + xy' = x^2$
    Ans: $y = C - x + \frac{1}{2}x^2$
14. $y' \cos x + y \sec x = \dfrac{1}{\cos x}, y(\pi) = 1/2$
    Ans: $1 - \dfrac{1}{2}e^{-\tan x}$
15. $y' + \dfrac{y}{x} = \dfrac{1}{x}$, y(1)=2   Ans: $\dfrac{1 + x}{x}$

Solve the Bernoulli Equations.

16. $y' + y = y^2$ Ans: $y = 1/(1 + Ce^x)$

17. $y' + y = y^3$ Ans: $y = (1 + Ce^{2x})^{-1/2}$
18. $y' + xy = \dfrac{x}{y}$ Ans: $y = \pm(1 + Ce^{-2x^3/3})^{1/2}$

Solve the homogeneous differential equations.

19. $y' = \dfrac{3x - 4y}{3x + 4y}$
20. $y' = 1 + \dfrac{y}{x}$
21. $y' = 1 + \dfrac{y}{x} - \left(\dfrac{y}{x}\right)^2$

22. Financial analysts currently suggest that the average middle class wage earner needs to accumulate approximately $2 million in savings over their work life to achieve a comfortable retirement. Assume that your graduate at age 22 and work for 50 years, to retire at age 72.

    (a) Suppose you deposit S dollars per month into a savings account with interest rate r percent per year. Write a differential equation for the total amount $A(t)$ accumulated.

    (b) Suppose you can afford to deposit $1000 per month. How long would it take at 1, 5, and 10 percent? What is the minimum interest rate you need to accumulate the two million dollars?

    (c) How much would you have to deposit every month to accumulate $2 million in 50 years at an interest rate of 1 percent per year? At a rate of 5%? At 10%?

23. Suppose you remove a pitcher of iced tea from the refrigerator at 40 deg. F and place it on a picnic table on a hot summer day at 98 deg. F. After ten minutes you measure the temperature of the iced tea and observe that it has increased to 50 deg. When will it reach 90 deg?

24. A population of bacteria is grown in culture. Initially there are 23 organisms. After 24 hours, 157 are observed. If the population grows at a rate proportional to the number present, how many will there be after 36 hours?

Appendix A

Algebraic Formulas

Algebraic Manipulation

$x + y = y + x$	Commutative Rule (addition)	(A.1)
$xy = yx$	Commutative Rule (addition)	(A.2)
$(x + y) + z = z + (y + z)$	Associative Rule (addition)	(A.3)
$(xy)z = x(yz)$	Associative Rule (multiplication)	(A.4)
$x(y + z) = xy + xz$	Distributive Rule	(A.5)

Fractions

$$\frac{x+y}{z} = \frac{x}{z} + \frac{y}{z} \quad \text{(A.6)} \qquad \frac{x}{p} + \frac{y}{q} = \frac{xp + yq}{pq} \quad \text{(A.7)}$$

$$\frac{x}{y} \cdot \frac{p}{q} = \frac{xp}{yq} \quad \text{(A.8)} \qquad \frac{x/y}{p/q} = \frac{xq}{py} \quad \text{(A.9)}$$

Differences of Powers

$$x^1 - y^1 = (x - y) \tag{A.10}$$
$$x^2 - y^2 = (x - y)(x + y) \tag{A.11}$$
$$x^3 - y^3 = (x - y)(x^2 + xy + y^2) \tag{A.12}$$
$$x^4 - y^4 = (x - y)(x^3 + x^2 y + xy^2 + y^3) \tag{A.13}$$

$$x^n - y^n = (x - y)(x^{n-1} + x^{n-2}y + x^{n-3}y^2 + \cdots \tag{A.14}$$
$$+ x^2 y^{n-3} + xy^{n-2} + y^{n-1})$$

Completing the Squares

$$ax^2 + bx + c = a\left(x^2 + \frac{b}{a}x\right) + c = a\left(x^2 + \frac{b}{a}x + \left(\frac{b}{2a}\right)^2 - \left(\frac{b}{2a}\right)^2\right) + c \tag{A.15}$$

$$= a\left(x + \frac{b}{2a}\right)^2 + c - \frac{b^2}{4a} \tag{A.16}$$

Quadratic Formula

The roots of $ax^2 + bx + c = 0$ are given by $x = \dfrac{-b \pm \sqrt{b^2 - 4ac}}{2a}$.

Binomial Theorem

$$(x+y)^n = \sum_{k=0}^{n} \binom{n}{k} x^k y^{n-k} = \sum_{k=0}^{n} \frac{n!}{k!(n-k)!} x^k y^{n-k} \qquad \text{(A.17)}$$

$$(x+y)^2 = x^2 + 2xy + y^2 \qquad \qquad \text{(A.18)}$$

$$(x+y)^3 = x^3 + 3x^2 y + 3xy^2 + y^2 \qquad \qquad \text{(A.19)}$$

$$(x+y)^4 = x^4 + 4x^3 y + 6^2 y^2 + 4xy^3 + y^4 \qquad \qquad \text{(A.20)}$$

$$(x+y)^n = \binom{n}{0} x^n y^0 + \binom{n}{1} x^{n-1} y^1 + \binom{n}{2} x^{n-2} y^2 + \cdots + \binom{n}{n-1} x^1 y^{n-1} + \binom{n}{n} x^0 y^n \qquad \text{(A.21)}$$

Pascals Triangle

The coefficients in the sum of $(x+y)^m$ are given by the mth row of Pascal's Triangle. Each number on the inside is the sum of the two numbers above it. The equation is given by the **Binomial Theorem**.

Equation of a Line

Let m be the slope of a line, a be its x intercept, b its y intercept, and (x_1, y_1) and (x_2, y_2) be any distinct points on the line, with $x_1 \neq x_2$.

$\dfrac{x}{a} + \dfrac{y}{b} = 1$	intercept-intercept form	(A.22)
$m = \dfrac{y_2 - y_1}{x_2 - x_1}$	slope of the line	(A.23)
$y = mx + b$	slope - intercept form	(A.24)
$y = y_1 + m(x - x_1)$	point-slope form	(A.25)

Roots and Exponents

$$x^n = \overbrace{x \cdot x \cdot x \cdots x}^{\text{repeated } n \text{ times}} \qquad (A.26)$$

$$y = x^n \iff x = y^{1/n} \qquad (A.27)$$

$$x^{1/n} = \sqrt[n]{x} \qquad (A.28)$$

$$x^{-n} = \frac{1}{x^n},\ (x \neq 0) \qquad (A.29)$$

$$\sqrt{\frac{x}{y}} = \frac{\sqrt{x}}{\sqrt{y}} \qquad (A.30)$$

$$\sqrt{xy} = \sqrt{x}\sqrt{y} \qquad (A.31)$$

$$x^{p-q} = \frac{x^p}{x^q} \qquad (A.32)$$

$$x^{p+q} = x^p \cdot x^q \qquad (A.33)$$

$$x^{p/q} = \sqrt[q]{x^p} \qquad (A.34)$$

$$(x^p)^q = x^{pq} \qquad (A.35)$$

$$(xy)^p = x^p y^p \qquad (A.36)$$

$$\left(\frac{x}{y}\right)^n = \frac{x^n}{y^n},\ (y \neq 0) \qquad (A.37)$$

Sums of Powers of Integers

$$1 + 2 + 3 + \cdots + n = \frac{n(n+1)}{2} \qquad (A.38)$$

$$1^2 + 2^2 + 3^2 + \cdots + n^2 = \frac{n(n+1)(2n+1)}{6} \qquad (A.39)$$

$$1^3 + 2^3 + 3^3 + \cdots + n^3 = \frac{n^2(n+1)^2}{4} \qquad (A.40)$$

Exponentials and Logarithms

$$y = e^x \iff x = \ln y \qquad (A.41)$$

$$\ln(xy) = \ln x + \ln y \qquad (A.42)$$

$$e^{\ln x} = \ln e^x = x \qquad (A.43)$$

$$\ln(x/y) = \ln x - \ln y \qquad (A.44)$$

$$e^{x+y} = e^x e^y \qquad (A.45)$$

$$\ln y^x = x \ln y \qquad (A.46)$$

$$e^{x-y} = \frac{e^x}{e^y} \qquad (A.47)$$

$$\log_b x = \frac{\ln x}{\ln b} \qquad (A.48)$$

Hyperbolic Functions and Identities

$$\cosh x = \tfrac{1}{2}(e^x + e^{-x}) \qquad (A.49)$$

$$\sinh x = \tfrac{1}{2}(e^x - e^{-x}) \qquad (A.50)$$

$$e^x = \cosh x + \sinh x \qquad (A.51)$$

$$e^{-x} = \cosh x - \sinh x \qquad (A.52)$$

$$\cosh^2 x - \sinh^2 x = 1 \qquad (A.53)$$

$$1 - \tanh^2 x = \operatorname{sech}^2 x \qquad (A.54)$$

$$\cosh(2x) = \cosh^2 x + \sinh^2 x \qquad (A.55)$$

$$\sinh 2x = 2 \sinh x \cosh x \qquad (A.56)$$

$$\operatorname{arcsinh}(x) = \ln\left(x + \sqrt{x^2 + 1}\right) \qquad (A.57)$$

$$\operatorname{arccosh}(x) = \ln\left(x + \sqrt{x^2 - 1}\right) \qquad (A.58)$$

Trigonometry

$$\sin\theta = \frac{\text{opposite}}{\text{hypotenuse}} \quad\text{(A.59)}$$

$$\cos\theta = \frac{\text{adjacent}}{\text{hypotenuse}} \quad\text{(A.60)}$$

$$\tan\theta = \frac{\sin\theta}{\cos\theta} = \frac{\text{opposite}}{\text{adjacent}} \quad\text{(A.61)}$$

$$\cot\theta = \frac{1}{\tan\theta} = \frac{\cos\theta}{\sin\theta} = \frac{\text{adjacent}}{\text{opposite}} \quad\text{(A.62)}$$

$$\sec\theta = \frac{1}{\cos\theta} = \frac{\text{hypotenuse}}{\text{adjacent}} \quad\text{(A.63)}$$

$$\csc\theta = \frac{1}{\sin\theta} = \frac{\text{hpotenuse}}{\text{opposite}} \quad\text{(A.64)}$$

Trigonometric Identities

$$\sin^2 x + \cos^2 x = 1 \quad\text{(A.65)}$$

$$\tan^2 x + 1 = \sec^2 x \quad\text{(A.66)}$$

$$1 + \cot^2 x = \csc^2 x \quad\text{(A.67)}$$

$$\cos(2x) = \cos^2 x - \sin^2 x \quad\text{(A.68)}$$

$$\sin(2x) = 2\sin x \cos x \quad\text{(A.69)}$$

$$\tan(2x) = \frac{2\tan x}{1 - \tan^2 x} \quad\text{(A.70)}$$

$$\sin^2 x = \frac{1}{2}(1 - \cos(2x)) \quad\text{(A.71)}$$

$$\cos^2 x = \frac{1}{2}(1 + \cos(2x)) \quad\text{(A.72)}$$

$$\sin(x \pm y) = \sin x \cos y \pm \cos x \sin y \quad\text{(A.73)}$$

$$\cos(x \pm y) = \cos x \cos y \mp \sin x \sin y \quad\text{(A.74)}$$

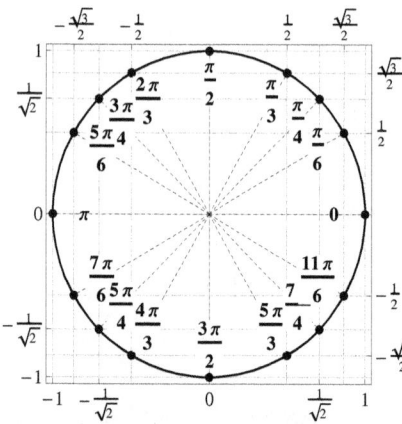

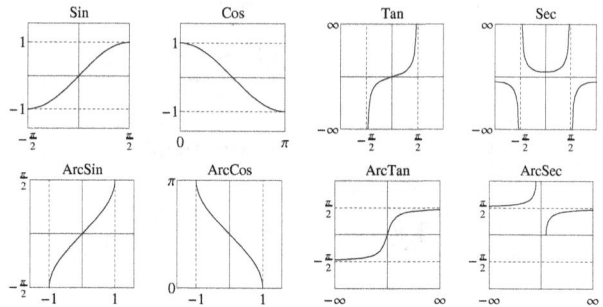

θ	0	$\frac{\pi}{6}$	$\frac{\pi}{4}$	$\frac{\pi}{3}$	$\frac{\pi}{2}$	$\frac{2\pi}{3}$	$\frac{3\pi}{4}$	$\frac{5\pi}{6}$	π
$\cos\theta$	1	$\frac{\sqrt{3}}{2}$	$\frac{1}{\sqrt{2}}$	$\frac{1}{2}$	0	$-\frac{1}{2}$	$-\frac{1}{\sqrt{2}}$	$-\frac{\sqrt{3}}{2}$	-1
$\sin\theta$	0	$\frac{1}{2}$	$\frac{1}{\sqrt{2}}$	$\frac{\sqrt{3}}{2}$	1	$\frac{\sqrt{3}}{2}$	$\frac{1}{\sqrt{2}}$	$\frac{1}{2}$	0
$\tan\theta$	0	$\frac{1}{\sqrt{3}}$	1	$\sqrt{3}$	∞	$-\sqrt{3}$	-1	$-\frac{1}{\sqrt{3}}$	0
$\sec\theta$	1	$\frac{2}{\sqrt{3}}$	$\sqrt{2}$	2	∞	-2	$-\sqrt{2}$	$-\frac{2}{\sqrt{3}}$	-1
$\csc\theta$	∞	2	$\sqrt{2}$	$\frac{2}{\sqrt{3}}$	1	$\frac{2}{\sqrt{3}}$	$\sqrt{2}$	2	∞
$\cot\theta$	∞	$\sqrt{3}$	1	$\frac{1}{\sqrt{3}}$	0	$-\frac{1}{\sqrt{3}}$	-1	$-\sqrt{3}$	∞

Appendix B

Table of Derivatives

$$\frac{d}{dx}(Cf(x)) = C\frac{d}{dx}f(x) \tag{B.1}$$

$$\frac{d}{dx}[f(x) \pm g(x)] = \frac{d}{dx}f(x) \pm \frac{d}{dx}g(x) \tag{B.2}$$

$$\frac{d}{dx}(f(x) \cdot g(x)) =$$
$$f(x) \cdot g'(x) + f'(x) \cdot g(x) \tag{B.3}$$

$$\frac{d}{dx}\frac{f(x)}{g(x)} = \frac{g(x)f'(x) - f(x)g'(x)}{g(x)^2} \tag{B.4}$$

$$\frac{d}{dx}f(g(x)) = f'(g(x))g'(x) \tag{B.5}$$

$$\frac{d}{dx}x^n = nx^{n-1} \tag{B.6}$$

$$\frac{d}{dx}f(x)^n = nf(x)^{n-1}f'(x) \tag{B.7}$$

$$\frac{d}{dx}\ln x = \frac{1}{x} \tag{B.8}$$

$$\frac{d}{dx}\ln f(x) = \frac{f'(x)}{f(x)} \tag{B.9}$$

$$\frac{d}{dx}e^x = e^x \tag{B.10}$$

$$\frac{d}{dx}e^{f(x)} = e^{f(x)}f'(x) \tag{B.11}$$

$$\frac{d}{dx}a^x = a^x \ln a \tag{B.12}$$

$$\frac{d}{dx}a^{f(x)} = a^{f(x)}f'(x)\ln a \tag{B.13}$$

$$\frac{d}{dx}x^x = x^x(1 + \ln x) \tag{B.14}$$

$$\frac{d}{dx}x^{f(x)} = x^{f(x)}\left(\frac{f(x)}{x} + f'(x)\ln x\right) \tag{B.15}$$

$$\frac{d}{dx}\sin x = \cos x \tag{B.16}$$

$$\frac{d}{dx}\cos x = -\sin x \tag{B.17}$$

$$\frac{d}{dx}\tan x = \sec^2 x \tag{B.18}$$

$$\frac{d}{dx}\cot x = -\csc^2 x \tag{B.19}$$

$$\frac{d}{dx}\sec x = \sec x \tan x \tag{B.20}$$

$$\frac{d}{dx}\csc x = -\csc x \cot x \tag{B.21}$$

$$\frac{d}{dx}\sin^{-1} x = \frac{1}{\sqrt{1 - x^2}} \tag{B.22}$$

$$\frac{d}{dx}\cos^{-1} x = -\frac{1}{\sqrt{1 - x^2}} \tag{B.23}$$

$$\frac{d}{dx}\tan^{-1} x = \frac{1}{x^2 + 1} \tag{B.24}$$

$$\frac{d}{dx}\cot^{-1} x = -\frac{1}{x^2 + 1} \tag{B.25}$$

$$\frac{d}{dx}\sec^{-1} x = \frac{1}{x\sqrt{x^2 - 1}} \tag{B.26}$$

$$\frac{d}{dx}\csc^{-1} x = -\frac{1}{x\sqrt{x^2 - 1}} \tag{B.27}$$

$$\frac{d}{dx}\sinh x = \cosh x \tag{B.28}$$

$$\frac{d}{dx}\cosh x = \sinh x \tag{B.29}$$

$$\frac{d}{dx}\tanh x = \operatorname{sech}^2 x \tag{B.30}$$

$$\frac{d}{dx}\coth x = -\operatorname{csch}^2 x \tag{B.31}$$

$$\frac{d}{dx}\operatorname{sech} x = -\tanh x \operatorname{sech} \tag{B.32}$$

$$\frac{d}{dx}\operatorname{csch} x = -\coth x \operatorname{csch} x \tag{B.33}$$

$$\frac{d}{dx}\sinh^{-1} x = \frac{1}{\sqrt{x^2+1}} \tag{B.34}$$

$$\frac{d}{dx}\cosh^{-1} x = \frac{1}{\sqrt{x^2-1}} \tag{B.35}$$

$$\frac{d}{dx}\tanh^{-1} x = \frac{1}{1-x^2} \tag{B.36}$$

$$\frac{d}{dx}\coth^{-1} x = -\frac{1}{1-x^2} \tag{B.37}$$

$$\frac{d}{dx}\operatorname{sech}^{-1} x = -\frac{1}{x\sqrt{1-x^2}} \tag{B.38}$$

$$\frac{d}{dx}\operatorname{csch}^{-1} x = -\frac{1}{|x|\sqrt{1+x^2}} \tag{B.39}$$

Appendix C

Table of Integrals

$$\int Cf(x)dx = C \int f(x)dx \tag{C.1}$$

$$\int (f \pm g)dx = \int fdx \pm \int gdx \tag{C.2}$$

$$\int x^n dx = \frac{x^{n+1}}{n+1}, n \neq -1 \tag{C.3}$$

$$\int \frac{1}{x}dx = \ln|x| \tag{C.4}$$

$$\int \ln x \, dx = x \ln x - x \tag{C.5}$$

$$\int x \ln x = \frac{1}{2}x^2 \ln x - \frac{1}{4}x^2 \tag{C.6}$$

$$\int e^x dx = e^x \tag{C.7}$$

$$\int xe^x \, dx = xe^x - e^x \tag{C.8}$$

$$\int x^2 e^x \, dx = e^x(x^2 - 2x + 2) \tag{C.9}$$

$$\int a^x dx = \frac{a^x}{\ln a} \tag{C.10}$$

$$\int xa^x dx = \frac{a^x(x \ln a - 1)}{(\ln a)^2} \tag{C.11}$$

$$\int \sin x dx = -\cos x \tag{C.12}$$

$$\int \cos x dx = \sin x \tag{C.13}$$

$$\int \tan x dx = \ln|\sec x| \tag{C.14}$$

$$\int \cot x dx = \ln|\sin x| \tag{C.15}$$

$$\int \sec x dx = \ln|\sec x + \tan x| \tag{C.16}$$

$$\int \csc x dx = -\ln|\csc x + \cot x| \tag{C.17}$$

$$\int \sec^2 x dx = \tan x \tag{C.18}$$

$$\int \csc^2 x dx = -\cot x \tag{C.19}$$

$$\int \sec x \tan x dx = \sec x \tag{C.20}$$

$$\int \csc x \cot x dx = -\csc x \tag{C.21}$$

$$\int \frac{dx}{\sqrt{1-x^2}} = \sin^{-1} x \tag{C.22}$$

$$\int \frac{dx}{1+x^2} = \tan^{-1} x \tag{C.23}$$

$$\int \frac{dx}{x\sqrt{x^2-1}} = \sec^{-1} x \tag{C.24}$$

$$\int \frac{dx}{\sqrt{x^2+1}} = \sinh^{-1} x \tag{C.25}$$

$$\int \frac{dx}{\sqrt{x^2-1}} = \cosh^{-1} x \tag{C.26}$$

$$\int \frac{dx}{1-x^2} = \tanh^{-1} x \tag{C.27}$$

$$\int \sinh x \; dx = \cosh x \tag{C.28}$$

$$\int \cosh x \; dx = \sinh x \tag{C.29}$$

$$\int \tanh x \; dx = \ln|\cosh x| \tag{C.30}$$

$$\int \coth x \; dx = \ln|\sinh x| \tag{C.31}$$

$$\int \operatorname{sech} x \; dx = 2\tan^{-1} \tanh \frac{x}{2} \tag{C.32}$$

$$\int \operatorname{csch} x \; dx = \ln \tanh \frac{x}{2} \tag{C.33}$$

$$\int \operatorname{sech}^2 x \; dx = \tanh x \tag{C.34}$$

$$\int \operatorname{csch}^2 x \; dx = -\coth x \tag{C.35}$$

$$\int \operatorname{sech} x \tanh x \; dx = -\operatorname{sech} x \tag{C.36}$$

$$\int \operatorname{csch} x \coth x \; dx = -\operatorname{csch} x \tag{C.37}$$

$$\int x \sin x \; dx = \sin x - x \cos x \tag{C.38}$$

$$\int x \cos x \; dx = \cos x + x \sin x \tag{C.39}$$

$$\int e^x \sin x \; dx = \frac{e^x}{2}[\sin x - \cos x] \tag{C.40}$$

$$\int e^x \cos x \; dx = \frac{e^x}{2}[\sin x + \cos x] \tag{C.41}$$

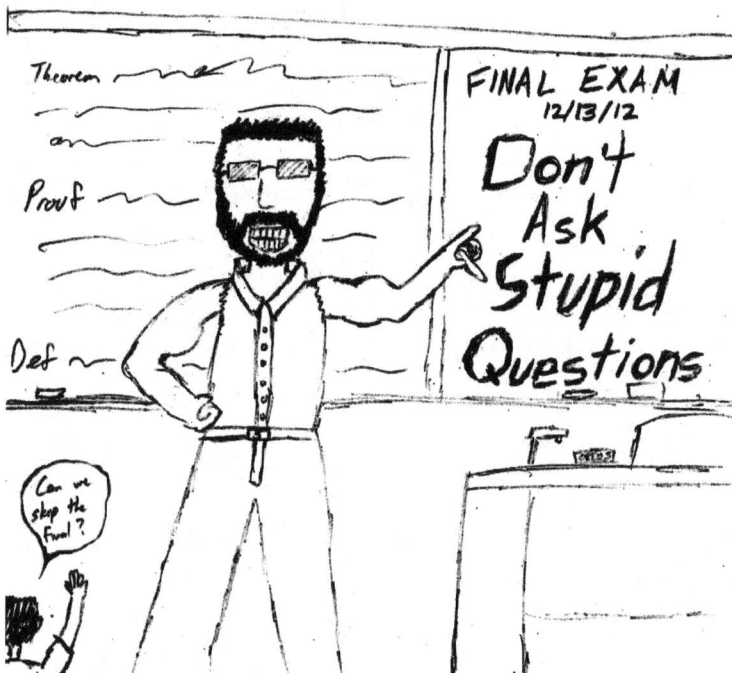

Appendix H

Help! An annotated bibliography.

Precalculus

Lippman, D & Rasmussen M. (2014). *Precalculus, An Investigation of Functions*. OpenTextBookStore. http://www.opentextbookstore.com If you need a little review of algebra, trigonometry, exponentials, logarithms, polynomials, plotting, functions, inverse functions, and that old precalc book of yours isn't doing it for you, try this one. Its free. Its good. You can get a nice bound copy for only around $15 too, if you prefer that to the pdf file.

The Spoon Feeders

Adams, C. C., Hass, J., & Thompson, A. (1998). *How to ace calculus: The streetwise guide*. New York: W.H. Freeman & Co. The title speaks for itself. Short, gives a few quick tricks.

Ayres, F., & Mendelson, E. (2013). *Schaum's outlines calculus*. New York: McGraw-Hill. Breaks everything down into its smallest bit. Doesn't really explain anything, just gives the formulas, then a few examples and solved problems.

Mendelson, E. (2009). *Schaum's outlines calculus: 3,000 solved problems in calculus*. New York: McGram-Hill. Just what the title says: a whole boatload of solved problems.

PatrickJMT (2014). *1,001 calculus practice problems for dummies*. Hoboken, N.J.: Wiley. Just what the title says it is. Problems and solutions. This one is by a guy who made about a zillion math videos.

Videos, Courseware, and MOOCS

Fowler, Jim & Barat Snapp. *Mooculus*. https://mooculus.osu.edu/ Video lectures and courseware, including open source textbook, from Ohio State University.

Jerison, David. *Single Variable Calculus: MIT Open CourseWare*. http://ocw.mit.edu/courses/mathematics/18-01-single-variable-calculus-fall-2006/video-lectures/. Video lectures of the class he taught at MIT in Fall 2007. Also includes 87MB of downloadable course material, such as PDF lecture notes.

Khan, Salman. *Khan Academy*. https://www.khanacademy.org/math/differential-calculus, also https://www.khanacademy.org/math/integral-calculus. The Khan Academy has everything.

PatrickJMT. http://patrickjmt.com/. Check the site. A bunch of stuff is listed.

More on Calculus

Boelkins, M., Austin, D., & Schlicker, S. (2013). *Active Calculus*. http://faculty.gvsu.edu/boelkinm/ Home/Open_Calculus.html. Calculus through a limited selection of examples and activities (Free PDF /CC BY-NC-SA License). Designed to accompany an inquiry-based learning class. Endorsed by the American Institute of Mathematics and Open Textbook Initiative.

Keisler, J. (1986)*Elementary Calculus: An Infinitesimal Approach*. Boston: Prindle, Weber & Schmidt. The only mainline calculus book to use an infinitesimal approach rather than limits, since it went out of print the author releases it open source as a pdf file at https://www.math.wisc.edu/~keisler/ calc.html. This book talks about infinitesimals – the stuff in chapter 8 – that you won't find in the other books like Stewart or Thomas

Strang, G. (1991). *Calculus* Wellesley, MA. Wellesley-Cambridge Press. http://ocw.mit.edu/resources/ res-18-001-calculus-online-textbook-spring-2005/textbook/. Scanned version of text released (Free/CC-SA) open source by MIT Open Courseware project. Make sure to get this one and not the second edition (which is not free). A traditional calculus book (like the doorstops) only much more concise. It has lots of exercises. Endorsed by the American Institute of Mathematics and Open Textbook Initiative.

Calculus for Biologists

Adler, FR (2013). *Modeling the Dynamics of Life: Calculus and Probability for Life Scientists, 3rd Ed.* Cengage: Belmont, CA. Just like the doorstops listed below, there are a lot of books advertised along the general theme of *Calculus for Biology Majors*. Probably trying to capitalize on all that pre-med tuition. Most of them take the view that all you have to do is talk slower and louder and the pre-meds will understand you. Unfortunately that's not the issue. We all know (but won't ever admit it, and you *certainly* didn't hear it from me) that typical pre-med students have higher IQs than typical calculus profs. So there is no way they are going to pay attention to a mathematicians pontificating on the virtues of the brachistochrone, because *it is absolutely irrelevant to them*. This book is different. It actually talks biology to biologists and presents the material in a subject appropriate manner. Most mathematicians won't understand it and won't use it for teaching because the biology is, well, just plain too hard for them. The fact of the matter is that its not rocket science – *rocket science is a whole lot easier than biology*. But this book actually makes the calculus understandable to biologists. If you're a confused bio major and really want to know what your calc teacher is blathering on so about, you might want to take a gander. Could induce sciatica, but that's not so unusual for biology books.

The Doorstops

Mostly aimed at the engineering/physics crowd, all of these books are the same. You can read any one of them. Despite what the web pages say about pedagogy and presentation they are just clones of one another. There is no spoon-feeding in any of these books - you have to read all the material and work the examples before you try to do the exercises. And some of the exercises are pretty tough.

Ellis, R., D. Gulick (2011) *Calculus with Concepts in Calculus, 6th Edition*. New York: Cengage. I've never actually used this book. I only mention it out of respect since I sat through four semesters of advanced calculus and analysis classes with Robert Ellis back in the stone age (the '70s) while he was writing the first edition. At the time he told us the book wasn't hard enough for us. A few years ago he told me that calculus had come to encompass his whole professional life. I hope it was worth the effort. 1204 pages. A little under half a GOM.

Larson, R., B. Edwards (2014). *Calculus*. Cengage. Part of a dynasty with lots of other changing authors I can't even begin to figure out. The publisher periodically bombards me with samples for no clear reason. The web site publishes side by side tables comparing this book to the others, explaining why its better. Of course nobody cares about any of the reasons listed. They are all directed at mythical P-Card auditors. 1280 pages, 0.7 GOM.

Stewart, J. (2012). *Calculus, Early Transcendentals, 7th Edition*. Belmont, CA: Thomson/ Brooks-Cole. The author provides a clear exposition. There are lots of good problems to work through.

There are seven different versions of this book; pick any one you like: *Calculus*; *Calculus, Early Transcendentals*; *Essential Calculus*; *Essential Calculus, Early Transcendentals*; *Calculus, Concepts and Contexts*; *Brief Applied Calculus*; *Calculus, Early Vectors*. At 1344 pages (about two-thirds a GOM[1]) it should hold even the heftiest door open. Be wary of sciatica.

Sullivan, M., K. Miranda (2014) *Calculus, Early Transcendentals.* Macmillan. This is a new one without much history so you won't be able to find any cheap rejects on the market. I only found out about it from the list on Larson's web page, so clearly Larson's publisher is scared, which means its probably pretty good. Its brand new so I haven't seen it and they haven't had the foresight to send me a review copy.

Weir, M. D., Hass, J., & Thomas, G. B. (2014). *Thomas' calculus: Early transcendentals*, 13th Edition. Boston: Pearson. Other versions include *Thomas' calculus, University Calculus: Early Transcendentals*, and *University Calculus Elements with Early Transcendentals*. Earlier editions were by G. B. Thomas and R. L. Finney, or just by G. B. Thomas. The BEST version of this book is the 4th edition (1969), though the 5th and 6th editions with Ross Finney as co-author weren't too shabby. It has been on a definite downhill spiral ever since. Grab one of the older editions for a few pennies plus shipping online. The exposition is clear and well written with lots of examples, but expect to have to work your way through the text.

Varberg, D. E., Purcell, E. J., & Rigdon, S. E. (2007). *Calculus, 9th Edition.* Upper Saddle River, N.J: Pearson Prentice Hall. Clearly yet concisely written; some students need more detail. Advertised by the publisher as "the shortest mainstream calculus book available – yet covers all the material needed by, and at an appropriate level for, students in engineering, science, and mathematics." Still more than half a GOM. Shoulda used thinner paper.

For Advanced Reading Only

Spivak, M. (2008). *Calculus, 4th Edition.* Cambridge: Cambridge University Press. Detailed. Pedantic. Constipating.

Apostol, T. M. (1967). *One-variable calculus, with an introduction to linear algebra.* Waltham, Mass.:Wiley. Same old, same old. Detailed. Pedantic. Constipating.

Courant, R., & John, F. (1999). *Introduction to calculus and analysis.* (3 Volumes). Berlin: Springer-Verlag. The first volume could be used in a very theoretical based version of this class. The writing is clear and easy to follow. The authors cover everything. The examples tend to be a bit on the lengthy side. The 1999 edition is a reprint of the 1965 Wiley edition. Best damn calculus book ever (IMHO). I mean, Courant, hey, they named a whole friggin math institute (at NYU) after him.

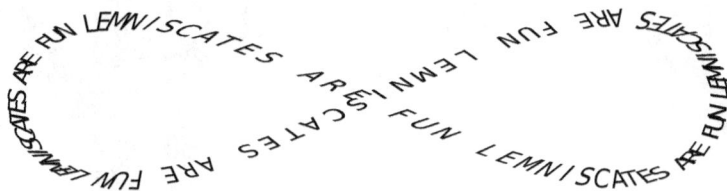

---

[1]One GOM is the weight of one gallon of milk, about 8.6 pounds.

Index

Illustration Credits

Except as indicated below all illustrations were produced by the author. Vector graphics were produced using Tikz[1] and/or Mathematica,[2] hand drawn images were processed with Inkscape,[3] and original photographs were processed using GIMP[4]. The spreadsheet images in chapter 21 were obtained from LibreOffice.[5] The entire manuscript was written and typeset in Ubuntu[6] Linux using LaTeX.[7]

Page	Title/Description	Credit
iv	Calculus Made Simple	Christopher Pyne
1	Facebook	Theodora Adde
2	Xtreme Mathematical Theory	Darrell Williams
2	Today We're Drawing Dinosaurs	Rheanna McKnight
3	Tardis (Photograph)	Zir / CC-by-2.5*
4	Integrals (Dr Shapiro)	Justin Meade
32	Royal Solver	Julia Arias de Liban
82	Yum, Cookies!	Setiawan Han
94,190	Super Derivative	Julia Arias de Liban
129	Function Machine	Bill Bailey*;**
138	Albert	Francisco Llamas
178	Importance of Cookies	Jennifer Mosher
206	Newton's Method	*Anonymous*
212	Integrals	Eduardo Sanchez
220	Brain Food (Donuts)	Patrick Basta
236	Bruce in the Box	Matthew Alegrete
262	Sir Integral	Julia Arias de Liban
270	Pink Floyd (Not)	Jesus Orozco
276	Integration by Parts	Hector Bonilla
284	Gang Sines	Arturo Menchaca
300	Pythagorean Theorem	Lennie Tran
324	Mushrooms	Andrew Freesh
326	Volumes by Cylindrical Shell	Luis Gomez
332	Trapezoid!?! We're not in geometry!	Dana Bocci
337	Math People are Crazy	Michelle Devost
344	Shapiro With His Mouth Open	Brianna Amador
360	Who Brought Cookies?	Kevin Callahan
376	Calculus with Cookies	Jose Tovar
388	Funny Guy With Cookies	Hovhannes Mikhitaryan
409	Stuff	Lawrence Milstein
430	Foundations of Pure and Utter Nonsense	Lorenzo Rulli
460	Don't Ask Stupid Questions	Erik Brinkhus
464, FC°	Derivatives Make Finding Slopes Simple	Julia Arias de Liban
Back Cover	Portrait of the Artist as a Young Man	Arno Babahekian

*Via Wikimedia.; **Released as public domain.; °FC = Front Cover.

[1] http://sourceforge.net/projects/pgf/
[2] http://www.wolfram.com/mathematica/
[3] http://www.inkscape.org/en/
[4] http://www.gimp.org/
[5] http://www.libreoffice.org/
[6] http://www.ubuntu.com/
[7] http://www.latex-project.org/

Sherwood Forest

The Sherwood Forest imprint evokes the image of Robin Hood, who, in some legends, hid in Sherwood Forest while fighting to help the poor, as they suffered under the oppressive regime of the medieval English aristocracy.

A statue of Robin Hood stands in front of Nottingham Castle, his legendary home. The castle dates to the 17th century; the statue was built in 1952.

Now as we enter the information age, modern university students suffer under the oppression of the expensive traditional publication model. Sherwood Forest Books aims to print low cost, affordable books in hard-copy and DRM-free electronic formats. In the logo, the roots of the tree sink down into the Earth, from which we all arise. The filament on the light bulb is a double helix, representing the DNA that binds all life on Earth and through which we grow, learn, interpret, and communicate our understanding of the world around us. This light of knowledge is spread through the easy and inexpensive dissemination of the printed word, like leaves on tree as they blow in the wind.